CW01203220

Public Health Significance of Urban Pests

EUROPE

Public Health Significance of Urban Pests

Xavier Bonnefoy
Helge Kampen
Kevin Sweeney

Abstract

The second half of the 20th century and the beginning of the 21st century have witnessed important changes in ecology, climate and human behaviour that favour the development of urban pests. Most alarmingly, urban planners are faced now with the dramatic expansion of urban sprawl, where the suburbs of our cities are growing into the natural habitats of ticks, rodents and other pests. Also, many city managers now erroneously assume that pest-borne diseases are relics that belong to the past.

All these changes make timely a new analysis of the direct and indirect impacts of present-day urban pests on health. Such an analysis should lead to the development of strategies to manage them and reduce the risk of exposure. To this end, WHO has invited international experts in various fields – pests, pest-related diseases and pest management – to provide evidence on which to base policies. These experts contributed to the present report by identifying the public health risk posed by various pests and appropriate measures to prevent and control them. This book presents their conclusions and formulates policy options for all levels of decision-making to manage pests and pest-related diseases in the future.

Keywords

PEST CONTROL - methods
INSECT CONTROL - methods
URBAN HEALTH
URBAN POPULATION
ENVIRONMENTAL EXPOSURE
CITY PLANNING
PUBLIC HEALTH
HEALTH POLICY

Address requests about publications of the WHO Regional Office for Europe to:

Publications
WHO Regional Office for Europe
Scherfigsvej 8
DK-2100 Copenhagen Ø, Denmark

Alternatively, complete an online request form for documentation, health information, or for permission to quote or translate, on the Regional Office web site (http://www.euro.who.int/pubrequest).

ISBN 978-92-890-7188-8

© World Health Organization 2008

All rights reserved. The Regional Office for Europe of the World Health Organization welcomes requests for permission to reproduce or translate its publications, in part or in full.

The designations employed and the presentation of the material in this publication do not imply the expression of any opinion whatsoever on the part of the World Health Organization concerning the legal status of any country, territory, city or area or of its authorities, or concerning the delimitation of its frontiers or boundaries. Dotted lines on maps represent approximate border lines for which there may not yet be full agreement.

The mention of specific companies or of certain manufacturers' products does not imply that they are endorsed or recommended by the World Health Organization in preference to others of a similar nature that are not mentioned. Errors and omissions excepted, the names of proprietary products are distinguished by initial capital letters.

All reasonable precautions have been taken by the World Health Organization to verify the information contained in this publication. However, the published material is being distributed without warranty of any kind, either express or implied. The responsibility for the interpretation and use of the material lies with the reader. In no event shall the World Health Organization be liable for damages arising from its use. The views expressed by authors, editors, or expert groups do not necessarily represent the decisions or the stated policy of the World Health Organization.

Graphic design: Pierre Finot
Text editor: Jerome M. Rosen

Contents

Foreword	VII
Executive summary	IX
Introduction	1
1. Allergic asthma	7
2. Cockroaches	53
3. House dust mites	85
4. Bedbugs	131
5. Fleas	155
6. Pharaoh ants and fire ants	175
7. Flies	209
8. Birds	239
9. Human body lice	289
10. Ticks	304
11. Mosquitoes	347
12. Commensal rodents	387
13. Non-commensal rodents and lagomorphs	421
14. Pesticides: risks and hazards	477
15. Integrated pest management	543
Annex 1. Abbreviations	563
Annex 2. Working Group	565

Foreword

Recent developments in pest-borne diseases, such as cases of West Nile fever in the United States of America and the spread of Lyme disease in both Europe and North America, have signalled strongly the crucial need to carefully assess the potential threat of urban pests to public and environmental health. Also, modern living conditions, urban sprawl and emerging changes in climate make the spread of pests and pest-borne diseases increasingly likely. The effects of these conditions and changes need to be properly monitored and understood. Moreover, the lesson learned from the outbreak of severe acute respiratory syndrome (SARS) is that modern forms of transport enable infected travellers to move quickly from one continent to another, with the ability to arrive at their destination before any symptoms appear. This same speed of travel also enables pests to spread freely and quickly from area to area in trucks, boats and planes. These factors, together with increasing concern about pathogens undergoing mutation and changing their host species and mode of transmission, need careful scientific evaluation.

This report considers the significance of the main urban pests and the medical conditions they create, as well as the resulting economic cost of the burden of disease, when data were available. It also proposes technical and policy options for governments that desire to implement adequate surveillance and contingency plans.

The report is based on contributions from international experts in the fields of pests, pest-borne diseases and pest management, invited by the WHO European Centre for Environment and Health, Bonn. The office acted as secretariat for the working group. WHO is very grateful for the contributions of these experts and believes that the recommendations in the report, if implemented, will reduce the health hazards caused directly and indirectly in Europe and North America by pests and unhealthy pest control practices.

Shortly before this report was finalized, one of the guiding spirits of the working group, Professor Marco Maroni, died suddenly and unexpectedly. He was an internationally renowned pesticide specialist and long-standing WHO advisor. He was not only the lead author of Chapter 14, on pesticide risks and hazards, but he also contributed significantly to the quality of the whole report, by constructive criticism and input during the meetings. Professor Maroni's charisma, devotion to public health, human qualities and constant availability to others (despite a fully booked agenda), along with his simplicity (although he was immensely knowledgeable), were recognized by all those who approached him. We shall miss his presence and deeply lament his death.

Dr Roberto Bertollini
Director, Special Programme on Health and Environment
WHO Regional Office for Europe

Executive summary

This report provides sound evidence that urban areas are being exposed increasingly to pests and (through them) to pest-related diseases. This multi-faceted problem of increasing exposure entails environmental, structural, institutional, regulative, managerial, financial, scientific and climatic aspects. Solutions that will better protect public health, by implementing improved pest and pest-related disease management, have been identified. These solutions address the need for legal action, education, institutional capacity building and research at international, national and local levels.

The conclusions drawn are based on the currently available evidence but it is important to understand that there are some major factors, such as the impact that climate change will have on landscapes, ecosystems and the future patterns of vector borne diseases, that are going to be very important in the future.

Climate change is particularly relevant because it is expected to alter not only the natural environment as a result of flooding or drought but also the urban environment as a result of changes in land use.

Only the future will tell to what degree these factors will affect the risk of pest-related diseases. Still, the conclusions drawn by this report will help and support national governments to understand better the public health relevance of urban pests and to be able to prepare the ground for increased technical capacity and ability for action.

The evidence reviewed in this book is based on a review of the current status of urban pests and health in Europe and North America, and mostly draws from scientific evidence produced and regulatory approaches developed within these countries. Still, the evidence and regulatory approaches described may be of use for a wide range of countries, depending on their national context.

Legal requirements

A fundamental requirement for implementing the right (and effective) preventive and control measures is having adequate legal requirements in place that allow appropriate ministries and agencies to take appropriate action and that provide them with the authority to take these actions.

Planning and construction

By destroying the borders between urban and rural environments, urban sprawl makes urban areas more susceptible to pests and the disease agents they carry. Since many zoonotic pathogens – that is, pathogens that can be transmitted to people from animals – are more likely to be transmitted between vectors and their reservoir hosts in rural environments, the risk of infection increases as rural amenities, such as woodland and recreational areas, are promoted. This increase in the risk of infection is due to the likelihood of inhabitants of inner-city areas coming into contact with such disease-bearing pests as ticks and rodents. Also, city planners and developers often seek to integrate (visually and ecologically) construction projects, such as housing developments, single buildings and recreational areas, with their natural surroundings; however, they often do so without considering the risk of increased pest infestation. This risk could be reduced by:

1. regulations about city planning, landscaping, design of recreational areas and the like taking into account the risks of pest infestation and disease transmission; and

2. construction regulations ensuring that new buildings are pest-proofed and do not create conditions conducive to pest infestation.

Responsibilities

Because pest management involves health, environmental and occupational factors, it is often difficult to decide which government department or agency should be responsible for its activities. At the local level, it is often unclear which body is responsible for pest prevention, surveillance and control. The following approach may help resolve the difficulty in decision-making and the lack of clarity.

3. A single government department should have the ultimate responsibility for supervising monitoring programmes and implementing pest management measures; this should be accompanied by the political will to implement programmes and measures.

4. With regard to pest management, adequate regulations should make clear the liabilities of contractors, building managers, homeowners, apartment occupants and local authorities.

Pesticide application

Potent pesticide products are often not only available to private individuals, but are also often misused by them, due to a lack of knowledge or expertise. In this case, pesticides may be applied when unnecessary, in wrong formulations, at wrong concentrations and in wrong amounts. Even if used correctly, pesticides still hold a risk for both human health and environmental health. They therefore require a technical risk–benefit analysis before being applied. The following measures may help improve this situation.

5. Although regulations that cover the sale and use of pesticides exist throughout Europe and North America, a stricter differentiation between professional and amateur products should be established and enforced, to prevent the general public from having access to products that need to be used only by trained and competent operators.

6. Through scientifically based risk assessments and proper approval processes, pesticide applications and the pesticides used should not pose an unacceptable risk to consumers, operators or the environment. Proper risk assessments should be required and carried out before pesticides are put on the market.

Notification, approval and public awareness

Notification

Because of differences among European Union (EU) Member States, the notification system in Europe is inconsistent. For example, Lyme borreliosis, the most frequent arthropod-borne disease in Europe, is reportable in some EU Member States but not in others. It is, therefore, extremely difficult to collate reliable epidemiological data. Also, where diseases are reportable, notification rules often differ from country to country, making it impossible to compare data. Finally, data are generally unavailable to the public, are not presented in easily accessible databases or are not offered in a user-friendly form. The following measures may help improve this situation.

7. At the international level, there should be an agreement on expanded and standardized notification requirements for pest-borne diseases, as well as other adequate mechanisms to collect and analyse data centrally and to make biological and epideiological data publicly available. Early notification, a clear requirement for developing adequate public health policies, should enable Member States to be properly informed.

Approvals

In addition to international differences in the requirements for the approval of pesticides, the complexity and cost of pesticide approvals are rising continually, which either currently prevents many companies from putting products on the market that could be more efficient and cheaper than the existing ones or results in acceptable products of minor use being withdrawn from the market. This makes it likely that future choices of the best available pesticide on the market for a particular application will be severely reduced by the economics of the approvals process. It also means that competition in the market for pesticides will be skewed towards large international companies able to afford to have their pesticides approved. As a result of this, the range of pesticides available will decrease, thereby reducing the options for treatment. Also, treatments of pests that are either of minor or new importance "will not be carried out", because it will be unprofitable to develop or obtain approval for pesticides for their control. The following is a step towards remedying this situation.

8. The prohibitive costs associated with obtaining pesticide approvals should be reconsidered and, when possible, lessened. This will allow for the competitive possibility of registering more efficient and cheaper pesticides and pesticides that fulfil treatment niches. Approval fees should not be inflated to cover unrelated needs.

Public awareness

Public information and education are fundamental to efficient and successful pest management, with respect to both preventive and control measures. Most people are unaware how their habits, their behaviour and changes in their homes can attract commensal pests and provide ideal living conditions for these pests to thrive. They are also largely unaware that pests may carry pathogens and that simple personal measures can be taken to avoid contact with pests. Moreover, they are largely unaware of how to handle pesticides. Thus, public information is not only a basic need, but it is also economically sound, because it contributes considerably to preventing pest infestations through private action. The following measure may help improve this situation.

9. Information should be developed for the public, to raise its awareness of how to protect itself through simple sanitary and behavioural measures. Such information should also familiarize them with how to best store and use pesticides, which would also minimize the risks associated with their storage and use.

Institutional capacities

Up-to-date data on the occurrence and distribution of pests and pest-related diseases are generally scarce (or even non-existent) in the EU. In the past, government departments and agencies dealt with pests and collected data. However, this form of activity has slowly (but substantially) been reduced (or even discontinued) by budget cuts. Although there has been a pest renaissance for several years, pertinent government agencies have not been upgraded or established anew with adequate staff, equipment and funds to act as surveillance units to collect epidemiological data. It is of general concern that in Europe there are neither national nor international institutions responsible for collecting vector-related information and coordinating pest control. The following measures may help improve this situation.

10. WHO Regional Office for Europe Member States, through coordinated efforts of their public health authorities, would benefit from: developing the capacity needed to identify pest-related risks in the urban environment (that is, identify pests and pest-borne diseases that occur at present or have the potential to occur); determining and recording the prevalence of various infections; and keeping track of existing reservoirs of host species and the geographical distribution of various pests and their transmission dynamics. They would also benefit from keeping an updated list of high-risk areas.

11. Governments of the European Region – as well as other countries – would benefit from ensuring that surveillance agencies and suitably educated staff are available. A well-trained public health force, available and prepared for pest and pest-related disease management, is needed to protect the public from the threats to health associated with urban pests. For example, it is needed at vulnerable sites, such as ports and airports. Educated specialists in such disciplines as medical entomology, medical zoology, toxicology, ecotoxicology and public health management are needed to: train pest managers; help develop control programmes, including strategies and pesticide use; reach agreements on action thresholds and defined control goals; and ensure that harmonious cooperation takes place between all the stakeholders involved, including government departments and agencies, local authorities, industry, consumer groups and the public.

12. At both the national and local levels, authorities in charge of vector-related information should be clearly identified. The role of partners, as well as mechanisms for coordinating partner efforts, should be defined and put in place. While there are European agencies that collect information on disease, there is a need for a similar organization that would collect information on vectors, because most data collection activities in this area are carried out at a local level nationally and no coordination exists.

Research

The study of the various chapters in this report (such as those that cover ticks, mosquitoes and fleas) generally demonstrates that while the biology and behaviour of the pests has been well studied, the epidemiology of the diseases they transmit, particularly in the case of newly emerging diseases, are poorly understood. Though the need for understanding exists, scientists specialized in the classical disciplines of medical zoology and medical entomology are becoming rare, as governments and universities progressively shift their limited financial resources to other fields. Because of this shift in resources, not only is research in the classical disciplines being neglected, but the knowledge that underpins it is also disappearing, slowly and irreversibly. Moreover, public health professionals and physicians are often overly strained when confronted with pests and emerging pest-borne diseases, and at universities there is rarely a specialist left who can be consulted.

The same is true of pest surveillance and pest control. Private pest management companies are becoming less and less (if at all) involved in research and development, and the pest management industry generally concentrates on products for which there are ready markets.

The following, the last item in this summary, is an important conclusion of this report.

13. Governments, public health programmes and the general public would benefit from encouraging, supporting and promoting pest-related scientific research. This would lead to refined knowledge of the biology, ecology and behaviour of pests and of the epidemiology of pest-borne diseases, which is urgently needed, as are more efficient and specific tools and active ingredients for pest surveillance and control.

Introduction

The second half of the 20th century and the beginning of the 21st have witnessed important changes in ecology, climate and human behaviour that favour the development of urban pests. Most alarmingly, urban planners are faced now with the dramatic expansion of urban sprawl, where the suburbs of our cities are growing into the natural habitats of ticks, rodents and other pests. Also, many city managers now erroneously assume that pest-borne diseases are relics that belong to the past.

Background

Changing times are accompanied by changing needs. Since the earliest of times, people have lived in communities, using caves in prehistoric times and dwellings built for their needs in more modern times. These communities strived to organize safe and secure settlements based on commonly agreed principles that would allow them to function in an orderly manner. Because the health and well-being of a community's inhabitants are vital parts of it functioning productively, specialized bodies were established to protect it from external and internal threats.

By the 19th century, the major threats to community health were recognized as coming from poor housing, poor management of sewage and drainage, foul air in industrialized towns, unsafe drinking water, and inadequate control of pests. Early environmental health practitioners fought to remedy these defects and spurred the founding of the environmental health movement. In the 20th century, engineering and construction techniques went a long way towards removing the problems of air pollution, sewage, drainage and poor water quality in cities and towns. At the same time, the development of pesticides that benefited public health made the control of pests much easier in increasingly urbanized areas. Subsequently, the new science of hazard (or risk) analysis filtered out a number of environmentally unacceptable products. Also, following major advances in medical research, antibiotics can now control most pest-borne diseases, while improved sanitation practices and immunization programmes have further reduced the adverse effects of infestation.

The development of cities and towns dramatically changed our lifestyle, especially our increased reliance on motorized transport, and as a result we are now confronted with problems caused by urban sprawl. As inner-city areas became crowded, degraded and (in some cases) unsafe, urban sprawl began with the more affluent residents moving to the new greener suburban areas, which were usually served by major transportation networks, such as railways and roads. These networks enabled the more advantaged people in a community to live in the desirable leafy suburbs while working in the city centres. Not only has this arrangement encouraged urban sprawl, but it has also changed the economic and health balance in the community. This general evolution has been accompanied by other macro-changes that affect the community, such as the globalization of national economies, the ease with which people move from country to country, the international exchange of goods and (more important) the first symptoms of global warming.

Since 1989, ministers responsible for health and the environment have met every five years to receive updates on the major environmental factors that adversely affect public health and to discuss them. During these discussions, they reviewed the options available to address problem areas and then agreed on common commitments for action.

In June 2004, at the Fourth Ministerial Conference on Environment and Health, in Budapest, Hungary, ministers of health and environment adopted a declaration that provides the political foundation for the present work. The declaration affirms that ministers have:

- recognized the importance of properly assessing the economic impacts of different levels of environmental degradation – in particular, the direct and indirect costs incurred by society in addressing diseases related to the environment;
- recognized that the existing housing stock, the lifestyles of our people, the immediate environment of dwellings and the social conditions of the inhabitants should all be considered when developing healthy and sustainable housing policies;
- recognized that preventing ill health and injury is infinitely more desirable and cost effective than trying to address the diseases;
- noted that large quantities of chemicals, currently produced and marketed with largely unknown effects on human health and the environment, constitute a potential risk to the people working with them, as well as to the general public; and
- recognized that delay in addressing a suspected health threat can have public health consequences.

Based on these premises, ministers have:

- recommended that the WHO Centre for Environment and Health should continue to provide Member States with evidence to support policy-making in environment and health;
- called for initiatives and programmes aimed at providing national and local authorities all over the Region with guidance on integrating health and environmental concerns into housing policies; and
- committed themselves to contributing to developing and strengthening housing policies that address the specific needs of the poor and the disadvantaged, especially with regard to children.

The present report has been prepared so that the ministers responsible for public health and the environment can better fulfil these commitments in an area of growing concern: the possible threat to public health that stems from urban pests and attempts to control them.

Scope of this report

The main purpose of this report is to identify approaches to urban pest prevention and control that beneficially reduce the impact of these pests on public health. Passive control through improved design and construction of our cities and our housing stock is certainly the most sustainable approach: when pests lack the conditions they need to breed, such as food, drink, warmth and safe harbourages, they simply cannot survive in an area. This very basic approach is valid for all pests. Unfortunately, suitable conditions often exist where we live, work and play, which means that pests usually can coexist perfectly in our environment and that passive measures have to be very specific to control either their presence or development.

Humankind's increasing desire to change its environment entails new risks from pests and the diseases with which they are associated. An example of this is the rise in tick-borne diseases. As cities expand and more houses are built on their wooded outskirts, people will be more exposed to tick-borne diseases, such as borreliosis (especially Lyme disease), tick-borne encephalitis and Rocky Mountain spotted fever. These severely disabling diseases have been able to spread over the past 30 years, (in part) because of our new lifestyles, despite the management techniques now available to control urban pests. These techniques have developed significantly, due to the discovery of new pesticides and new application methods.

Chapter 14 of this report reviews the main health risks associated with pesticides. It also provides examples of models designed to estimate human exposure to pesticides used in pest control activities. The science in this field is developing quickly.

Regulations are also evolving to make pest control less hazardous. As this report makes clear, good pest control cannot be achieved through the sole use of chemicals. Chapter 15 gives a detailed description of the principles and basic techniques of *integrated pest management*; it is the key concept that supports sustainable pest management practices and should be enshrined in national regulations that deal with pest control.

This report discusses many urban pests, including such emerging ones as non-commensal rodents. However, non-vector-borne health threats associated with pets, especially cats (and cat allergens), have not been considered in the report. Similarly, feral cats and dogs are not reviewed, because techniques for controlling them are very specific and (in many countries) are seldom used by pest control managers. This is also why Chapter 9, on lice, deals mainly with body lice, which are usually handled only by pest control staff in cases of emerging outbreaks. Moreover, this chapter touches very briefly on head lice, because they are usually controlled by medical or paramedical teams in schools, kindergartens and day-care centres or by care providers at home.

This report focuses on the urban pest management challenges faced by developed countries in Europe and North America, its primary audience are public health authorities at international, national and local levels. However, universities and other research bodies, the pest management industry, urban planners, architects, nongovernmental organizations, and consumer groups from the WHO European and North American regions will find information of considerable relevance to their daily practices in the report. Although evidence from the other regions of the world has not been reviewed in the context of this report, there may still be relevant issues and elements that could be applicable depending on the national context. Nevertheless, the information contained in this report should not be extrapolated to these regions without careful consideration of the relevant literature that deals with their specifics.

The editors and authors of this report would be grateful for any comments and questions about it. The scope of this report is finite, because it is not possible to deal either with all pests present in the urban environment or with all possible diseases carried or transmitted by pests.

The process

At a meeting held in London, 26–28 June 2002, a steering committee (see Annex 2) was established. It agreed on the scope and purpose of this report, a workplan and a schedule. The institutions to be consulted were identified, as were the authors of the individual chapters (see Annex 2). The basis for selecting authors was their credentials – their contributions to peer-reviewed literature. To provide the most accurate epidemiological evidence from both the North American continent and the WHO European Region, a team of at least two authors, one from each geographical region, was created where it was believed to be relevant.

The authors were requested to adhere to the following guidance, which was decided upon at the first meeting of the steering committee. Based on existing literature, authors were invited to review both the distribution of urban pests and the prevalence of diseases associated with them. In preparing their chapters, authors were asked to address in some detail, existing management techniques, with special emphasis on the application of integrated pest management. The literature on the health aspects of the use of pesticides, as well as methods for evaluating human exposure and assessing the risk of pesticides used during pest control activities, was also to be reviewed.

This report does not attempt to be a systematic review of the literature per se, but it does attempt to use the most objective criteria to select the literature cited. In choosing literature on which to base the individual chapters, grey literature was not automatically excluded when it was the only source available to support hypotheses or to report on anecdotal, but significant events. With regard to the literature cited, it is important to note that the authors of the different chapters were asked to carefully evaluate this literature. In line with this, each author was asked to provide the criteria he or she used to include or reject a reference.

At a meeting of the steering committee, held in Bonn, 9–10 May 2005, the initial drafts produced by the authors were examined. At this meeting, further guidance was provided to the authors: to avoid overlaps, to include or delete paragraphs on specific subjects and to ensure consistency.

The final drafts were reviewed by the steering committee, at a meeting held in London, 23–24 April 2006. In addition to the members of the steering committee and the authors, a representative of the European Centre for Disease Prevention and Control attended the meeting. Comments at the meeting were provided to the authors, who then submitted their final work to WHO within four weeks of the end of the meeting.

WHO then sent each chapter to peer reviewers (see Annex 2). These reviewers were recommended both by the authors and by the WHO Regional Office for Europe or WHO headquarters and were neither involved in the steering committee nor in the writing process. Where appropriate, the comments and suggestions of the peer reviewers have been incorporated in the final report.

1. Allergic asthma

Matthew S. Perzanowski, Ginger L. Chew,
Rob C. Aalberse and Frederic de Blay

Summary

Asthma is a major urban disease and a substantial burden from the standpoint of both the quality of life for the many suffering from the disease and the economics of health care. The global increase in the prevalence of asthma in the last half of the 20th century has affected urban communities in many countries disproportionately. Asthma is an allergic disease for more than 50% of adults and 80% of children. The evidence for a relationship between allergic asthma and domestic exposure to cockroaches, mice and dust mites is strong. These pests are common in urban environments, especially impoverished communities, and play a significant role in the pathogenesis of urban asthma.

While exposure to allergens is necessary to develop an allergic response, the thresholds of exposure needed for an allergic immune response and for exacerbation of asthma symptoms are unclear. Recent studies would suggest that the differences observed among studies on the quantity of allergen exposure that relates to disease might be due to the influence of concurrent exposures that alter the immunological response (such as bacterial products) and to genetic factors that affect an individual's susceptibility to produce immunoglobulin E antibodies to an allergen. Despite uncertainties in quantitative thresholds, studies have demonstrated a clear relationship between increasing domestic exposure to allergens from cockroaches, mice and dust mites and an increased risk of allergic sensitization and severe asthma.

Given the relationship between exposure and disease, removing these pests and their allergens is a logical tactic for preventing disease and reducing symptoms, but current methods need improved efficacy. Results from studies on avoiding allergens suggest cautious expectations about the ease with which long-term clinically relevant allergen reduction can be accomplished. Also, public health practitioners should be aware that focusing on one aspect of asthma control will not necessarily result in an improvement in the prevalence of asthma at the community level. A multifaceted approach to controlling allergic asthma consists of using controller medications, avoiding irritants (such as ozone and outdoor air pollution) and avoiding allergens. These more broadly defined asthma interventions, which include tailoring allergen reductions to an individual's specific allergy and providing education about effective methods for sustained integrated pest management, may be effective in reducing the burden of asthma in urban communities.

1.1. Allergic asthma and urban environmental factors

In industrialized countries, asthma has emerged as one of the most common chronic diseases of childhood, and in the United States of America and other countries it is the leading cause of hospital visits for children. WHO has estimated that 300 million people worldwide have asthma (WHO, 2006). The prevalence of asthma varies from less than 1% in rural Africa, to 7–20% in western Europe, to as high as 25–40% in some cities in the United States and suburban Australia (Fig. 1.1; Peat et al., 1995; ECRHS, 1996; Yemaneberhan et al., 1997; Ng'ang'a et al., 1998; Ronmark et al., 1998; Gupta et al., 2004; Nicholas et al., 2005). Asthma in the urban environment has been linked to allergic sensitization – in particular, to pests in the indoor environment, including cockroaches, rodents and dust mites. Other diseases are related to allergic sensitization, including allergic rhinitis and atopic dermatitis. These diseases and asthma often occur in the same individuals, especially in childhood. Also, some pests, such as ticks and midges, can induce anaphylactic reactions (Elston, 2004). While these other allergic responses to pests are important in the urban environment, the scope of this chapter is limited to allergic asthma.

1.1.1. The association between asthma and allergy

Asthma, as defined by the Global Initiative for Asthma (GINA; originally established by WHO and the National Institutes of Health in the United States), is:

> ... a chronic inflammatory disorder of the airways in which many cells and cellular elements play a role. Chronic inflammation causes an associated increase in airway hyperresponsiveness that leads to recurrent episodes of wheezing, breathlessness, chest tightness and coughing, particularly at night or in the early morning.

These episodes are usually associated with variable airflow obstruction that is often reversible, either spontaneously or with treatment (WHO, 2003).

Asthma is an allergic disease for more than 50% of adults and 80% of children (WHO, 2003). In an urban environment, sensitization to pests, including rodents, cockroaches and dust mites, is common among asthmatics. In the United States, the National Cooperative Inner-City Asthma Study (NCICAS) found that 77% of mild or moderate asthmatics 4–9 years of age were sensitized to at least one of the allergens tested, including a high prevalence of sensitization to cockroach and mouse allergens (Kattan et al., 1997). Also, a similar study in Atlanta, Georgia, found that 80% of mild or moderate asthmatic children had a positive allergy skin test to at least one allergen, primarily from cockroaches and dust mites (Carter et al., 2001). Recently a study from inner-city New York City found sensitization to mice, cockroaches and dust mites was common (about 15%) in children as young as 2 years old (Miller et al., 2001a). However, rates of sensitization to both dust mites and cockroaches vary between cities and ethnic groups within cities (Stevenson et al., 2001).

The *atopic march* is a term that describes the process whereby an individual who is genetically predisposed to allergy is exposed to an antigen, becomes sensitized and develops an allergic disease (Fig. 1.1; von Mutius, 1998). This sensitization is the result of an immunoglobulin E (IgE) antibody-mediated hypersensitivity reaction to the specific allergen (antigen). In this reaction, mast cells are stimulated and release histamine. This leads to a complex inflammatory reaction that involves multiple inflammatory cell types. The diseases allergic rhinitis, atopic eczema and allergic asthma are all associated with immediate hypersensitivity to allergens.

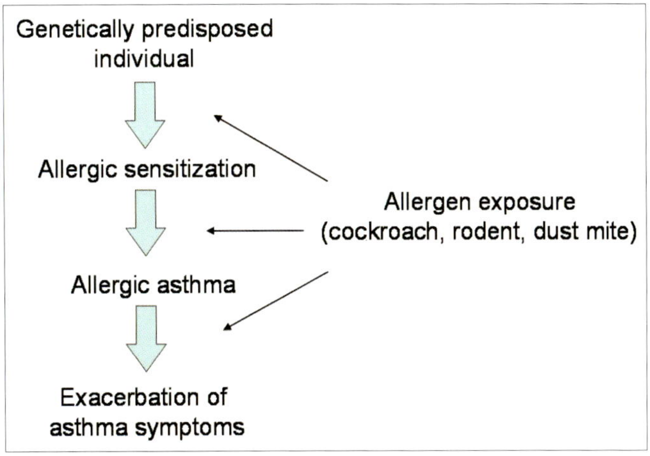

Fig. 1.1. Allergen exposure and the atopic march to asthma

1.1.1.1. Risk of developing allergic sensitization

For this discussion, atopy is defined as an individual's tendency to "produce IgE antibodies in response to ordinary exposure to low doses of allergens" (WHO, 2003). Allergic sensitization can be assessed by two common methods: a skin test, in which an allergen extract is placed on the skin and the skin is pricked (Bousquet & Michel, 1983), and a measurement of allergen-specific IgE in the serum. In most communities without endemic parasitic infections, the correlation between skin tests and serum IgE test results is strongly positive (Bousquet & Michel, 1983; Eriksson, 1989). However, not all sensitized individuals, as judged either by a skin test or serum IgE, have noticeable allergic symptoms.

Exposure to an allergen is essential for the development of allergic sensitization. However, the quantity of allergen necessary for sensitization is not known and may depend on the individual and the allergen itself. For sensitization to dust mites, there does appear to be a dose–response relationship, with atopically predisposed individuals exposed to more allergen being more likely to become sensitized (Platts-Mills et al., 2001). In communities where most homes have concentrations greater than 10 μg of dust mite allergen per gram of dust collected (μg/g), the response of a majority of the allergic individuals sensitized to mites appears to plateau, likely indicating a population-level saturation in prevalence (Marks, 1998). The allergic response to cockroach allergens also seems to show a dose–response relationship, with increased exposure associated with increased symptoms (Rosenstreich et al., 1997). For exposure to cat and dog allergens, there appears to be a protective effect of high-dose exposure early in life, with those children exposed to the highest levels of pet allergen – that is, those living with pets – less likely to become sensitized to pets (Ronmark et al., 1998; Hesselmar et al., 1999; Platts-Mills et al., 2001; Ownby, Johnson & Peterson, 2002; Perzanowski et al., 2002). However, this view is contradicted by several studies that report an increase in the development of sensitization among pet owners (Lau et al., 2000; Melen et al., 2001; Almqvist et al., 2003).

Several determinants have been identified as risk factors for allergy. There is a genetic component, which has been observed in studies of monozygotic and dizygotic twins (Marsh, Meyers & Bias, 1981; Ownby, 1990; Duffy, Mitchell & Martin, 1998; Borish, 1999). In general, studies seem to show that a child with one atopic parent is twice as likely as a child with non-atopic parents to develop atopy, and a child with two atopic parents is four times as likely (WHO, 2003). Compared with rural life, urban life has been shown to be associated with an increase in atopy (Braback et al., 1994; Ronmark et al., 1999). However, the variation in risk between urban and rural communities are difficult to attribute, since lifestyle differences can include many determinants, such as exposure to animals, increased exposure to bacteria, housing type and physical activity, all of which have been evaluated independently as risk factors for allergy (Crater & Platts-Mills, 1998). While many studies have identified exposure to environmental tobacco smoke (ETS) as a risk for developing asthma, the relationship between exposure to ETS and allergic sensitization does not seem as clear (Rylander et al. 1993; Chang et al., 2000; Gold, 2000; Ronmark et al., 1998).

The first few years of life are thought to be an important period in the development of sensitization. Several studies have even suggested that prenatal exposure to allergens may influence the immune response (Warner et al., 2000; Miller et al., 2001b). The development of sensitization, as judged by skin test or serum IgE level, has been shown to occur during the early years of life through the teenage years; however, exposure to new allergens as an adult can lead to the development of new sensitizations (Dowse et al., 1985; Sporik et al., 1990).

Exposure to some substances, referred to as *adjuvants*, can influence the development of sensitization to environmental allergens. Two examples of environmental exposure that are currently being studied for their potentially relevant immunomodulatory effects are bacterial cell wall components and products of diesel combustion.

In the 20th century, increased cleanliness at the community level has generally resulted in lower exposure to bacteria early in life. That this increased cleanliness is responsible for a more allergic population is part of the proposed *hygiene hypothesis*. Measuring endotoxin, a lipopolysaccharide from Gram-negative bacteria, has been used to assess exposure to bacteria. Experiments that use animal models have shown that exposure to high levels of endotoxin, in conjunction with an allergen, leads to a non-IgE response, while low doses of endotoxin and exposure to allergens result in an IgE response (Eisenbarth et al., 2002). Cross-sectional studies have shown that rural children with high levels of endotoxin in their bedroom had a lower prevalence of allergic sensitization than those with low exposure (von Mutius et al., 2000; Braun-Fahrlander et al., 2002). Also, recent birth cohort studies from two American cities, Boston (Phipatanakul et al., 2004) and New York (Perzanowski et al., 2006), have found exposure to endotoxin in the home to have a modest inverse relationship to the allergic skin disease eczema in the first two years of life.

In urban environments, exposure to diesel exhaust particulates is common. Components of diesel particulates can shift an allergen-specific immune response to an enhanced IgE response. This has been demonstrated in people, where nasal challenges with diesel particulates increased the IgE antibodies to ragweed allergen (Diaz-Sanchez et al., 1997). Due to difficulties in measuring diesel in epidemiological studies, few studies have dealt with its associations with asthma. Higher levels of particulate matter and proximity to traffic, which have been used as proxies for exposures that would include diesel exhaust particulates, have been associated with asthma and other respiratory symptoms in several studies (Riedl & Diaz-Sanchez, 2005).

1.1.1.2. Risk factors for developing asthma

As already mentioned, a primary risk factor for developing asthma is allergic sensitization (Platts-Mills et al., 1997). Allergy to dust mites, cockroaches, mice, cats, dogs and the fungus *Alternaria* have all been shown to be significantly associated with asthma (Call et al., 1992; Peat et al., 1994; Sporik et al., 1995; Custovic et al., 1996a; Halonen et al., 1997; Rosenstreich et al., 1997; Perzanowski et al., 1998; Ronmark et al., 1998; Phipatanakul et al., 2000a; Matsui et al., 2004a). In fact, a strong association between asthma and allergy has been reported in virtually all studies in westernized communities (Platts-Mills et al., 1997). When evaluating asthma at an early age (before the age of 6 years), sensitization to an inhaled allergen is a strong risk factor for asthma symptoms that persist into later childhood (Martinez et al., 1995). As with allergic sensitization, having a family history of asthma also contributes to the risk of becoming asthmatic (Bracken et al., 2002). Exposure to ETS in the home is also associated with asthma (Gold, 2000). Moreover, having two or more siblings in day care and frequent respiratory infections early in life have been shown to be protective against asthma, and these findings are sighted as evidence for the hygiene hypothesis (von Mutius et al., 1999; Ball et al., 2000). Furthermore, exposure to ozone has been associated with exacerbating asthma (d'Amato et al., 2005). Psychosocial stress has also been associated with asthma (Wright, Rodriguez & Cohen, 1998). Finally, there also may be dietary and physical-activity components to the development of asthma (Crater & Platts-Mills, 1998; Sprietsma, 1999; Shore & Fredberg, 2005).

The current hypothesis for the genesis of asthma in children involves a complex interaction, in the first years of life, between the two arms of the immune system: the innate and adaptive pathways. Viruses and other immunostimulatory substances may modulate the immune system's response to non-pathogenic allergens, such as those from dust mites, cockroaches and mice. This occurs at a critical point in lung growth and development in children, and effects of this immune response could have permanent effects on lung function (Holt, Upham & Sly, 2005).

1.1.1.3. Exposure to allergens is a risk for exacerbating asthma

Exposure to allergens to which individuals are sensitized can exacerbate the severity of asthma and trigger asthma attacks. A study of inner-city asthmatic children from several cities in the United States found that those asthmatic children who were sensitized to cockroaches and exposed to higher levels of cockroach allergen in the home had more frequent asthma symptoms and hospital admissions for asthma (Rosenstreich et al., 1997). Therefore, reducing exposure to allergens has been a major goal of asthma intervention studies, the efficacy of which is discussed in section 1.7.

1.1.2. The global asthma epidemic

The global epidemic of asthma, which occurred at the end of the 20th century, has been well documented and appears to parallel changes in industrialized countries (Evans et al., 1987; Sears & Beaglehole, 1987). Results from studies around the world clearly demonstrate an overall global increase in the prevalence of asthma; however, some countries have reported a plateau or even a decrease over the past decade (Eder et al., 2006) The frequency and occurrence of asthma varies among countries and communities within countries, but in most industrialized nations about 7–10% of the population are affected. Australia, New Zealand and urban communities in the United States, however, have reported 25–40% of their population affected (Peat et al., 1995; ISAAC, 1998; Nicholas et al., 2005). A study of conscripts from Finland (where military service is compulsory) over the years 1926–1989 showed an increase in asthma, from less than 0.1% to 1.8%, among 18-year-old men, with the increase starting in the 1960s (Haahtela et al., 1990). In Charleston, South Carolina, admissions for asthma at the university hospital increased 20-fold among African-American males over a 40 year period, starting in 1960 (Crater et al., 2001). This association with industrialized countries can be observed cross-sectionally in low-income countries as well. In Kenya, the rural prevalence of paediatric asthma was found to be 2%, increasing to 6% in the more urbanized villages and to 10% in Nairobi (Ng'ang'a, 1996; Odhiambo et al., 1998).

Much research effort has been focused on understanding the reasons for the global increase in the prevalence of asthma; however, no clear cause has been demonstrated. The central question is: what has changed as a result of (or in parallel with) industrialization over the last several decades to cause an increase in asthma? In response to the question, there are many hypotheses and propositions.

An initial proposition was that the reported increase in disease was not a real increase; instead, it was simply a change in the way asthma was diagnosed or was an increased awareness of the disease. While this may add to the overall increase in the prevalence of asthma, a true increase in the disease is widely accepted and has been documented in the literature (Evans et al., 1987; Sears & Beaglehole, 1987; Lundback, 1998).

Outdoor air pollution has been hypothesized as a cause for the increase in the prevalence of asthma (Samet, 1995). Air pollution is known to be associated with the exacerbation of asthma symptoms (Sandstrom, 1995); increased ozone has been associated with decreased lung function (Kinney et al., 1989); and exposure to diesel exhaust particulates has been associated with increased allergic inflammation (Nel et al., 1998; Sydbom et al., 2001).

However, one argument against air pollution causing the increase in asthma relates to timing. The rise in the prevalence of asthma has not necessarily occurred in parallel with an increase in outdoor air pollution, since environmental regulations in the 1970s have led to a decrease in overall air pollution in the United States. Specifically, a report by Pope and colleagues (2002) showed a moderate decrease in fine particles, with an aerodynamic diameter smaller than 2.5 μm (designated $PM_{2.5}$), in many cities over the past 20 years. Moreover, studies from Germany have shown an association between air pollution and bronchitis, but not asthma (von Mutius et al., 1994; Braun-Fahrlander et al., 1997).

Several hypotheses focus on the increase in asthma being caused by changes in lifestyle. Trends in conserving energy in buildings in the last half of the 20th century have resulted in people living in houses with less natural ventilation. This has led to an increase in indoor air humidity and to a more stable climate for supporting dust mite growth (Platts-Mills & de Weck, 1989). Increases in the use of carpets, a common dust mite reservoir, have also contributed to the increase in dust mite allergens (Platts-Mills & de Weck, 1989). In areas of the world currently reporting some of the highest prevalences of asthma, such as suburban Australia, the United Kingdom and inner-city areas of the southern United States, allergy to dust mites is one of the greatest risk factors for having asthma (Call et al., 1992; Peat et al., 1994; ISAAC, 1998; Crater et al., 2001). Thus, an increase in exposure to dust mites has been cited as a possible driver of the increase in asthma. However, similar prevalence rates of asthma have been reported in areas of the world where, because of low humidity, dust mites cannot live, such as northern Sweden and Los Alamos in the United States (Sporik et al., 1995; Ronmark et al., 1999), and such rates have also been reported in cities in the north-eastern United States, such as New York and Boston, where the prevalence of asthma is high, but where dust mite allergy does not contribute greatly to the profile of allergic sensitization (Rosenstreich et al., 1997).

The so-called hygiene hypothesis, which recently gained support, is another theory for the increase in the prevalence of asthma. Originally proposed by Strachan to explain increases in allergic rhinitis in the latter part of the 20th century, this hypothesis has been extended to explain the increase in asthma (Strachen, 1989). It proposes that a cleaner living environment for children in 20th century developed countries, including less exposure to viruses and bacteria, has led to an overall shift in the immune response to the more allergic T-helper cell 2. Decreased exposure to bacteria, both in the environment and in the gut, has been proposed as a factor that contributes to the increase in the prevalence of asthma (von Mutius et al., 2000; Bjorksten et al., 2001).

With respect to the relationship between asthma and pests, centuries ago exposure to allergens of cockroaches, mice and rats was invariably associated with pathogens coming from the human environment. Those exposed to these pests were also exposed to infectious pathogens. Now, this link between pests and pathogens in the urban environment has been greatly disrupted, and pests might induce only allergies (Tatfeng et al., 2005).

Decreased exposure to bacteria in early life may have contributed to the increase in allergy that occurred in the late 19th century in the United Kingdom, when hay fever became prevalent (Emanuel, 1988). In the United States, a cross-sectional study of adults found a lower prevalence of hay fever among older adults. This was associated with an increased prevalence of infections, possibly suggesting a change in the prevalence of hay fever associated with hygiene in the 20th century (Matricardi et al., 2002). Examined as an issue of timing, contact with farm animals (a significant source of microbial exposure) for the majority of individuals in the United States ended more than a century ago. While some studies have found higher bacterial endotoxin levels in homes with pets, a recent study of inner-city residents (n = 301) did not find a significant association (Heinrich et al., 2001; Park et al., 2001; Wickens et al., 2003; Perzanowski et al., 2006).

Conversely, it is debatable whether increased cleanliness resulted in the post-1950s increase in asthma. To correspond with the increase in asthma, the proposed change to cleaner living would have to have occurred within one generation. While it seems doubtful that parents of today's asthmatic children grew up in much dirtier environments than those of their offspring, Matricardi, Bouygue & Tripodi (2002) have proposed a theory whereby the increase in inner-city asthma is associated with increased hygiene, brought on by delayed class-driven urbanization in urban poor communities. Although many of the recent studies are compelling, the relevance of the hygiene hypothesis to the asthma epidemic is still not well established scientifically.

Recently the trends towards a more sedentary lifestyle in industrialized countries have been scrutinized as a possible cause of the increase in asthma (Crater & Platts-Mills, 1998). Technological advances in indoor entertainment, such as television, became common in households in developed countries in the 1960s, which along with today's cable television, video games, computers and video players, have resulted in children today spending a larger proportion of their time in sedentary activities than did prior generations. A study of asthmatic children in inner-city Atlanta, Georgia, found that 82% of the children had a television in their bedroom (Carter et al., 2001). Sedentary entertainment could have the additive effect of both lack of exercise and longer exposure to indoor allergens from sources such as dust mites, cockroaches, cats and dogs, which are known to be strongly associated with asthma (Crater & Platts-Mills, 1998).

It is likely that causes for the increase in asthma include multifactorial features of westernized countries, with some contributing more than others in different communities. What is clear is that there must be some commonality in the causes of the increase observed in different communities, given the similar time frames of the increases (Evans et al., 1987; Sears & Beaglehole, 1987; Aberg, 1989; Haahtela et al., 1990; Crater & Platts-Mills, 1998; ISAAC, 1998).

1.1.3. Public health impact of urban asthma

The public health impact of asthma in the urban environment is substantial, with the prevalence in some communities estimated at close to one in three children (Peat et al., 1995; Nicholas et al., 2005). While mortality from asthma is low, the day-to-day burden for those with asthma is substantial, and the economic costs to society are high. As more countries develop urban centres and adopt western lifestyles and methods of building residential environments, the importance of understanding the effect of these changes on urban pests and how they relate to health will only increase.

Findings from several studies show that the prevalence of childhood asthma in an urban population could range from 8% to 22% and that the prevalence of allergy among asthmatic children varies by community (50–80%) (Martinez et al., 1995; Kattan et al., 1997; Ronmark et al., 1999; Lau et al., 2000; Carter et al., 2001). Among these atopic children in an urban environment, many are allergic to urban pest allergens that can exacerbate symptoms (Rosenstreich et al., 1997). Therefore, exposure to urban pests could affect 4–17% (for example, from 50% of 8% to 77% of 22%) of children living in an urban environment. One study in North America found that for every (United States) dollar spent, asthma disease management has been shown to save US$ 3–4, by decreasing visits to emergency departments (Rossiter et al., 2000). Overall, the long-term economic and quality-of-life benefits of decreasing the number of missed school and work days are difficult to assess, but they are surely positive.

1.2. Assessing allergen exposure

1.2.1. Questionnaire assessment

Several investigators have examined worldwide associations between home characteristics, reported by residents, and dust mite allergen levels (Chan-Yeung et al., 1995; Munir et al., 1995; Platts-Mills et al., 1997; Tobias et al., 2004). In warm and humid environments, the concentration of dust mite allergens is expected to be high in soft furnishings throughout the year (Zhang et al., 1997). In more temperate climates, a seasonal variation has been observed (Platts-Mills et al., 1987; Miyazawa et al., 1996; Chew et al., 1999a), and several characteristics have been associated with exposure to dust mite allergens: altitude, type of building, and presence and type of carpeting (Wickman et al., 1991; Arlian, 1992; Harving, Korsgaard & Dahl, 1993; van Strien et al., 1994, 2002; Chew et al., 1998; Wickens et al., 2001; Basagana et al., 2002; Mihrshahi et al., 2002; Matheson et al., 2003). In general, factors that lead to large dust mite populations will lead to high concentrations of the allergen. However, some evidence exists for the passive transfer of dust mite allergens from dust-mite-hospitable environments to inhospitable environments, such as automobiles (Neal, Arlian & Morgan, 2002).

In homes, a visual assessment is possible for cockroaches and mice, but not for microscopic dust mites. A resident, a trained inspector, or both can make visual assessments for cockroaches or mice. In the NCICAS, 937 homes were evaluated for cockroach and rodent infestation by residents and trained evaluators (Crain et al., 2002). The percentage of residents that reported problems with mice or rats in the past 12 months was higher than that observed by the inspector (40% versus 9%), and the percentage of residents that reported problems with cockroaches in the past 12 months was higher than that observed by the inspector (58% versus 15%. In another study, mouse allergen (>1.6 µg/g) had a significant association with resident reports of rodent infestation (odds ratio (OR) = 3.38, $P < 0.05$), but not with inspector observations (Cohn et al., 2004). In a study of inner-city children in Baltimore, 41% of homes had substantial amounts of a major cockroach allergen, Bla g 1 (> 1 unit/g), but only 18% had a cockroach infestation identified by a home inspector (Matsui et al., 2003). Although residents might not accurately report cockroaches or rodents in their homes for a variety of reasons, the studies above suggest that residents might be better informed than trained inspectors with regard to previous infestations that could lead to high levels of allergen in the dust.

Because of the range of housing stock across the world, some home characteristics are not consistently associated with mouse or cockroach allergens (Phipatanakul et al., 2000b; Rauh et al., 2002; Stelmach et al., 2002a, 2002b; Chew et al., 2003; Cohn et al., 2004; Matsui et al., 2005). For example, the United States National Survey of Lead and Allergens in Housing, which included information from buildings in 75 locations, found that the level of mouse allergens was higher in high-rise buildings (five storeys or more) than in low-rise apartments (one to four storeys) (Cohn et al., 2004). This finding is not directly applicable to some cities, such as New York City, where the majority of the housing in low-income neighbourhoods is greater than five storeys. In fact, shorter apartment buildings (fewer than eight storeys in New York City) were more likely to have high levels of mouse allergen in the kitchen and bed dust than taller high-rise buildings) (OR = 10 and 6.25, for kitchen and bed dust, respectively) (Chew et al., 2003). This highlights the importance of considering the geographic factors that influence allergen levels within the home. However, some predictors, such as resident reports of mice or cockroaches, are consistently associated with high levels of their respective allergens (Chew et al., 1998; Phipatanakul et al., 2000a; Rauh et al., 2002; Cohn et al., 2004; Matsui et al., 2004a). The use of a standardized questionnaire, such as that employed by the large analysis and review of European housing and health status (LARES) project, can enable the survey of larger populations and comparisons across study sites when allergen measurements are not feasible, but care must be taken in the interpretation, for the reasons noted in this section (WHO Regional Office for Europe, 2006).

1.2.2. Pest counts

Before immunoassays were developed, exposure to dust mites was assessed by trained acarologists who identified and counted dust mites in house dust. Subsequently, mite counts were shown to correlate with allergen concentrations in the dust (Platts-Mills et al., 1992). Other investigators have made measurements of faecal pellets (which contain guanine) and found significant associations with dust mite allergens (Tovey et al., 1981; van Bronswijk et al., 1989), but the development of allergen assays has enabled cost- and time-effective assessments of exposure that do not depend on the expertise of acarologists. The main problem with bypassing the identification step is that many environments have more than one type of dust mite (Arlian et al., 1992; Montealegre et al., 1997; Warner et al., 1998), and the use of a single dust mite allergen immunoassay (see section 1.2.3) could preclude finding meaningful etiological agents of allergic disease.

Correlations between cockroaches and cockroach-allergen levels in urban homes have been observed; a major cockroach allergen, Bla g 2 (from the German cockroach, *Blatella germanica*), was positively correlated with cockroaches collected on sticky traps in bedrooms ($r = 0.63$, $P < 0.001$) and in kitchens ($r = 0.53$, $P < 0.001$) (McConnell et al., 2003). A study of low-income suburban homes found that the number of cockroaches collected on sticky traps correlated best with the allergens in dust sampled two months after trap collection (Mollet et al., 1997). Of note is that cockroach allergen was still detectable in kitchen and bed dust samples, even when no cockroaches were detected.

1.2.3. Allergen assays

Serum IgE antibodies from allergic individuals have been used to identify the relevant (allergenic) proteins from allergen sources. Enzyme-linked immunosorbent assays

(ELISAs) were subsequently developed to measure these relevant proteins (from cockroaches, mice, rats and dust mites) in dust and air samples (Chapman et al., 1987; Pollart et al., 1991a; Renstrom et al., 1997; Ferrari et al., 2004). The ELISAs differ somewhat depending on which allergen is being measured, but most use a sandwich method with primary monoclonal capture antibodies bound to a plastic surface. The quantity of an allergen-specific secondary antibody is detected by a colorimetric enzymatic reaction. These immunoassays demonstrate a high degree of sensitivity and specificity, with nanogram quantities of protein detectable with little cross-reactivity between allergens. Since ELISAs involve an antibody or antigen interaction, they are susceptible to interference from contaminants in the dust samples (Woodfolk et al., 1994; Chew et al., 1999b). With the antibodies used in the ELISA, rapid screens have been developed and enable a quick assessment of allergen levels in the home or workplace without the use of major laboratory facilities. The results from the rapid test are not as quantitative as those obtained by ELISA, but they do correlate well with those obtained by ELISA (Chapman, Tsay & Vailes, 2001; Lau et al., 2001).

1.2.4. Dust sampling

Dust collected from floors and furniture has been used to assess exposure in many studies (Gelber et al., 1993; Platts-Mills et al., 1997). Typically, a vacuum cleaner modified with a dust collection device mounted on the nozzle is used. Sometimes, the dust is sieved to remove large particles, which would not become airborne easily. The dust is extracted in a buffered solution with a detergent and assayed for allergens by the ELISA, guanine or rapid screen methods (van Bronswijk et al., 1989; Tsay et al., 2002). This provides a relatively easy technique for assessing allergens in the dust of individual homes.

1.2.5. Airborne sampling

Airborne dust can be collected on filters, extracted and assayed for allergens by ELISA. Samples have been collected at high flow rates for short periods of time (less than an hour) and at low flow rates for longer periods of time (a week). Small pumps worn on a belt can be used to better assess an individual's exposure. One important consideration in airborne sampling is causing a disturbance in the room being sampled, because allergens from different animals travel on particles of different sizes, which affect how easily they become airborne and how quickly they settle out of the air.

1.2.6. Comparison of various methods of allergen exposure assessment

Identifying a method that accurately assesses what has been deposited in a person's airway over a lifetime is extremely difficult. Even a study that assesses exposures in the homes of individuals from birth on a regular basis and then assesses other exposure locations, such as day-care centres and schools, will not completely determine lifetime exposure. The reasons for this are as follow.

Moderate temporal fluctuations occur in allergen concentrations in homes, even on a month-to-month basis (Chew et al., 1999a). Therefore, a sample collected once (or even once a year) may not truly represent the fluctuations in exposure to allergens seen over the time period between samples.

Peak exposures to some allergens – for example, to cat allergens for non-cat owners – may occur in locations outside of the home (Custovic et al., 1996b).

Due to the lack of large temporal variations in allergen concentrations within one home (except in some intervention studies), apportioning the risk from exposure to allergens that occurred early versus late in life is difficult.

When considering how accurately a dust sample reflects the allergen that deposits in a lung, the allergen under study must be considered. The particles that mite and cockroach antigens travel on are not easily made airborne; therefore, correlations between dust and undisturbed airborne sampling are poor (Platts-Mills et al., 1997; Tovey et al., 1981). Although, cockroach allergens can be carried on particles with a wide range of sizes, depending on the amount of air disturbed (de Blay et al., 1997; de Lucca et al., 1999), most of the Bla g 1 allergens tend to be associated with particles greater than 10 μm in aerodynamic diameter. The recent development of intranasal samplers has detected Bla g 1 in residences with a low level of disturbed air (de Lucca et al., 1999), but most research has shown that a considerable disturbance is necessary to measure airborne cockroach allergens (Swanson, Agarwal & Reed, 1985; de Blay et al., 1997). Exposure to dust mite allergens most likely occurs in close contact with the reservoirs; allergen is probably inhaled either when a person's head is close to bedding or when children are playing on carpets (Custovic et al., 1999). Vacuum cleaning can also increase exposure (de Blay et al., 1991, 1997).

To assess exposure to dust mite and cockroach allergens, dust sampling is probably a better method than airborne sampling. Since dust mite allergens are not measurable without disturbance, the dose–response relationship between exposure and sensitization has been stronger when the exposure was assessed with a settled dust sample than with an airborne sample (Sporik et al., 1990; Platts-Mills et al., 1997). The fact that cockroach allergen is associated mainly with larger particles might in part explain why studies have found that relationships among exposure, sensitization and asthma are most strongly associated with exposure in bedrooms, where residents spend prolonged periods of time (Sarpong et al., 1996; Rosenstreich et al., 1997; Gruchalla et al., 2005).

Previous studies have found variability between the quantity of pet allergen in airborne and settled dust, due in large part to the ease with which the particles on which cat and dog allergens become airborne. Cat and dog allergens have the ability to travel on smaller particles and can be detected in air even three hours after disturbance (Luczynska et al., 1990; de Blay et al., 1991; Custovic et al., 1996b, 1997). Therefore, previous correlations between airborne samples and dust samples were often poor. Recently, though, a good correlation between a major cat allergen (Fel d 1), measured in the air and in the dust, was demonstrated when the airborne samples were collected over a 24-hour period (Custis et al., 2003). It is likely that the longer sampling period better reflects an average exposure. Karlsson and colleagues (2002) collected settled dust in a Petri dish over a period of a week and found a good correlation with personal air sampling, and this may provide an alternative to long-term airborne sampling. These findings could be extended to rodent allergens, which travel on particles with similar aerodynamic properties (Ohman et al., 1994).

1.3. Cockroaches and asthma

1.3.1. Allergic sensitization to cockroach allergens

In the 1960s and 1970s, researchers found that sensitivity to cockroaches was common among atopic populations (Bernton & Brown, 1967; Kang et al., 1979; Kang & Chang, 1985). Specifically, sensitivity to crude extracts of cockroach allergens among allergic individuals ranged from 20% in Boston to 53% in Chicago. Not all cockroach allergen extracts are the same, and this in part explains some of the variation in the prevalence of allergy to cockroaches in the earlier studies. German cockroach whole body extracts correlate well with faecal extracts (r = 0.88, P< 0.001) (Musmand et al., 1995). Specific cockroach antigens that elicit an IgE response in a majority of cockroach-allergic individuals have been identified. Pollart and colleagues (1991b) found that twice as many cockroach allergic asthmatics had IgE antibodies against Bla g 2 (an allergen from German cockroaches), compared with those against Bla g 1 (an allergen found in several different types of cockroaches). Furthermore, an estimated 60–70% of cockroach-sensitive individuals have IgE antibodies to the allergens Bla g 4 and Bla g 5 (Arruda et al., 2001).

1.3.2. Cockroach sensitization and asthma

In many areas of the world, sensitization to cockroach allergens has been associated with asthma (Platts-Mills et al., 1997; Eggleston, 2001). In inner cities in the United States, more asthmatic children with cockroaches reported or observed in their home were sensitized to cockroach allergens than those without exposure to cockroaches (75% versus 53%, P<0.01) (Crain et al., 2002). A study of urban, suburban and rural atopic patients found a higher prevalence of cockroach sensitivity among patients with a primary diagnosis of asthma (49.6%) than among those with a primary diagnosis of allergic rhinitis (30.3%) (Garcia et al., 1994). Also in this study, no difference was observed between the prevalence of sensitivity to cockroach allergens in urban and rural environments. This finding is in contrast to previous studies in Poland and the United States, which reported a high degree of sensitization to cockroach allergens among urban dwelling asthmatics (Matsui et al., 2003; Stelmach et al., 2002b). Nonetheless, sensitivity to cockroach allergens appears to be important in the pathogenesis of asthma in areas with cockroaches.

1.3.3. Asthma and the exposure of sensitized individuals to cockroaches

Exposure to cockroach allergens in early life has been associated with recurrent asthmatic wheezing in children with a family history of atopy (Litonjua et al., 2001). This was also observed in the first year of life, when allergy-specific IgE to inhaled allergens typically cannot be detected, suggesting a potential role for a non-allergic mechanism of airway inflammation due to exposure to cockroach allergens (Gold et al., 1999). Also, the combination of exposure and sensitization to cockroach allergens has been shown to increase rates of hospitalization, missed school days and days with wheezing among asthmatic children (Rosenstreich et al., 1997). Moreover, recent evidence suggests that the combination of exposure and sensitization to cockroach allergens is a stronger risk factor for asthma morbidity in the inner cities of the United States than is exposure and sensitization to allergens produced by dust mites, cats or dogs (Gruchalla et al., 2005).

1.4. Rodents and asthma

1.4.1. Sensitization and symptoms

Most of what is known about allergic reactions to mice and rats originates from occupational studies. Symptoms of allergy, including rhinitis, conjunctivitis and asthma, have been reported by workers in animal laboratories, notably those that work with rats and mice (Taylor, Longbottom & Pepys, 1977; Renstrom et al., 1994; Hollander et al., 1997; Lieutier-Colas et al., 2002). In the 1970s, British researchers reported that five people who worked with laboratory animals developed asthma within two weeks to two years of starting to work with mice and rats (Taylor, Longbottom & Pepys, 1977). Within a year of developing asthma, all five workers experienced asthma symptoms after only a few minutes of exposure to the animals. These studies were among the first to show that urinary proteins from rats and mice, rather than fur, elicited stronger positive skin prick tests and bronchial hyperreactivity. In a multi-site study of 650 people who worked in animal laboratories in the Netherlands, Sweden and the United Kingdom, researchers found that 9.7% of the workers were sensitized to rat urinary allergens (Heederik et al., 1999). They also found a higher risk of sensitization to rats associated with high exposure to rat allergens.

Recently, researchers realized that mouse and rat allergens might also contribute to the development or exacerbation (or both) of childhood allergic asthma. Polish researchers found that 61% of inner-city children exposed to detectable levels of mouse allergen were skin prick positive to these allergens, whereas only 14% of children with levels below the limit of detection were sensitized to mice (Stelmach et al., 2002a). In 2000, a study of 499 children in the United States with asthma found that 18% were allergic to mouse allergen and that those with exposure to Mus m 1 (an allergen in mouse urine) > 1.6 µg/g in kitchen dust were more likely (OR = 2.2) to become sensitized to the mouse allergen than those with a lower level of exposure (Phipatanakul et al., 2000a). Among the same cohort of children, allergy to rats was also prevalent (21%), and those with sensitization and exposure to rat allergens experienced more unscheduled medical visits, hospitalizations and days with diminished activity due to asthma (Perry et al., 2003). While the terms *urban* and *suburban* are somewhat subjective, it is of interest to report that mouse allergens measured in suburban homes in the United States have also been associated with sensitization to mouse allergen, as judged by a skin prick test (Phipatanakul et al., 2005). Specifically, quartile increases in bed dust Mus m 1 (median: 0.76 µg/g; interquartile range: 0.16–3.20 µg/g) were associated with a greater likelihood of having a positive skin prick test (OR = 1.4). This finding suggests that even low levels of mouse allergen can pose a risk of developing allergic sensitization. Furthermore, Polish researchers found that children of workers in animal laboratories were also allergic to mice and rats, suggesting that passive transfer of rodent allergens from work to the home might be a biologically relevant exposure pathway (Krakowiak, Szulc & Gorski, 1999).

1.4.2. Exposure

1.4.2.1. Size characteristics of airborne mouse and rat allergens

Occupational studies are the source of most of the size characterization of particles that bear rodent allergens. Dutch researchers conducted ambient and personal air sampling in seven laboratory animal facilities and found that rat and mouse allergens were contained mainly on particles larger than 5.8 µm in aerodynamic diameter (Hollander et al., 1998). In a French study of 12 rat breeding facilities, different tasks resulted in varying levels of exposure to allergens (Lieutier-Colas et al., 2001). Rat n 1 is an allergen from rats, and changing cages and feeding the rats were associated with an average level of 91.1 ng Rat n 1 per m^3 of air compared with an average level of 0.4 ng Rat n 1 per m^3 of air for office duties. Both the Dutch and French researchers found that the greater the number of animals in a room, the higher the allergen level. In the United Sates, 13 rooms of a major animal handling facility were sampled for airborne mouse allergen (Mus m 1), and the particles were size fractionated (Ohman et al., 1994). Among the rooms without mice, most of the Mus m1 was contained on 0.4–10-µm-diameter particles. However, most of the Mus m 1 in rooms with a high density of mice was found on particles larger than 10 µm in aerodynamic diameter.

1.4.2.2. Residential exposures

Mouse allergens have been measured in dust and, to some extent, air samples from homes in urban and suburban environments where children reside. In some studies, Mus m 1 was measured with a monoclonal antibody ELISA; in others, mouse urinary proteins (MUP), which include Mus m 1, were measured with polyclonal antibody-based ELISAs (Phipatanakul et al., 2000a; Chew et al., 2003; Cohn et al., 2004), but the two allergen measurements were highly correlated ($r = 0.96$, $P < 0.001$) (Chew et al., 2003). Mus m 1 levels in bed dust were as high as 294 µg/g, with a median level of 0.5 µg/g in a multi-site study of asthmatic children living in inner cities across the United States (Phipatanakul et al., 2000a). In a national survey of American homes (urban, suburban and rural), the median MUP level in bed dust was 0.25 µg/g (Cohn et al., 2004). The median MUP level was 0.5 µg/g for bed dust in low-income New York City homes (Chew et al., 2003), which was slightly higher than that (median = 0.23 µg/g) for Polish inner city homes (Stelmach et al., 2002a). Mouse allergen levels have been found to be consistently higher in kitchen dust than in bed dust. The level of mouse allergen in the air of homes where children reside is usually lower than that in occupational settings; however, the levels can be relatively high (Chew, Correa & Perzanowski, 2005; Matsui et al., 2005). While one study observed that mouse allergen levels were lower among cat owners, all studies found associations between problems reported by residents with rodents and increased levels of mouse allergens. Reasons for differences in home characteristics being predictive of mouse allergen levels are explained in greater detail in section 1.2.1.

Unlike mouse allergens, rat allergens in house dust have not been associated positively with sensitization (Perry et al., 2003; Phipatanakul et al., 2005). The reason for this disparity could be that significant rat allergen exposure might also occur outside of the home, such as in subways, schools and restaurants, and that home measurement of this allergen might not be representative of total exposure. Although the same could be said for mouse allergen, it appears that a significant exposure to mouse allergen does arise from homes. Nevertheless, high levels of mouse allergen (≥ 2 µg/g) were found in 78% of New York City school classrooms during at least one sampling season (Chew, Correa & Perzanowski, 2005) and ranged from 0.3 µg/g to 118 µg/g in 12 urban Baltimore schools (Amr et al., 2003).

1.5. Dust mites and asthma

1.5.1.Sensitization and asthma

The evidence for a causal role of exposure to allergens in the pathogenesis of perennial asthma has been better established for dust mite allergens than any other inhaled allergen (Platts-Mills et al., 1992). In communities where exposure to dust mites is prevalent, dust mite allergens are often among the most common allergens to which atopic individuals are sensitized (Platts-Mills et al., 1997). A birth cohort study by Sporik and colleagues (1990) found that children exposed to high levels of dust mite allergens in their homes in the first year of life were more likely to develop sensitization to dust mite allergens and that those exposed to the highest levels were the most likely to develop asthma. This established a dose–response relationship between exposure to dust mite allergens and sensitization. This relationship was also observed in a German study (Lau et al., 2000; Illi et al. 2006). Sensitization to dust mites has been observed to be strongly associated with asthma in many areas of the world, including Australia, the United Kingdom, the mid-Atlantic and southern United States and especially New Zealand, which reports some of the highest residential dust mite allergen concentrations in the world (Sporik et al., 1990; Peat et al., 1994; Squillace et al., 1997; Wickens et al., 1997). Sensitization to the storage mites of the genera *Tyrophagus, Glycyphagus, Lepidoglyphus* and *Blomia* has also been associated with allergic disease (van Hage-Hamsten & Johansson, 1998). Further evidence for the causality of exposure to dust mite allergens exacerbating asthma comes from studies that have shown successful improvement in asthma patient symptoms when exposure to dust mite allergens were reduced. The efficacy of avoiding allergens is discussed in detail in section 1.7.

1.5.2. Factors associated with exposure to dust mite allergens

Dust mites are found in the bedding, pillows, mattresses, carpets and upholstered furniture of homes, where they feed on scales from human skin (Tovey et al., 1981). They absorb water from the air, and therefore require a relative humidity above 50% to survive (Arlian, 1992). This restricts the environments inhabitable by mites and is a major controlling factor in the geographical distribution of dust mites. Reducing indoor air humidity has recently been used with some success as a mite eradication strategy (Arlian, Neal & Vyszenski-Moher, 1999; Arlian et al., 2001). In major cities of the north-eastern United States, multifamily buildings are often overheated during the winter, which leads to low humidity air indoors and precludes the proliferation of dust mites (Chew et al., 1999a). Mite genera vary geographically, with *Dermataphagoides* being more common in temperate climates and *Blomia* being found only in tropical climates (Arlian, Morgan & Neal, 2002).

1.6. Proposed thresholds of exposure to allergens associated with allergic asthma

1.6.1. Studies examining thresholds of exposure relevant to disease

Prior exposure to an allergen is necessary for an immediate hypersensitivity immune response; however, the quantity of allergen necessary for sensitization appears to depend on the allergen in question, concurrent exposure to other substances that stimulate the immune system and the genetic predisposition of the individual being exposed. In examining thresholds of exposure to pests or (more specifically) allergens from those pests that lead to allergic disease, it is first necessary to consider allergic sensitization and symptoms of allergic diseases as separate outcomes. The level of allergens in homes necessary for allergic sensitization – that is, development IgE antibodies – is probably less than is necessary for precipitating asthma symptoms (Platts-Mills et al., 1992). For example, for dust mite allergens, the proposed threshold of exposure for sensitization is fivefold lower than that for asthma symptoms (Sporik et al., 1990).

The strongest scientific evidence for a threshold level of exposure to an allergen exists for dust mite allergens; however, the levels proposed by Sporik and colleagues (1990) have been debated by subsequent studies (Warner et al., 1996). Due to the design of studies conducted to date, mostly cross-sectional and lacking controls (non-asthmatics), the thresholds for developing sensitization to cockroach and mouse allergens have not been well established. Recent studies suggest that the differences observed between studies that evaluated the risk of exposure to mite allergens may likely be due to the influence of concurrent exposures that alter the immunological response and genetic factors that affect an individual's susceptibility to produce IgE antibodies in response to an allergen (see subsection 1.1.1.1) (Kleeberger & Peeden, 2005).

1.6.1.1. Dust mites

Based on findings from a prospective birth cohort in 1990, Sporik and colleagues proposed a threshold level of the dust mite allergen Der p 1 (from the house dust mite *Dermatophagoides pteronyssinus*) sufficient for sensitization (2 µg/g) and a threshold for an increased risk of asthma (10 µg/g) (Sporik et al., 1990). However, subsequent studies have proposed that exposure to much lower concentrations is sufficient for sensitization in a sizeable fraction of the population (Munir et al., 1995; Warner et al., 1996). Exposure to dust mite allergens in the home does appear to have a dose–response relationship, with those individuals exposed to more allergen being more likely to become sensitized to mite allergens, especially among those with atopic mothers (Platts-Mills et al., 2001; Cole Johnson et al., 2004; Brussee et al., 2005; Kerkhof et al., 2005). In communities where dust mite allergen in most homes is greater than 10 µg/g, the population response appears to plateau, leading to the majority of allergic individuals being sensitized to mite, likely indicating saturation (Marks, 1998).

As discussed in subsection 1.6.1, the response to exposure to dust mite allergens, as well as to other allergens, is likely influenced by environmental exposures and the genetic predisposition of the individual, and this can explain differences in thresholds reported from different studies. A recent study in a birth cohort in Manchester, England, found that increasing exposure to dust mite allergens was associated with an increased risk of sensitization, but that effect was dampened if the children were also exposed to higher levels of endotoxin. However, the modifying effect of endotoxin was only observed in children with a specific genetic polymorphism in a gene for an endotoxin binding receptor (Simpson et al., 2006). From this example, it is clear that defining allergen exposure thresholds is complicated by other environmental exposures and by individual genetic susceptibilities.

1.6.1.2. Cockroaches

To date, exposure thresholds for developing IgE to cockroach allergens have not been well established. However, two studies examined exposure to cockroach allergens and its relationship to the risks of developing asthma and asthma symptoms. In a prospective study of children with allergic parents who lived in the Greater Boston area (urban and suburban), those exposed to between 0.05 unit/g and 2 units/g of cockroach allergen (Bla g 1 or Bla g 2) were eight times more likely to develop asthma than those exposed to less than 0.05 unit/g (Litonjua et al., 2001). Those children exposed to 2 units/g or more were even more likely to develop asthma. A study of inner-city asthmatics in the United States found that asthmatics who were sensitized to cockroach allergens and were exposed to more than 8 units/g of allergen (Bla g 1) had more severe asthma symptoms (Rosenstreich et al., 1997).

1.6.1.3. Rodents

Population-based studies on exposure to mouse allergens, sensitization and asthma are limited to cross-sectional studies. One study of rat allergens found that sensitization was common among inner-city asthmatics (22%); however, it was not associated with rat allergen measured in the house dust at the same time (Perry et al., 2003). In the same cohort of asthmatic children, investigators found that those children exposed to greater than the median level of mouse allergen in the kitchen (Mus m 1 >1.6 µg/g) were twice as likely to be sensitized to mouse allergen (Phipatanakul et al., 2000a). A study conducted in a suburban community where exposure to mouse allergen was substantially lower (median Mus m 1: 0.02 µg/g in the bedroom) also found an increasing risk of sensitization with increasing exposure (OR in quartile of exposure) (Matsui et al., 2004a). A larger effect was seen in an occupational study of exposure to airborne mouse allergens and sensitization to them (OR = 1.7 for each increase in quintile of exposure) (Matsui et al., 2004b). A recent study of the homes of inner-city asthmatics found that many homes had airborne mouse allergen levels in the range of those found in occupational settings (Matsui et al., 2005). Two caveats for application of these thresholds are that the studies do not take into account past exposure to allergens – that is, cross-sectional associations – and that individuals with allergen levels below these thresholds were also sensitized. Although specific monoclonal antibodies have been developed for detecting rat allergens in environmental samples (Ferrari et al., 2004), establishing the levels of rat allergen associated with development of sensitization may be hindered by the cross-reactivity of human IgE antibodies to mouse and rat allergens (Spitzauer, 1999; Renstrom et al., 2001).

1.7. Public health recommendations

Asthma and other allergic diseases are typically treated with medications to control current symptoms, prevent future exacerbations, or both. Reduction of clinical symptoms is also successfully achieved for some individuals who use allergen-specific immunotherapy (Abramson, Puy & Weiner, 2003). In the field of asthma research, an emerging strategy for reducing or preventing symptoms has been to reduce exposure to allergens in the home. However, findings on the efficacy of current allergen-avoidance strategies are conflicting. Table 1.1 summarizes current levels of scientific evidence in clinical trials of allergen avoidance (Shekelle et al., 1999).

A previous report by WHO defined prevention with respect to asthma and allergic disease in the following terms (WHO, 2003). Primary prevention is the "prevention of immunological sensitization (i.e., the development of IgE antibodies)". Secondary prevention is "preventing the development of an allergic disease" (such as atopic eczema, atopic dermatitis, allergic rhinitis or allergic asthma) after the onset of sensitization, but before the development of clinical symptoms. Tertiary prevention is the "treatment of asthma and allergic diseases".

1.7.1. Avoiding allergens to decrease asthma symptoms (tertiary prevention)

A successful demonstration of how avoiding exposure to allergens can reduce asthma symptoms occurs annually when asthmatics allergic to pollen have fewer symptoms outside of the pollen season (Custovic & Wijk, 2005). Specifically ragweed pollen can fluctuate by orders of magnitude, and during the non-peak season allergic individuals have fewer asthma symptoms and also produce less allergen-specific IgE (Creticos et al., 1996). The reduction of domestic allergens seems a logical strategy against asthma for the following three reasons (Creticos et al., 1996; Platts-Mills et al., 1997).

Asthma is strongly associated with sensitization to inhaled allergens found in the home.

An increase in the severity of asthma among sensitized individuals is observed with increasing domestic levels of most of these allergens.

Asthma symptoms and IgE decrease for pollen allergic asthmatics outside of the pollen season.

However, many efforts at tertiary prevention that have attempted to reduce allergen exposure in the home environment have been ineffective in reducing symptoms, especially when a single allergen or single exposure site was targeted (O'Connor, 2005).

The longest history of studies of tertiary intervention to reduce indoor allergens exists for dust mites. Initial studies with small numbers of patients achieved successful improvement of asthma symptoms, by removing the patient to a geographical location or hospital room without dust mites (Platts-Mills et al., 1982; Boner et al., 1985). With varying success, subsequent strategies have focused on reducing mites and their allergens in homes (Marks, 1998; O'Connor, 2005). The method most commonly used is encasement of mattresses and pillows in an allergen impenetrable fabric.

Recently, there has been considerable debate in the literature on asthma about the efficacy of dust mite allergen avoidance, especially with regard to encasement-only strategies (Platts-Mills, 2004; Custovic & Wijk, 2005; O'Connor, 2005; Schmidt & Gotzsche, 2005). In 2004, an update of a meta-analysis evaluated the efficacy of control measures for house dust mites in reducing asthma (Gotzsche et al., 2004). This report concluded that current methods of avoiding dust mite allergens were ineffective and should not be recommended. However, subsequent criticisms of the meta-analysis charged it with a narrow inclusion criteria and a lack of sufficient control for studies that demonstrated a reduction in allergen exposure (Platts-Mills, 2004; Custovic & Wijk, 2005; O'Connor, 2005). The authors of the meta-analysis also reported a positive reference bias in narrative reviews on the subject – that is, reviewers preferentially citing studies with positive effects (Schmidt & Gotzsche, 2005). Recently a well-designed placebo control trial was conducted in which dust mite mattress encasements were distributed through a physician's office as a single avoidance measure – that is, no other avoidance measures were advocated (Woodcock et al., 2003). This study found no statistically significant improvement among the patients exposed to avoidance measures relative to the placebo controls. What seems clear from the meta-analysis and the recent placebo control trial is that targeting single allergens or single domestic exposure locations for allergen avoidance is not sufficient to reliably reduce symptoms (O'Connor, 2005).

Although some recent studies have shown that allergen interventions can effectively reduce mouse and cockroach allergens, allergen reduction below levels that are thought to be clinically important may not be sustainable (Eggleston et al., 1999; Gergen et al., 1999; Williams, Reinfired & Brenner, 1999; McConnell et al., 2003). The challenge of reducing exposure to cockroach and rodent allergens in urban homes is formidable (see Chapter 2, on cockroaches). Unlike dust mites, for which infestations are typically localized in bedding, carpeting and soft furniture, cockroaches and mice are more mobile, and their infestations are more dispersed. In urban locales, a special challenge is posed by infestations that may span an entire multifamily apartment building (Chew et al., 1998, 2006; Eggleston et al., 1999; Hynes et al., 2004). A study by Eggleston and colleagues (2005) of inner-city asthmatics, which included cockroach abatement for infested households, resulted in a significant decrease in cockroach allergen in the abatement group, while the control group had no change. While the abatement group had significantly fewer daytime asthma symptoms than the controls at the end of the one-year trial, no other measures of asthma symptoms, lung function or visits for acute attacks of asthma were significantly different among the members of the abatement group compared with controls. The authors speculated that lack of significant improvements might have been due to inclusion criteria that did not cover more severe asthmatics and did not require that the patients were allergic to cockroaches. Also, the reduction of cockroach allergens by 40% may not have resulted in low enough exposure for clinical relevance.

When evaluating strategies for avoiding allergens, several important features of the relationship between allergen exposure and asthma must be considered. First, as a popula-

Table 1.1. Summary of evidence: domestic allergen exposure, asthma and preventative strategies

<table>
<tr><th rowspan="2">Study type</th><th rowspan="2">Outcome of interest</th><th colspan="3">Source of allergens</th></tr>
<tr><th>Dust mites</th><th>Cockroaches</th><th>Rodents</th></tr>
<tr><td rowspan="3">Association between exposure to domestic allergens and allergic sensitization and asthma</td><td>Sensitization</td><td>High exposure is associated with development of sensitization in prospective studies</td><td colspan="2">High exposure is associated with sensitization in cross-sectional studies.</td></tr>
<tr><td>Development of asthma</td><td>Sensitization is associated with the development of asthma.[a]</td><td colspan="2">Sensitization is associated with asthma.</td></tr>
<tr><td>Asthma exacerbation</td><td colspan="2">Current exposure is associated with more severe asthma among sensitized individuals.</td><td>–</td></tr>
<tr><td>Efficacy of current methods for reducing allergens in the home</td><td>Allergen reduction</td><td>Ib. Effective: allergen impermeable bed covers and replacing carpets with hardwood floor.

IIb. Effective: wash bedding in hot water and use of high-efficiency particulate air vacuum cleaners (Custovic & Wijk 2005).</td><td colspan="2">Ib. Effective, but difficult to maintain: integrated pest management, which includes education of families to maintain reduced allergen levels.</td></tr>
<tr><td rowspan="4">Efficacy of avoiding domestic allergens for preventing asthma and asthma exacerbation</td><td>Primary prevention (development of atopy)</td><td>Ia. Not effective: allergen impermeable bed covers alone.

Ib. Potentially effective: studies have been of short duration or included interventions in addition to allergen avoidance, making the contribution of allergen reduction difficult to determine, so more studies with longer duration and follow-up to older ages are needed.</td><td>Not tested</td><td>Not tested</td></tr>
<tr><td>Secondary prevention (development of asthma after atopy)</td><td>Studies to date have focused on primary and tertiary prevention</td><td>Not tested</td><td>Not tested</td></tr>
<tr><td rowspan="2">Tertiary prevention (reduction of asthma symptoms)</td><td>Ia. Not effective: mattress or pillow encasement alone.</td><td></td><td></td></tr>
<tr><td colspan="3">Ib. Effective: multifaceted, allergen- (patient-)specific, comprehensive avoidance. Results are from one study, so more studies of this type are needed to verify findings.</td></tr>
</table>

[a] Dust mite birth cohort studies have shown conflicting evidence on whether dust mite allergen exposure is a risk factor for the development of asthma (Sporik et al., 1990; Lau et al., 2000).

Note. Categories of clinical evidence based on Shekelle et al. (1999) are shown for allergen avoidance: *Ia.* = evidence from a meta-analysis of randomized controlled trials (RCT); *Ib.* = evidence from at least one RCT; *IIa.* = evidence from at least one trial without randomization; *IIb.*= evidence from at least one other type of quasi-experiment; *III.* = evidence from non-experimental descriptive studies; and *IV.*= evidence from expert committee reports or opinions or clinical experience or respected authorities. Although there are no III or IV levels of evidence referenced in the table, they are part of the Shekelle scale and are included in this note, so readers will not erroneously think that level II is the least rigorous level of evidence.

tion, allergic asthmatics are sensitized to a variety of allergens, and reduction of allergens to which an individual is not sensitized is unlikely to yield any significant changes in their clinical symptoms. Therefore, the use of clinical allergy testing to determine the allergens to which an asthmatic is sensitized, is recommended (National Heart, Lung and Blood Institute, 1997; WHO, 2003; Johansson & Haahtela, 2004; O'Connor et al., 2004). A study of low-income urban asthmatics found that patients were often not evaluated for sensitization or did not receive any education in avoiding allergens (Busse, Wang & Halm, 2005).

A second important feature of the relationship between allergen exposure and asthma is the sensitization of many asthmatics to multiple allergens, and reduction of exposure to only one of these allergens may prove ineffective in reducing symptoms, due to continued exposure to other relevant allergens that exacerbate the disease. As O'Connor and colleagues (2004) point out in a review of allergen-avoidance strategies, exposure to other asthma triggers, such as ETS, should also be included in avoidance strategies, since they may perpetuate symptoms. Therefore, avoidance measures should be global, by including all of the allergens to which an individual is sensitized and exposed, as well as other respiratory irritants.

A third important feature of the relationship between allergen exposure and asthma is the occurrence of exposures in multiple locations; therefore, targeting one location for allergen reduction may not be sufficient. For example, while encasing mattresses and pillows to reduce exposure to dust mite allergens is the most common strategy, dust mites are also found in the bedding, carpets, upholstered furniture and curtains. Thus, more comprehensive avoidance strategies include washing bedding weekly in hot water to kill live mites and remove allergens, removing carpets and fabrics, and reducing the amount of upholstered furniture (Eggleston, 2005). Custovic & Wijk (2005) summarized the scientific evidence for mite-allergen-avoidance strategies and reported that while some methods have level I and II evidence,[1] the majority have not been tested in randomized, placebo-control trials and are only expert opinions. An additional problem with evaluating comprehensive interventions is that truly blinding the participants to multiple changes in their homes is impossible and could lead to a bias in evaluating changes in participants, as compared with controls.

The fourth important feature of the relationship between allergen exposure and asthma is the unknown magnitude of allergen reduction required to have a clinical effect. Few, if any, studies resulted in the complete removal of allergens from a domestic environment. Reductions can be evaluated as a percentage of the reduction or as a reduction below a threshold. Given that an individual's response is also likely to be due to underlying genetic factors and other concurrent exposures, it is difficult to know which is more relevant.

[1] Categories of clinical evidence can be divided into levels, where *level I* is evidence from a meta-analysis of randomized controlled trials or evidence from at least one RCT and *level II* is evidence from at least one trial without randomization or evidence from at least one other type of quasi-experiment.

While many of the previous allergen-avoidance studies have not demonstrated clinical efficacy, a recent study with a more comprehensive intervention had positive findings. Published in 2004, a large environmental allergen-avoidance study of asthmatic children living in several inner-city communities in the United States did find an improvement in asthma symptoms for the group that avoided allergens, as compared with the control group (Morgan et al., 2004). Children with moderate to severe asthma, only half of whom reported taking medication to control their asthma, were enrolled. Specifically, the children in the allergen-avoidance group had fewer symptoms than controls in both the year of allergen avoidance and the subsequent year. The children in the allergen-avoidance group also had significantly fewer complications due to asthma than did controls. An aim of the study was to provide parents with the knowledge, skills, motivation and equipment for comprehensive, sustained environmental remediation. Within the group that avoided allergens (as well as within the control group), a significant decrease in symptom days, unscheduled visits to a health facility and hospitalizations was found for those children whose cockroach and dust mite allergens were reduced more than 50%. This study may be particularly relevant in that it was conducted in impoverished urban communities. While there is enthusiasm for the findings of this study, the message overall seems to be that to be clinically effective, avoidance must be tailored to the allergens to which an individual is allergic, must include all of the relevant allergens and exposures and must include a successful education component that empowers families to keep allergen levels low.

Some studies provided asthma or environmental caseworkers (or both) to help families mitigate allergies and asthma. Although the results were promising (de Blay et al., 2003; Laforest et al., 2004; Krieger et al., 2005; Perez et al., 2006), the duration of the benefits is limited by the study duration, which in some cases was only 12 months. Reducing the severity of asthma also relies on comprehensive asthma management, including controller medications, which work over a period of time to treat the underlying inflammation of the airways. Caseworker management of allergen-avoidance measures seems particularly compelling, given the difficulties with allergen-avoidance strategies administered through a clinic (Busse, Wang & Halm, 2005). Further studies with similar avoidance strategies are needed to confirm these findings, but home-based allergen avoidance conducted by trained staff (rather than advice in a doctor's office) may lead to more effective and prolonged avoidance.

Another important factor to consider in the efficacy of allergen avoidance is the benefit relative to the cost, which has not been evaluated for most studies. The effective multifaceted American study described previously found that the intervention cost was US$27 per additional symptom-free day (Kattan et al., 2005); however, health care costs would be expected to vary greatly by country and should decrease over time, if allergen reduction is sustained. Future studies should include cost–benefit analyses of allergen avoidance.

In summary, based on the evidence that exposure to allergens is associated with asthma symptoms in allergic individuals and that, for asthmatics allergic to pollen, symptoms decrease with a substantial reduction (orders of magnitude) in the pollen allergen, domestic allergen reduction for the tertiary prevention of asthma is compelling. However, the efficacy of methods currently used for allergen reduction and whether they should be recommended for asthmatic individuals are controversial (O'Connor, 2005). While single-allergen, single-site allergen-avoidance strategies (such as the use of allergen impermeable mattress covers) do not seem effective on their own, recent evidence has shown that multifaceted domestic avoidance measures, which include targeting multiple allergens and multiple sites, have some benefit among children who live in urban environments.

1.7.2. Primary and secondary prevention

The goals of primary and secondary prevention of allergic asthma are, respectively, preventing the onset of allergic sensitization and preventing the development of asthma after having developed allergic sensitization (WHO, 2003). As with tertiary prevention, the scientific basis for the hypothesis that avoiding allergens is an effective primary prevention measure is that exposure to allergens (or a certain threshold amount of those allergens) is necessary for developing sensitization. Therefore, reducing exposure to allergens before the development of sensitization should, in theory, prevent the onset of sensitization and thus allergic asthma.

Far fewer studies have focused on primary or secondary prevention than on tertiary prevention. Several recent and ongoing well-designed cohort studies with varying levels of intervention intensity have reported mixed results on the efficacy of primary prevention. Some studies suggest that *in utero* exposure to allergens or other modulators of the developing immune and respiratory systems may play a role in the development of asthma, indicating the potential need for measures that avoid them during pregnancy (Wright, 2004). Thus, many of the longitudinal studies on primary prevention have included prenatal interventions. It is important to note that some of the studies have reported results at ages 2 and 3 years only, an age before asthma can be reliably diagnosed. More conclusive results on primary and secondary prevention will be available when the children in these cohorts are older. Given the relevance of primary and secondary prevention to this chapter, the findings to date are discussed in detail below. These studies have not focused specifically on secondary prevention, except by including means to measure both sensitization and asthma. In theory, a study could be ineffective in demonstrating primary prevention, where children become sensitized, but effective in secondary prevention, where atopic children do not become asthmatic.

1.7.2.1. Limited avoidance of dust mite allergens

To date, primary prevention studies of domestic environments have focused on avoiding dust mite allergens, and not on cockroach or rodent allergens. The approaches of the randomized, controlled trials described in the following subsections (1.7.2.2 and 1.7.2.3) vary from simply furnishing mattress impermeable covers to multifaceted methods for eradicating exposure to dust mites in sites throughout the home. A study in the Netherlands that only used mattress covers found that children in the group with covers (called the active group) had fewer episodes of night cough without a cold (one symptom of developing asthma) at age 2 years than did controls (called the placebo group), but no differences in sensitization or other symptoms at ages 2 or 4 years (Koopman et al., 2002; Corver et al., 2006). One complication reported by the authors of this study was an unexplained lower dust mite allergen level (2- to 10-fold) in the cohort (active and placebo) than reported previously in the Netherlands (Brunekreef et al., 2005). A multinational European study that used mattress covers and education on mite allergen reduction also found no difference in sensitization or symptoms indicative of developing asthma at age 2 years (Horak et al., 2004). It is important to note that many of the homes still contained carpets (a reservoir for dust mites) at the follow-up assessment. Also, mite allergen levels were not measured, so the effect of avoiding exposure to allergens could not be evaluated. Two other European studies, using mattress encasement and education, have attempted primary prevention of sensitization in older, high-risk children. Children in the active group of both a cohort of toddlers and preschool children (mean age: 3 years) and a cohort of 5–7-year olds who were at risk of developing sensitization (atopic parent or sensitized to other aeroallergens) were less likely than controls to develop sensitization to dust mite allergens (Arshad et al., 2002; Tsitoura et al., 2002). Both of these studies demonstrated effective primary prevention of sensitization, but only over a short period of time. Further follow-up at later ages is required to determine the long-term efficacy.

1.7.2.2. Comprehensive avoidance of dust mite allergens

Dust mite allergen levels were lowered for the active group in more stringent environmental interventions. In a Manchester, England, intervention study, homes received allergen impermeable covers for both the mother's bed and the child's bed, a high filtration vacuum cleaner, vinyl flooring in the child's bedroom (after removing carpets), bedding that was washed weekly in hot water and washable toys. At age 3 years, there was no significant difference in the symptoms reported, but airway resistance was significantly better in the active group. Unexpectedly, the children in the active group were significantly more likely than the children in the control group to be sensitized to dust mite allergens at age 3 years (Woodcock et al., 2004).

A less stringent intervention study in Australia also reduced allergen levels significantly (Marks et al., 2006). Dust mite avoidance measures in the active group included allergen impermeable covers, weekly washing of bedding in hot water and education about dust mite avoidance. At age 5 years, there was no difference in the prevalence of asthma, eczema or atopy between the active and placebo groups. The authors cited several reasons for the failure of the intervention, including that while allergen levels were significantly reduced in the active group, they may not have been low enough to be clinically relevant. They also discussed the possible need for a multifaceted intervention, as was discussed in the previous section on tertiary prevention.

1.7.2.3. Multifaceted primary prevention studies

From the studies discussed in the previous two sections, it appears that reducing exposure to allergens alone may not be a sufficient avoidance measure for primary prevention – at least for outcomes in the early years of life. A large multifaceted trial in Canada included control of house dust mites in multiple locations in the home, a reduction in exposure to pet allergens, avoidance of exposure to ETS, encouragement of extended (4 months) breastfeeding and the introduction of partially hydrolysed formula as a breast-milk substitute. While there was no difference in atopy by age 2 years, those children in the active group were significantly less likely to have *possible or probable asthma* (Becker et al., 2004). At age 7 years, the children in the active group were significantly less likely than children in the placebo group to have asthma (diagnosed by a study physician) (Chan-Yeung et al., 2005). While lung function testing did not reveal a difference in the prevalence of bronchial hyperreactivity between the active and control groups, the active children were significantly less likely to have bronchial hyperreactivity with symptoms of wheezing. Another study, a controlled trial from the Isle of Wight, used allergen avoidance and either the mother maintained a diet with a low level of allergens while breastfeeding or fed the child hydrolysed formula (Arshad, Bateman & Matthews, 2003). This relatively small study reported that at the age of 8 years the children in the active group were significantly less likely than the children in the control group to wheeze, have nocturnal cough or be atopic (skin test positive to several food and inhaled allergens).

In summary, to date the studies on primary prevention of asthma, by avoiding allergens, do not allow for a conclusion on the efficacy of this approach. As with tertiary prevention, dust mite covers alone do not appear to be sufficient (Koopman et al., 2002; Horak et al., 2004). In one study, stringent dust mite control in the domestic environment resulted in better lung function, but increased atopy at age 3 years, a finding that must be examined in greater detail when the children reach an age at which a reliable diagnosis of asthma can be made (Woodcock et al., 2004). Multifaceted interventions, which include extended breastfeeding, delayed introduction of allergic foods and reduced exposure to ETS, appear to offer some protection (Arshad, Bateman & Matthews, 2003; Chan-Yeung et al., 2005). While exclusive breastfeeding and reduced exposure to ETS also can be recommended for health reasons other than asthma (Arshad, 2005), the relative contribution of allergen avoidance in these types of studies is difficult to determine. Continued follow-up of the children in these studies at a later age and more comprehensive studies are needed to better understand the potential long-term efficacy of allergen avoidance as a component of primary prevention of asthma.

1.7.3. Targeting housing conditions of high-risk groups

In light of the high prevalence of asthma, it seems reasonable to focus public health efforts on urban communities, especially impoverished communities. The ramifications of gentrification of these neighbourhoods are myriad and include displacement of low-income communities to fringe areas that could have equally poor or worse housing conditions (Kennedy & Leonard, 2001; Higgins, Wakefield & Cloutier, 2005). Certainly, a holistic approach is necessary to improve housing conditions without marginalizing those at risk of developing asthma.

1.8. Conclusions

Asthma is a major disease of the urban environment and a substantial burden from the standpoint of both the quality of life for the many suffering from the disease and the economics of health care. The global increase in the prevalence of asthma in the last half of the 20th century has disproportionately affected urban communities in many countries. The evidence that relates asthma and domestic exposure to cockroaches, mice, and dust mites is clear. These pests are common in urban environments and play a significant role in the pathogenesis of urban asthma. Removing these pests and their allergens is a logical tactic for preventing disease and reducing symptoms, but this tactic needs improved efficacy. Results from studies on avoiding allergens suggest cautious expectations about the ease with which long-term clinically relevant allergen reductions can be accomplished. However, the burden of asthma in urban communities may be effectively reduced through more broadly defined asthma interventions that include allergen reduction tailored to an individual's specific allergy, education about effective methods for sustained integrated pest management and general education about asthma.

References[2]

Aberg N (1989). Asthma and allergic rhinitis in Swedish conscripts. *Clinical and Experimental Allergy*, 19:59–63.

Abramson MJ, Puy RM, Weiner JM (2003). Allergen immunotherapy for asthma. *Cochrane Database Systematic Reviews*, 4:CD001186 (DOI:10.1002/14651858.CD001186; http://www.mrw.interscience.wiley.com/cochrane/clsysrev/articles/CD001186/frame.html, accessed 21 January 2007).

Almqvist C et al. (2003). Direct and indirect exposure to pets – risk of sensitization and asthma at 4 years in a birth cohort. *Clinical and Experimental Allergy*, 33:1190–1197.

Amr S et al. (2003). Environmental allergens and asthma in urban elementary schools. *Annals of Allergy, Asthma & Immunology*, 90:34–40.

Arlian LG (1992). Water balance and humidity requirements of house dust mites. *Experimental & Applied Acarology*, 16:15–35.

Arlian LG, Morgan MS, Neal JS (2002). Dust mite allergens: ecology and distribution. *Current Allergy and Asthma Reports*, 2:401–411.

Arlian LG, Neal JS, Vyszenski-Moher DL (1999). Reducing relative humidity to control the house dust mite *Dermatophagoides farinae*. *Journal of Allergy and Clinical Immunology*, 104:852–856.

Arlian LG et al. (1992). Prevalence of dust mites in the homes of people with asthma living in eight different geographic areas of the United States. *The Journal of Allergy and Clinical Immunology*, 90:292–300.

Arlian LG et al. (2001). Reducing relative humidity is a practical way to control dust mites and their allergens in homes in temperate climates. *The Journal of Allergy and Clinical Immunology*, 107:99–104.

Arruda LK et al. (2001). Cockroach allergens and asthma. *The Journal of Allergy and Clinical Immunology*, 107:419–428.

Arshad SH (2005). Primary prevention of asthma and allergy. *The Journal of Allergy and Clinical Immunology*, 116:3–14; quiz 5.

[2] This chapter is offered as an overview of allergic asthma and its association with urban pests. The authors selected studies according to topic (keyword searches: cockroach, dust mite, mouse, rat, asthma, allergy, prevention), with attention given to inclusion of studies, when available, conducted in urban residential environments. A more stringent systematic literature review of prevention for allergy and asthma was published by WHO (2003).

Arshad SH, Bateman B, Matthews SM (2003). Primary prevention of asthma and atopy during childhood by allergen avoidance in infancy: a randomised controlled study. *Thorax*, 58:489–493.

Arshad SH et al. (2002). Prevention of sensitization to house dust mite by allergen avoidance in school age children: a randomized controlled study. *Clinical and Experimental Allergy*, 32:843–849.

Ball TM et al. (2000). Siblings, day-care attendance, and the risk of asthma and wheezing during childhood. *The New England Journal of Medicine*, 343:538–543.

Basagana X et al. (2002). Domestic aeroallergen levels in Barcelona and Menorca (Spain). *Pediatric Allergy and Immunology*, 13:412–417.

Becker A et al. (2004). The Canadian asthma primary prevention study: outcomes at 2 years of age. *The Journal of Allergy and Clinical Immunology*, 113:650–656.

Bernton HS, Brown H (1967). Cockroach allergy II: the relation of infestation to sensitization. *Southern Medical Journal*, 60:852–855.

Bjorksten B et al. (2001). Allergy development and the intestinal microflora during the first year of life. *The Journal of Allergy and Clinical Immunology*, 108:516–520.

Boner AL et al. (1985). Pulmonary function and bronchial hyperreactivity in asthmatic children with house dust mite allergy during prolonged stay in the Italian Alps (Misurina, 1756 m). *Annals of Allergy*, 54:42–45.

Borish L (1999). Genetics of allergy and asthma. *Annals of Allergy, Asthma & Immunology*, 82:413–424; quiz 424–426.

Bousquet J, Michel FB (1983). In vivo methods for the study of allergy: skin tests, techniques and interpretation. In: Middleton E et al., eds. *Allergy principles and practice*, 3rd ed. St. Louis, Mosby, 573–589.

Braback L et al. (1994). Atopic sensitization and respiratory symptoms among Polish and Swedish school children. *Clinical and Experimental Allergy*, 24:826–835.

Bracken MB et al. (2002). Genetic and perinatal risk factors for asthma onset and severity: a review and theoretical analysis. *Epidemiologic Reviews*, 24:176–189.

Braun-Fahrlander C et al. (1997). Respiratory health and long-term exposure to air pollutants in Swiss schoolchildren. SCARPOL Team. Swiss Study on Childhood Allergy and Respiratory Symptoms with Respect to Air Pollution, Climate and Pollen. *American Journal of Respiratory and Critical Care Medicine*, 155:1042–1049.

Braun-Fahrlander C et al. (2002). Environmental exposure to endotoxin and its relation to asthma in school-age children. *The New England Journal of Medicine*, 347:869–877.

Brunekreef B et al. (2005). La mano de DIOS...was the PIAMA intervention study intervened upon? *Allergy*, 60:1083–1086.

Brussee JE et al. (2005). Allergen exposure in infancy and the development of sensitization, wheeze, and asthma at 4 years. *The Journal of Allergy and Clinical Immunology*, 115:946–952.

Busse PJ, Wang JJ, Halm EA (2005). Allergen sensitization evaluation and allergen avoidance education in an inner-city adult cohort with persistent asthma. *The Journal of Allergy and Clinical Immunology*, 116:146–152.

Call RS et al. (1992). Risk factors for asthma in inner city children. *The Journal of Pediatrics*, 121:862–866.

Carter MC et al. (2001). Home intervention in the treatment of asthma among inner-city children. *The Journal of Allergy and Clinical Immunology*, 108:732–737.

Chang MY et al. (2000). Salivary cotinine levels in children presenting with wheezing to an emergency department. *Pediatric Pulmonology*, 29:257–263.

Chan-Yeung M et al. (1995). House dust mite allergen levels in two cities in Canada: effects of season, humidity, city and home characteristics. *Clinical and Experimental Allergy*, 25:240–246.

Chan-Yeung M et al. (2005). The Canadian Childhood Asthma Primary Prevention Study: outcomes at 7 years of age. *The Journal of Allergy and Clinical Immunology*, 116:49–55.

Chapman MD, Tsay A, Vailes LD (2001). Home allergen monitoring and control – improving clinical practice and patient benefits. *Allergy*, 56:604–610.

Chapman MD et al. (1987). Monoclonal immunoassays for major dust mite (*Dermatophagoides*) allergens, Der p I and Der f I, and quantitative analysis of the allergen content of mite and house dust extracts. *The Journal of Allergy and Clinical Immunology*, 80:184–194.

Chew GL, Correa JC, Perzanowski MS (2005). Mouse and cockroach allergens in the dust and air in northeastern United States inner-city public high schools. *Indoor Air*, 15:228–234.

Chew GL et al. (1998). Limitations of a home characteristics questionnaire as a predictor of indoor allergen levels. *American Journal of Respiratory and Critical Care Medicine*, 157:1536–1541.

Chew GL et al. (1999a). Monthly measurements of indoor allergens and the influence of housing type in a northeastern US city. *Allergy*, 54:1058–1066.

Chew GL et al. (1999b). The effects of carpet fresheners and other additives on the behaviour of indoor allergen assays. *Clinical and Experimental Allergy*, 29:470–477.

Chew GL et al. (2003). Distribution and determinants of mouse allergen exposure in low-income New York City apartments. *Environmental Health Perspectives*, 111:1348–1351.

Chew GL et al. (2006). Determinants of cockroach and mouse exposure and associations with asthma among families and the elderly living in New York City public housing. *Annals of Allergy, Asthma & Immunology*, 97:502–513.

Cohn RD et al. (2004). National prevalence and exposure risk for mouse allergen in US households. *The Journal of Allergy and Clinical Immunology*, 113:1167–1171.

Cole Johnson C et al. (2004). Family history, dust mite exposure in early childhood, and risk for pediatric atopy and asthma. *The Journal of Allergy and Clinical Immunology*, 114:105–110.

Corver K et al. (2006). House dust mite allergen reduction and allergy at 4 yr: follow up of the PIAMA-study. *Pediatric Allergy and Immunology*, 17:329–336.

Crain EF et al. (2002). Home and allergic characteristics of children with asthma in seven U.S. urban communities and design of an environmental intervention: the Inner-City Asthma Study. *Environmental Health Perspectives*, 110:939–945.

Crater DD et al. (2001). Asthma hospitalization trends in Charleston, South Carolina, 1956 to 1997: twenty-fold increase among black children during a 30-year period. *Pediatrics*, 108:E97.

Crater SE, Platts-Mills TA (1998). Searching for the cause of the increase in asthma. *Current Opinion in Pediatrics*, 10:594–599.

Creticos PS et al. (1996). Ragweed immunotherapy in adult asthma. *The New England Journal of Medicine*, 334:501–506.

Custis NJ et al. (2003). Quantitative measurement of airborne allergens from dust mites, dogs, and cats using an ion-charging device. *Clinical and Experimental Allergy*, 33:986–991.

Custovic A, Wijk RG (2005). The effectiveness of measures to change the indoor environment in the treatment of allergic rhinitis and asthma: ARIA update (in collaboration with GA(2)LEN). *Allergy*, 60:1112–1125.

Custovic A et al. (1996a). Exposure to house dust mite allergens and the clinical activity of asthma. *The Journal of Allergy and Clinical Immunology*, 98:64–72.

Custovic A et al. (1996b). Domestic allergens in public places. II: Dog (Can f 1) and cockroach (Bla g 2) allergens in dust and mite, cat, dog and cockroach allergens in the air in public buildings. *Clinical and Experimental Allergy*, 26:1246–1252.

Custovic A et al. (1997). Aerodynamic properties of the major dog allergen Can f 1: distribution in homes, concentration, and particle size of allergen in the air. *American Journal of Respiratory and Critical Care Medicine*, 155:94–98.

Custovic A et al. (1999). Relationship between mite, cat, and dog allergens in reservoir dust and ambient air. *Allergy*, 54:612–616.

d'Amato G et al. (2005). Environmental risk factors and allergic bronchial asthma. *Clinical and Experimental Allergy*, 35:1113–1124.

de Blay F et al. (1991). Airborne dust mite allergens: comparison of group II allergens with group I mite allergen and cat-allergen Fel d I. *The Journal of Allergy and Clinical Immunology*, 88:919–926.

de Blay F et al. (1997). Dust and airborne exposure to allergens derived from cockroach (*Blattella germanica*) in low-cost public housing in Strasbourg (France). *The Journal of Allergy and Clinical Immunology*, 99:107–112.

de Blay F et al. (2003). Medical Indoor Environment Counselor (MIEC): role in compliance with advice on mite allergen avoidance and on mite allergen exposure. *Allergy*, 58:27–33.

de Lucca SD et al. (1999). Measurement and characterization of cockroach allergens detected during normal domestic activity. *The Journal of Allergy and Clinical Immunology*, 104:672–680.

Diaz-Sanchez D et al. (1997). Combined diesel exhaust particulate and ragweed allergen challenge markedly enhances human in vivo nasal ragweed-specific IgE and skews cytokine production to a T helper cell 2-type pattern. *Journal of Immunology*, 158:2406–2413.

Dowse GK et al. (1985). The association between *Dermatophagoides* mites and the increasing prevalence of asthma in village communities within the Papua New Guinea highlands. *The Journal of Allergy and Clinical Immunology*, 75:75–83.

Duffy DL, Mitchell CA, Martin NG (1998). Genetic and environmental risk factors for asthma: a cotwin-control study. *American Journal of Respiratory and Critical Care Medicine*, 157:840–845.

ECRHS (1996). Variations in the prevalence of respiratory symptoms, self-reported asthma attacks, and use of asthma medication in the European Community Respiratory Health Survey (ECRHS). *The European Respiratory Journal*, 9:687–695.

Eder W, Ege MJ, von Mutius E (2006). The asthma epidemic. *New England Journal of Medicine*, 355:2226–2235.

Eggleston P (2001). Methods and effectiveness of indoor environmental control. *Annals of Allergy, Asthma & Immunology*, 87 (Suppl. 3):44–47.

Eggleston PA (2005). Improving indoor environments: reducing allergen exposures. *The Journal of Allergy and Clinical Immunology*, 116:122–126.

Eggleston PA et al. (1999). Removal of cockroach allergen from inner-city homes. *The Journal of Allergy and Clinical Immunology*, 104:842–846.

Eggleston PA et al. (2005). Home environmental intervention in inner-city asthma: a randomized controlled clinical trial. *Annals of Allergy, Asthma & Immunology*, 95:518–524.

Eisenbarth SC et al. (2002). Lipopolysaccharide-enhanced, toll-like receptor 4-dependent T helper cell type 2 responses to inhaled antigen. *The Journal of Experimental Medicine*, 196:1645–1651.

Elston DM (2004). Prevention of arthropod-related disease. *Journal of the American Academy of Dermatology*, 51:947–954.

Emanuel MB (1988). Hay fever, a post industrial revolution epidemic: a history of its growth during the 19th century. *Clinical Allergy*, 18:295–304.

Eriksson NE (1989). Total IgE influences the relationship between skin test and RAST. *Annals of Allergy*, 63:65–69.

Evans R 3rd et al. (1987). National trends in the morbidity and mortality of asthma in the US. Prevalence, hospitalization and death from asthma over two decades: 1965–1984. *Chest*, 91(Suppl.):65S–74S.

Ferrari E et al. (2004). Environmental detection of mouse allergen by means of immunoassay for recombinant Mus m 1. *The Journal of Allergy and Clinical Immunology*, 114:341–346.

Garcia DP et al. (1994). Cockroach allergy in Kentucky: a comparison of inner city, suburban, and rural small town populations. *Annals of Allergy*, 72:203–208.

Gelber LE et al. (1993). Sensitization and exposure to indoor allergens as risk factors for asthma among patients presenting to hospital. *The American Review of Respiratory Disease*, 147:573–578.

Gergen PJ et al. (1999). Results of the National Cooperative Inner City Asthma Study (NCICAS) environmental intervention to reduce cockroach allergen exposure in inner city homes. *The Journal of Allergy and Clinical Immunology*, 103:501–506.

Gold DR (2000). Environmental tobacco smoke, indoor allergens, and childhood asthma. *Environmental Health Perspectives*, 108 (Suppl. 4):643–651.

Gold DR et al. (1999). Predictors of repeated wheeze in the first year of life: the relative roles of cockroach, birth weight, acute lower respiratory illness, and maternal smoking. *American Journal of Respiratory and Critical Care Medicine*, 160:227–236.

Gotzsche PC et al. (2004). House dust mite control measures for asthma. *Cochrane Database of Systematic Reviews*, 4:CD001187 (DOI:10.1002/14651858.CD001187).

Gruchalla RS et al. (2005). Inner City Asthma Study: relationships among sensitivity, allergen exposure, and asthma morbidity. *The Journal of Allergy and Clinical Immunology*, 115:478–485.

Gupta R et al. (2004). Burden of allergic disease in the UK: secondary analyses of national databases. *Clinical and Experimental Allergy*, 34:520–526.

Haahtela T et al. (1990). Prevalence of asthma in Finnish young men. *British Medical Journal*, 301:266–268.

Halonen M et al. (1997). Alternaria as a major allergen for asthma in children raised in a desert environment. *American Journal of Respiratory and Critical Care Medicine*, 155:1356–1361.

Harving H, Korsgaard HH, Dahl J (1993). House-dust mites and associated environmental conditions in Danish homes. *Allergy*, 48:106–9.

Heederik D et al. (1999). Exposure-response relationships for work-related sensitization in workers exposed to rat urinary allergens: results from a pooled study. *The Journal of Allergy and Clinical Immunology*, 103:678–684.

Heinrich J et al. (2001). Pets and vermin are associated with high endotoxin levels in house dust. *Clinical and Experimental Allergy*, 31:1839–1845.

Hesselmar B et al. (1999). Does early exposure to cat or dog protect against later allergy development? *Clinical and Experimental Allergy*, 29:611–617.

Higgins PS, Wakefield D, Cloutier MM (2005). Risk factors for asthma and asthma severity in nonurban children in Connecticut. *Chest*, 128:3846–3853.

Hollander A et al. (1997). Exposure of laboratory animal workers to airborne rat and mouse urinary allergens. *Clinical and Experimental Allergy*, 27:617–626.

Hollander A et al. (1998). Determinants of airborne rat and mouse urinary allergen exposure. *Scandinavian Journal of Work, Environment & Health*, 24:228–235.

Holt PG, Upham JW, Sly PD (2005). Contemporaneous maturation of immunologic and respiratory functions during early childhood: implications for development of asthma prevention strategies. *The Journal of Allergy and Clinical Immunology*, 116:16–24; quiz 5.

Horak F Jr et al. (2004). Effect of mite-impermeable mattress encasings and an educational package on the development of allergies in a multinational randomized, controlled birth-cohort study – 24 months results of the Study of Prevention of Allergy in Children in Europe. *Clinical and Experimental Allergy*, 34:1220–1225.

Hynes HP et al. (2004). Investigations into the indoor environment and respiratory health in Boston public housing. *Reviews on Environmental Health*, 19:271–289.

Illi S et al. (2006). Perennial allergen sensitisation early in life and chronic asthma in children: a birth cohort study. *Lancet*, 368:763–770.

ISAAC Steering Committee (1998). Worldwide variation in prevalence of symptoms of asthma, allergic rhinoconjunctivitis and atopic eczema: The International Study of Asthma and Allergies in Childhood (ISAAC) Steering Committee. *Lancet*, 351:1225–1232.

Johansson SG, Haahtela T (2004). World Allergy Organization Guidelines for Prevention of Allergy and Allergic Asthma. Condensed Version. *International Archives of Allergy and Immunology*, 135:83–92.

Kang B, Chang JL (1985). Allergenic impact of inhaled arthropod material. *Clinical Reviews in Allergy*, 3:363–375.

Kang B et al. (1979). Cockroach cause of allergic asthma. Its specificity and immunologic profile. *The Journal of Allergy and Clinical Immunology*, 63:80–86.

Karlsson AS et al. (2002). Evaluation of Petri dish sampling for assessment of cat allergen in airborne dust. *Allergy*, 57:164–168.

Kattan M et al. (1997). Characteristics of inner-city children with asthma: the National Cooperative Inner-City Asthma Study. *Pediatric Pulmonology*, 24:253–262.

Kattan M et al. (2005). Cost-effectiveness of a home-based environmental intervention for inner-city children with asthma. *The Journal of Allergy and Clinical Immunology*, 116:1058–1063.

Kennedy M, Leonard P (2001). *Dealing with neighborhood change: a primer on gentrification and policy choices*. Washington, DC, The Brookings Institution Center on Urban and Metropolitan Policy (http://www.brookings.edu/es/urban/gentrification/gentrification.pdf, accessed 29 October 2006).

Kerkhof M et al. (2005). The effect of prenatal exposure on total IgE at birth and sensitization at twelve months and four years of age: The PIAMA study. *Pediatric Allergy and Immunology*, 16:10–18.

Kinney PL et al. (1989). Short-term pulmonary function change in association with ozone levels. *The American Review of Respiratory Disease*, 139:56–61.

Kleeberger SR, Peden D (2005). Gene-environment interactions in asthma and other respiratory diseases. *Annual Review of Medicine*, 56:383–400.

Koopman LP et al. (2002). Placebo-controlled trial of house dust mite-impermeable mattress covers: effect on symptoms in early childhood. *American Journal Respiratory and Critical Care Medicine*, 166:307–313.

Krakowiak A, Szulc B, Gorski P (1999). Allergy to laboratory animals in children of parents occupationally exposed to mice, rats and hamsters. *The European Respiratory Journal*, 14:352–356.

Krieger JW et al. (2005). The Seattle-King County Healthy Homes Project: a randomized, controlled trial of a community health worker intervention to decrease exposure to indoor asthma triggers. *American Journal of Public Health*, 95:652–659.

Laforest L et al. (2004). Association between asthma control in children and loss of workdays by caregivers. *Annals of Allergy, Asthma & Immunology*, 93:265–271.

Lau S et al. (2000). Early exposure to house-dust mite and cat allergens and development of childhood asthma: a cohort study. Multicentre Allergy Study Group. *Lancet*, 356:1392–1397.

Lau S et al. (2001). Comparison of quantitative ELISA and semiquantitative Dustscreen for determination of Der p 1, Der f 1, and Fel d 1 in domestic dust samples. *Allergy*, 56:993–995.

Lieutier-Colas F et al. (2001). Difference in exposure to airborne major rat allergen (Rat n 1) and to endotoxin in rat quarters according to tasks. *Clinical and Experimental Allergy*, 31:1449–1456.

Lieutier-Colas F et al. (2002). Prevalence of symptoms, sensitization to rats, and airborne exposure to major rat allergen (Rat n 1) and to endotoxin in rat-exposed workers: a cross-sectional study. *Clinical and Experimental Allergy*, 32:1424–1429.

Litonjua AA et al. (2001). Exposure to cockroach allergen in the home is associated with incident doctor-diagnosed asthma and recurrent wheezing. *The Journal of Allergy and Clinical Immunology*, 107:41–47.

Luczynska CM et al. (1990). Airborne concentrations and particle size distribution of allergen derived from domestic cats (*Felis domesticus*). Measurements using cascade impactor, liquid impinger, and a two-site monoclonal antibody assay for Fel d I. *The American Review of Respiratory Disease*, 141:361–367.

Lundback B (1998). Epidemiology of rhinitis and asthma. *Clinical and Experimental Allergy*, 28 (Suppl. 2):3–10.

Marks GB (1998). House dust mite exposure as a risk factor for asthma: benefits of avoidance. *Allergy*, 53:108–114.

Marks GB et al. (2006). Prevention of asthma during the first 5 years of life: A randomized controlled trial. *The Journal of Allergy and Clinical Immunology*, 118:53–61.

Marsh DG, Meyers DA, Bias WB (1981). The epidemiology and genetics of atopic allergy. *The New England Journal of Medicine*, 305:1551–1559.

Martinez FD et al. (1995). Asthma and wheezing in the first six years of life. The Group Health Medical Associates. *The New England Journal of Medicine*, 332:133–138.

Masoli M et al. (2004). The global burden of asthma: executive summary of the GINA Dissemination Committee report. *Allergy*, 59:469–478.

Matheson MC et al. (2003). Residential characteristics predict changes in Der p 1, Fel d 1 and ergosterol but not fungi over time. *Clinical and Experimental Allergy*, 33:1281–1288.

Matricardi PM, Bouygue GR, Tripodi S (2002). Inner-city asthma and the hygiene hypothesis. *Annals of Allergy, Asthma & Immunology*, 89(Suppl. 1):69–74.

Matricardi PM et al. (2002). Hay fever and asthma in relation to markers of infection in the United States. *The Journal of Allergy and Clinical Immunology*, 110:381–387.

Matsui EC et al. (2003). Cockroach allergen exposure and sensitization in suburban middle-class children with asthma. *The Journal of Allergy and Clinical Immunology*, 112:87–92.

Matsui EC et al. (2004a). Mouse allergen exposure and mouse skin test sensitivity in suburban, middle-class children with asthma. *The Journal of Allergy and Clinical Immunology*, 113:910–915.

Matsui EC et al. (2004b). Mouse allergen exposure and immunologic responses: IgE-mediated mouse sensitization and mouse specific IgG and IgG4 levels. *Annals of Allergy, Asthma & Immunology*, 93:171–178.

Matsui EC et al. (2005). Airborne mouse allergen in the homes of inner-city children with asthma. *The Journal of Allergy and Clinical Immunology*, 115:358–363.

McConnell R et al. (2003). Cockroach counts and house dust allergen concentrations after professional cockroach control and cleaning. *Annals of Allergy, Asthma & Immunology*, 91:546–552.

Melen E et al. (2001). Influence of early and current environmental exposure factors on sensitization and outcome of asthma in pre-school children. *Allergy*, 56:646–652.

Mihrshahi S et al. (2002). Predictors of high house dust mite allergen concentrations in residential homes in Sydney. *Allergy*, 57:137–142.

Miller RL et al. (2001a). Increased cockroach and mouse-IgE level at age 2 years in a highly exposed inner-city cohort. *American Journal of Respiratory and Critical Care Medicine*, 163:A603.

Miller RL et al. (2001b). Prenatal exposure, maternal sensitization, and sensitization in utero to indoor allergens in an inner-city cohort. *American Journal of Respiratory and Critical Care Medicine*, 164:995–1001.

Miyazawa H et al. (1996). Seasonal changes in mite allergen (*Der* I and *Der* II) concentrations in Japanese homes. *Annals of Allergy, Asthma & Immunology*, 76:170–174.

Mollet JA et al. (1997). Evaluation of German cockroach (Orthoptera:Blattellidae) allergen and seasonal variation in low-income housing. *Journal of Medical Entomology*, 34:307–311.

Montealegre F et al. (1997). Identification of the domestic mite fauna of Puerto Rico. *Puerto Rico Health Sciences Journal*, 16:109–116.

Morgan WJ et al. (2004). Results of a home-based environmental intervention among urban children with asthma. *The New England Journal of Medicine*, 351:1068–1080.

Munir AK et al. (1995). Mite allergens in relation to home conditions and sensitization of asthmatic children from three climatic regions. *Allergy*, 50:55–64.

Musmand JJ et al. (1995). Identification of important allergens in German cockroach extracts by sodium dodecylsulfate-polyacrylamide gel electrophoresis and Western blot analysis. *The Journal of Allergy and Clinical Immunology*, 95:877–885.

Neal JS, Arlian LG, Morgan MS (2002). Relationship among house-dust mites, Der 1, Fel d 1, and Can f 1 on clothing and automobile seats with respect to densities in houses. *Annals of Allergy, Asthma & Immunology*, 88:410–415.

Nel AE et al. (1998). Enhancement of allergic inflammation by the interaction between diesel exhaust particles and the immune system. *The Journal of Allergy and Clinical Immunology*, 102:539–554.

Ng'ang'a LW (1996). *The epidemiology of childhood asthma in Kenya: objective markers in rural and urban school children* [Ph.D. thesis]. Montreal, McGill University.

Ng'ang'a LW et al. (1998). Prevalence of exercise induced bronchospasm in Kenyan school children: an urban-rural comparison. *Thorax*, 53:919–926.

National Heart, Lung and Blood Institute (1997). *Guidelines for the diagnosis and management of asthma: clinical practice guidelines*. Bethesda, Maryland, National Institutes of Health, National Heart, Lung and Blood Institute (NIH Publication No. 97–4051; http://www.nhlbi.nih.gov/guidelines/asthma/asthgdln.pdf, accessed 29 October 2006).

Nicholas SW et al. (2005). Addressing the childhood asthma crisis in Harlem: the Harlem Children's Zone Asthma Initiative. *American Journal of Public Health*, 95:245–249.

O'Connor GT (2005). Allergen avoidance in asthma: what do we do now? *The Journal of Allergy and Clinical Immunology*, 116:26–30.

O'Connor GT et al. (2004). Airborne fungi in the homes of children with asthma in low-income urban communities: The Inner-City Asthma Study. *The Journal of Allergy and Clinical Immunology*, 114:599–606.

Odhiambo JA et al. (1998). Urban-rural differences in questionnaire-derived markers of asthma in Kenyan school children. *The European Respiratory Journal*, 12:1105–1112.

Ohman JL Jr et al. (1994). Distribution of airborne mouse allergen in a major breeding facility. *The Journal of Allergy and Clinical Immunology*, 94:810–817.

Ownby DR (1990). Environmental factors versus genetic determinants of childhood inhalant allergies. *The Journal of Allergy and Clinical Immunology*, 86:279–287.

Ownby D, Johnson CC, Peterson EL (2002). Exposure to dogs and cats in the first year of life and risk of allergic sensitization at 6 and 7 years of age. *The Journal of the American Medical Association*, 288:963–972.

Park JH et al. (2001). Predictors of airborne endotoxin in the home. *Environmental Health Perspectives*, 109:859–864.

Peat JK et al. (1994). Asthma severity and morbidity in a population sample of Sydney schoolchildren: Part II – Importance of house dust mite allergens. *Australian and New Zealand Journal of Medicine*, 24:270–276.

Peat JK et al. (1995). Prevalence and severity of childhood asthma and allergic sensitisation in seven climatic regions of New South Wales. *The Medical Journal of Austria*, 163:22–26.

Perez M et al. (2006). The impact of community health worker training and programs in NYC. *Journal of Health Care for the Poor and Underserved*, 17(Suppl. 1):26–43.

Perry T et al. (2003). The prevalence of rat allergen in inner-city homes and its relationship to sensitization and asthma morbidity. *The Journal of Allergy and Clinical Immunology*, 112:346–352.

Perzanowski MS et al. (1998). Association of sensitization to Alternaria allergens with asthma among school-age children. *The Journal of Allergy and Clinical Immunology*, 101:626–632.

Perzanowski MS et al. (2002). Effect of cat and dog ownership on sensitization and development of asthma among preteenage children. *American Journal of Respiratory and Critical Care Medicine*, 166:696–702.

Perzanowski MS et al. (2006). Endotoxin in inner-city homes: associations with wheeze and eczema in early childhood. *The Journal of Allergy and Clinical Immunology*, 117:1082–1089.

Phipatanakul W et al. (2000a). Mouse allergen. II. The relationship of mouse allergen exposure to mouse sensitization and asthma morbidity in inner-city children with asthma. *The Journal of Allergy and Clinical Immunology*, 106:1075–1080.

Phipatanakul W et al. (2000b). Mouse allergen. I. The prevalence of mouse allergen in inner-city homes. The National Cooperative Inner-City Asthma Study. *The Journal of Allergy and Clinical Immunology*, 106:1070–1074.

Phipatanakul W et al. (2004). Endotoxin exposure and eczema in the first year of life. *Pediatrics*, 114:13–18.

Phipatanakul W et al. (2005). Predictors of indoor exposure to mouse allergen in urban and suburban homes in Boston. *Allergy*, 60:697–701.

Platts-Mills TA (2004). Allergen avoidance. *The Journal of Allergy and Clinical Immunology*, 113:388–391.

Platts-Mills TAE, de Weck A (1988). Dust mite allergens and asthma: a worldwide problem. International Workshop report. *Bulletin of the World Health Organization*, 66:769–780.

Platts-Mills TA et al. (1982). Reduction of bronchial hyperreactivity during prolonged allergen avoidance. *Lancet*, 2:675–678.

Platts-Mills TA et al. (1987). Seasonal variation in dust mite and grass-pollen allergens in dust from the houses of patients with asthma. *The Journal of Allergy and Clinical Immunology*, 79:781–791.

Platts-Mills TA et al. (1992). Dust mite allergens and asthma: report of a second international workshop. *The Journal of Allergy and Clinical Immunology*, 89:1046–1060.

Platts-Mills TA et al. (1997). Indoor allergens and asthma: report of the Third International Workshop. *The Journal of Allergy and Clinical Immunology*, 100:S2–S24.

Platts-Mills T et al. (2001). Sensitisation, asthma, and a modified Th2 response in children exposed to cat allergen: a population-based cross-sectional study. *Lancet*, 357:752–756.

Pollart SM et al. (1991a). Identification, quantification, purification of cockroach allergens using monoclonal antibodies. *The Journal of Allergy and Clinical Immunology*, 87:511–521.

Pollart SM et al. (1991b). Environmental exposure to cockroach allergens: analysis with monoclonal antibody-based enzyme immunoassays. *The Journal of Allergy and Clinical Immunology*, 87:505–510.

Pope CA 3rd et al. (2002). Lung cancer, cardiopulmonary mortality, and long-term exposure to fine particulate air pollution. *Journal of the American Medical Association*, 287:1132–1141.

Rauh VA et al. (2002). Deteriorated housing contributes to high cockroach allergen levels in inner-city households. *Environmental Health Perspectives*, 110 (Suppl. 2):323–327.

Renstrom A et al. (1994). Prospective study of laboratory-animal allergy: factors predisposing to sensitization and development of allergic symptoms. *Allergy*, 49:548–852.

Renstrom A et al. (1997). A new amplified monoclonal rat allergen assay used for evaluation of ventilation improvements in animal rooms. *The Journal of Allergy and Clinical Immunology*, 100:649–655.

Renstrom A et al. (2001). Working with male rodents may increase risk of allergy to laboratory animals. *Allergy*, 56:964–970.

Riedl M, Diaz-Sanchez D (2005). Biology of diesel exhaust effects on respiratory function. *The Journal of Allergy and Clinical Immunology*, 115:221–228; quiz 229.

Ronmark E et al. (1998). Asthma, type-1 allergy and related conditions in 7- and 8-year-old children in northern Sweden: prevalence rates and risk factor pattern. *Respiratory Medicine*, 92:316–324.

Ronmark E et al. (1999). Different pattern of risk factors for atopic and nonatopic asthma among children – report from the Obstructive Lung Disease in Northern Sweden Study. *Allergy*, 54:926–935.

Rosenstreich DL et al. (1997). The role of cockroach allergy and exposure to cockroach allergen in causing morbidity among inner-city children with asthma. *The New England Journal of Medicine*, 336:1356–1363.

Rossiter LF et al. (2000). The impact of disease management on outcomes and cost of care: a study of low-income asthma patients. *Inquiry*, 37:188–202.

Rylander E et al. (1993). Parental smoking and other risk factors for wheezing bronchitis in children. *European Journal of Epidemiology*, 9:517–526.

Samet JM (1995). Asthma and the environment: do environmental factors affect the incidence and prognosis of asthma? *Toxicology Letters*, 82–83:33–38.

Sandstrom T (1995). Respiratory effects of air pollutants: experimental studies in humans. *The European Respiratory Journal*, 8:976–995.

Sarpong SB et al. (1996). Socioeconomic status and race as risk factors for cockroach allergen exposure and sensitization in children with asthma. *The Journal of Allergy and Clinical Immunology*, 97:1393–1401.

Schmidt LM, Gotzsche PC (2005). Of mites and men: reference bias in narrative review articles: a systematic review. *Journal of Family Practice*, 54:334–338.

Sears MR, Beaglehole R (1987). Asthma morbidity and mortality: New Zealand. *The Journal of Allergy and Clinical Immunology*, 80:383–388.

Shekelle PG et al. (1999). Clinical guidelines: developing guidelines. *British Medical Journal*, 318:593–596.

Shore SA, Fredberg JJ (2005). Obesity, smooth muscle, and airway hyperresponsiveness. *The Journal of Allergy and Clinical Immunology*, 115:925–927.

Simpson A et al. (2006). Endotoxin exposure, CD14, and allergic disease: an interaction between genes and the environment. *American Journal of Respiratory and Critical Care Medicine*, 174:386–392.

Spitzauer S (1999). Allergy to mammalian proteins: at the borderline between foreign and self? *International Archives of Allergy and Immunology*, 120:259–269.

Sporik R et al. (1990). Exposure to house-dust mite allergen (Der p I) and the development of asthma in childhood. A prospective study. *The New England Journal of Medicine*, 323:502–507.

Sporik R et al. (1995). Association of asthma with serum IgE and skin test reactivity to allergens among children living at high altitude. Tickling the dragon's breath. *American Journal of Respiratory and Critical Care Medicine*, 151:1388–1392.

Sprietsma JE (1999). Modern diets and diseases: NO-zinc balance. Under Th1, zinc and nitrogen monoxide (NO) collectively protect against viruses, AIDS, autoimmunity, diabetes, allergies, asthma, infectious diseases, atherosclerosis and cancer. *Medical Hypotheses*, 53:6–16.

Squillace SP et al. (1997). Sensitization to dust mites as a dominant risk factor for asthma among adolescents living in central Virginia. Multiple regression analysis of a population-based study. *American Journal of Respiratory and Critical Care Medicine*, 156:1760–1764.

Stelmach I et al. (2002a). The prevalence of mouse allergen in inner-city homes. *Pediatric Allergy and Immunology*, 13:299–302.

Stelmach I et al. (2002b). Cockroach allergy and exposure to cockroach allergen in Polish children with asthma. *Allergy*, 57:701–705.

Stevenson LA et al. (2001). Sociodemographic correlates of indoor allergen sensitivity among United States children. *The Journal of Allergy and Clinical Immunology*, 108:747–752.

Strachan DP (1989). Hay fever, hygiene, and household size. *British Medical Journal*, 299:1259–1260.

Swanson M, Agarwal M, Reed C (1985). An immunochemical approach to indoor aeroallergen quantification with a new volumetric air sampler: studies with mite, roach, cat, mouse, and guinea-pig antigens. *The Journal of Allergy and Clinical Immunology*, 76:724–729.

Sydbom A et al. (2001). Health effects of diesel exhaust emissions. *The European Respiratory Journal*, 17:733–746.

Tatfeng YM et al. (2005). Mechanical transmission of pathogenic organisms: the role of cockroaches. *Journal of Vector Borne Diseases*, 42:129–134.

Taylor AN, Longbottom JL, Pepys J (1977). Respiratory allergy to urine proteins of rats and mice. *Lancet*, 2:847–849.

Tobias KR et al. (2004). Exposure to indoor allergens in homes of patients with asthma and/or rhinitis in southeast Brazil: effect of mattress and pillow covers on mite allergen levels. *International Archives of Allergy and Immunology*, 133:365–370.

Tovey ER et al. (1981). The distribution of dust mite allergen in the houses of patients with asthma. *The American Review of Respiratory Disease*, 124:630–635.

Tsay A et al. (2002). A rapid test for detection of mite allergens in homes. *Clinical and Experimental Allergy*, 32:1596–1601.

Tsitoura S et al. (2002). Randomized trial to prevent sensitization to mite allergens in toddlers and preschoolers by allergen reduction and education: one-year results. *Archives of Pediatrics & Adolescent Medicine*, 156:1021–1027.

van Bronswijk JE et al. (1989). Evaluating mite (Acari) allergenicity of house dust by guanine quantification. *Journal of Medical Entomology*, 26:55–59.

van Hage-Hamsten M, Johansson E (1998). Clinical and immunologic aspects of storage mite allergy. *Allergy*, 53:49–53.

van Strien RT et al. (1994). Mite antigen in house dust: relationship with different housing characteristics in The Netherlands. *Clinical and Experimental Allergy*, 24:843–853.

van Strien RT et al. (2002). Mite and pet allergen levels in homes of children born to allergic and nonallergic parents: the PIAMA study. *Environmental Health Perspectives*, 110:A693–A698.

von Mutius E (1998). The influence of birth order on the expression of atopy in families: a gene environment interaction? *Clinical and Experimental Allergy*, 28:1454–1456.

von Mutius E et al. (1994). Prevalence of asthma and atopy in two areas of West and East Germany. *American Journal of Respiratory and Critical Care Medicine*, 149:358–364.

von Mutius E et al. (1999). Frequency of infections and risk of asthma, atopy and airway hyperresponsiveness in children. *The European Respiratory Journal*, 14:4–11.

von Mutius E et al. (2000). Exposure to endotoxin or other bacterial components might protect against the development of atopy. *Clinical and Experimental Allergy*, 30:1230–1234.

Warner AM et al. (1996). Childhood asthma and exposure to indoor allergens: low mite levels are associated with sensitivity. *Pediatric Allergy and Immunology*, 7:61–67.

Warner A et al. (1998). Environmental assessment of *Dermatophagoides* mite-allergen levels in Sweden should include Der m 1. *Allergy*, 53:698–704.

Warner JA et al. (2000). Prenatal origins of allergic disease. *The Journal of Allergy and Clinical Immunology*, 105:S493–S498.

WHO (2003). *Prevention of allergy and allergic asthma*. Geneva, World Health Organization (document number: WHO/NMH/MNC/CRA 03.2; http://whqlibdoc.who.int/hq/2003/WHO_NMH_MNC_CRA_03.2.pdf, accessed 27 October 2006).

WHO (2006). *GARD launch, Beijing, 28 March: fact sheet for media*. Geneva, World Health Organization (http://www.who.int/respiratory/gard/GARD_Fact_Sheet.doc, accessed 31 October 2006).

WHO Regional Office for Europe (2006). *The LARES project (Large Analysis and Review of European Housing and Health Status)*. Copenhaen, WHO Regional Office for Europe (www.euro.who.int/housing/activities/20020711_1, accessed 26 October 2006).

Wickens K et al. (1997). Determinants of house dust mite allergen in homes in Wellington, New Zealand. *Clinical and Experimental Allergy*, 27:1077–1085.

Wickens K et al. (2001). The importance of housing characteristics in determining Der p 1 levels in carpets in New Zealand homes. *Clinical and Experimental Allergy*, 31:827–835.

Wickens K et al. (2003). Determinants of endotoxin levels in carpets in New Zealand homes. *Indoor Air*, 13:128–135.

Wickman M et al. (1991). House dust mite sensitization in children and residential characteristics in a temperate region. *The Journal of Allergy and Clinical Immunology*, 88:89–95.

Williams LW, Reinfired P, Brenner RJ (1999). Cockroach extermination does not rapidly reduce allergen in settled dust. *The Journal of Allergy and Clinical Immunology*, 104:702–703.

Woodcock A et al. (2003). Control of exposure to mite allergen and allergen-impermeable bed covers for adults with asthma. *The New England Journal of Medicine*, 349:225–236.

Woodcock A et al. (2004). Early life environmental control: effect on symptoms, sensitization, and lung function at age 3 years. *American Journal of Respiratory and Critical Care Medicine*, 170:433–439.

Woodfolk JA et al. (1994). Chemical treatment of carpets to reduce allergen: a detailed study of the effects of tannic acid on indoor allergens. *The Journal of Allergy and Clinical Immunology*, 94:19–26.

Wright AL (2004). The epidemiology of the atopic child: who is at risk for what? *The Journal of Allergy and Clinical Immunology*, 113:S2–S7.

Wright RJ, Rodriguez M, Cohen S (1998). Review of psychosocial stress and asthma: an integrated biopsychosocial approach. *Thorax*, 53:1066–1074.

Yemaneberhan H et al. (1997). Prevalence of wheeze and asthma and relation to atopy in urban and rural Ethiopia. *Lancet*, 350:85–90.

Zhang L et al. (1997). Prevalence and distribution of indoor allergens in Singapore. *Clinical and Experimental Allergy*, 27:876–885.

2. Cockroaches

Michael K. Rust

Summary

Cockroaches are one of the most significant pests found in apartments, homes, food-handling establishments, hospitals and health care facilities worldwide. Indoor species, especially the German cockroach, exploit conditions associated with high-density human populations and impoverished living conditions. In some areas of the United Kingdom and Europe, the oriental and brownbanded cockroach may also be present. Poor sanitation, disrepair of a structure and clutter contribute to large populations of cockroaches. In these situations, their medical importance requires the implementation of aggressive integrated pest management (IPM) programmes. To minimize the likelihood of insecticide resistance in cockroach populations and human exposure to insecticides, strategies that include baiting and built-in pest control should be adopted.

Cockroach species that are found outdoors, such as the American, smokybrown and oriental cockroach, require IPM programmes that focus on altering and removing suitable habitats and that use baits and possibly even biological control options. The primary focus is directed outdoors to prevent them from gaining access to structures.

2.1. Introduction

Of the 3500–4000 species of cockroaches only about 50 have been reported as pests of human structures and dwellings worldwide (Cochran, 1999). Many of them are only occasional intruders or unwanted guests in commerce and do not pose a serious threat of becoming established indoors. Of the 69 species recorded from North America, 24 species are invasive, including all the major pest species (Atkinson, Koehler & Patterson, 1991). In fact, all of the major domiciliary pest species of cockroaches in North America and Europe are invasive and have relied on human activities and commerce to spread throughout the world. This review will focus on the five pest species most commonly found in urban settings in Canada, Europe and the United States: the German cockroach (*Blattella germanica*), the American cockroach (*Periplaneta americana*), the oriental cockroach (*Blatta orientalis*), the brownbanded cockroach (*Supella longipalpa*) and the smokybrown cockroach (*Periplaneta fuliginosa*). A four-year survey of a military installation in North Carolina found the following species in housing and food preparation areas: German cockroaches (84.8% of the observations), American cockroaches (14.7%), brownbanded cockroaches (0.5%), oriental cockroaches (0.2%), and smokybrown cockroaches (0.1%) (Wright & McDaniel, 1973). A survey of 219 hospitals in the United Kingdom revealed the following: oriental cockroaches only (58%); German cockroaches only (0.5%); oriental and German cockroaches (4.5%); oriental, German and American cockroaches (0.5%); and oriental, German and brownbanded cockroaches (0.5%) (Baker, 1990). The average ratio of cockroach pest species, German cockroach, brownbanded cockroach, and oriental cockroach, in Zurich, Switzerland, was, respectively, 17.45 to 3.75 to 1.00 for the years 1994–1998 (Landau, Muller & Schmidt, 1999).

Other species – such as the Australian cockroach (*Periplaneta australasiae*), brown cockroach (*Periplaneta brunnea*), Asian cockroach (*Blattella asahinai*) and the Turkestan cockroach (*Blatta lateralis*) – will at times, in specific localities or tropical conditions, become a significant problem. Health concerns with these species, however, are either unknown or of much less importance in temperate climates and will not be addressed.

Some important life history and biological parameters are summarized in Table 2.1 for the five most important cockroach pest species. Details of the biology and life histories of these species have been published in several sources; these include the German cockroach (Cornwell, 1968; Ebeling, 1975; Appel, 1995; Ross & Mullins, 1995), the American cockroach (Cornwell, 1968; Ebeling, 1975; Ross & Mullins, 1995), the oriental cockroach (Cornwell, 1968; Ebeling, 1975), the smokybrown cockroach (Cornwell, 1968; Appel & Smith, 2002), and the brownbanded cockroach (Cornwell, 1968; Ebeling, 1975). The following information about these species will emphasize information published since these reviews.

2.1.1. German cockroach

Once thought to have originated from Africa, the German cockroach probably spread from an area in East and South-East Asia (Roth, 1985). It is now a cosmopolitan pest because of changes in human travel, commerce and the urban environment. The advent of heating and cooling systems in buildings and of indoor cooking is probably a major contributor to its rapid spread and importance as an indoor pest throughout the world. Studies in Shanghai, China, support this view. Because Shanghai households are not well heated in the winter, lack cooking facilities and have inadequate amounts of food and water for German cockroaches, less than 6% of them are infested with cockroaches (Robinson, 1996b). The German cockroach, however, is the main pest species in Shanghai restaurants with heating sources.

The German cockroach is associated with indoor human activity, rarely being encountered outdoors. Alexander, Newton & Crowe (1991) found it in about 6% of outdoor situations in the United Kingdom. In public housing in France, it is the most common species found indoors (Rivault & Cloarec, 1997), but it is less likely to be found in single-family dwellings. It is typically found in areas associated with food preparation and storage, such appliances as stoves and refrigerators, and refuse containers. Areas that provide dark harbourage, such as under a stove or refrigerator and in cracks and crevices in close proximity to food and water, are ideal sites for German cockroaches. Rivault & Cloarec (1995) found that some environmental factors, such as population density, type of structure and cleanliness, partly explain the presence or absence of the German cockroach, concluding that it was a complex and multifactorial problem. Whyatt and colleagues (2002) also found that the quality of housing was a predictor of whether pests were sited and treated. However, less is known about the specific preferred microhabitats and the micro-distribution within structures (Appel, 1995).

Adult German cockroaches are about 1.3–1.6 cm long and pale brown to tan. The two black stripes on the prothorax – the anterior part of the insect's thorax, which bears its first pair of legs – help to easily distinguish this species from the similar sized brownbanded

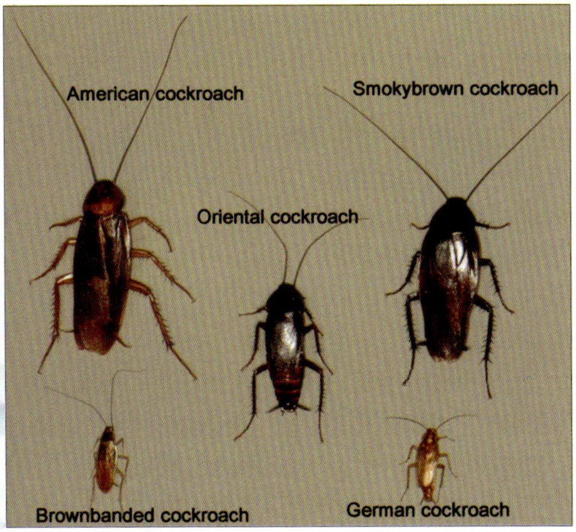

Fig. 2.1. Adult male cockroaches of the five most important urban pest species

Source: Photo by D.-H. Choe.

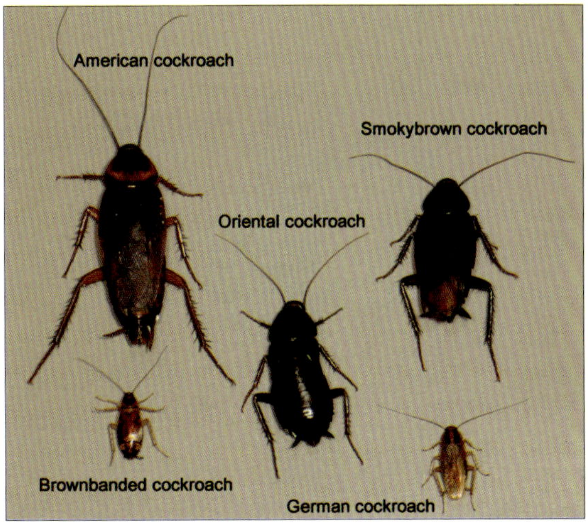

Fig. 2.2. Adult female cockroaches of the five most important urban pest species

Source: Photo by D.-H. Choe.

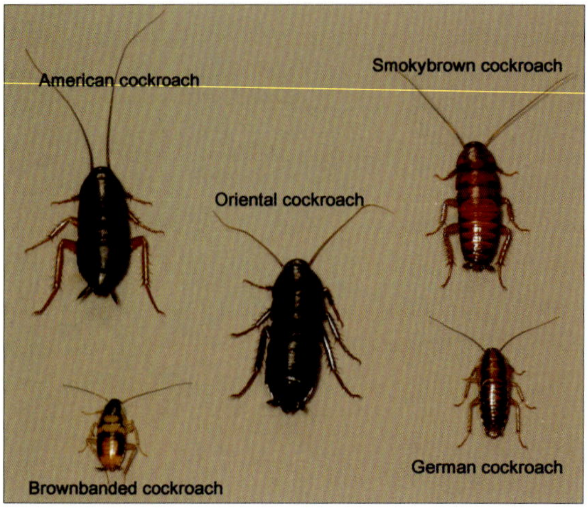

Fig. 2.3. Nymphal cockroaches of the five most important urban pest species

Source: Photo by D.-H. Choe.

cockroach (Fig. 2.1–2.3). The development of nymphs occurs rapidly, and three to four generations are possible each year (Table 2.1). The female carries the ootheca (egg case) until it is almost ready to hatch. These empty egg capsules are frequently scattered around infested structures, aiding in its identification (Fig. 2.4).

In recent years, studies have demonstrated the important role of German cockroach faeces in human allergies and asthma. Stejskal (1997) found that the mean density of faecal material decreased in the following order: shelter, edges of arena, area around the edges and remainder of the open area. The deposition of faeces may play some role in the orientation of cock-

Table 2.1. Life history and biological parameters of important domiciliary cockroaches

Species (common name)	**Indoors/ outdoors**	**Harbourage preferences**	**Temperature preferences** (°C)	**Adult size** (mm)	**Adult lifespan** (days)	**Generation interval** (days)	**Number of oothecae and size** (mm)
Blattella germanica (German cockroach)	Indoors	Food preparation areas, kitchens, food storage areas, bathrooms, areas combining warmth and moisture	20–26.7	10–15	♀ 153 ♂ 128	♀ 41 ♂ 40	~36 eggs (8)
Periplaneta americana (American cockroach)	Indoors/ oudoors/	Sewer systems, steam tunnels, zoological parks, greenhouses, areas that combine heat and moisture	24–31	34–40	♀ 125–706 ♂ 125–362	♀150–450	~18 eggs (8)
Blatta orientalis (oriental cockroach)	Indoors/ outdoors	Dense vegetation, water meter boxes, crawl spaces under structures/basements, cellars, areas that combine damp and cool conditions	20–29	25–30	♀ 34–181 ♂ 112–160	♀ 216 ♂ 185	~16 eggs (10)
Periplaneta fuliginosa (smokybrown cockroach)	Indoors/ outdoors	Trees, under logs, stones and flower pots, sewer systems, greenhouses	15–30	25–38	♀ 218 ♂ 215	♀ 320 ♂ 388	~26 eggs (11–14)
Supella longipalpa (brownbanded cockroach)	Indoors	Elevated closets, storage areas, cupboards, animal rearing facilities, warm areas	26–30	11–14	♀ 60 ♂ 115	♀ 56 ♂ 54	16 eggs (4–5)

Source: Cornwell (1968); Ebeling (1975); Appel & Smith (2002).

roaches and their aggregation around harbourages and as sources of allergens. Adult males, non-gravid females and gravid females produce an average of 9.6, 9.1 and 2.7 faecal pellets a day, respectively. Adult females consistently produce more allergen Bla g 1 in their faeces than do males (Gore & Schal, 2005): in their lifetime, adult females will produce about 25 000 to 50 000 units of the allergen, compared with about 2000 to 3000 units for males. Mated non-gravid females and adult males are extremely mobile, moving about 75 m in five days (Demark & Bennett, 1995). This may have tremendous implications for the management of allergens within structures. It also highlights the problem of trying to remove faecal material from indoor environments.

Fig. 2.4. Oothecae of the five most important urban cockroach species

Source: Photo by D.-H. Choe.

Studies on the feeding behaviour of first instars (immature insects in their first nymphal developmental stage) suggest that facultative coprophagy (feeding on faecal matter) permits them to molt into second instars with minimal foraging (Kopanic et al., 2001). The first instars survived significantly longer and also gained more nutrients than did first instars not fed faeces. Survival of the second instars did not increase when fed faecal matter. In addition, the quality of adult diet affects the nutritional content of faecal matter and the development of first instars. More important, feeding on anal secretions (proctodeal feeding) by first instars facilitated the horizontal transfer of slow-acting pesticides, such as hydramethylnon (see subsection 2.6.2.1, on baits). This data clearly indicates that slow-acting pesticides should be recommended in baiting programmes.

Numerous researchers have shown that the aggregation pheromone produced by cockroaches to attract other cockroaches has potential uses in pest-control programmes, because this multi-component pheromone will cause immature cockroaches (nymphs) to aggregate and will arrest their activity on faecal deposits. The exact characteristics of the pheromone remain unknown, but it holds promise as an aid to help reduce pesticide use.

Cockroach traps have been baited with various food attractants and so-called pheromones. Traps have never demonstrated control of German cockroaches. Their effectiveness depends on their placement and not on attractive semiochemicals (chemicals that transfer a signal from one organism to another). Nojima and colleagues (2005) reported the structure of the volatile sex pheromone blattellaquinone of the German cockroach. Field tests indicated that males were attracted to sticky traps baited with synthetic blattellaquinone. This may be a new promising attractant for monitoring traps, increasing their effective range.

2.1.2. American cockroach

The American cockroach is probably of tropical African origin, where it lives both inside and outside of structures. Adult males and females will readily fly, especially in warm tropical conditions. The adult female's ability to survive 90 days with water and no food, the deposition of oothecae and extended nymphal developmental periods probably contributed to their spread in maritime commerce before the age of steam powered ships. The American cockroach is now cosmopolitan and a significant pest in tropical and subtropical climates.

Indoors, the American cockroach will inhabit food preparation and storage areas, but they are not as common as German cockroaches. In buildings, they are found around steam and heating pipes and in areas associated with high temperature and humidity (Appel, 1995). They can be troublesome in greenhouses, damaging plants and feeding on pest insects. Outdoors American cockroaches frequent palm trees and vegetation around structures (Roth, 1981). Their close association with latrines, cesspools, sewers and dumps increases the likelihood of them transmitting human pathogens. In the arid south-western United States, they are commonly found in sewer systems.

Adult male and female American cockroaches are among the largest cockroach species found in urban settings. They are about 4 cm long on average and are reddish brown except for a pale brown band around the edge of the prothorax. The cerci – that is, paired appendages on the rear-most segments, which serve as sensory organs – are extremely long and slender (Fig. 2.1–2.3). In the field, the nymphs of American and oriental cockroaches are difficult to separate without a hand lens, and the initial identification often depends on the habitat where they are collected.

Outdoor surveys that use traps can be biased and depend on the species in the area. Appel (1984) has shown that American cockroaches are more likely to be trapped than are smokybrown cockroaches. However, traps with nymphal smokybrown cockroaches repel American cockroaches, resulting in an under-representation of American cockroaches in those areas of sympatry – that is, geographical areas where the different species of cockroach overlap without interbreeding – where smokybrown cockroaches predominate.

2.1.3. Oriental cockroach

The oriental cockroach is probably indigenous to North Africa, inhabiting climates that combine summer heat and moderate winter temperatures. Oriental cockroaches are a major domiciliary pest in England and are found in northern cities of the United States and throughout Europe. In the United Kingdom, the favourite localities in structures for oriental cockroaches are the cellar, boiler room and heating ducts (Cornwell, 1968). Surveys of pest control service technicians throughout the United Kingdom indicated that oriental cockroaches were found indoors (in all questionnaires) and outdoors about 60% of the time (Alexander, Newton & Crow, 1991). In the Czech Republic and in Slovakia, the prevalence of oriental cockroaches has been declining, probably because of changes in building practices (Stejskal & Verner, 1996). In Hungary, apartments older than 60 years were infested with oriental cockroaches, whereas new apartments (less than 10 years old) were infested with German cockroaches. Oriental cockroaches have also been reported in sewers in Germany (Pospischil, 2004) and Hungary (Bajomi, Kis-Varga & Bánki, 1993). In Germany, they are also pests in public baths, bakeries and breweries (Pospischil, 2004). In the south-western United States, they are primarily found outside structures, around water meter boxes and woodpiles and under uplifted concrete walks (Cornwell, 1968). Adult populations peak in late June and July. Oriental cockroaches have been observed feeding on garbage, dead insects, slugs, bird droppings and turf grass, but areas where pets are fed outdoors frequently become heavily infested. In central California, they were found in 7.1% of sewers inspected, whereas they are rarely encountered in sewers in southern California (Rust et al., 1991).

Cold hardiness probably contributes to its northern distribution. Oriental cockroaches are not limited by temperature throughout much of the United Kingdom and western Europe, if it can avoid short-term exposures to extremely low temperatures (le Patourel, 1993). Outdoor populations may survive in urban areas if heated buildings provide attractive harbourages. During a two-week acclimation to 10 °C, none of the nymphs successfully molted. Acclimation, however, did result in increased survival at low temperatures. Critical temperatures for formation of ootheca, viability and successful hatch are between 15 °C and 20 °C (le Patourel, 1995). If the temperature is reduced to around 15 °C, the populations can be arrested and will decline.

Oriental cockroaches remain close to preferred harbourages and existing refuge (Mielke, 1996). Nymphs are most likely to initiate movement to new harbourages, depending on the distance. Increased temperature (29–38 °C) and relative humidity between 48% and 70% decreases activity and increases the radius of movement. In the eastern United States, oriental cockroaches remained near preferred harbourages, aggregating around crawl spaces and vents under structures and with only 2% moving indoors (Thoms & Robinson, 1986). Consequently, locating refuge and aggregation sites is essential to control them.

2.1.4. Smokybrown cockroach

The smokybrown cockroach is commonly found in the south-eastern United States, except in central and southern Florida where it is replaced by the Australian cockroach (Atkinson, Koehler & Patterson, 1991). It is also found outdoors in some coastal areas of California. The lower and upper temperature limits on their development are 15 °C and 35 °C, respectively (Benson, Zungoli & Smith, 1994). The distribution could extend to all subtropical areas of the world, and isolated infestations could survive in any modern city (Appel & Smith, 2002).

Adults are dark shiny tan to ebony and 25–38 mm long (Fig. 2.1 and Fig. 2.2). The unpatterned pronotum – that is, the upper or dorsal surface of the prothorax – is distinguishable among the other species of pestiferous *Periplaneta*. Both sexes have full wings, unlike female oriental and Turkestan cockroaches that are brachypterous – that is, having very short or rudimentary wings. Early instars are easily recognized by distinctive white-banded antennae, white mesonotum (the dorsal portion of the mesothorax) and whitish first two abdominal nota (the dorsal parts of the first two abdominal segments). Compared with other outdoor species, the ootheca is large (Fig. 2.4).

Outdoors, smokybrown cockroaches are likely to be found in woodpiles, bark, leaf mulch, tree holes, planters and utility vaults. They feed on fruit, dead insects, worms and pet food. They will infest water-meter boxes, but are rarely found in sewer systems. Smokybrown cockroaches will occasionally invade structures; however, sustained indoor breeding populations are rare. The proximity of preferred habitats close to homes is responsible for domestic infestations (Brenner & Pierce, 1991). Catches in indoor traps positively correlate with outdoor catches (Smith et al., 1995). Preferred indoor harbourages include empty spaces of porch and carport ceilings, exterior walls of furnace rooms and empty spaces of water damaged walls (Appel & Smith, 2002).

Like the oriental cockroach, the smokybrown cockroach has a restricted small outdoor home range (less than 280 m^2) and restricted movements near harbourages, such as under decks and within debris that offers dark hiding places (Appel & Rust, 1985). Traps or baits in these areas dramatically reduce populations, eliminating the need for perimeter spraying.

The incidence of smokybrown cockroaches in urban habitats has led to the development of a cockroach habitat index (Smith et al., 1995). Such factors as the number of trees, pets, residents, age of the house and the obvious cockroach harbourages were the most important indicators of infestation. Such items as mulched and bushy landscapes next to retaining walls and sheds are especially attractive. The index can help target sites for a comprehensive IPM programme, including insecticide treatment and habitat modification.

The volatile sex pheromone of the smokybrown cockroach has been isolated and identified as periplanone D (Takahashi et al., 1995). It elicits low activity in American cockroach, in *Periplaneta japonica* and in oriental cockroach males. Periplanone D may prove useful in monitoring traps, especially since smokybrown cockroaches are less likely to be trapped than American cockroaches.

2.1.5. Brownbanded cockroach

The brownbanded cockroach is an occasional indoor pest, especially in such heated structures as animal rearing facilities, apartments and homes in which temperatures are higher than normal human comfort (Ebeling, 1975). First reported in Germany in 1954, it has only been periodically reported as a problem (Mielke, 1995; Pospischil, 2004). It has been only sporadically reported in Budapest, Hungary (Stejskal & Verner, 1996). Also, rarely encountered in the United Kingdom, it has been reported with German cockroaches in only 0.5% of 219 hospitals surveyed (Baker, 1990). However, recent reports from professionals in pest control suggest that it is becoming a more important pest in the United Kingdom. Additional surveys and careful identification of indoor pest species are needed.

Little research has been conducted on this species, in part because it is infrequently encountered. Liang and colleagues (1998) reported that the synthetic sex pheromone supellapyrone is highly attractive to males in the field, increasing male trap catches 6–28 times. This might provide a useful monitoring tool in situations where brownbanded cockroaches are a problem.

2.2. Health hazards

2.2.1. Surveys and public opinion

Cockroaches typically rank as one of the most common and objectionable insects encountered by homeowners, especially in low-income housing. In a survey of 315 inner-city and low-income women in New York City, 66% of them reported seeing cockroaches (Whyatt et al., 2002). Greater than 75% of apartment tenants considered cockroaches a serious problem (Wood et al., 1981). In three cities in the eastern United States – Baltimore, Roanoke, and Norfolk – surveys of tenants of public housing indicated that 83% felt that cockroaches were a serious problem (Zungoli & Robinson, 1984). Cockroaches were more important than other negative factors in these buildings, such as leaky faucets, broken windows and trash in the yard. Only the presence of mice was considered worse. In a similar study of London residents, more than 80% of the residents from uninfested apartments felt that cockroach infestations were worse than poor security, dampness, poor heating and poor repair (Majekodunmi, Howard & Shah, 2002). Again, only infestations of mice were considered worse. Residents with cockroach infestations were slightly more tolerant of the problem than residents that did not have cockroaches. Only 2% of the respondents mentioned asthma or allergies as a potential health concern associated with cockroaches. A survey by Davies, Phil & Peltranovic (1986) of apartment residents in Toronto, Canada, indicated that about 50% of them had cockroach infestations, 89% considered them a health hazard and 94% considered them a source of anxiety. In a household survey in Kentucky, 63% of the respondents listed seeing 0 to 1 cockroach as their tolerance threshold. Less than 10% of the respondents would tolerate seeing more than five cockroaches (Potter & Bessin, 1998).

In addition to the direct health problems associated with cockroaches (such as allergic responses, transport of pathogenic organisms and contamination of food), improper applications of insecticides and a heavy reliance on aerosols and the application of liquid sprays to surfaces may create potential human exposure problems. More than 90% of the pesticides applied in apartments are directed at cockroaches (Whyatt et al., 2002). Fischer and colleagues (1999) reported several cases in sensitive areas, such as schools and health care facilities, where house dust contained detectable amounts of the pesticides pyrethrum, piperonyl butoxide, chlorpyrifos and cyfluthrin. Three insecticides, acephate, chlorpyrifos and propetamphos, were detected on both target and non-target surfaces in schools treated for cockroach infestations (Williams et al., 2005). In apartments infested with German cockroaches, air samples of the residences ($n = 60$) found that more than four pesticides used for cockroach control were present. In 18 of 60 cases, eight different pesticides were detected in the air, clearly suggesting an over-reliance on chemical measures to control cockroaches in these housing units. No explanation was given for the possible source of the dichlorodiphenyltrichloroethane (DDT) and chlordane found in air samples (Whyatt et al., 2002).

Data collected throughout the United States suggest that pesticide applications in schools may produce acute illness among school employees and students (Alarcon et al., 2005). The authors indicate that these are "albeit mainly of low severity and with relatively low incidence rates". Of the 2593 cases of illness examined, about 35% resulted from the insecticides. Of the 406 cases with more detailed information, 69% were associated with pesticides used in schools. The most common active ingredients reported were diazinon, chlorpyrifos, and malathion.

2.2.2. Allergy and asthma overview

In recent years, cockroach pest management has focused on the association between asthma and the presence of cockroach allergens (see Chapter 1 of this report). Educational intervention and attempts to lower the source of cockroach allergens resulted in about a 60% reduction in cockroaches in the intervention group, compared with the non-intervention group (McConnell et al., 2005). Although allergen loads in bedding were reduced by these efforts, kitchen levels remained high.

Even low numbers of cockroaches can produce significant amounts of allergen. Over their lifetime, adult female German cockroaches can produce 25 000 to 50 000 units (Gore & Schal, 2005). In spite of reductions in cockroach numbers, the amount of allergen or cockroach dust often remains for longer than 6 months, even with aggressive cleaning (Eggleston, 2003). In summary, best pest management strategies seem to significantly reduce allergen, but not to below the disease threshold (8 U/g of house dust) (Katial, 2003). Clearly, additional research and new approaches are needed.

Numerous studies have shown the association between (and potential significance of) cockroaches in lower-income households and asthma among children (Brenner, 1995; Baumholtz et al., 1997). Rosenstreich and colleagues (1997) write that "exposure to cockroach allergen has an important role in causing morbidity due to asthma among inner city children".

2.2.3. Food contamination and disease transmission

Cockroaches present a potential health problem to people and their companion animals. Brenner (1995) and Baumholtz and colleagues (1997) have provided extensive reviews of literature on the pathogens associated with cockroaches, including such pathogens as viruses, bacteria, fungi and molds. Table 2.2 is a summary and update of the important pathogenic bacteria, viruses and fungi reported for cockroaches. Fathpour, Emtiazi & Ghasemi (2003) collected German, American and brownbanded cockroaches from hospitals, houses and poultry sheds in Iran. Of the 80 cockroaches tested, about 70% were contaminated with *Salmonella* spp., many of which were resistant to antibacterial drugs.

Indirectly, cockroaches may affect human health by transmitting disease to agricultural products that ultimately end up in the human food supply. In the past 20 years, the presence of German cockroaches has increased dramatically in Czech and Slovak dairies (Stejskal & Verner, 1996). Oriental cockroaches have been a problem in pig farms for

Table 2.2. List of pathogenic microbes isolated from cockroaches

Bacteria

Alcaligenes faecalis
Bacillus subtilis
Campylobacter enteritis
Campylobacter jejuni
Clostridium novyi
Clostridium perfringens
Enterobacter aerogenes
Escherichia coli (*B. orientalis*, Auer, Asperger & Bauer, 1994; *B. germanica*, Tarry & Lucas, 1977)
Klebsiella pneumoniae
Listeria monocytogenes (*B. orientalis*, Hechmer & van Driesche, 1996)
Mycobacterium leprae
Nocardia spp.
Proteus mirabilis
Proteus morganii
Proteus rettgeri
Proteus vulgaris
Pseudomonas aeruginosa
Salmonella spp. (*B. germanica*, *P. americana*, *S. longipalpa*, Rosenstreich et al., 1997)
Salmonella bareilly
Salmonella bovismorbificans
Salmonella bredeney
Salmonella enterica serotype Oranienburg
Salmonella enterica serotype Panama
Salmonella enteritidis (*B. orientalis*, Auer, Asperger & Bauer, 1994)
Salmonella newport
Salmonella paratyphi B
Salmonella typhimurium (*B. germanica*, *P. americana, B. orientalis*, Zurek & Schal, 2004)
Serratia marcescens
Shigella dysenteriae
Staphylococcus aureus (*B. orientalis*, Auer, Asperger & Bauer, 1994)
Streptococcus faecalis
Streptococcus pyogenes
Vibrio spp.
Yersinia pestis

Fungi and moulds

Alternaria spp.
Aspergillus niger
Aspergillus flavus
Aspergillus fumigatus
Candida krusei
Candida parapsilosis
Candida tropicalis
Cephalosporium acremonium
Cladosporium spp.
Fusarium spp.
Geotrichum candidum
Mucor spp.
Penicillium spp.
Rhizopus spp.
Trichoderma viride
Trichosporon cutaneum

Helminths

Ancylostoma duodenale
Ascaris lumbricoides
Ascaris spp.
Enterobius vermicularis
Hymenolepis spp.
Necator americanus
Trichuris trichiura

Protozoans

Entamoeba histolytica
Giardia spp.

Viruses

Poliomyelitis

Source: Compiled from Brenner (1995); more recent citations (shown in parentheses) are included.

decades, and they may spread porcine parvoviruses (Tarry & Lucas, 1977). In addition, German cockroaches may serve as an important mechanical vector of porcine verotoxigenic *Escherichia coli*. As a result, Zurek & Schal (2004) recommend the incorporation of cockroach IPM into disease prevention and control programmes in the pig-farming industry. Kopanic and colleagues (1994) reported that American cockroaches collected at feed mills and poultry hatcheries were positive for *Salmonella* spp., raising concerns about infecting hatched chicks and farm flocks. Fischer and colleagues (2003) reported that nymphs of oriental cockroaches were capable of transmitting paratuberculosis (*Mycobacterium avium* ssp. *paratuberculosis*) and avian tuberculosis (*Mycobacterium avium* ssp. *avium*) in their faeces. Isolates from the cockroaches and their faeces were virulent to chickens.

American cockroaches, feeding on the faecal matter of opossums that carried *Sarcocystis falcatula*, were potential carriers to non-American psittacine birds, especially cockatoos and cockatiels. The birds contracted the disease by either eating infected cockroaches or possibly contaminated food (Clubb & Frenkel, 1992).

Even though numerous studies have demonstrated the ability of cockroaches to pick up and later excrete or transfer pathogens, definitive evidence that cockroaches are vectors for human disease is still lacking. However, the prevalence of cockroaches near human and animal wastes, human food, and human environments creates sufficient concern about their role as vectors. This potential health threat necessitates the control of cockroaches in food handling areas, hospitals, animal-rearing facilities, zoos and human residences.

The incidence of cockroaches in commercial food-handling establishments exceeds 50%. In New York City, 53% of the 18 000 food establishments inspected in 1976 had insect infestations (Dupree, 1977). In a random survey of 100 commercial food-handling establishments in Los Angeles, 62 were infested with German cockroaches. All of them had professional pest control service (Rust & Reierson, 1991). It is likely that as many as 70% of all food-handling establishments have cockroach infestations.

2.3. Public costs

2.3.1. Cost of health-related conditions

In the United States, there were more than 21 000 pest control companies in 1997, with an estimated annual income of US$ 4.5 billion. Some 300 000 retail food outlets, 500 000 commercial restaurants and kitchens, and 70 000 hotels and motels were under service contract (Potter & Bessin, 1998). In 2004, some 19 000 pest control companies generated US$ 6.5 billion – a 6% increase over revenue generated in 2003 (Curl, 2004). The importance of cockroaches for the pest control industry in the United States has declined in recent years, but cockroaches still represent 22% of service sales (Anonymous, 2002; Curl, 2004). Possibly US$ 1 billion are spent on professional services in the United States each year to control cockroaches.

The costs of medical problems associated with cockroaches are difficult to estimate. Asthma affects 15 million Americans, approximately a third being under 8 years of age (Benson & Marano, 1998). Children allergic to cockroach allergen and exposed to high levels had a 3.4 times higher rate of hospitalization for asthma than other children. This group also had 78% more unscheduled visits to health care providers because of asthma. They also missed significantly more days of school than did other children (Rust & Reierson, 1991). An estimate of the national economic burden of asthma in the United States in 2000 is US$ 14.5 billion (Krieger et al., 2002). The actual medical and societal costs associated with cockroach-related asthma might be a considerable portion of that cost.

2.3.2. Cost of control and management

The costs of cockroach control services vary greatly, depending on the pest species and locality. Because it is generally assumed that cockroach IPM programmes cost more than conventional pest control services, a comparison of conventional programmes with IPM programmes is insightful. In 1998, in the north-eastern United States, conventional pest control services cost US$ 65 an hour, whereas IPM services were US$ 80 an hour (Rambo, 1998). However, it is assumed that IPM should actually reduce the frequency of visits and consequently the labour costs in the long term. Williams and colleagues (2005) reported that after all costs were considered, conventional service was US$ 8.57 per unit and IPM was US$ 7.49 per unit. In their study, cockroach infestations were very low in schools, and costs would probably have increased if they had been more severe.

Brenner and colleagues (2003) set up an IPM-based programme for low-income households in New York City, involving monitoring, baiting, cleaning and structural repairs. The costs for IPM were US$ 46–69 per unit in the first year and US$ 24 per unit in the following year. In comparison, conventional chemical controls cost US$ 24–46 per unit and involved no repairs or structural modifications to the apartments. In IPM programmes, the number of cockroach infestations declined by 50% over six months.

In public housing in Portsmouth, Virginia, the costs of conventional crack-and-crevice treatments with sprays and dusts were compared with vacuuming, baits and insect growth regulators (IGRs) for controlling German cockroaches (Miller & Meek, 2004). The average costs for IPM and conventional treatments were US$ 4.06 and US$ 1.50 per unit, respectively. After eight months, cockroach populations decreased about 80% in IPM units, compared with a 300% increase with conventional treatments.

In 2006, in the United States, the cost of treating an apartment was about US$ 150 and buildings typically cost US$ 1200 or more to treat. Commercial accounts, such as restaurants, typically cost US$ 250 a month to treat.

2.4. Impact of poverty

The physical and sanitary conditions of a dwelling greatly affect the likelihood of cockroach infestations, especially of German cockroaches. Bradman and colleagues (2005) reported on a study of pregnant Latina women and their children in 644 homes in an agricultural community in Salinas Valley, California. These were mostly multiple dwelling units that were characterized by high residential densities (39% had more than 1.5 persons per room). In the United States, only 3% of Hispanic households and 0.5% of all households experience this level of crowding. About 60% of the Salinas Valley households had cockroaches and 32% had rodents. The chances of having cockroaches increased with the presence of peeling paint, water damage and high residential density.

In New York City, the frequency of cockroach sightings and allergens is related directly to the level of housing problems and level of disrepair (Rauh, Chew & Garfinkel, 2002). Indicators of disrepair include holes in ceilings or walls, peeling paint, water damage, leaking pipes, and lack of gas or electricity in the past six months. In a study of asthmatics from Connecticut and Massachusetts, low socioeconomic indicators and minority status were associated with a high likelihood of cockroach allergens in house dust (Leaderer et al., 2002). Elevated cockroach allergen levels increased as the number of families living in habitations increased.

In addition to the disrepair and poor sanitary conditions associated with poverty that promote cockroach infestations, inner-city children are exposed to heavy applications of pesticides (Landrigan et al., 1999). In 1997, the number of gallons of chlorpyrifos applied in New York City exceeded the total number of pesticides applied in any other county in that state. In housing projects in the East Harlem section of New York City, chlorpyrifos, cyfluthrin and lesser amounts of bendiocarb were applied on a monthly basis. The use of illegal pesticides, such as aldicarb, Chinese chalk, and methyl parathion, is another problem encountered in impoverished neighbourhoods.

The quality of the indoor environment is especially important, because 75% of children between the ages of 5 and18 years spend 16 hours or more in the home (Bonnefoy et al., 2003). Consequently, conditions that contribute to cockroach infestations and control measures to eradicate them will greatly impact children living under conditions of poverty.

2.5. Impact of new housing technology

Changes in housing have clearly had an effect on the prevalence and species of cockroach infesting structures. In Hungary, for example, changes in construction practices, common ventilation systems, false ceilings, wall coverings, and heating and sewage pipes have contributed to German cockroaches becoming the main pest species in structures (Bajomi, Kis-Varga & Bánki, 1993). Similarly, in food-processing plants in the Czech Republic and Slovakia, the upsurge of German cockroaches has been attributed in part to central heating (Stejskal & Verner, 1996). The increase in kitchens in different types of buildings, such as clubs, cinemas, offices and shops and all types of factories after the Second World War has been cited as the cause of an increase in German cockroach infestations. Cornwell (1968) discounted the increase in central heating as a major reason for its spread in the United Kingdom.

Building practices, such as hollow-wall voids, drop ceilings and voids under cabinets, attics and built-in appliances, provide suitable harbourage for cockroaches. The elimination of these harbourages is the primary goal of so-called built-in pest control or insect proofing. The use of inorganic dusts to eliminate cockroaches in harbourages and voids has long been advocated and is a successful means of controlling German cockroaches (Ebeling, 1975). Inorganic dusts are preferred, because they retain their insecticidal activity as long as the dust deposits remain intact and do not clump or cake. Repellent dusts, such as silica aerogel, are typically applied at the time of construction to prevent cockroaches from establishing themselves in wall and sub-cabinet voids. Non-repellent dusts, such as boric acid, are applied to existing infestations to provide remedial control. Non-repellent dust, however, will not scatter existing infestations. Non-repellent dusts should be routinely reapplied when flats are being refurbished to accommodate new occupants. One important advantage of built-in treatments is that the dusts are applied in areas not readily accessible to people and pets. Contrary to common belief, relative humidity does not strongly affect the toxicity of most insecticidal dusts, especially inorganic dusts. In fact, as boric acid dusts and silica aerogel dust plus synergized pyrethrin are wetted, their toxicity actually increases.

In many cities throughout Europe and the United States, formerly state-owned and managed housing projects are being converted into privately owned dwellings. These conversions present special problems in providing IPM programmes, because of the inability to inspect, monitor and treat all units within a building complex. In these units, inorganic dusts should be reapplied when flats are renovated between occupants. Community action plans and education programmes will be important in implementing IPM under these conditions. Also, cooperation is essential to ensure that all flats are treated.

2.6. IPM of cockroaches

Basic IPM programmes to control cockroaches were initiated in the 1980s and 1990s (Robinson & Zungoli, 1995). Unfortunately, as Robinson (1996a) notes, "In spite of the need, the potential benefits, and sufficient development time, the concept of IPM has not been developed fully for household and structural pests." Since these initial attempts, most of the research has focused on the reduction of cockroaches and allergens in structures. Some pest control companies have incorporated baits into their cockroach control programme and simply renamed the service IPM.

IPM programmes to control cockroaches are poorly understood. Even though 67.8% of pest management professionals (PMPs) thought that IPM was necessary to carry out their pest control mission, less than 25% felt that the average pest control professional understood what IPM is. The PMPs also thought that there was no universal or clear definition of IPM in the industry (Anonymous, 1995). The challenge is to provide effective and economical uses of pesticides and alternative technologies that control and eliminate pests in the living space (Robinson, 1996a). The programmes must be economically and aesthetically acceptable and must address specific attitudes of the target audience if they are to be successful (Robinson & Zungoli, 1995).

Outdoor IPM programmes have drawn less attention. A checklist of factors that contribute to smokybrown cockroach infestations, such as the age of the structure, the presence of woodpiles, railroad ties and tool sheds, and the number of pets, has been developed to identify structures and sites to be treated. By incorporating habitat removal, granular and gel bait treatments, and some spot sprays, greater than 80% reductions were achieved with less total insecticide used, compared with conventional perimeter sprays. In addition, these programmes were able to reduce the number of visits that the pest-management technician made to the structure. Additional research and refinements of IPM programmes indicate that baiting consistently provided excellent reductions, and combination treatments that targeted different habitats were especially effective. The application of insecticide to targeted sites is the most cost-effective and environmentally sound approach to control smokybrown cockroach infestations (Smith et al., 1998; Appel & Smith, 2002).

Among the goals of IPM programmes are reducing the amount of insecticide applied and reducing unnecessary treatments. Careful monitoring identifies areas that do not need to be treated. In their two-year control programme based on tenant observations, Rivault & Cloarec (1997) estimated that more than 50% of flats were unnecessarily treated. Monitoring of schools with traps revealed that a number of them were unnecessarily treated for German cockroaches (Williams et al., 2005). Also, the actual amount applied to control the cockroaches can be reduced. About 6 g of insecticide were applied per unit in the IPM approach, compared with about 140 g in conventional treatments (Miller & Meek, 2004). Greene & Breisch (2002) reported a 93% reduction in the total amount of insecticide applied after the adoption of an IPM programme (Table 2.3). Similarly, Williams and colleagues (2005) reported that 99.9% less insecticide was applied in their IPM programmes.

In addition to using less insecticide, IPM programmes incorporate alternative strategies. In a pilot project in Toronto, Canada, most tenants showed improved knowledge about IPM concepts and selected non-spray chemical treatments, such as baits, resulting in reductions in the numbers of cockroaches in about 67% of the units (Campbell et al., 1999). In Seattle, Washington, a comprehensive project was implemented to improve indoor environmental quality for children with asthma (Krieger et al., 2002). To control the cockroaches – in addition to treating with bait – daily cleaning, structural repairs, and the removal of clutter, food and sources of standing water were initiated.

2.6.1. Cockroach control: sanitation practices

The removal of potential clutter, debris, harbourage sites and food sources is frequently stressed in cockroach control programmes. Demonstrating that these efforts have a significant impact on cockroach control is extremely difficult and few well-designed field studies have been conducted. Lee & Lee (2000) reported that increased sanitation in homes resulted in better bait performance (0.5% chlorpyrifos bait) at 12-week post-treatment evaluations of American cockroach infestations. When German and American cockroaches are provided alternative food along with toxic bait in laboratory studies, the time required to kill cockroaches is consistently longer than with unfed cockroaches (Appel, Eva & Sims, 2005).

Table 2.3. Effect of IPM on insecticide applications and cockroach control

Service category	Before IPM (1988)	After implementing IPM (1999)
Pesticide-treatment requests (total)	14 659	954
Spray application requests	2661	0
Cockroaches requests only	10 647	733
Pesticide applications (%)	99.6	60.3
Total pesticides (g)	4426	305.8

Note. Control services considered pertain to buildings under the United States General Services Administration in the National Capitol Region (Washington, DC, Maryland and Virginia).

Source: Adopted from Greene & Breisch (2002).

Sanitation is only a part of a more comprehensive IPM programme to remove preferred cockroach harbourages and resources, and cleaning alone will not reduce cockroach populations. In conjunction with structural repairs and built-in pest control, improved sanitation will stress cockroaches, making them more susceptible to baits and crack-in-crevice treatments.

2.6.2. Cockroach control: pesticide applications

2.6.2.1. Baits

The development of baits has revolutionized cockroach control, especially against the German cockroach (Reierson, 1995). This new paradigm has been widely accepted by both the professional pest control community and the public. Nearly 69% of homeowners preferred baits to sprays, and 33.5% thought that baits were the least hazardous method of control (Potter & Bessin, 1998).

Operational factors are also important to the success of baiting programmes (Reierson, 1995). Durier & Rivault (2003) found that applying numerous discrete droplets of bait were better than one large spot of bait against high densities of German cockroaches; at low densities, a single spot of bait was adequate. Providing many bait stations helped reduce aggression among cockroaches and allowed greater access to bait in their lab study. In addition to greater numbers of bait placements, the location of the bait station may be extremely important. Appel (2004) found that bait consumption decreased if it was placed on surfaces treated with repellents, such as some pyrethroid insecticides. When repellents were applied to solid cockroach baits, there was decreased consumption. This did not occur on gel baits.

Baiting outdoors often presents special problems, because of rainfall, irrigation and high humidity. Thoms & Robinson (1987a) reported that hydramethylnon bait molded in high humidity conditions in outdoor bait trials against oriental cockroaches, but still managed 69.3–97.8% reductions in cockroaches. Gel baits that contained fipronil and imidacloprid provided 96–99% control of American cockroaches in sewers for 6 months or longer.

Resurgence of cockroaches appeared to result from a loss of bait and not degradation of active ingredient (Reierson et al., 2005).

Baits are an important component of cockroach IPM programmes. However, a common misconception is that the use of baits alone is IPM. Nothing could be farther from the truth. Effective IPM involves a systematic approach and process (see Chapter 15 of this report).

2.6.2.2. Crack-and-crevice treatments

The method of applying insecticides greatly affects its bioavailability to pests and people (Rust, 1995, 2001). Applications of small amounts of non-repellent insecticides to cracks and crevices are effective against such insects as cockroaches and silverfish, because these insects prefer to hide in small dark spaces. Non-repellent insecticides are more likely to be encountered by cockroaches than repellent insecticides that flush them out of hiding areas. Small crevices are ideal sites to place gels baits, dusts and liquid sprays.

Crack-and-crevice treatments will quickly reduce the numbers of foraging cockroaches, but will not eliminate breeding populations in voids and outdoor sites. These spot treatments are most effective when applied in conjunction with structural repairs, void treatments and sanitation. Spot treatments to cracks and crevices reduce the amount of insecticide applied indoors and also reduce the likelihood of exposing people and pets to insecticides (Rust, 2001). These treatments are designed to be preventive and should only be made when visual monitoring or traps indicate the presence of cockroaches.

2.6.2.3 General surface treatments

Indoors, baiting and other less pervasive methods have largely replaced the use of spot, baseboard and surface sprays. Nevertheless, perimeter and spot treatments have a place in IPM, especially in treating peridomestic species outdoors, where the application of bait, structural modifications or built-in controls is problematic. The activity of insecticides against eight pest species, including American, oriental and smokybrown cockroaches, is (in order of most active to least active) as follows: pyrethroid, carbamate and organophosphate (Valles, Koehler & Brenner, 1999). Encapsulated diazinon and chlorpyrifos applied to perimeters of homes and in crawl spaces provided about 95% reductions of oriental cockroaches for at least 12 weeks (Thoms & Robinson, 1987a).

The use of IGRs as surface sprays or in total release aerosols has always been of interest, because of their low toxicity to mammalians. However, because of their modest efficacy against cockroaches, they have never gained acceptance. In a recent study, noviflumuron sprays applied in infested apartments provided up to 90% reductions of German cockroaches at four weeks post-treatment (Ameen et al., 2005).

Also, in the control of German cockroaches, residual sprays of chlorfenapyr provided 80% reductions in infested apartments at eight weeks post-treatment (Ameen, Kaakeh & Bennett, 2000). Chlorfenapyr is a pro-insecticide – that is, it is only activated to its toxic form in the cockroach's body – that shows promise as an alternative treatment for field strains with high levels of mixed function oxidase, resulting in insecticide resistance.

These indoor perimeter applications are designed to provide preventive control, and they should only be used when visual monitoring or traps indicate the presence of cockroaches. Special precautions should be taken to reduce the likelihood of people, pets and non-target organisms being exposed.

2.6.2.4. Alternative strategies

There will always be situations and sensitive areas where the use of insecticides is unfeasible, impractical or prohibited. Cockroach infestations in equipment such as cash registers, computers, televisions and radios or in such places as museums, health care facilities, airplanes, ships and vehicles might be difficult to treat with conventional strategies. Alternative approaches have been successful, but have been used on a limited basis.

For use in museums, a simple procedure that uses special polyester bags, nitrogen gas and an oxygen scavenger to produce anoxic atmospheres (less than 0.1% oxygen) readily kills cockroaches (Rust et al., 1996). Adult American, German, and brownbanded cockroaches are killed with 6-hour exposures, whereas nymphs of all three species are killed within 24 hours. The most resistant stage is the ootheca. The American, German, and brownbanded cockroach oothecae require exposures of 120, 24, and 72 hours, respectively, to produce 100% mortality.

Most household insect pests are extremely sensitive to high temperatures. At 51.7°C, a 30-minute exposure kills 100% of adult male German cockroaches (Rust & Reierson, 1998). In field studies, it was possible to control German cockroaches by heating food-handling areas in buildings to 46°C for 45 minutes (Zeichner et al., 1996). The heat treatment successfully reduced chemical treatments by 60%. Even though it was expensive, the treatment was considered a success because of the repeated failures with conventional chemical controls prior to heating.

2.7. Insecticide resistance

2.7.1. Physiological insecticide resistance

As long as insecticides are an integral part of pest management programmes, the potential for development of physiological resistance to them will be significant, especially for the control of German cockroaches. Also, insecticide resistance in other cockroach species has appeared in recent reports. German cockroaches have developed resistance to many of the organophosphates, carbamates and pyrethroids extensively used against them (Cochran, 1995). Resistance to some insecticides, such as pyrethrins, bendiocarb, diazinon and chlorpyrifos, was widespread in 1989, but resistance to such pyrethroids as permethrin, cyfluthrin and cypermethrin was found in only a few strains. Of 30 strains collected in the field from three continents, 15 had a more than twofold resistance to pyrethroids and 12 were also resistant to chlorpyrifos and propoxur (Hemingway et al., 1993). By 1998, at least one gene for resistance to cypermethrin was common in strains collected in the field. Of 57 strains in which the gene frequencies were determined, 31 were resistant (Cochran, 1998). A full range of resistance mechanisms, including esterase- and oxidase-based metabolism and nerve insensitivity, was found in these strains.

Cross-resistance continues to be an obstacle to finding alternative treatments when resistance develops. Holbrook and colleagues (2003) found that about 13 of 19 strains collected in the field showed some level of resistance to fipronil. Also, there was a direct relationship to resistance to cyclodienes. However, when fipronil was fed in baits, there was more than enough active ingredient to kill even the most tolerant individuals in their study. Even though the use of dieldrin in Denmark was discontinued over 30 years ago, resistance to it persists. Kristensen, Hansen & Vagn Jensen (2005) reported four strains collected in the field resistant to dieldrin and cross-resistant to fipronil. The mutation confers about 15-fold resistance to fipronil.

Strategies that rotate the use of different insecticides have often been recommended to counteract resistance. Applications of organophosphates may be most effective when used before pyrethroids in a rotational programme. Rotating on each generation and using a third insecticide or mixture of insecticides may be the best strategy to prevent high-level resistance to any one compound. The importance of developing effective rotational schemes diminished with the advent of baits in control programmes. However, the appearance of the cross-resistance to fipronil and behavioral resistance to baits has made this an important topic again.

2.7.2. Behavioural resistance

Even though numerous studies have shown the existence of physiological insecticide resistance in cockroaches, there have been few examples of behavioral resistance. In one such example, a strain of German cockroach was shown to have an aversion to feeding on glucose in the diet and, consequently, did not consume toxic baits containing glucose (Silverman & Bieman, 1993). Subsequently, strains collected in the field in the Republic of Korea and the United States were also shown to have this glucose aversion (Silverman & Ross, 1994).

These studies suggest that the rotation of baits with different active ingredients will not be enough to prevent reduced performance. It will be necessary to also monitor food preferences and make necessary changes to prevent avoidance. This is extremely important, because of the central and pivotal role that baits play in cockroach IPM programmes.

2.8. Biological control

The incorporation of natural predators, parasitoids (insects whose larvae are parasites that eventually kill their hosts) and pathogens to control peridomestic (undomesticated animals that, nevertheless, live in close proximity to humans) cockroaches has been a long-term goal of IPM (Suiter, 1997). The use of biological predators is especially attractive in sensitive situations where insecticide applications may be inappropriate, such as animal rearing facilities, zoos, sewers, and greenhouses. Lebeck (1991) has provided an excellent review of the hymenopterous (insects having two pairs of membranous wings and an ovipositor specialized for stinging or depositing eggs) natural enemies of cockroaches and those species most likely to be used in biological control. The most promising parasitoids are the cockroach wasp (*Aprostocetus hagenowii*), against *Periplaneta* spp., and the encyrtid wasp *Comperia merceti* (family Encyrtidae), against the brownbanded cockroach. Some transitory predator wasps (family Evaniidae), such as the emerald cockroach wasp (*Ampulex compressa*) are difficult to rear, and their large size probably makes them unacceptable candidates in indoor settings. Thoms & Robinson (1987a) reported that 60% of the homeowners killed the evaniid parasitoid *Prosevania punctata* when they encountered it. Outdoor releases may be more appropriate for the evaniid parasitoids, because of their size and wasp-like behaviour (Lebeck, 1991).

The efficacy of evaniid parasitoids used to control cockroaches is mixed. The evaniid parasitoid *P. punctata* and the cockroach wasp were collected at a field site in Virginia from oothecae of oriental cockroaches (Thoms & Robinson, 1987b). None of the oothecae placed inside the structure and observed was parasitized, even though adult *P. punctata* were regularly collected indoors. In the study, about 51% of oothecae recovered from outdoor sites were parasitized, 15% by *P. punctata* and 36% by cockroach wasps. Thoms and Robinson postulated that the cockroach wasp may be more efficient because the development time is 24–64 days, compared with 37–337 days for *P. punctata*. Also, about 30–93 adult cockroach wasps emerge per ootheca compared with 1 adult for *P. punctata*. Moreover, the cockroach wasp is smaller and less conspicuous than *P. punctata*. In another study, releases of cockroach wasps in sewers did not become established or provide control of American cockroaches (Reierson et al., 2005).

Releases of the encyrtid wasp *C. merceti* in indoor facilities to control the brownbanded cockroach have probably been the best-studied effort of the biological control of cockroaches. In an insect rearing facility, Coler, Van Driesche & Elkinton (1984) reported increased parasitism by *C. merceti* and decreased brownbanded cockroach trap catches. Initially cockroach populations were low, and cockroaches were added to increase populations, resulting in 70–90% parasitism of oothecae. Cockroach populations increased for a while and then decreased to low levels. For studies of low-level populations of brownbanded cockroaches, augmentation may be necessary. In follow-up studies, Hechmer & Van Driesche (1996) reported that two populations of the parasitoid *C. merceti* had maintained themselves on two separate indoor brownbanded cockroach populations without supplemental releases for 10 years. Levels of parasitism ranged from 36% to 93%. Lebeck (1991) believes that *C. merceti* has pest management potential, especially with overwhelming releases.

Even though some reviews suggest that fungi and nematodes may be promising biological control agents (Suiter, 1997), desiccation, spore viability and avoidance of bait stations have been a problem. Koehler, Patterson & Martin (1992) found that the time required to kill 50% of the cockroaches exposed to *Steinernema carpocapsae* nematodes was related inversely to the moisture of their preferred habitats. German and brownbanded cockroaches were the most susceptible. American, oriental and smokybrown cockroaches were less susceptible. Nematodes, however, require moisture and the delivery of viable nematodes in relatively dry habitats of German and brownbanded cockroaches is problematic.

2.9. Conclusions

Cockroaches present an unnecessary threat to public health, especially in multifamily dwellings and commercial food-handling establishments. IPM programmes must be designed to accommodate the pest species and each situation. No two programmes will be identical; however, they all should incorporate the following six steps.

- It is essential to *identify* the pest species and locations where indoor and outdoor infestations are breeding or gaining access to structures. The use of commercial traps is an important first step in determining the extent and severity of the problem. Traps can locate sites that need corrective measures. Traps have never been shown to be effective in controlling cockroaches, especially German cockroaches. Trap counts provide a quantitative mechanism upon which to base treatments and evaluate the success of the IPM programme. Trapping can also reduce unnecessary treatments and thereby reduce the amount of insecticides applied.

- The development of a *community action plan* is necessary, so that the tenants, landlords and proprietors actively participate in the IPM programme. The tenant's assistance and cooperation in removing clutter, food and water sources and in permitting access to their flats is essential. For example, treating infested appliances is essential to prevent cockroaches from moving between flats when tenants move. The other participants, such as landlord or caretaker, have responsibility for maintaining and repairing the structure and surrounding property and for providing adequate trash removal. Also, personnel responsible for repairing structural deficiencies and for treating cockroach infestations must provide tenants and landlords with progress reports and positive feedback.

- Control strategies should involve: prevention (built-in pest control); elimination of potential pest harbourages, such as clutter and cracks and crevices and voids; and *prevention* of cockroach movement across common pipes and conduits. The application of non-repellent dusts to voids should be repeated when flats are refurbished between occupancies. In addition to improving living conditions, structural repairs help reduce potential cockroach harbourages and movement within buildings. *Cleaning* and *sanitation* can be important in removing harbourage sites and sources of food and water. This is especially important in multifamily dwellings, where cockroaches can rapidly spread and the tenants do not feel directly responsible for the problem.

- When necessary, applications of insecticidal sprays, dusts and gel baits should be applied to cracks, crevices and voids where cockroaches harbour. Non-repellent insecticides should be used to avoid scattering cockroaches and slow-acting baits might be used indoors whenever possible to improve the control of early instars. When possible, containerized baits should be used, especially in extremely sensitive situations, such as schools and health care facilities. Applications should be made to minimize potential exposure to people and pets. Insecticides should only be applied where warranted, by monitoring with traps or visual inspections.

- In some situations, large blocks of flats or entire buildings will need to be treated. It is essential, therefore, that the IPM action plan embraces both the tenants and landlords. Failing to treat all of the units will leave potential refuges for reinfestation. Unsuccessful results are discouraging and encourage future noncompliance by tenants and landlords.

- Post-treatment evaluations are essential to determine if the IPM action plan has corrected the problems and controlled the cockroaches. These evaluations need to be shared with tenants and landlords to ensure continued cooperation and support. Input from the tenants and landlords allows the treatment team to evaluate and alter the action plan as needed. The programme must be economically and aesthetically acceptable and must address the needs of the target audience if is to be successful.

Acknowledgements

A special thanks to Mr. Dong-Hwan Choe for the photographs of the cockroaches and to Mary Rust for her editorial comments.

References[1]

Alarcon WA et al. (2005). Acute illnesses associated with pesticide exposure at schools. *The Journal of the American Medical Association*, 294:455–465.

Alexander JB, Newton J, Crowe GA (1991). Distribution of oriental and German cockroaches, *Blatta orientalis* and *Blattella germanica* (Dictyoptera), in the United Kingdom. *Medical and Veterinary Entomology*, 5:395–402.

Ameen A, Kaakeh W, Bennett G (2000). Integration of chlorfenapyr into a management program for the German cockroach (Dictyoptera: Blattellidae). *Journal of Agricultural and Urban Entomology*, 17:135–142.

Ameen A et al. (2005). Residual activity and population effects of noviflumuron for German cockroach (Dictyoptera: Blattellidae) control. *Journal of Economic Entomology*, 98:899–905.

Anonymous (1995). IPM survey results. *Pest Control Technology*, 23:18.

Anonymous (2002). Pest services drive profits. Pest Control [web site] (http://www.pestcontrolmag.com/pestcontrol/article/articleDetail.jsp?id=31025, accessed 19 July 2006).

Appel AG (1984). Intra- and interspecific trappings of two sympatric peridomestic cockroaches (Dictyoptera: Blattidae). *Journal of Economic Entomology*, 87:1027–1032.

Appel AG (1995). *Blattella* and related species. In: Rust MK, Owens JM, Reierson DA, eds. *Understanding and controlling the German cockroach*. New York, Oxford University Press:1–19.

Appel AG (2004). Contamination affects the performance of insecticidal baits against German cockroaches (Dictyoptera: Blattellidae). *Journal of Economic Entomology*, 97:2035–2042.

Appel AG, Eva MJ, Sims SR (2005). Toxicity of granular ant bait formulations against cockroaches (Dictyoptera: Blattellidae and Blattidae). *Sociobiology*, 46:65–72.

Appel AG, Rust MK (1985). Outdoor activity and distribution of the smokybrown cockroach, *Periplaneta fuliginosa* (Dictyoptera: Blattidae). *Environmental Entomology*, 14:669–673.

[1] The literature cited was selected according to the following criteria. (1) The information used in this chapter has been published in part in about five major reviews. Additional articles cited have appeared since 1995. (2) Technical refereed papers were selected whenever available and appropriate. (3) As papers that deal with the situation in Europe are not as common as those that deal with the conditions in the United States, less visible publications and Internet references are sometimes cited.

Appel AG, Smith LM 2nd (2002). Biology and management of the smokybrown cockroach. *Annual Review of Entomology*, 47:33–55.

Atkinson TH, Koehler PG, Patterson RS (1991). Catalogue and atlas of the cockroaches (Dictyoptera) of North America North of Mexico. *Miscellaneous Publications of the Entomological Society of America*, 78:1–86.

Auer B, Asperger H, Bauer J (1994). Zur Bedeutung der Schaben als Vektoren pathogener Bakterien. *Archiv für Lebensmittelhygiene, Fleisch-, Fisch- und Milchhygiene*, 45:89–93.

Bajomi D, Kis-Varga A, Bánki G (1993). Cockroach control and maintenance of cockroach-free conditions in housing estates of panel construction in Budapest. *International Pest Control*, 35:39–43.

Baker LF (1990). Cockroach incidence in English hospitals and a model contract. In: Robinson WH, ed. *Proceedings of the National Conference on Urban Entomology.* College Park, MD, University of Maryland:120.

Baumholtz MA et al. (1997). The medical importance of cockroaches. *International Journal of Dermatology*, 36:90–96.

Benson EP, Zungoli PA, Smith LM 2nd (1994). Comparison of developmental rates of two separate populations of *Periplaneta fuliginosa* (Dictyoptera: Blattidae) and equations describing development, preoviposition, and oviposition. *Environmental Entomology*, 23:979–986.

Benson V, Marano MA (1998). Current estimates from the National Health Interview Survey, 1995. *Vital Health Status*, 10:1–428.

Bonnefoy XR et al. (2003). Housing and health in Europe: preliminary results of a Pan-European study. *American Journal of Public Health*, 93:1559–1563.

Bradman A et al. (2005). Association of housing disrepair indicators with cockroach and rodent infestations in a cohort of pregnant Latina women and their children. *Environmental Health Perspectives*, 113:1795–1801.

Brenner BL et al. (2003). Integrated pest management in an urban community: a successful partnership for prevention. *Environmental Health Perspectives*, 111:1649–1653.

Brenner RJ (1995). Economics and medical importance of German cockroaches. In Rust MK, Owens JM, Reierson DA, eds. *Understanding and controlling the German cockroach*. New York, Oxford University Press: 77–92.

Brenner RJ, Pierce, RR (1991). Seasonality of peridomestic cockroaches (Blattoidea: Blattidae): mobility, winter reduction, and effect of traps and baits. *Journal of Economic Entomology*, 84:1735–1745.

Campbell ME et al. (1999). A program to reduce pesticide spraying in the indoor environment: evaluation of the 'Roach Coach' project. *Canadian Journal of Public Health*, 90:277–281.

Clubb SL, Frenkel JK (1992). *Sarcocystis falcatula* of opossums: transmission by cockroaches with fatal pulmonary disease in psittacine birds. *Journal of Parasitology*, 78:116–124.

Cochran DC (1995). Insecticide resistance. In: Rust MK, Owens JM, Reierson DA, eds. *Understanding and controlling the German cockroach*. New York, Oxford University Press:171–192.

Cochran DC (1998). Prevalence of the cypermethrin resistance gene in field-collected populations of the German cockroach. *Resistant Pest Management*, 10:18–21.

Cochran DG (1999). Cockroaches: their biology, distribution and control. Geneva, World Health Organization (document number: WHO/CDS/CPC/WHOPES/99.3; http://whqlibdoc.who.int/hq/1999/WHO_CDS_CPC_WHOPES_99.3.pdf, accessed 20 July 2006).

Coler RR, Van Driesche RG, Elkinton JS (1984). Effect of an oothecal parasitoid, *Comperia merceti* (Compere) (Hymenoptera: Encyrtidae), on a population of the brown-banded cockroach (Orthoptera: Blattellidae). *Environmental Entomology*, 13:603–606.

Cornwell PB (1968). *The cockroach. Vol. 1*. London, Hutchinson & Co.

Curl G (2004). *A strategic analysis of the U.S. structural pest control industry – the 2004 season*. Mendham, NJ, Specialty Products Consultants.

Davies K, Phil D, Petranovic T (1986). Survey of attitudes of apartment residents to cockroaches and cockroach control. *Journal of Environmental Health*, 49:85–88.

Demark JJ, Bennett GW (1995). Adult German cockroach (Dictyoptera: Blattellidae) movement patterns and resource consumption in a laboratory arena. *Journal of Medical Entomology*, 32:241–248.

Dupree R (1977). New York City problems and one solution. In: Westphal K, ed. *Proceedings of the 1977 seminar on cockroach control*. New York, New York State Department of Health, 4–6.

Durier V, Rivault C (2003). Improvement of German cockroach (Dictyoptera: Blattellidae) population control by fragmented distribution of gel baits. *Journal of Economic Entomology*, 96:1254–1258.

Ebeling W (1975). *Urban entomology*. Berkeley, CA, Division of Agricultural Sciences, University of California.

Eggleston PA (2003). Cockroach allergen abatement: the good, the bad, and the ugly. *Journal of Allergy and Clinical Immunology*, 112:265–267.

Fathpour H, Emtiazi G., Ghasemi E (2003). Cockroaches as reservoirs and vectors of drug resistant *Salmonella* spp. *Fresenius Environmental Bulletin*, 12:724–727.

Fischer AB et al. (1999). Pest control in public institutions. *Toxicology Letters*, 107:75–80.

Fischer OA et al. (2003). Nymphs of the oriental cockroach (*Blatta orientalis*) as passive vectors of causal agents of avian tuberculosis and paratuberculosis. *Medical and Veterinary Entomology*, 17:145–150.

Gore JC, Schal C (2005). Expression, production and excretion of Bla g 1, a major human allergen, in relation to food intake in the German cockroach, *Blattella germanica*. *Medical and Veterinary Entomology*, 19:127–134.

Greene A, Breisch NL (2002). Measuring integrated pest management programs for public buildings. *Journal of Economic Entomology*, 95:1–13.

Hechmer A, van Driesche RG (1996). Ten year persistence of a non-augmented population of brownbanded cockroach (Orthoptera: Blattellidae) parasitoid, *Comperia merceti* (Hymenoptera: Encyrtidae). *Florida Entomologist*, 79:77–79.

Hemingway J et al. (1993). Pyrethroid resistance in German cockroaches (Dictyoptera: Blattellidae): resistance levels and underlying mechanisms. *Journal of Economic Entomology*, 86:1631–1638.

Holbrook GL et al. (2003). Origin and extent of resistance to fipronil in the German cockroach, *Blattella germanica* (L.) (Dictyoptera: Blattellidae). *Journal of Economic Entomology*, 96:1548–1558.

Katial RK (2003). Cockroach allergy. *Immunology and Allergy Clinics of North America*, 23:483–499.

Koehler PG, Patterson RS, Martin WR (1992). Susceptibility of cockroaches (Dictyoptera: Blattellidae, Blattidae) to infection by *Steinernema carpocapsae*. *Journal of Economic Entomology*, 85:1184–1187.

Kopanic Jr RJ, Sheldon BW, Wright CG (1994). Cockroaches as vectors of *Salmonella*: laboratory and field trials. *Journal of Food Protection*, 57:125–132.

Kopanic Jr RJ et al. (2001). An adaptive benefit of facultative coprophagy in the German cockroach *Blattella germanica*. *Ecological Entomology*, 26:154–162.

Krieger JK et al. (2002). The Seattle-King County healthy homes project: implementation of a comprehensive approach to improving indoor environmental quality for low-income children with asthma. *Environmental Health Perspectives*, 110(Suppl. 2):311–322.

Kristensen M, Hansen KK, Vagn Jensen K-M (2005). Cross-resistance between dieldrin and fipronil in German cockroach (Dictyoptera: Blattellidae). *Journal of Economic Entomology*, 98:1305–1310.

Landau I, Muller G, Schmidt M (1999). The Urban Pest Advisory Service of Zurich (Switzerland) and the situation of some selected pests. In: Robinson WH, Rettich F, Rambo GW, eds. *Proceedings of the Third International Conference of Urban Pests*, 19-22 July 1999, Prague, Czech Republic. Hronov, Czech Republic, Grafické Závody:67-72 (http://www.stadt-zuerich.ch/internet/ugz/home/fachbereiche/schaedlingsbekaempfung/publikationen.ParagraphContainerList.ParagraphContainer0.ParagraphList.0013.File.pdf/sb_paper_prag.pdf, accessed 09 January 2007).

Landrigan PJ et al. (1999). Pesticides and inner-city children: exposures, risks, and prevention. *Environmental Health Perspectives*, 107(Suppl. 3):431–437.

Leaderer BP et al. (2002). Dust mite, cockroach, cat, and dog allergen concentrations in homes of asthmatic children in the northeastern United States: impact of socioeconomic factors and population density. *Environmental Health Perspectives*, 110:419–425.

Lebeck LM (1991). A review of the hymenopterous natural enemies of cockroaches with emphasis on biological control. *Entomophaga*, 36:335–352.

Lee C-Y, Lee L-C (2000). Influence of sanitary conditions on the field performance of chlorpyrifos-based baits against American cockroaches, *Periplaneta americana* (L.) (Dictyoptera: Blattidae). *Journal of Vector Ecology*, 25: 218–221.

le Patourel GNJ (1993). Cold-tolerance of the oriental cockroach, *Blatta orientalis*. *Entomologia Experimentalis et Applicata*, 68:257–263.

le Patourel G (1995). Effect of environmental conditions on progeny production by the oriental cockroach *Blatta orientalis*. *Entomologia Experimentalis et Applicata*, 74:1–6.

Liang D et al. (1998). Field and laboratory evaluation of female sex pheromone for detection, monitoring, and management of brownbanded cockroaches (Dictyoptera: Blattellidae). *Journal of Economic Entomology*, 91:480–485.

Majekodunmi A, Howard MT, Shah V (2002). The perceived importance of cockroach [*Blatta orientalis* (L.) and *Blattella germanica* (L.)] infestation to social housing residents. *Journal of Environmental Health Research*, 1: 27–34.

McConnell R et al. (2005). Educational intervention to control cockroach allergen exposure in homes of hispanic children in Los Angeles: results of the La Casa study. *Clinical and Experimental Allergy*, 35:426–433.

Mielke U. (1995). Die Verbreitung des Hygieneschädlings Braunbandschabe (*Supella longipapla* [Fabricius, 1798]) (Blattodea, Blattidae) in Deutschland. *Anzeiger für Schädlingskunde, Pflanzenschutz, Umweltschutz*, 68:187–189.

Mielke U (1996). Erfolgreiche Bekämpfung der Orientalischen Schabe (*Blatta orientalis*) (L.) (Blattodea, Blattidae) an Hand eines Beispieles aus der Praxis. *Anzeiger für Schädlingskunde, Pflanzenschutz, Umweltschutz*, 69:106–108.

Miller DM, Meek F (2004). Cost and efficacy comparison of integrated pest management strategies with monthly spray insecticide applications for German cockroach (Dictyoptera: Blattellidae) control in public housing. *Journal of Economic Entomology*, 97:559–569.

Nojima S et al. (2005). Identification of the sex pheromone of the German cockroach, *Blattella germanica*. *Science*, 307:1104–1106.

Pospischil R (2004). Schaben – Blattodea. Ein Beispiel für die Einschleppung und Einbürgerung von Insekten in Europa. *Mitteilungen der Deutschen Gesellschaft für Allgemeine und Angewandte Entomologie*, 14:93–100.

Potter MF, Bessin RT (1998). Pest control, pesticides, and the public: attitudes and implications. *American Entomologist*, 44:142–147.

Rambo G (1998). Developing reduced cost methods of IPM. *Pest Control Technology*, 26:74.

Rauh VA, Chew GL, Garfinkel RS (2002). Deteriorated housing contributes to high cockroach allergen levels in inner-city housing. *Environmental Health Perspectives*, 110(Suppl. 2):323–327.

Reierson DA (1995). Baits for German cockroach control. In: Rust MK, Owens JM, Reierson DA, eds. *Understanding and controlling the German cockroach*. New York, Oxford University Press:231–265.

Reierson DA et al. (2005). Control of American cockroaches (Dictyoptera: Blattidae) in sewer systems. In: Lee C-Y, Robinson WH, eds. *Proceedings of the Fifth International Conference on Urban Pests*. Singapore, P&Y Design Network:141–148.

Rivault C, Cloarec A (1995). Limits of insecticide cockroach control in council flats in France. *Journal of Environmental Management*, 45:379–393.

Rivault C, Cloarec A (1997). Outcome of insecticide control of cockroaches (Dictyoptera: Blattellidae) in public housing in France. *Journal of Environmental Management*, 51:187–197.

Robinson WH (1996a). Integrated pest management in the urban environment. *American Entomologist*, 42:76–78.

Robinson WH (1996b). *Urban entomology: insect and mite pests in the human environment*. London, Chapman & Hall.

Robinson WH, Zungoli PA (1995). Integrated pest management: an operational view. In: Rust MK, Owens JM, Reierson DA, eds. *Understanding and controlling the German cockroach*. New York, Oxford University Press:345–359.

Rosenstreich DL et al. (1997). The role of cockroach allergy and exposure to cockroach allergen in causing morbidity among inner-city children with asthma. *The New England Journal of Medicine*, 336:1356–1363.

Ross MH, Mullins DE (1995). Biology. In: Rust MK, Owens JM, Reierson DA, eds. *Understanding and controlling the German cockroach*. New York, Oxford University Press:21–47.

Roth LM (1981). Introduction. In: Bell WJ, Adiyodi KG, eds. *The American cockroach*. London, Chapman and Hall:1–14.

Roth LM (1985). A taxonomic revision of the genus *Blattella* Caudell (Dictyoptera, Blattaria: Blattellidae). *Entomologica Scandinavica Supplement*, 22:1–221.

Rust MK (1995). Factors affecting control with insecticides. In: Rust MK, Owens JM, Reierson DA, eds. *Understanding and controlling the German cockroach*. New York, Oxford University Press:149–170.

Rust MK (2001). Insecticides and their use in urban structural pest control. In: Kreiger RI, ed. *Handbook of pesticide toxicology. Vol. 1*, 2nd ed. San Diego, CA, Academic Press:243–250.

Rust MK, Reierson DA (1991). Chlorpyrifos resistance in German cockroaches (Dictyoptera: Blattellidae) from restaurants. *Journal of Economic Entomology*, 84:736–740.

Rust MK, Reierson DA (1998). Use of extreme temperatures in urban insect pest management. In: Hallman GJ, Denlinger DL, eds. *Temperature sensitivity in insects and application in integrated pest management*. Boulder, CO, Westview Press:179–200.

Rust MK et al. (1991). Control of American cockroaches (Dictyoptera: Blattidae) in sewers. *Journal of Medical Entomology*, 28:210–213.

Rust MK et al. (1996). The feasibility of using modified atmospheres to control insect pests in museums. *Restaurator*, 17:43–60.

Silverman J, Bieman DN (1993). Glucose aversion in the German cockroach, *Blattella germanica*. *Journal of Insect Physiology*, 39:925–933.

Silverman J, Ross MH (1994). Behavioral resistance of field-collected German cockroaches (Blattodea: Blattellidae) to baits containing glucose. *Environmental Entomology*, 23:425–430.

Smith LM 2nd et al. (1995). Model for estimating relative abundance of *Periplaneta fuliginosa* (Dictyoptera: Blattidae) by using house and landscape characteristics. *Journal of Economic Entomology*, 88:307–319.

Smith LM 2nd et al. (1998). Comparison of conventional and targeted insecticide application for control of smokybrown cockroaches (Dictyoptera: Blattidae) in three urban areas of Alabama. *Journal of Economic Entomology*, 91:473–479.

Stejskal V (1997). Distribution of faeces of the German cockroach, *Blattella germanica*, in a new refuge. *Entomologia Experimentalis et Applicata*, 4:201–205.

Stejskal V, Verner PH (1996). Long-term changes of cockroach infestations in Czech and Slovak food-processing plants. *Medical and Veterinary Entomology*, 10:103–104.

Suiter DR (1997). Biological suppression of synanthropic cockroaches. *Journal of Agricultural Entomology*, 14:259–270.

Takahashi S et al. (1995). Isolation and biological activity of the sex pheromone of the smokybrown cockroach, *Periplaneta fuliginosa* Serville (Dictyoptera: Blattidae). *Applied Entomology and Zoology*, 30:357–360.

Tarry DW, Lucas M (1977). Persistence of some livestock viruses in the cockroach *Blatta orientalis*. *Entomologia Experimentalis et Applicata*, 22:200–202.

Thoms EM, Robinson WH (1986). Distribution, seasonal abundance, and pest status of the oriental cockroach (Orthoptera: Blattidae) and an evaniid wasp (Hymenoptera: Evaniidae) in urban apartments. *Journal of Economic Entomology*, 9:431–436.

Thoms EM, Robinson WH (1987a). Insecticide and structural modification strategies for management of oriental cockroach (Orthoptera: Blattidae) populations. *Journal of Economic Entomology*, 80:131–135.

Thoms EM, Robinson WH (1987b). Potential of the cockroach oothecal parasite *Prosevania punctata* (Hymenoptera: Evaniidae) as a biological control agent for the oriental cockroach (Orthoptera: Blattidae). *Environmental Entomology*, 16:938–944.

Valles SM, Koehler PG, Brenner RJ (1999). Comparative insecticide susceptibility and detoxification enzyme activities among pestiferous Blattodea. *Comparative Biochemistry and Physiology. Part C, Pharmacology, Toxicology & Endocrinology*, 124:227–232.

Whyatt R et al. (2002). Residential pesticide use during pregnancy among a cohort of urban minority women. *Environmental Health Perspectives*, 110:507–514.

Williams GM et al. (2005). Comparison of conventional and integrated pest management programs in schools. *Journal of Economic Entomology*, 98:1275–1283.

Wood FE et al. (1981). Survey of attitudes and knowledge of public housing residents toward cockroaches. *Bulletin of the Entomological Society of America*, 27:9–13.

Wright CG, McDaniel HC (1973). Further evaluation of the abundance and habitat of five species of cockroaches on a permanent military base. *Florida Entomologist*, 56:251–254.

Zeichner BC et al. (1996). The use of heat for control of chronic German cockroach infestations in food service facilities – fresh start. In: Wildey KB, ed. *Proceedings of 2nd International Conference of Insect Pests in the Urban Environment*, Edinburgh, Scotland, 7–10 July 1996:507–513.

Zungoli PA, Robinson WH (1984). Feasibility of establishing an aesthetic injury level for German cockroach pest management programs. *Environmental Entomology*, 13:1453–1458.

Zurek L, Schal C (2004). Evaluation of the German cockroach (*Blattella germanica*) as a vector for verotoxigenic *Escherichia coli* F18 in confined swine production. *Veterinary Microbiology*, 101:263–267.

3. House dust mites

David Crowther and Toby Wilkinson

Summary

Less than a millimetre in size, house dust mites are found worldwide, primarily in human dwellings. They themselves are harmless, but they give rise to potent allergens associated with several diseases, notably asthma. The major component of their diet is scales from human skin (dander), which is in plentiful supply in a typical home – for example, in mattresses and bedding, carpets, and upholstered furniture. However, mites are also dependent on both temperature and relative humidity, and they cannot survive if hygrothermal conditions are unfavourable. The observed variability in mite populations and allergen levels, both between and within different regions, can be related, at least in part, to the variability in hygrothermal conditions found within dwellings. In turn, these conditions are affected by the complex interaction of such factors as climate, type of building construction (particularly in relation to the standards of ventilation and insulation provided) and occupant behaviour (in relation to moisture production, as well as to heating and ventilating habits).

Modifying the domestic hygrothermal environment in an appropriate manner thus has considerable potential to be an effective method of controlling house dust mites. Many issues are involved, however, and there are several other methods of control, including various forms of cleaning (particularly steam cleaning), temperature control (such as electric blankets), physical methods (such as barrier fabrics) and pesticides. Whether the aim is to prevent mite infestation from occurring or to control an infestation once it has occurred, it is essential to adopt an integrated approach.

3.1. Introduction

House dust mites (HDMs) are microscopic organisms that live where scales from human skin, the major component of their diet, accumulate. Although they can be found in schools, offices and other workplace environments, they are more commonly found in dwellings, where a variety of habitats, particularly bedding and carpets, can provide favourable conditions in terms of:

- material properties
- relative security from disturbance, competition and predation
- food supply (human skin scales or dander)
- temperature and relative humidity.

The physical spaces provided by modern mattresses (whether of spring or foam construction), duvets, pillows (whether feather or synthetic), carpets and upholstered furniture are well suited to support mite development. The interconnected air gaps and interstices provided are:

- the right size for them to crawl into
- dark (they are photophobic)
- difficult to extract them from (due to their ability to cling to fibres)
- close to sources of the food they require.

Given these conditions, their distribution within the built environment depends primarily on temperature and relative humidity. Since HDMs are unable to drink liquid water, they absorb the water they need from the atmosphere. They thus need high levels of relative humidity to survive.

HDMs do not attack, bite or transmit disease to people, nor are they a sign of poor hygiene. However, their faecal pellets contain a number of potent allergens known to trigger and possibly cause allergic disease, such as perennial rhinitis, eczema and, most important, asthma. Although the HDM has not historically been considered to be an urban pest, the increasing prevalence of these diseases, especially asthma, justifies its consideration in this context.

3.1.1. Biology and ecology

3.1.1.1. Origins and natural history

Mites are arthropods within the class Arachnida. HDM is primarily used to describe three species of mite, *Dermatophagoides pteronyssinus*, *Dermatophagoides farinae* and *Euroglyphus maynei*.[1] The allergens these mites produce are probably the most important allergens

[1] Unlike with many other urban pests, a sound system of common names does not exist for HDMs. For example, the names "American mite" and "European mite" are rarely used by acarologists, because they are misleading – neither species is exclusive to the region named and both have a worldwide distribution. Therefore the system of nomenclature based on binomial Latin names is used for HDM species throughout this chapter.

associated with asthma worldwide (Tovey, 1992). In temperate climates, *D. pteronyssinus* and *D. farinae* are the most abundant species (Arlian et al., 1998). *E. maynei* is common, but under-researched; some evidence, however, shows that it can be clinically important (Mumcuoglu, 1977; van Hage-Hamsten & Johansson, 1989).

HDMs belong to the family Pyroglyphidae. Mites belonging to this family are normally associated with birds, mammals and stored products (van Bronswijk, 1981). It is likely that HDMs were originally nest dwellers that moved to occupy peoples' beds at the time of the first settlements (Colloff, 1998). HDMs (and mites in general) are very difficult to identify, largely due to their small size. A number of guides that specialize in mites found in dust exist – for example, by Fain (1990) and by Colloff & Spieksma (1992).

3.1.1.2. Life cycle

The life-cycles of HDMs consist of five major stages: egg, larva, protonymph, tritonymph and adult. Development time from egg to adult under optimal conditions, 25°C and 75% relative humidity, takes about 25 days. The speed of HDM development depends largely on temperature (see subsection 3.1.3.3).

As the HDMs progress through each stage of life, they become larger and display more taxonomic features. Each active stage is separated by a quiescent phase that lasts for a third to a half the time of the preceding active period (Arlian, Rapp & Ahmed, 1990). *D. farinae* is able to survive adverse conditions, particularly low relative humidity, by forming a prolonged, so-called drought-resistant quiescent protonymph stage. The quiescent protonymph of *D. farinae* consumes 28.5 times less oxygen per hour than an active protonymph. It has a water exchange half-life of 160 days, compared with just 20 hours for an active protonymph and 28 hours for an adult (Arlian et al., 1983).

Current research indicates that *D. pteronyssinus* does not form a drought-resistant quiescent protonymph stage similar to *D. farinae*. Arlian, Rapp & Ahmed (1990) found that quiescent *D. pteronyssinus* protonymphs were not formed even at low humidities, although they suggested that a combination of unfavourable conditions may trigger a quiescent protonymph stage. De Boer & Kuller (1997) observed that the quiescent phase of *D. pteronyssinus* was relatively brief and to their knowledge, not prolonged. De Boer, Kuller & Kahl (1998) reported that in the carpeted ground floor of Dutch houses, *D. pteronyssinus* adults remained active throughout the winter.

At 25°C and 75% relative humidity, the difference between the development time of males and females is non-significant (Arlian, Rapp & Ahmed (1990). The lifespan of adult males, however, was found to be 77 days, compared with 45 days for unmated adult females and 31 days for mated adult females. During their adult lifespan, *D. pteronyssinus* females have been shown to produce between 40 and 80 eggs (Colloff, 1987). *D. farinae* females are able to live considerably longer than *D. pteronyssinus* females (100 and 31 days, respectively). The reproductive period and number of eggs produced were similar for both species. After death, the bodies of HDMs dried thoroughly, allowing them to persist in the environment for a considerable period of time before breaking down.

3.1.1.3. Food

As well as skin scales, pollen, spores of microorganisms, fungal mycelia and bacteria have all been found in the gut of *Dermatophagoides* spp. (van Bronswijk, 1981). A person typically produces 0.5–1.0 g of dead skin a day, although a large proportion of this will be removed via the cleaning of clothes and bed sheets and the vacuuming of carpets. Food is not normally considered to be a constraint on HDM population growth, although its scarcity after washing or vacuuming could conceivably impede HDM population growth and allergen production (de Boer, 1998).

It is thought that freshly shed skin scales may be dry and indigestible and that mites prefer older skin scales with a higher moisture content (Maunder, 1990). Also, fungal growth on skin scales can affect their nutritional value. Van Bronswijk (1981) reported that *D. pteronyssinus* grew better on skin that had previously been partially digested by the fungus *Aspergillus amstelodami* than skin with no fungal treatment. However, Douglas & Hart (1989) concluded that no experimental studies provided definite evidence of this, although they did show that small quantities of the fungus *Aspergillus penicillioides* may be of nutritional value. De Saint Georges-Gridelet (1987) reported that the fungus *Aspergillus repens* also increases the nutritional value of skin scales to the mites by reducing the fat content and adding vitamins B and D.

Mites kept in the laboratory are normally fed yeast-based food, which is often supplemented with different types of animal protein. Experiments indicate that the type of food used can affect the rate of HDM growth (Hart et al., 2007). This can have a significant impact on the comparability of laboratory trials.

3.1.1.4. Limiting factors

In most situations, HDM populations are primarily limited by temperature and relative humidity (van Bronswijk, 1981). HDM habitats, such as carpets and especially mattresses, provide vast amounts of space for these mites. HDMs and food are more commonly found near the top surfaces of spring and foam mattresses (de Boer & Kuller, 1994; Hay, 1995). However, HDMs may be present throughout solid foam mattresses, providing food and suitable hygrothermal conditions are available (de Boer & Kuller, 1994). Food quality is also likely to have an effect on mites. Arlian (1992) found that feeding rates are affected by the moisture content of food; this could have implications for nutritional uptake and the speed of reproduction. The carrying capacity of an artificial yeast- and liver-based mite food has been found to be about 12 000 mites per gram of culture (Wilkinson et al., 2002). Food is only likely to become a limiting factor when HDMs experience prolonged favourable hygrothermal conditions. The food quality is also likely to vary considerably between habitats; carpet dust, for example, contains considerably more grit than mattress dust. Although fungi provide nutritional benefits to HDMs, at very high levels of relative humidity they also can kill mites, either by producing toxins or as a result of the physical effect that a large quantity of fungi has on the HDM habitat (Arlian et al, 1998; van Asselt, 1999). The impact of predation is also thought to be minimal in the domestic environment (see section 3.3.8).

3.1.1.5. Habitats within the home

HDMs occupy a wide range of habitats in the home, including mattresses, bedding, carpets, soft furnishings, gaps in laminate floors, clothes and toys. It is difficult to generalize about where HDMs are most abundant within houses, since population size may be influenced by a number of factors, including variations in microclimatic conditions. Mattresses and carpets, however, are generally considered to be the main living and breeding grounds for HDMs. The mattress forms a complex hygrothermal environment, containing areas that are more and less favourable to mites than room conditions (Pretlove et al., 2001). Mattresses are not always the dominant HDM habitat. Arlian, Bernstein & Gallagher (1982) found a living room carpet in Ohio, which supported a population of HDMs (primarily *D. farinae*) seven times larger than that found in the mattresses. Other soft furnishings may also support HDMs. Mitchell and colleagues (1969) found that mite density was greater in frequently used furniture than in pieces rarely used. This is likely to be due to the greater number of skin scales supplying food for HDMs and the higher moisture content of furniture used regularly.

D. pteronyssinus and *D. farinae* show behavioural differences that enable them to occupy different niches within the same habitat. Van Bronswijk (1981) reported findings from a study by Wharton that noted that although *D. pteronyssinus* and *D. farinae* occupied the same niche, *D. farinae* tended to crawl on top of the substrate while *D. pteronyssinus* remained beneath it. Generally, HDMs are found closer to the surface of the mattress, although de Boer (1990a) observed that HDMs are also present deep inside a mattress, not just near the surface.

3.1.2. Distribution in Europe and North America

HDMs inhabit a wide geographical range. Their distribution is well understood in Europe and North America, although comprehensive studies in other parts of the world are limited. *D. farinae* is generally the most dominant species in North America and other continental regions with prolonged dry weather, while *D. pteronyssinus* is abundant in temperate areas with constantly higher humidity levels, such as the United Kingdom (Platts-Mills & Chapman, 1987).

Within mattresses in the United Kingdom, *D. pteronyssinus* is usually the most abundant mite species, followed by *E. maynei*, although there is some variation in the proportions of these two species (Rao et al., 1975; Blythe, Williams & Smith, 1974). In general, studies have recorded no or very few *D. farinae* in samples and speculate that conditions within the United Kingdom may be too cool and damp for the survival of this species (Blythe, Williams & Smith, 1974; Abbott, Cameron & Taylor, 1981; Hart & Whitehead, 1990; Wanner et al., 1993).

Walshaw & Evans (1987) stated that there was a strong correlation between the density of *E. maynei* within a mattress and social class, increasing numbers of mites being found as social class declined. However they did not examine this relationship with *D. pteronyssinus*. Colloff (1991a) also showed that *E. maynei* occurred more frequently in samples from homes that he assessed as being damp.

In other areas of Europe and in Australia and New Zealand, *D. pteronyssinus* is usually the dominant species (Blythe, 1976; Crane et al., 1998; Dharmage et al., 1999), although this is not always the case. Solarz (1997) found that in mattress dust samples from the Czech Republic, *D. farinae* was the dominant species, representing 62.7% of the total mite population, while *D. pteronyssinus* comprised 28.8% and *E. maynei* 1.4%.

In contrast to Europe and Australasia, *D. farinae* is generally the most abundant species of HDM in the United States of America (Wharton, 1976; Arlian, Bernstein & Gallagher, 1982), although *D. pteronyssinus* is also present. Mulla and colleagues (1975) found that in California *D. pteronyssinus* was more common than *D. farinae*. In the homes they examined, 29% supported *D. pteronyssinus* alone, compared with 21% supporting *D. farinae* alone and the remainder supported both species. The houses they surveyed were all within 24 km of the Pacific Ocean. They noted that *D. pteronyssinus* numbers were greatest near the coast while *D. farinae* numbers increased in abundance further inland (see section 3.2.2).

3.1.3. House dust mite physiology

A total of 72% male and 74% female HDM weight is water (Arlian, 1975). *D. pteronyssinus* females weigh 5.8 ± 0.2 µg, males being considerably smaller, weighing 3.5 ± 0.2 µg. HDMs have a high ratio of surface area to volume, which means they risk becoming dehydrated. Critically, HDMs are unable to drink liquid water; instead, they absorb the water they need from the air, so that they need high levels of relative humidity to survive.

3.1.3.1. Water balance

HDMs obtain their water in four ways, through:

1. ingestion with food
2. production of metabolic water from the oxidation of carbohydrates and fats
3. passive absorption, through the outer surface of the body
4. active absorption from unsaturated air, via the supracoxal gland.

Water is simultaneously lost in four ways:

1. by transpiration or evaporation to ambient air from the supracoxal gland
2. by evaporation through the permeable surfaces of the body
3. in digestive secretions and reproductive fluids
4. during other functions, such as feeding, excretion and oviposition (Arlian, 1992).

Of the four ways HDMs obtain water, first, the moisture content of the food that mites eat tends to be in equilibrium with ambient air. Arlian (1977) reported that above the critical equilibrium humidity (CEH; see subsection 3.1.3.2), the feeding rate increased for both *D. pteronyssinus* and *D. farinae*; below the CEH, water gained by feeding was less than 9% of that transpired. The results of the study indicate that the moisture content of the food only has a small role to play in providing mites with the water they require.

Second, Arlian (1977) also concluded that HDMs are unlikely to obtain significant amounts from the oxidization of carbohydrates and fats and, third, passive absorption through the outer surface of the body does occur, but only slowly (Arlian & Veselica, 1982). Finally, the active uptake of water is facilitated by a pair of glands, the supracoxal glands. These glands secrete a hygroscopic salt solution that contains sodium chloride and potassium chloride (Wharton, Duke & Epstein, 1979); the secretion flows down a duct, eventually reaching the HDM mouth. Providing the relative humidity exceeds the CEH, the salt solution experiences a net gain in water. Wharton, Duke & Epstein (1979) found that the rapid uptake of water took place after three hours, when they moved dehydrated *D. farinae* from a low relative humidity environment and placed them in a high relative humidity environment.

Of the four ways HDMs lose water, first, when relative humidity is below the CEH, the salt solution dries out, eventually blocking the gland and thus preventing further water loss (Wharton, Duke & Epstein, 1979). Arlian & Wharton (1974) observed that, in *D. farinae* it took 14 hours for transpiration from the supracoxal gland to become negligible. Second, mites and mite eggs are able to lose water through the outer surface of the body, but only slowly. Increasing temperature, however, increases the permeability of the cuticle and results in a more rapid water loss (Arlian & Veselica, 1982). Third, digestive secretions and reproductive fluids are only likely to account for a small proportion of the loss. Finally, at low relative humidity, HDMs generally feed less, and their reproductive rate is reduced (Arlian, 1992). Also, their rate of excretion and oviposition is reduced.

3.1.3.2. CEH

CEH is very important in understanding how HDMs survive, especially in relation to their water balance. CEH is most commonly defined as the relative humidity below which mites are unable to maintain their water balance and lose water more rapidly than they can gain it. Arlian & Veselica (1981a) found that CEH for *D. farinae* varies with temperature. CEH also appears to vary according to the state of hydration of the HDMs (Arlian & Veselica, 1981a). At 25°C, Arlian & Veselica found the CEH for *D. farinae* to be 58%, which was significantly lower than the CEH of 70% found previously by Larson (1969). Arlian & Veselica (1981b) suggested that this was probably due to differences in experimental methods, their mites being dehydrated for significantly longer than Larson's.

A similar situation was reported for *D. pteronyssinus*. Arlian (1975) found the CEH at 25°C to be 73%, using mites that had not been pre-dehydrated. De Boer & Kuller (1997) and then De Boer alone (1998), using pre-dehydrated mites, found the CEH at 16°C to be between 56–58% relative humidity and at 20°C to be between 58–60% relative humidity. When HDMs are exposed to brief daily spells of elevated humidity above CEH, they can survive in otherwise detrimental conditions, as Arlian, Neal & Bacon (1998) found for *D. farinae* and de Boer, Kuller & Kahl (1998) found for *D. pteronyssinus*. This has significant implications for their survival and for preventive environmental control measures.

3.1.3.3. Influence of temperature

In the domestic environment, the rate of HDM population growth is related to temperature (Arlian et al., 1983; Colloff, 1987). Temperature primarily affects the rate of HDM growth, although it also plays a vital role (together with the moisture content of the air) in determining relative humidity, which affects HDM survival. Extremes of temperature can have detrimental effects on HDM population growth. High temperatures of 51°C for 6 hours or 45°C for 24 hours (both at 60% relative humidity) will cause the death of *D. pteronyssinus* (Kinnaird, 1974). *D. pteronyssinus* is relatively resistant to low temperatures; a six-hour exposure to temperatures of –15°C results in only 60% mortality (compared with 100% mortality in *D. farinae*). Increasing the time of exposure to cold also causes an increase in mortality, with only 15% of mites surviving after a three-week exposure at –1°C (van Bronswijk, 1981). However, a small number of *D. pteronyssinus* eggs wrapped in a duvet with an insulation value of 1.35 $m^2 \cdot K/W$ (13.5 togs) were able to survive being frozen for 48 hours in a conventional –18°C freezer (T. Wilkinson, unpublished data, 2002). This has implications for HDM control (see section 3.3.5).

Arlian & Dippold (1996) examined the effect of temperature on the life-cycle of *D. farinae* and compared it with a similar study on *D. pteronyssinus* (Arlian, Rapp & Ahmed, 1990). They found that *D. pteronyssinus* was able to develop under a far wider range of temperatures than *D. farinae*.

Increasing the temperature from 23°C to 35°C has been shown to accelerate development by 2.3 times (Arlian, Rapp & Ahmed, 1990). Despite this, the greatest population growth for *D. pteronyssinus* occurs at 25°C and 75% relative humidity, since higher temperatures also affect other survival factors. At 35°C, for example, both fecundity and longevity were significantly reduced, compared with 23°C (Arlian, Rapp & Ahmed, 1990). As temperature increases, CEH also increases (Arlian & Veselica, 1981a).

3.2. Impact of the domestic environment

Van Bronswijk (1981) described how the domestic environment evolved over time in ways that favoured the HDM. At the turn of the 19th and 20th centuries, house dust on floors and in furniture and mattresses was crowded with arthropods, both dust feeders and their predators. This was due to a plentiful supply of organic dust (from the use of straw and horsehair in furniture and mattresses, for example) and high relative humidities. Despite this biological abundance, however, there is no record of *Dermatophagoides* spp. being present at this time. By the 1930s, domestic environments were warmer, drier and cleaner, but relative humidity and organic dust levels were still high enough to support a range of arthropods, usually in high numbers. According to Bronswijk (1981), there is some evidence that HDMs were present, but only in small numbers. Since then, the trend towards warmer, drier and cleaner domestic environments has continued, to the extent that there are now few other arthropod survivors in house dust apart from HDMs, which can thus enjoy an assured source of food (human skin scale – the principal remaining organic component of house dust) without hindrance from competitors or predators. Only if relative humidities in domestic environments continued to decline would their future be endangered. Unfortunately, for reasons that will be explained, this has not happened and the modern home is consequently providing an ecological niche for HDMs that, without intervention, is currently encouraging their proliferation.

3.2.1. Impact of the hygrothermal environment

As explained in section 3.1, HDMs need a particular combination of temperature and relative humidity to survive, even if food is plentiful. However, these ideal hygrothermal conditions are by no means universal in dwellings. Several studies have found marked differences in mite numbers or allergen concentrations (or both) between dwellings, ranging from high to non-existent, even within the same region. Moreover, these differences can be related to differences in indoor climate (Voorhorst, Spieksma & Varekamp, 1969; Korsgaard, 1983a, 1983b; Hart & Whitehead, 1990; Kuehr et al., 1994; van Strien et al., 1994; Sundell et al., 1995). Outside the field of building science, it is not generally appreciated quite how much hygrothermal conditions can vary between dwellings, variations that make all the difference between whether HDMs survive or not. Indeed, many appear to believe that HDMs are ubiquitous and found in all homes within a region, when this is rarely the case.

The three principal factors that affect hygrothermal conditions within a dwelling are:

1. climate – that is, hygrothermal conditions outside the home;

2. building construction – in relation to, for example, airtightness, insulation standard and heating provision; and

3. occupant behaviour – in relation to, for example, moisture production from cooking, washing and drying and the extent to which windows are opened for ventilation.

3.2.2. Impact of climate

Climate alone can account for some (but not all) of the observed differences in mite concentrations and asthma prevalence between countries and regions. As summarized by Arlian (1989), the percentage of mite-positive homes in a geographical area tends to be determined by average annual outdoor relative humidity. Within the United States, nearly 100% of homes are mite-positive in humid areas, such as New Orleans, while less than 10% are mite-positive in drier areas, such as Denver. Similarly, Colloff (1991b) reviewed several studies that found mite concentrations and mite hypersensitivity to be higher in coastal locations than in continental interiors. In addition, mites and asthma have been found to be less prevalent at high altitudes (Vervloet et al., 1982; Charpin et al., 1988) and in colder climates (Munir, 1998).

In general, mite proliferation is likely to be lowest in geographical areas with cold winters (whether due to high latitude or high altitude) where homes are heated. This is because cold air cannot contain as much moisture as warm air and, during winter, warm moist air from inside a dwelling is continually exchanged (assuming adequate ventilation) with cooler air that contains less moisture. When this cooler, dryer air is heated, the relative humidity inside the dwelling falls. The extent to which it falls depends on the coldness of the outside air, the level of ventilation and the temperature to which the thermostat controlling the heating is set: the colder the outside air, the more the ventilation and the warmer the indoor temperature, the lower the resultant relative humidity. Once relative humidity falls below about 50%, HDM populations start to dwindle (Arlian et al., 1998, 2001).

Unlike winter, the higher outdoor temperatures of summer limit the ability to reduce mite populations in this way. As a result, marked seasonal variations are typically observed in HDM populations, with peaks in late summer and autumn, when indoor conditions are most favourable for proliferation, and troughs in late winter and spring, when conditions are least favourable (Voorhorst, Spieksma & Varekamp, 1969; van Bronswijk, 1981; Arlian et al., 1983; Platts-Mills et al., 1987). However, the precise timing of these peaks and troughs tends to vary with climatic conditions from year to year and also tends to differ from region to region (Colloff, 1991b). Voorhorst, Spieksma & Varekamp (1969) showed that while this seasonal variation is easiest to see in damp homes, it can also be detected in dry homes, but here the peaks are orders of magnitude lower. Although a few mites may survive to take advantage of the favourable conditions of summer or autumn, if the conditions of winter and spring are dry enough too few of them will survive to cause medical problems.

Several acarologists – such as van Bronswijk (1981) and Arlian and colleagues (2001) – have suggested that HDM populations might be controlled by exploiting the natural seasonal culling effect and modifying the hygrothermal environment within homes during the critical winter months. Korsgaard (1979, 1983b), in particular, suggested that, if carried out regularly, this might lead to permanent reductions in mite populations (and even eradication). He was also clear about the key to such reductions being adequate winter ventilation. Thus successful culling can only be achieved if both (a) winters are cold or dry enough and (b) the standard of ventilation in winter is sufficiently high. In other words, even in cold winter regions, the natural culling of mites can be overridden by inadequate ventilation, allowing mite populations to survive and even prosper. This helps to explain the variation in the mite numbers found in homes within cold winter regions.

In winter, continental interiors are typically drier than coastal areas, making the seasonal culling of mites easier to achieve. The critical factor is the amount of moisture contained in the air brought in from outside to ventilate a dwelling. This can be low, either because the outside air temperature is low or because the air is dry due to other geographical factors. In cold winter regions, whether coastal or not, the laws of physics limit the amount of moisture in the air during this period. In regions with mild or warm winters, however, air sufficiently dry to allow seasonal culling is more likely to occur in continental interiors than by the coast, which is burdened by the additional moisture in the sea air.

The natural seasonal culling of mites is most difficult to achieve in humid tropical and subtropical regions, where there is no part of the year when the outside air is either cold or dry enough. This helps to explain the high mite concentrations and asthma prevalence in such cities as Sydney, Singapore and Caracas (Colloff, 1991b). On the other hand, Arlian (1989) reported considerable variation in mite density in humid regions, for reasons that are not yet understood. Colloff (1991b) reported that mites in climates where conditions are nearly ideal all year round appear to be more susceptible to the effects of minor hygrothermal fluctuations than mites accustomed to temperate climates. It is also possible that they suffer more from competition and predation in such constantly humid climes.

The association between mite concentration and the variation in hygrothermal environment within dwellings is thus primarily relevant to geographical areas that experience several months a year with either cold or dry (or both) outdoor conditions. Fortunately, this still accounts for a high proportion of those affected worldwide by mite-related illness.

3.2.3. Impact of building construction

Although seasonal culling of mites is possible in many parts of the world, whether or not it is achieved in practice depends on several other factors, including those that relate to building construction. In addition to climate, two aspects affect the hygrothermal environment within a dwelling:

1. the integrity of the building envelope and its ability to keep out excessive moisture
2. the ventilation and heating system it provides.

3.2.3.1. The building envelope

The presence of excessive moisture is almost certainly the cause of more building defects than any other single factor and, in the worst cases, can adversely affect the health of its occupants (Eldridge, 1976). The Committee on Damp Indoor Spaces and Health (2004) summarized the extent to which this is recognized as a significant public health problem. The various ways in which the building envelope can fail to keep out moisture are described by Singh (1994), who provided many illustrations of how excess moisture can get into (or arise within) and accumulate within a building; in many cases, this occurs because of simple failures of maintenance, such as broken roof tiles, damaged water pipes, spillages and overflowing cisterns. If not dealt with adequately, it is clear that the resulting increase in humidity is likely to lead to higher HDM proliferation. Other examples relate to common mistakes in design or construction, such as the excessive use of impermeable membranes that do not allow moisture trapped within the building envelope to evaporate outwards.

A particular problem of the building envelope is how its design responds to moisture from the ground. Several studies have found that relative humidity and HDM proliferation are higher in dwellings where the ground floor consists of a concrete slab in direct contact with the ground (Wickman et al., 1994, for example). With this type of construction, it is often difficult to achieve the comprehensive seal required to prevent moisture penetration and rising damp. At the same time, it can be sufficiently impermeable to trap indoor moisture – for example, from leaks, accidental spillage or condensation. If the floor covering is absorbent – a carpet, for example – it can act as a reservoir, leading to long-term dampness and high relative humidities. If the concrete slab is not adequately insulated, especially around the outer edges, the floor will also tend to be significantly colder than the rest of the room, raising relative humidity locally and increasing the risk of condensation. Even if floor level temperatures at the edges are too low to support mite population growth, the dampness created will raise relative humidity within the room as a whole.

One would expect building envelope failures to increase with building age, and some studies have found an association between dwelling age and HDM concentrations (Voorhorst, Spieksma & Varekamp, 1969; Hart & Whitehead, 1990). However, there appear to be few studies that relate building envelope failures to dwelling type.

3.2.3.2. The ventilation and heating system

A building's ventilation and heating system can significantly affect the hygrothermal environment within it and hence, HDM population growth. Since the Second World War, new dwellings, aided by modern technology, have generally become more airtight (Mage & Gammage, 1985) and with rising fuel prices, householders have become ever more energy conscious. As a result, ventilation standards have fallen and since water vapour is continuously produced in the home, indoor relative humidity levels have tended to rise. At the same time, expectations of thermal comfort have risen, with occupants relying more on heating systems to provide warmth in winter than on clothing, as before. Studies have shown that human beings are insensitive to gradual changes in relative humidity (McIntyre, 1980, for example), so that the higher levels of relative humidity resulting from inadequate ventilation tend to go unnoticed. The perceived need for ventilation has thus fallen and householders are increasingly reluctant to ventilate and lose expensively heated air. Furthermore, adequate ventilation in many modern homes can only be provided by the conscious act of opening a vent or window, which, given this reluctance, tends not to happen, especially in the crucial winter months when the need for it, in terms of culling mites, is greatest. The combination of more airtight dwellings and lower ventilation standards are often suggested as principal reasons for the rise in the prevalence of asthma in cold winter countries (Harving et al., 1991; Wickman et al., 1991; Løvik, Gaarder & Mehl, 1998).

In contrast to modern dwellings, older dwellings tend to be leakier and much less airtight. In this way, adequate background ventilation is provided involuntarily by various means – for example, through the multiple cracks around loose-fitting doors and windows and open chimneys, which are very efficient ventilation stacks, even with no fire. Because of far less tolerance than before of uncontrolled air movements – so-called draughts – older housing is being gradually rehabilitated. This is significant, because older housing constitutes the largest fraction of the housing stock in most high- and middle-income countries. The need for adequate background ventilation should thus be recognized for all types of construction.

The health benefits of adequate ventilation are obvious. They relate to reduced exposure not only to mite allergens, but also to other indoor airborne pollutants that would otherwise accumulate. Although the health benefits of higher standards of insulation are similarly obvious, in this case the effect on mite populations is not so clear-cut. The time it takes a HDM to develop from an egg to an adult increases rapidly as room temperature falls below 23°C, thereby significantly slowing population growth even when room relative humidity is high (van Bronswijk, 1981). Raising room temperature thus tends to shorten egg-to-adult development time and to favour mite population growth. On the other hand, raising room temperature, assuming the moisture content of the room air stays constant, has the simultaneous effect, by the laws of physics, of lowering room relative humidity, which is unfavourable for HDM population growth. The two effects thus tend to cancel each other. Modelling studies suggest that the favourable effect on mite growth of the rise in room temperature that results from improved insulation and heating systems tends to be outweighed by the unfavourable effect of the fall in relative humidity (Pretlove et al., 2005). This is to be welcomed, since it means that modifying the hygrothermal environment without sacrificing the health benefits of providing affordable warmth can potentially control mite populations.

The key is the provision of adequate ventilation. Although this necessarily involves some loss of energy, this can be lessened in some cases by technological means (see section 3.3.2). However, even without such active interventions, studies have shown that ventilation heat loss can be relatively modest (Marsh, 1996). Adequate ventilation is thus not necessarily incompatible with energy efficiency.

3.2.4. Occupant behaviour

As used here, *occupant behaviour* refers to how householders use their homes. This factor is far more significant than is generally realized and contributes to the large variations in hygrothermal conditions found in different households, even in identically constructed and located dwellings. To begin with, typical household moisture production can vary from less than 7 to more than 14 litres a day (Garratt & Nowak, 1991), according to:

- the number of occupants and how much of the day they spend at home
- their moisture producing activity, mainly washing, cooking and bathing.

The many householders, particularly in the United Kingdom, that still hang wet laundry up to dry indoors illustrates the last point. Hygrothermal conditions are then affected by:

- the extent to which windows are kept tightly shut in winter to conserve heat
- whether internal doors (especially to kitchen and bathroom) are kept open or shut
- the temperature at which the thermostat is set and the number of hours heating is on.

Occupant behaviour can often override the influence of building construction or climate. Such behaviour can be simply the result of varying personal preferences or acquired habits (whether consciously or unconsciously applied), but it can also be the result of socioeconomic and cultural influences. Moreover, there is often confusion in householders' minds as to what constitutes so-called healthy behaviour. For example, excessively dry air can irritate sensitized respiratory airways, so that patients or their parents may seek to raise relative humidity to alleviate symptoms, without being aware that this increases the risk of mite proliferation. As Platts-Mills and colleagues (1996) observed, occupant behaviour, in relation to the effects on hygrothermal conditions within buildings, is becoming an increasingly complex topic to study.

One specific factor is worth examining in more detail with reference to indoor relative humidity. It often tends to be high in low-income households, due to a low level of ventilation being maintained to preserve heat, but at the same time indoor temperatures tend to be low, due to the lack of affordable warmth (Boardman, 1991). As noted in subsection 3.2.3.2, low temperatures inhibit mite population growth and if indoor temperatures are kept low enough for long enough, this is likely to prevent mites from proliferating, even though the relative humidity is high. The fact that mite growth is inhibited by low room temperatures may partly explain the different results obtained in studies of childhood asthma prevalence and socioeconomic status. While some studies have found a positive association (Cesaroni et al., 2003; Almqvist, Pershagen & Wickman, 2005), others, particularly in the United Kingdom where low bedroom temperatures have until recently been common, have not (Prescott-Clarke & Primatesta, 1998). With rising living standards and wider access to affordable warmth, indoor temperatures in low-income housing can be expected to rise. In general, as noted earlier, this will be beneficial, but in the coldest households it may remove the inhibiting effect of low temperatures on mite population growth. Unless moisture production in such households is curtailed simultaneously, near ideal conditions for mite proliferation could be created – in which case, a clearer association between childhood asthma prevalence and socioeconomic status is likely to become evident.

3.2.5. Furniture and furnishings

Furniture and furnishings have some influence on hygrothermal conditions within the home, although far less than climate, building construction and occupant behaviour. Upholstery, carpets and soft fabrics tend to be highly hygroscopic, absorbing moisture when the relative humidity is high and releasing it when the relative humidity is low. When they are present in large quantities, this property tends to smooth out the peaks and troughs in relative humidity levels within the dwelling, although the overall effect on HDM populations is likely to be marginal. For more information on furnishings, including carpets, see section 3.3.6.

3.2.6. Overall effect of the housing environment on health or illness

The role housing plays in supporting public health has been generally acknowledged, together with the perception that a healthy house is one that is warm, dry, light and airy. Unfortunately, by the turn of the 21st century, providing these desirable attributes has become more difficult than expected and for reasons this chapter has attempted to clarify, conditions have been created that favour widespread HDM proliferation. While modern beds, carpets, upholstery and soft toys make ideal habitats for mites, the key factor in this development is likely to be the hygrothermal conditions that modern dwellings now provide.

However, a common misconception needs to be addressed. Many people, even within building physics, initially assume that hygrothermal conditions in mite habitats, such as beds or carpets, are mostly independent of room conditions, so that the latter are largely irrelevant. In beds, for example, it is assumed that people provide independent and long-lasting sources of warmth and moisture that do not dissipate significantly when they get out of bed. However, experiments using volunteers sleeping in fully instrumented beds in a laboratory have shown that this assumption is incorrect (Pretlove et al., 2001). Before the sleeper gets into bed, hygrothermal conditions in the mattress are the same as room conditions. These then change quickly when the sleeper gets into bed and almost as quickly when the sleeper gets out of bed eight hours later. This return to room conditions is delayed only slightly by leaving bed coverings in place. In other words, for most of the 16 hours that the bed is typically unoccupied, hygrothermal conditions within the bed are the same as room conditions. This means that room conditions do indeed play a major role in determining whether mite populations grow or decline.

In large parts of the world, cold or dry winters allow for the seasonal culling that can control mite populations. To the extent to which such culling does not occur, it is most likely related to the housing environment – that is, the combined effect of occupant behaviour and building construction.

Attempts have been made recently to estimate the economic burden of disease of asthma and other allergic disease. For the United Kingdom, for example, Gupta and colleagues (2004) reported that 39% of children and 30% of adults have been diagnosed with one or more atopic conditions and that the direct National Health Service cost for managing them is estimated at more than £1 billion annually. For the United States, the direct annual cost of asthma was estimated to be US$ 9.4 billion and the indirect cost (such as missed school and work days) was estimated to be US$ 4.6 billion (NHLBI, 2002). Because of the multifactorial nature of its causation, it is difficult to estimate the fraction of this vast burden that can be attributed specifically to the housing environment, but there is little doubt that it has been an important factor in the increased proliferation of HDMs.

3.3. Methods of house dust mite control

Professional pest controllers and public health specialists conduct very little HDM control; the majority of this work is done by members of the public. The primary aim of any control programme should be to prevent HDM infestations of buildings from occurring in the first place. A variety of products, available in shops and over the Internet, can help manage HDM populations effectively. In any HDM control programme it is essential to consider three aspects:

1. the prevention of HDM infestation;

2. the control of HDMs, if present; and

3. the control or removal of their allergens, which are extremely persistent in the environment.

It is therefore essential that members of the public be advised as to how to integrate the available control methods to best effect.

Unfortunately, published data are unavailable on some specific methods of HDM or allergen control. Also, control measures are often integrated in large scale trials; while integrating the techniques available is the best way of reducing HDM populations and the allergens they produce, it makes it difficult to evaluate the individual contribution of a specific control measure. Moreover, there are large differences in study designs, individuals targeted and methods used to monitor outcomes. Not all such trials report positive clinical outcomes. Tovey (1992) reported that the most successful trials usually ran for a longer period of time and included the removal of carpets, the encasement of mattresses and vigorous attention being given to controlling all dust.

3.3.1. Dust mite and allergen inspection and detection methods

3.3.1.1. Sampling methods

HDMs are very difficult to extract from their various habitats. They do not live just on the surface of these habitats, but also deep down inside them. This can be at least partially attributed to their being photophobic (van Bronswijk, 1981). HDMs are also very good at clinging to the substrate in which they live; each of their eight (six in the case of larvae) legs possesses a sucker, a number of spikes and a hook. Bischoff, Fischer & Liebenberg (1992) reported that vacuuming a 1 m^2 area of carpet for two minutes may only remove 10% of the HDM population.

The methods used to collect HDMs seldom give identical results and vary in a number of ways, such as the power of the vacuum cleaner used, the area sampled and the length of time for which the samples are taken, in addition to the type of collection device used. This variation can make the results of different studies difficult to compare. Although different sampling methods generally give slightly different results, Twiggs and colleagues (1991) found two different collection techniques that yielded similar results.

Nearly without exception, vacuum cleaners are now used to remove dust, and brushing has become redundant, as vacuum cleaning is more efficient at removing mites (Blythe, Williams & Smith, 1974; Abbott, Cameron & Taylor, 1981). HDMs can also be extracted by destructive sampling, which can provide accurate estimates of HDM populations, but can be time consuming and cannot always be used in field situations.

3.3.1.1.1. Mite extraction from dust

Live mites can be removed from dust and small quantities of materials by exploiting their natural response to changes in light and humidity. Also, extracting them with heat onto an adhesive film can remove 65% of live mites present (Hill, 1998). Colloff (1991b) applied mild heat to the top of a Petri dish that contained HDMs and an adhesive film. The aim of this method is to stimulate mite movement while not significantly reducing relative humidity, thus increasing the likelihood of the HDMs coming into contact with the film. The application of heat can also be used to extract mites from precollected dust samples. Van Bronswijk (1973) found a Tullgren funnel, which uses heat applied to the upper layers of a collected mass to slowly dry it out, extracted 40–60% of live HDMs. These sampling methods only collect live, active mites, not dead or immobile life forms. It is also possible to extract HDMs from dust by using the flotation–suspension technique. This generally involves placing dust samples in a saturated salt solution (sodium chloride); the mites float to the surface and the majority of the dust sinks to the bottom (Hart & Fain, 1987).

3.3.1.1.2. Sampling site variations

The location sampled may also influence the number of mites recovered. For example, it is known that dust distribution in mattresses, and therefore the presence of HDMs, is influenced by the pattern of seams and buttons. Blythe (1976) sampled a 10 cm by 10 cm section of mattress with a seam running across the centre and found that over 80% of the mite population was collected from within 0.5 cm of the seam. This highlights the difficulty of estimating how representative a sample is of the total reservoir of dust in the object being sampled.

Estimates of population size based on numbers of mites per unit weight of dust are also difficult to compare between different habitats. Abbott, Cameron & Taylor (1981), for example, reported that the surfaces of innerspring mattresses and foam mattresses yielded a similar number of mites: 746 and 706 mites per gram of dust, respectively. However, when the number of mites/m^2 was examined, innerspring mattresses contained 2489 mites, compared with only 720 mites for foam mattresses. This was due to the mass of dust collected from the innerspring mattresses being over three times greater than that from foam mattresses. Also, floor dust, which generally contains such material as sand and grit, is heavier than dust collected from a mattress, and comparison of such data will give misleading results. Blythe (1976) recorded 276 mites per 0.1 gram of dust in a mattress and 16.9 mites per 0.1 gram of dust in a carpet. However, when he expressed this as the number of mites per 100 cm^2, the carpet had higher numbers: 32, compared with 9.7 in the mattress. The number of mites per unit area is, therefore, a more satisfactory measurement of mites collected in different locations, although this still does not take into account the substrate thickness.

3.3.1.1.3. Allergen

Each HDM faecal pellet contains a number of different allergens (Arlian, 1991), Der p 1 being the one most studied. In the first international review of HDM allergens, Platts-Mills & de Weck (1989) proposed two threshold levels for HDM allergen: 2 µg of Der p 1 per gram of dust from mattresses and carpets to be regarded as a level at which sensitization can take place, and 10 µg of Der p 1 per gram as a level above which the development of acute asthma in sensitized individuals can take place. Although these are currently taken to be the generally accepted thresholds, measurement of airborne allergens is likely to give more realistic exposure levels. However, Custovic & Chapman (1998) stated that too few reliable data sets are available to produce benchmark figures and that the measurement of HDM allergen concentration in dust reservoirs should still be regarded as the best validated index of exposure and quantification. As with mites, there is also a debate as to whether or not to measure allergen levels in terms of Der p 1 per gram of dust or Der p 1 per m^2. Custovic & Chapmen (1998) recommend that both should be reported.

For further information see Chapter 1 of this report.

3.3.2. Modifying environmental conditions

As explained in section 3.2, HDMs can be controlled through winter culling in climates that allow it. Controlling mites in this way has two major advantages.

1. Hygrothermal conditions that are unfavourable to mite population growth can often be created by relatively minor adjustments to heating and ventilation methods or occupant behaviour. In such cases, the control of mites can be achieved inexpensively and without the use of acaricides (agents that kill ticks and mites).

2. Controlling mites by environmental means has potential both as a curative measure for alleviating symptoms and as a preventive measure – that is, before sensitization occurs.

Intervention studies have demonstrated that reducing humidity within dwellings, typically by using mechanical ventilation or dehumidifiers, does lead to reduced mite numbers and allergen levels, compared with controls (Harving et al., 1991; McIntyre, 1992; Harving, Korsgaard & Dahl, 1994; Wickman et al., 1994; Cabrera et al., 1995; Emenius, Egmar & Wickman, 1998; Warner et al., 2000). However, some other studies have found no or very little benefit (Fletcher et al., 1996; Niven et al., 1999; Hyndman et al., 2000).

As explained earlier, hygrothermal conditions in the home are in constant flux, affected by several continuously changing and interacting factors. The results of the studies that reported no beneficial effects from reducing humidity may thus have been due to atypical conditions or to a combination of unforeseen factors. With respect to climate, for example, de Boer (2000) observed that temperatures in an untypically mild winter in the Netherlands may not be low enough for long enough to result in significant mite deaths. On the other hand, during an untypically cool summer, they may not be high enough to result in significant mite population growth. De Boer and colleagues (1997, 1998) also demonstrated that mite populations can survive and prosper under conditions that on average are unfavourable, provided that they are exposed to favourable conditions for a few hours daily. This very important finding, confirmed by Pike, Cunningham & Lester (2005), introduces another level of complexity into an already complex situation and illustrates the difficulty of attempting to predict the outcome of specific interventions.

While controlling mites by environmental means is a potentially attractive option, the most appropriate way of achieving such control in practice is not always clear. One strategy is to set a target for the maximum indoor absolute humidity, the moisture content of the air (normally measured in grams of water vapour per kilogram of dry air). At the 1987 WHO Workshop at Bad Kreuznach, Germany, it was recommended that the absolute humidity in the home should be kept below 7g/kg (Platts-Mills & de Weck, 1989). This recommendation followed the suggestion by Korsgaard that mite populations can be controlled if winter absolute humidity values are kept below this target level (Korsgaard, 1979, 1983b). Thus, maintaining the indoor temperature at, say, 21°C during the winter heating season and keeping absolute humidity below 7g/kg (by means of ventilation and controlling moisture production) will ensure that the relative humidity is kept below 45% – that is, well below the level HDMs need to prosper. For heating engineers, it is then straightforward to calculate the ventilation rate required to keep absolute humidity at 7g/kg for any given combination of outdoor absolute humidity, dwelling volume, number of occupants and moisture production rate. As an example, Korsgaard (1983b) showed that for a typical case the ventilation rate required would be 0.75 air changes per hour (the normal way of expressing ventilation rates, meaning 75% of the volume of air in the dwelling is being replaced every hour).

Unfortunately, this strategy has proved to be too simple. As Colloff (1994) and Lowe (2000) have pointed out, it is based on the assumption that bedroom temperatures are continuously maintained at around 21°C for 24 hours a day. While this may be justified for some regions, such as Scandinavia, it is not justified for regions such as the United Kingdom. At 7g/kg, the low bedroom temperatures typically found in the United Kingdom lead to relative humidity levels that are high enough to risk mite infestation. This helps to explain the different results obtained in studies where the absolute humidity has been kept below 7g/kg in winter. While some studies found this strategy to be effective in reducing mite infestations (Vervloet et al., 1991, in France, for example), other studies did not find it to be effective or else found a poor correlation between absolute humidity and HDM populations (Raw et al., 1998, in the United Kingdom, for example). Although the 7g/kg threshold for absolute humidity is relevant to some situations, it lacks general applicability.

Another approach is to use computer-modelling techniques for simulating hygrothermal conditions in the home. In this way, all the various interacting factors that affect them can be taken into account, including fluctuating conditions. To simulate energy use in buildings, building physicists have developed, over the last three decades, a range of validated computer models that predict indoor temperature and relative humidity. Using such predictions, the next step is to simulate conditions within mite habitats, such as beds and carpets. The advent of small cheap sensors and loggers has catalysed interest in this area (Cunningham, 1996, 1998; Lowe, 2000; Pretlove et al., 2001, 2005). The final step is then to simulate how habitat conditions affect mite populations (Crowther et al., 2006; Biddulph et al., 2007).

Preliminary results of this modelling effort are promising and tend to confirm that the considerable variation in mite numbers found both seasonally and between similarly located households can indeed be explained and simulated using computer modelling techniques. Sensitivity analyses also show that factors that relate to a building and its occupants have more effect than other factors, such as the physical properties of the bedding, and that relatively small changes can make a substantial difference in hygrothermal conditions in mite habitats. This confirms the potential for controlling mite populations by environmental means.

Modelling techniques thus make it possible to investigate which modifications to home environments have most impact on mite populations, for any given climatic region, housing type and pattern of occupant behaviour. In this way, it will be possible to establish the most effective, energy efficient and socially acceptable ways of achieving HDM control by modifying environmental conditions, where this is feasible.

3.3.3. Constructional changes to improve conditions

In advance of such modelling studies, it is still worth discussing, in general terms, changes in building construction that improve conditions.

3.3.3.1. Improved maintenance and the avoidance of excess moisture

According to Eldridge (1976):

> Defects in buildings and building materials are often said to be caused by the weather, especially when severe or unusual conditions have been experienced. However, careful diagnosis will demonstrate that faulty design, the wrong choice of materials or faults on site are usually the root cause, the weather only providing the appropriate conditions for the failure to occur.

New building requirements and ways of using buildings can also give rise to unforeseen consequences. Even though regularly updated guides for diagnosing and avoiding building defects are published in most high- and middle-income countries, and the importance of avoiding excess moisture is generally well recognized, many obstacles to making progress persist. Elaborating on this point, the Committee on Damp Indoor Spaces and Health (2004) called attention to the lack of sufficient information on which to base quantitative recommendations, as well as to institutional, social and economic factors that tend to hinder the widespread adoption of technical measures and practices that could improve the situation.

3.3.3.2. Improved insulation and ventilation standards

As suggested earlier, providing affordable heating and improved insulation standards is likely to have the beneficial effect of lowering bedroom relative humidity levels and thus reducing mite populations, provided that ventilation standards are maintained. This proviso needs to be stressed, and it is vital that insulation standards are improved in conjunction with measures that enable and encourage householders to achieve good ventilation.

Householders can do a great deal to modify hygrothermal conditions, to reduce mite population growth, both by controlling moisture production and by being aware of the need to ventilate adequately in winter. For example, the spread of moisture vapour to other rooms can be restricted by keeping kitchen and bathroom doors closed, as well as by drying clothes only in rooms that can be closed and well ventilated. The use of thresholds and automatic door closers would undoubtedly also be beneficial. Above all, householders need to be provided with ways of ventilating their home that are both effective and easy to use. Some windows, for example, do not allow sufficient flexibility or range of opening positions, so that it is difficult to achieve the desired level of ventilation or to change it easily in response to varying external conditions. Trickle vents, such as those in window frames, can often improve the situation, by allowing more precise control of incoming air.

3.3.3.3. Mechanical solutions

One method for which success has been claimed is mechanical ventilation with heat recovery (MVHR), whereby the heat from the outgoing warm stale air (such as from bathrooms and kitchens) is recovered and transferred to the cold incoming fresh air. Higher ventilation rates can thus be achieved for less energy loss, thus encouraging householders to raise their ventilation standards. Several studies have found that when householders use MVHR installations correctly – that is, on 24 hours a day at a high enough volume setting to achieve at least 0.5 air changes an hour – both mite-population and allergen levels have fallen as a result of improved ventilation (Harving et al., 1991; McIntyre, 1992; Warner et al., 2000).

For this method to work, however, both outgoing and incoming air need to be ducted, with an electric fan in each direction, and the dwelling needs to be relatively airtight, to maximize the proportion of ventilated air that passes through the system. This last requirement makes the method more relevant to new housing than to older housing, where airtightness is more difficult to achieve. The amount of ductwork required is another obstacle for use in refurbished houses. Other possible problems relate to the correct positioning of inlet and outlet air grilles and to the need for regular cleaning of input air ductwork, as well as filters. Although electric fans are becoming quieter, the energy consumption and noise of two fans is also likely to attract the attention of householders. In one study, it was found that the system was turned off overnight in 50% of the dwellings with MVHR, thereby significantly reducing its effectiveness and the standard of ventilation (McIntyre, 1992).

MVHR is thus a somewhat complex *high tech* solution. Even if properly installed, it puts a considerable onus on householders to use it correctly and to keep it well maintained. Moreover, the energy apparently saved is offset by the energy used (and carbon dioxide emissions produced) by the two electric fans. The system thus makes most sense in regions with very cold winters and hydroelectric power, such as northern Scandinavia.

Besides whole-house MVHR systems, small single-room versions are available. In these systems, the two fans, filters and grilles are all combined in one compact unit that can be installed in an external wall. With virtually no ductwork, cleaning and maintaining, these systems are less problematic. However, Htut and colleagues (1996) found that the unit could only be operated 24 hours a day at the lowest setting without producing unacceptable noise at night. At this setting, the unit did reduce humidity and mite numbers in an occupied bedroom (compared with a control), but not sufficiently to effect a permanent reduction.

Positive pressure ventilation (PPV) is another solution. With PPV, a large quantity of air is sucked in, using a fan and ductwork, to put the whole dwelling under positive pressure. Unlike MVHR, the stale air simply leaks out of the dwelling through cracks or trickle vents in individual rooms, making strict airtightness less of a requirement. On the other hand, although PPV is a less complicated and more feasible option in refurbishment projects, there are similar concerns: maintaining clean ducts and filters and choosing to leave the system running for 24 hours a day, as required.

Finally, the evidence on the effect of air conditioning is conflicting. Some studies have found a beneficial effect on mite numbers and allergen levels (Lintner & Brame, 1993; van Strien et al., 2004), while others have not (Chan-Yeung et al., 1995; Chew et al., 1998). When considering the suitability of air-conditioning as a possible solution for controlling mites, its energy requirements should be taken into account.

Smaller scale mechanical systems that modify environmental conditions, such as portable dehumidifiers and bed heaters, are discussed in sections 3.3.5 and 3.3.8.

3.3.3.4. Non-mechanical passive ventilation systems

The most important non-mechanical passive ventilation system is passive stack ventilation (PSV) which, like a traditional chimney, makes use of the stack effect – that is, buoyant warm air rising to the cold outside. The system consists of a 10–15 cm-diameter tube (typically polyvinyl chloride) installed in the ceiling of kitchens and bathrooms and extending to vents in the roof ridge. An advantage of PSV systems is that the stack effect works best in winter, when the temperature difference between the inside and outside is greatest, which is also when the need to continuously remove warm moist air from kitchens and bathrooms is greatest. Following a pilot study with a test house fitted with both MVHR and PSV systems, used alternately, Palin and colleagues (1993) reported that the use of PSV was as energy efficient as the use of MVHR in achieving the same conditions of comfort. PSV can also be used in conjunction with so-called supply air windows, which preheat the incoming air, using heat that would otherwise be lost through the window glazing (McEvoy & Southall, 2002).

Architects and engineers have devised a variety of effective ways to achieve good ventilation at low or reasonable energy cost and without sacrificing thermal comfort. Householders are thereby encouraged to ventilate sufficiently well in the crucial winter period to keep HDM populations under control. Appropriate technology can undoubtedly play a helpful role, but perhaps the greatest need is to put more emphasis on educating the public about how to use the buildings they live in most efficiently. To this end, it is to be hoped that computer modelling will in due course increase our knowledge of how buildings are best used to achieve a healthy environment.

3.3.4. Cleaning

Although rigorous household cleaning, with the exception of steam cleaning, will not eliminate or significantly reduce HDM populations, it will remove allergens. Many cleaning methods will not remove dirt and allergens from deep within a carpet or mattress, although they will remove allergens from the surface where it is most likely to come in contact with people. Cleaning will also reduce the amount of food (human skin scales) available to HDMs, thus potentially reducing the size of future HDM populations.

3.3.4.1. Home disinfectants

Cleaning of hard surfaces is generally conducted using some form of disinfectant diluted with water. In laboratory experiments, Schober and colleagues (1987) found that a number of different household disinfectants, some containing acaricides, were able to kill

HDMs. However, the mites were soaked in the solutions for a considerable period of time, which does not necessarily reflect real life conditions, where mites may only be exposed to the disinfectants for a short period of time or else come into contact with them after the water has evaporated.

3.3.4.2. Washing

HDM allergens are extremely soluble in water, enabling them to be removed from items during washing. Watanabe and colleagues (1995) found that washing blankets with a hot (55°C) soap solution reduced Der p 1 levels by an average of 97% and Der p 2 levels by an average of 91%. McDonald & Tovey (1992) found washing bedding at 55°C killed all mites present, while reducing the temperature to 50°C killed only half of them. Most washing machines have washing cycles at 40°C, 60°C and 90°C; temperatures of 60°C or more are therefore recommended. If the bedding has to be washed at low temperatures, it is possible to add special products to the wash to kill the mites (McDonald & Tovey, 1993; Bischoff et al., 1998). In addition to bedding, clothing can also be a source of allergens (Tovey, Mahmic & McDonald, 1995).

3.3.4.3. Dusting

Dusting is an effective method of removing house dust and therefore allergens; damp dusting is particularly effective. No specific studies have been reported, although Tovey (1992) found that a vigorous cleaning regime in addition to a number of other measures generally is important in the success of clinical trials.

3.3.4.4. Dry cleaning

Kniest, Liebenberg & Bischoff (1989) reported that all HDMs were killed by dry cleaning in perchloroethylene. Watanabe and colleagues (1995) found that dry cleaning of blankets with perchloroethylene reduced levels of Der p 1 by a mean of 69% and Der p 2 by an average of 54%. These levels of allergen reduction are significantly less than would be achieved through washing at 55°C.

3.3.4.5. Vacuuming

Householders primarily use vacuuming to remove dirt and dust from carpets and other floor coverings, but in this way they are also able to reduce the quantity of allergen within the home. However, this can take a substantial period of time (de Boer, 1990b) and is best achieved by regular vacuuming (Munir, Einarsson & Dreborg, 1993). The concentration of airborne allergens has been found to increase significantly after vacuuming with a standard vacuum cleaner, thus increasing the risk of exposure to these allergens. This can be minimized when vacuum cleaners are fitted with high efficiency filters (Kalra et al., 1990).

Vacuuming an area has been shown to decrease the number of mites present (Hill, 1998), although it does not remove all of them. Bischoff, Fischer & Liebenberg (1992) reported that vacuuming a 1 m^2 area of carpet for two minutes may only remove 10% of the HDM population. Hay (1995) vacuumed the surface area of a spring mattress and recorded a population density of 3–46 living mites per m^2, three orders of magnitude lower than the estimate of 8,200 m^2 26,800 living mites per m^2 obtained by extracting mites from a core taken from the upper 1.5 cm of the same mattress. This demonstrates not only the inefficiency of vacuuming, but also that HDMs are not restricted to the mattress surface, where vacuuming is likely to be most effective.

3.3.4.6. Steam cleaning

Steam can be used as an alternative to insecticides. It can be extremely effective at controlling HDMs and their allergens and can be used frequently, provided the room is subsequently well ventilated. Steam does not, however, have residual activity. In laboratory experiments, Colloff, Taylor & Merrett, (1995) found that steam cleaning controlled 100% of HDMs in samples of carpeting. Also, no subsequent development of HDM populations occurred, indicating that the cleaning also controlled the HDM eggs, which are normally more resistant to extremes of temperature than the mites themselves. Under field conditions, steam cleaning also resulted in a mean reduction in the measurable Der p 1 concentration of 86.7%. The mean temperature measured in the carpets during treatment was 103.4°C, decreasing to the background temperature in 20 minutes. After treatment, the humidity was found to remain at saturation point for 25 minutes before falling to background levels after 140 minutes. It should be noted that treatment of an entire room is likely to cause humidity levels to become elevated for a more prolonged period of time and that it is essential to ventilate both during and after treatment, to minimize both mite survival and the potential for the growth of mould.

3.3.4.7. Carpet cleaning

Carpet cleaning is one of the few professional anti-HDM, anti-allergen services available to householders. The devices used to clean carpets range from small domestic units to powerful truck mounted units. These machines have the potential to remove allergens and if they use high temperatures, to kill mites. The inclusion of an acaricide in the cleaning solution also provides the potential to control HDMs.

Although the use of these cleaners has not been widely studied, they are considered to be effective. De Boer (1990b) found that the shampooing and wet cleaning of carpets resulted in a significant reduction of allergens and habitat deterioration (see also subsection 3.3.4.5). A domestic carpet cleaner has also been found to reduce Der p 1 levels in carpets by 70% (Thompson et al., 1991). Wassenaar (1988), however, found an increase in HDM populations after cleaning. After wet cleaning or shampooing, it is essential that the carpet is thoroughly dried; after the carpets are cleaned, it also is essential that houses be properly aired.

3.3.5. Temperature control

3.3.5.1. Autoclaving

De Boer (1990b) found that autoclaving samples of carpet reduced allergen levels to below detectable limits, effectively providing 100% control. Autoclaving will also eliminate HDM populations. While autoclaving is very effective, it is not a practical technique to use in the domestic environment, although it could be used in hospitals.

3.3.5.2. Steam cleaning

Information about the effectiveness of steam cleaning can be found in subsection 3.3.4.6.

3.3.5.3. Electric blankets

De Boer & van der Geest (1990) observed reductions in HDM populations of between 19% and 84% over a 10-week period in heated areas of a mattress. Areas of the mattress that had been heated by electric blankets had significantly fewer mites than unheated areas. This indicates that electric blankets are able to suppress, but not completely eliminate, HDM populations. Using a combination of electric blankets and vacuuming, Mosbech, Korsgaard & Lind, (1988) also observed significant reductions in allergens. They suggested electric blankets are best used to help prevent mattresses from becoming infested, rather than to combat pre-existing mattress infestations.

3.3.5.4. Bed heaters

A number of devices are currently being developed to introduce hot air, at varying temperatures, into a mattress. As they are able to heat up mattresses, they have the potential, over several days, to control HDMs by reducing the humidity within the mattress. However, no work on their effectiveness has been published to date.

3.3.5.5. Freezing

Laboratory and field studies indicate that the use of liquid nitrogen as a freezing agent, combined with vacuuming, is effective at reducing HDM populations when compared with vacuum cleaning alone (Colloff, 1986). Freezing toys in a conventional freezer can also kill the mites (Nagakura et al., 1996). While it is true that freezing will control HDMs, we have found (T. Wilkinson, unpublished observations, 2002) that 100% control of HDM populations can only be observed after 72 hours, although 98.3% mortality can be achieved after 24 hours. It appears that some eggs, which later hatch, survive the freezing process. Freezing does not, however, remove allergens from treated items. It should therefore be combined with washing, in the case of smaller items, such as toys and pillows; it should also be combined with such actions as vacuuming, in the case of such large items as mattresses. At present, the use of freezing to treat carpets seems impractical. Although freezing leaves behind no residues, this does mean that it has no residual activity (see also subsection 3.1.3.3).

3.3.5.6. Sunlight

In some countries, rugs and bedding are commonly aired outdoors. Tovey & Woolcock (1994) found that exposing carpets to direct sunlight killed all mites within three hours, by creating microclimates hostile to HDMs – with peaks in temperature of 55°C and relative humidity as low as 24% within the carpets. Allergens, however, were not affected significantly. The effectiveness of these experiments was in part due to the hot sunny weather in which they were conducted. Tovey (1992) reported that the ambient conditions during the day were 30°C and 60% relative humidity. While extremely effective, the use of this technique will generally be confined to the summer months, which coincide with the period of maximum HDM growth, and will not be as applicable in some cooler climates.

Some products and services are available to consumers that incorporate an ultraviolet light, but there is no scientific information available to ascertain the effectiveness of this technique. However, it is unlikely to be effective, since the ultraviolet light will not focus on one spot for a prolonged period of time and a considerable proportion of the HDMs will be shaded from the light by their environment.

3.3.6. Physical control methods

3.3.6.1. Habitat modification

HDMs inhabit a diverse range of habitats within the home, and by removing or modifying these habitats it is possible to reduce the potential for mite population development. Habitat modification can also make rooms easier to clean, thus aiding the removal of HDM allergens.

3.3.6.2. Carpets

Physically, carpets provide an excellent habitat for HDMs, particularly near the edges of beds, chairs and sofas where they are showered with skin scales (Colloff, 1998). Platts-Mills and colleagues (1996) suggested that fitted carpets, together with increased indoor temperatures and decreased ventilation, are among the housing-related changes that have increased the prevalence and severity of asthma. Carpets are one of the major habitats of HDMs and can contain the largest reservoir of mite allergens in the house (Tovey, 1992). However, carpets favour a somewhat different hygrothermal environment than the rest of the room. Van Bronswijk (1981) cites a 1966 study by Leupen & Varekamp that shows that they tend to be cooler and damper. This is particularly the case where carpets are laid directly onto a concrete and screed ground floor. Hygrothermally, carpets are distinct from beds and upholstery in that they do not generally benefit from proximity to human warmth and moisture. Studies have shown that long or loose-pile carpets tend to harbour more mites and allergens than short-pile carpets or hard floors (Arlian, 1989, for example).

Carpets in homes are typically replaced with hard flooring, such as wood, tile and laminate. When properly fitted, these hard surfaces are inhospitable to HDMs and, by removing one of their major habitats, it is possible to reduce the number of HDMs and therefore the amount of allergen produced. Hard floors are also more readily cleaned than soft floors. Mulla and colleagues (1975) found that vacuum cleaning removed nearly all of the mites and allergens from hard floors. Because dust may become more easily airborne from a hard floor than from a carpet; it is essential that hard floors be cleaned on a regular basis.

3.3.6.3. Soft furnishings

Allergen levels in soft furnishings are often similar to those found in beds (Tovey, 1992). Upholstered furniture allows the mites to penetrate deep below the surface, and such furniture is difficult to cover with barrier covers. Leather and vinyl furniture is less likely to be colonized by HDMs than conventional soft furnishing and can also be cleaned more easily. Also, in terms of the materials used, it has generally been found that synthetic fibres do not have any significant inherent benefit over natural fibres (Wickman et al., 1994; Hallam et al., 1999).

3.3.6.4. Barrier fabrics

Barrier bedding materials can be a major benefit in combating HDMs. They can act as a physical barrier to the mites themselves; on a new mattress, the barrier prevents mites from entering and therefore colonizing it. Also, the barrier stops mite faecal material (already present in an old mattress) from escaping into the atmosphere, where it can be inhaled. Fabrics with a pore size of 10 µm or less can effectively block the faecal pellets (Vaughan et al., 1999). In addition, these fabrics prevent the dead skin cells upon which the mites feed from penetrating the mattress. Thus, the mites and their allergens are effectively contained within the mattress, and the mites are cut off from a continuing food supply.

A number of trials of barrier bedding have highlighted the clinical benefits of this type of intervention. As a result of its use, atopic infants who received barrier bedding did not become sensitized to HDM allergens (Nishioka, Yasueda & Saito, 1998). Also, airway hyperresponsiveness improved significantly after six months with barrier bedding in place (van der Heide et al., 1997), and symptom scores in patients with barrier bedding improved after a year.

Barrier fabrics should be fitted not only to mattresses, but also to pillows and duvets. It is also essential that barrier fabrics be properly constructed. An effective fabric can lose a significant amount of its efficacy as a result of poor quality zippers and stitching, which can let through allergens. It is also recommended that zippers be covered with flaps and that the barrier provides 100% cover. Moreover, it is important to select a breathable fabric, to prevent the risk of mould growing within the mattress and to minimize discomfort from sweating; in general, woven fabrics are longer lasting than other types. For barrier fabrics to be effective, it is also essential for them to be properly cleaned and for the rest of the bedding to be washed regularly, so that it remains free of allergens.

3.3.7. Pesticides

Acaricides or insecticides can be used to control HDMs. They are typically applied as surface treatments or impregnated into fibres and fabrics used in the construction of mattresses, soft furnishings and carpets. Acaricides can kill mites, providing they are applied correctly, although they generally have very little effect on HDM allergen levels.

For controlling HDMs, benzyl benzoate is the most commonly used acaricide, although an increasing number of products use pyrethroids, such as permethrin. In vitro studies normally show acaricidal products to be highly effective, causing rapid HDM death (Hart, Guerin & Nolard, 1992; Hyden et al., 1992). However, Colloff and colleagues (1992) reported that the high level of mortality observed in the laboratory from the use of acaricides cannot be simply reproduced in the home. Also, Tovey & Marks (1999) highlighted the importance of ensuring that acaricides made available to the public are not only subjected to in vitro studies, but that they also have been properly field tested in domestic environments. Moreover, De Boer (1998) questioned the ability of acaricides to penetrate deeply into upholstered furniture and mattresses, thus reducing their potential efficacy. In such a situation, it may be more effective to make bedding and upholstery with materials that have been impregnated with acaricides. Therefore, studies that simply place mites onto fabrics impregnated with insecticides or that apply insecticides directly to mites should only be considered as being the first step in determining their potential in the field.

An effective acaricide will only remove the source of the allergen, the HDMs; it will not remove or denature any allergen present. Colloff and colleagues (1992) advise that acaricide use must be followed by an intensive and thorough vacuuming of the treated surfaces to remove allergens. De Boer, van de Hoeven & Stapel (1995) found that samples of carpet treated with lindane showed only a very small decrease in Der p 1 levels over 18 months, although Der p 2 decreased by a slightly larger amount. The use of tannic acid, in combination with an acaricide, has been found to reduce allergen levels more effectively than the use of acaricides alone (Green et al., 1989). Alternatively, steam cleaning could be conducted just prior to the application of an acaricide. Also, Cameron (1997) reported that benzyl benzoate reduced mite allergen levels in carpet dust more than in mattress dust, which is likely due to carpets being cleaned more frequently than mattresses.

In Europe and elsewhere, there is a shift away from the use of insecticides, particularly in the domestic environment. Colloff (1986) discussed the risk of using acaricides or insecticides in the presence of atopic individuals, because of possible sensitization, although he did state that no occurrence of toxicity was reported in people due to these treatments. More recently, in agreement with the United States Environmental Protection Agency (EPA) (2000), a company removed a benzyl benzoate based acaricidal product from the market, as a small percentage of consumers reacted adversely to a fragrance used in its formulation. Problems reported included asthma attacks, respiratory problems, burning sensations and skin irritation. This case highlights the care needed in formulating products for use by people with allergies. Many of the other control methods described in this chapter can be used instead of acaricides. If acaricides are used, it is essential that they be applied correctly and in such a way as to minimize any direct contact with people during and after treatment.

3.3.8. Other methods

3.3.8.1. Biological control

Since HDMs are not the only arthropods living in house dust, it is likely that there is some interaction with other species. Potential predators of HDMs include silverfish, dust lice, pseudoscorpions and other predatory mites. For example, van Bronswijk (1981) cited a 1971 study by McGarth that found that the adults and juveniles of *Cheyletus aversor* eliminated a well-established culture of *D. farinae* within 20 days. However, this particular species of *Cheyletus* has not been reported as naturally occurring within house dust. Similarly, van Bronswijk and colleagues (1971) found that, while two predator mite species that can be present in house dust (*Glycyphagus destructor* and *Acarus siro*) were able to control *D. pteronyssinus* populations under laboratory conditions, in nature these species live in different habitats and are thus unlikely to interact in real conditions.

Typically, cheyletid mites found in house dust (such as *Cheyletus malaccensis*) constitute only a small percentage of the total number of mites, and they are present mainly in carpets, rather than mattresses (Rao et al., 1975). In comparison with HDMs, their numbers are also lower than would be expected in a predator–prey population in equilibrium (Colloff, 1991b); therefore, they are not suitable contenders for biological control. Also, previous studies have suggested that in the absence of suitable prey, *Cheyletus* mites may not only bite people in self-defence, but they may also feed on their body fluids (Yoshikawa, 1980; Htut, 1994).

3.3.8.2. Air filters

The use of air filtration devices that incorporate high efficiency particulate air (HEPA) filters have the ability to filter or remove airborne allergens from the atmosphere, thus reducing allergen exposure to individuals in the room. These filters have been found to be of clinical benefit in a number of trials (Zwemer & Karibo, 1973; Villaveces, Rosengren & Evans, 1997). However, most HDM allergens do not remain airborne unless disturbed (Custis et al., 2003). Custis and colleagues (2003) found that an ion-charging device was able to remove HDM allergens from the air after disturbance, although their experiments were not designed to test the ability of this device to clean air. Colloff and colleagues (1992) concluded that HEPA filters were more effective than electrostatic filters and that air filtration should not be conducted in isolation from other forms of control.

3.3.8.3. Anti-allergy sprays

No published data are available on these treatments, which generally work by binding the allergen to the fibres of the article. Some products claim efficacy levels of up to 75% in reducing airborne allergens from a treated article. The effectiveness of these products may well be short lived.

3.3.8.4. Antimicrobial treatments

Recently, there has been a revived interest in the use of fungicides to control HDMs. This can be at least partially attributed to the reluctance of the general public to use insecticides or acaricides within their homes. Antimicrobial treatments are primarily thought to retard HDM growth, by preventing fungal development on skin scales, thus reducing the nutritional value of the food available to HDMs (see subsection 3.1.1.3). Studies using the fungicide Natamycin have reported mixed results (Bronswijk et al., 1987). However, de Saint Georges-Gridelet (1988) found the treatment to be at least somewhat effective, which may have been caused by higher application rates. Some products can be applied topically or applied to carpets and soft furnishings prior to their sale. Fibres used in bedding materials can also incorporate an antimicrobial agent onto or into their structure. Little is published in scientific journals on the effectiveness of these treatments, although laboratory experiments conducted by private research laboratories have shown them to be effective in reducing the rate of HDM population growth. It should be noted that not all antifungal products have the ability to control HDMs.

3.3.8.5. Dehumidifiers

Custovic and colleagues (1995) investigated the use of portable dehumidifiers in the control of HDMs and mite allergens. They found that a single portable dehumidifier placed centrally in a house was unable to reduce humidity to a level capable of retarding HDM population growth, although it was found to reduce condensation. It is likely that a more powerful dehumidifier will have a greater effect, although dehumidifiers can be noisy. However, switching them off at night can create brief spells of elevated humidity that are sufficient to allow HDM growth and survival (de Boer, Kuller & Kahl, 1998).

3.4. Conclusions

The association between exposure to HDM allergens and disease has prompted considerable research over the past several decades. Much is now known about HDM biology, physiology and ecology. The different and varying factors that affect HDM population growth are gradually being elucidated, as are the efficacies of different methods of control. Nevertheless, there is still much that is not known. For example, the currently available data sets that describe how hygrothermal conditions affect HDM life processes tend to be inconsistently gathered and incomplete, providing only partial coverage of the full range of hygrothermal conditions experienced by mites in real habitats. In particular, data on the effect of fluctuating conditions is lacking. Similarly, little is known about allergen production at different combinations of temperature and relative humidity or about mite movement, either within a habitat or in the form of mite migration between different habitats within a dwelling. On a broader scale, not enough is known about the size and distribution of HDM populations and allergen reservoirs at local, regional, national and international levels. Moreover, the surveys that have been carried out have not always monitored hygrothermal conditions adequately.

The need for continuing research is thus urgent. In the meantime, the following is suggested for climate, building construction, occupant behaviour, control methods, medical practitioners and other areas.

3.4.1. Climate

There are two suggestions with respect to climate.

1. In regions with suitable winter climates – that is, cold or dry or both – public health campaigns should make householders more aware that correctly implemented heating and ventilation can suppress HDM populations in all but certain times of year, but is particularly effective during the winter months, the winter culling thus achieved significantly reducing the risk of mite infestation.

2. Special training should be given to health professionals that make home visits – for example, to pregnant women and mothers with newly born children, who are especially vulnerable to mite allergen exposure – to provide advice about how best to environmentally control HDMs and achieve winter culling of mites, as well as other mite control measures.

3.4.2. Building construction

There are three suggestions with respect to building construction.

Public funding for reducing the number of dwellings with raised humidity levels due to low maintenance standards or to designer construction faults should be continued. Also, the possibility of more stringent building codes that minimize dampness rising in ground floors and basements should be considered.

Throughout the building industry (from legislative controls to design and construction), more emphasis should be placed on the means of providing adequate ventilation in winter and throughout the year through the use of suitable technology and improved window design. For both new and refurbished dwellings, thermal comfort and airtightness must not be achieved at the expense of adequate ventilation.

Particular priority should be given to the extraction of excess moisture from bathrooms and kitchens. Both rooms should always have doors capable of isolating them from the rest of the dwelling, and the use of automatic door closers should be encouraged.

3.4.3. Occupant behaviour

There are five suggestions with respect to occupant behaviour.

1. Public health campaigns should inform householders of the extent to which they themselves can influence humidity levels in the home.

2. Particular emphasis should be placed on the desirability of:

    a. ventilating and closing doors while cooking, and bathing or showering; and

    b. drying clothes only in a room with good ventilation and a door that can be closed (if not outside or in a tumble dryer).

3. Advice to allergic individuals should emphasize the importance of allergen-avoidance methods – for example, a stringent cleaning regime and the use of barrier covers.

4. More cleaning should be encouraged, especially late autumn cleaning and spring cleaning.

5. The use of fitted carpets should be discouraged, particularly in children's bedrooms, on solid ground floors and in basements. Also, it is essential that the hard floor that replaces the carpet be cleaned frequently.

3.4.4. Control methods

There are five suggestions with respect to control methods.

1. Education and guidelines are essential to enable people to integrate the methods of control and habitat modification available.

2. More in-depth research into individual control methods is needed.

3. As with many pesticides, further work may need to be done to establish the long-term effects on human health of using acaricides to control HDMs.

4. Alternatives to insecticides or acaricides should be considered in all control programmes.

5. Vacuum cleaners, especially in the homes of people with HDM and other allergies, should be fitted with HEPA filters.

3.4.5. Medical practitioners

There are two suggestions intended for medical practitioners.

1. More doctors should be able to conduct allergy tests, to identify patients whose asthma is caused or triggered by HDM or other allergens.

2. More advice needs to be available on allergen avoidance, control of HDMs and modification of the hygrothermal environment to prevent dwellings from being infested in the first place.

3.4.6 Other

There are two suggestions that do not fit in any of the above categories.

1. More research is needed into the relationship between levels of airborne allergens and asthma.

2. Further research is needed to fully establish the most effective methods of environmental control.

References[2]

Abbott J, Cameron J, Taylor B (1981). House dust mite counts in different types of mattresses, sheepskins and carpets, and a comparison of brushing and vacuuming collection methods. *Clinical Allergy*, 11:589–595.

Almqvist C, Pershagen G, Wickman M (2005). Low socioeconomic status as a risk factor for asthma, rhinitis and sensitization at 4 years in a birth cohort. *Clinical and Experimental Allergy*, 35:612–618.

Arlian LG (1975). Water exchange and effect of water vapour activity on metabolic rate in the dust mite *Dermatophagoides*. *Journal of Insect Physiology*, 21:1439–1442.

Arlian LG (1977). Humidity as a factor regulating feeding and water balance of the house dust mites *Dermatophagoides farinae* and *D. pteronyssinus* (Acari: Pyroglyphidae). *Journal of Medical Entomology*, 14:484–488.

Arlian LG (1989). Biology and ecology of house dust mites, *Dermatophagoides* spp. and *Euroglyphus* spp. *Immunology and Allergy Clinics of North America*, 9:339–356.

Arlian L (1991). House-dust-mite allergens: a review. *Experimental & Applied Acarology*, 10:167–186.

Arlian LG (1992). Water balance and humidity requirements of house dust mites. *Experimental & Applied Acarology*, 16:15–35.

Arlian LG, Bernstein IL, Gallagher JS (1982). The prevalence of house dust mites, *Dermatophagoides* spp., and associated environmental conditions in homes in Ohio. *The Journal of Allergy and Clinical Immunology*, 69:527–532.

Arlian LG, Dippold JS (1996). Development and fecundity of *Dermatophagoides farinae* (Acari: Pyroglyphidae). *Journal of Medical Entomology*, 33:257–260.

Arlian LG, Neal JS, Bacon SW (1998). Survival, fecundity and development of *Dermatophagoides farinae* (Acari: Pyroglyphidae) at fluctuating relative humidity. *Journal of Medical Entomology*, 35:962–966.

Arlian LG, Rapp CM, Ahmed SG (1990). Development of *Dermatophagoides pteronyssinus* (Acari: Pyroglyphidae). *Journal of Medical Entomology*, 27:1035–1040.

[2] In selecting the references, priority was given to peer-reviewed papers published in internationally recognized journals, followed by books or reports published by recognized experts or institutions, and then by peer-reviewed papers in published proceedings of international conferences. In one or two cases, reference is made to a PhD dissertation. When reference is made to unpublished material, this is clearly noted. Within each category, priority is given to research based on larger (rather than smaller) sample sizes.

Arlian LG, Veselica M (1981a). Effect of temperature on the equilibrium body water mass in the mite *Dermatophagoides farinae*. *Physiological Zoology*, 54:393–399.

Arlian LG, Veselica M (1981b). Re-evaluation of the humidity requirements of the house dust mite *Dermatophagoides farinae* (Acari: Pyroglyphidae). *Journal of Medical Entomology*, 18:351–352.

Arlian LG, Veselica M (1982). Relationship between transpiration rate and temperature in the mite *Dermatophagoides farinae*. *Physiological Zoology*, 55:344–354.

Arlian LG, Wharton G (1974). Kinetics of water active and passive components of water exchange between the air and a mite, *Dermatophagoides farinae*. *Journal of Insect Physiology*, 20:1063–1077.

Arlian LG et al. (1983). Seasonal population structure of house dust mites, *Dermatophagoides* spp. (Acari: Pyroglyphidae). *Journal of Medical Entomology*, 20:99–102.

Arlian LG et al. (1998). Population dynamics of the house dust mites *Dermatophagoides farinae, D. pteronyssinus* and *Euroglyphus maynei* (Acari: Pyroglyphidae) at specific relative humidities. *Journal of Medical Entomology*, 35:46–53.

Arlian LG et al. (2001). Reducing relative humidity is a practical way to control dust mites and their allergens in homes in temperate climates. *The Journal of Allergy and Clinical Immunology*, 107:99–104.

Biddulph P et al. (2007). Predicting the population dynamics of the house dust mite *Dermatophagoides pteronyssinus* (Acari: Pyroglyphidae) in response to a constant hygrothermal environment using a model of the mite life cycle. *Experimental & Applied Acarology*, 41:61–86.

Bischoff ERC, Fischer A, Liebenberg B (1992). Assessment of mite numbers: new methods and results. *Experimental & Applied Acarology*, 16:1–14.

Bischoff ERC et al. (1998). Mite control with low temperature washing – II. Elimination of living mites on clothing. *Clinical and Experimental Allergy*, 28:60–65.

Blythe ME (1976). Some aspects of the ecological study of the house dust mites. *British Journal of Diseases of the Chest*, 70:3–31.

Blythe ME, Williams JD, Smith JM (1974). Distribution of pyroglyphid mites in Birmingham with particular reference to *Euroglyphus maynei*. *Clinical Allergy*, 4:25–33.

Boardman B (1991). *Fuel poverty: from cold homes to affordable warmth*. London, Belhaven Press.

Cabrera P et al. (1995). Reduction of house dust mite allergens after dehumidifier use.

The Journal of Allergy and Clinical Immunology, 95:635–636.

Cameron MM (1997). Can house dust mite-triggered atopic dermatitis be alleviated using acaricides? *The British Journal of Dermatology*, 137:1–8.

Cesaroni G et al. (2003). Individual and area-based indicators of socioeconomic status and childhood asthma. *The European Respiratory Journal*, 22:619–624.

Chan-Yeung M et al. (1995). House dust mite allergen levels in two cities in Canada: effects of season, humidity, city and home characteristics. *Clinical and Experimental Allergy*, 25:240–246.

Charpin D et al. (1988). Asthma and allergy to house-dust mites in populations living in high altitudes. *Chest*, 93:758–761.

Chew GL et al. (1998). Limitations of a home characteristics questionnaire as a predictor of indoor allergen levels. *American Journal of Respiratory and Critical Care Medicine*, 157:1536–1541.

Colloff MJ (1986). Use of liquid nitrogen in the control of house dust mite populations. *Clinical Allergy*, 16:41–47.

Colloff MJ (1987). Effects of temperature and relative humidity on development times and mortality of eggs from laboratory and wild populations of the European house-dust mite *Dermatophagoides pteronyssinus* (Acari: Pyroglyphidae). *Experimental & Applied Acarology*, 3:279–289.

Colloff MJ (1991a). Population studies on the house dust mite, *Euroglyphus maynei* (Cooreman, 1950) (Pyroglyphidae). In: Schuster R, Murphy PW, eds. *The Acari: reproduction, development and life history strategies*. London, Chapman & Hall:497–505.

Colloff MJ (1991b). Practical and theoretical aspects of the ecology of house dust mites in relation to the study of mite mediated allergy. *Review of Medical and Veterinary Entomology*, 79:611–630.

Colloff MJ (1994). Dust mite control and mechanical ventilation: when the climate is right. *Clinical and Experimental Allergy*, 24:94–96.

Colloff MJ (1998). Distribution and abundance of dust mites within homes. *Allergy*, 53(Suppl. 48):24–27.

Colloff MJ, Spieksma FT (1992). Pictorial keys for the identification of domestic mites. *Clinical and Experimental Allergy*, 22:823–830.

Colloff MJ, Taylor C, Merrett TG (1995). The use of domestic steam cleaning for the control of house dust mites. *Clinical and Experimental Allergy*, 25:1061–1066.

Colloff MJ et al. (1992). The control of allergens of dust mites and domestic pets: a position paper. *Clinical and Experimental Allergy*, 22(Suppl. 2):1–28.

Committee on Damp Indoor Spaces and Health (2004). *Damp indoor spaces and health*. Washington, DC, National Academies Press.

Crane J et al. (1998). A pilot study of the effect of mechanical ventilation and heat exchange on house-dust mites and Der p 1 in New Zealand homes. *Allergy*, 53:755–762.

Crowther D et al. (2006). A simple model for predicting the effect of hygrothermal conditions on populations of house dust mite *Dermatophagoides pteronyssinus* (Acari: Pyroglyphidae). *Experimental & Applied Acarology*, 39:127–148.

Cunningham M (1996). Controlling dust mites psychrometrically – a review for building scientists and engineers. *Indoor Air*, 6:249–258.

Cunningham M (1998). Direct measurements of temperature and humidity in dust mite microhabitats. *Clinical and Experimental Allergy*, 28:1104–1112.

Custis NJ et al. (2003). Quantitative measurement of airborne allergens from dust mites, dogs, and cats using and ion-charging device. *Clinical and Experimental Allergy*, 33:986–991.

Custovic A, Chapman M (1998). Risk levels for mite allergens. Are they meaningful? *Allergy*, 53(Suppl. 48):71–76.

Custovic A et al. (1995). Portable dehumidifiers in the control of house dust mites and mite allergens. *Clinical and Experimental Allergy*, 25:312–316.

de Boer R (1990a). Effect of heat treatments on the house dust mite *Dermatophagoides pteronyssinus* and *D. farinae* (Acari: Pyroglyphidae) in a mattress-like polyurethane foam block. *Experimental & Applied Acarology*, 9:131–136.

de Boer R (1990b). The control of house dust mite allergens in rugs. *The Journal of Allergy and Clinical Immunology*, 86:808–814.

de Boer R (1998). Reflections on the control of mites and mite allergens. *Allergy*, 53(Suppl. 48):41–46.

de Boer R (2000). Explaining house dust mite infestations on the basis of temperature and air humidity measurements. In: Siebers R et al., eds. *Proceedings of mites, asthma and domestic design III*. Wellington, New Zealand, November 1997, Wellington Asthma Research Group, Wellington School of Medicine, Otago University:13–19.

de Boer R, Kuller K (1994). House dust mites (*Dermatophagoides pteronyssinus*) in mattresses: vertical distribution. In: Sommeijer R, Franke P, eds. *Proceedings of the Netherlands*

Entomological Society (NEV), Amsterdam, 1994. (Experimental and Applied Entomology 5:129–130).

de Boer R, Kuller K (1997). Mattresses as a winter refuge for house-dust mite populations. *Allergy*, 52:299–305.

de Boer R, Kuller K, Kahl O (1998). Water balance of *Dermatophagoides pteronyssinus* (Acari: Pyroglyphidae) maintained at brief daily spells of elevated air humidity. *Journal of Medical Entomology*, 35:905–910.

de Boer R, van der Geest LP (1990). House dust mite (Pyroglyphidae) populations in mattresses, and their control by electric blankets. *Experimental & Applied Acarology*, 9:113–122.

de Boer R, van de Hoeven WA, Stapel SO (1995). The decay of house dust mite allergens, Der p 1 and Der p 2, under natural conditions. *Clinical and Experimental Allergy*, 25:765–770.

de Saint Georges-Gridelet D (1987). Vitamin requirements of the European house dust mite, *Dermatophagoides pteronyssinus* (Acari: Pyroglyphidae), in relation to its fungal association. *Journal of Medical Entomology*, 24:408–411.

de Saint Georges-Gridelet D (1988). Optimal efficacy of a fungicide preparation, Natamycin, in the control of the house dust mite, *Dermatophagoides pteronyssinus*. *Experimental & Applied Acarology*, 4:63–72.

Dharmage S et al. (1999). Residential characteristics influence Der p 1 levels in homes in Melbourne, Australia. *Clinical and Experimental Allergy*, 29:461–469.

Douglas AE, Hart BJ (1989). The significance of the fungus *Aspergillus penicilloides* to the house dust mite *Dermatophagoides pteronyssinus*. *Symbiosis*, 7:105–116.

Eldridge H (1976). *Common defects in buildings*. London, Her Majesty's Stationery Office.

Emenius G, Egmar A, Wickman M (1998). Mechanical ventilation protects one-storey single-dwelling house against increased air humidity, domestic mite allergens and indoor pollutants in a cold climate region. *Clinical and Experimental Allergy*, 28:1389–1396.

EPA (2000). Pesticide Product recalled due to asthma concerns. In: *Annual report 2000 – Office of Pesticide Programs*. Washington, DC, United States Environmental Protection Agency:26 (http://www.epa.gov/oppfead1/annual/2000/2000annual.htm#ch6, accessed 2 April 2007).

Fain A. (1990). Morphology, systematics and geographical distribution of mites responsible for allergic disease in man. In: Fain A, Guerin B, Hart B, eds. *Mites and allergic disease*. Varennes-en-Argonne, France, Allerbio:11–152.

Fletcher AM et al. (1996). Reduction in humidity as a method of controlling mites and mite allergens: the use of mechanical ventilation in British domestic dwellings. *Clinical and Experimental Allergy*, 26:1051–1056.

Garratt J, Nowak F (1991). *Tackling condensation: a guide to the causes of, and remedies for, surface condensation and mould in traditional housing*. Watford, United Kingdom, Building Research Establishment (BRE Report 174).

Green et al. (1989). Reduction of house dust mites and mite allergens; effects of spraying carpets and blankets with Allersearch DMS, an acaricide combined with an allergen reducing agent. *Clinical and Experimental Allergy*, 19:203–207.

Gupta R et al. (2004). Burden of allergic disease in the UK: secondary analyses of national databases. *Clinical and Experimental Allergy*, 34:520–526.

Hallam C et al. (1999). Mite allergens in feather and synthetic pillows. *Allergy*, 54:407–408.

Hart BJ, Fain A (1987). A new technique for isolation of mites exploiting the difference in density between ethanol and saturated NaCl: qualitative and quantitative studies. *Acarologia*, 28:251–254.

Hart BJ, Guerin B, Nolard N (1992). In vitro evaluation of acaricidal and fungicidal activity of the house dust mite acaricide, Allerbiocid. *Clinical and Experimental Allergy*, 22:923–928.

Hart B, Whitehead L (1990). Ecology of house dust mites in Oxfordshire. *Clinical and Experimental Allergy*, 20:203–209.

Hart B et al. (2007). Reproduction and development of laboratory and wild house dust mites (*Dermatophagoides pteronyssinus* (Acari: Pyroglyphidae)) and their relation to the natural dust ecosystem. *Journal of Medical Entomology* (in press).

Harving H, Korsgaard J, Dahl R (1994). House dust mite exposure reduction in specially designed, mechanically ventilated "healthy" homes. *Allergy*, 49:713–718.

Harving H et al. (1991). House dust mite allergy and anti-mite measures in the indoor environment. *Allergy*, 46(Suppl. 11):33–38.

Hay D (1995). An *in situ* coring technique for estimating the population size of house dust mites in their natural habitat. *Acarologia*, 36:341–345.

Hill M (1998). Quantification of house-dust-mite populations. *Allergy*, 53(Suppl. 48):18–23.

Htut T (1994). A case study of bite reactions in man and domestic dust samples that implicate the house dust mite predator *Cheyletus malaccensis* Oudemans (Acari: Cheyletidae).

Indoor Environment, 3:103–107.

Htut T et al. (1996). A pilot study on the effect of one room MVHR units on HDM populations and Der p 1 in laboratory simulated bedrooms. *International Journal of Environmental Health Research*, 6:301–313.

Hyden ML et al. (1992). Benzyl benzoate moist powder; investigation of acaricidal activity in cultures and reduction of dust mite allergens in carpets. *The Journal of Allergy and Clinical Immunology*, 89:536–45.

Hyndman SJ et al. (2000). A randomized trial of dehumidification in the control of house dust mite. *Clinical and Experimental Allergy*, 30:1172–1180.

Kalra S et al. (1990). Airborne house dust mite antigen after vacuum cleaning. *Lancet*, 336:449.

Kinnaird CH (1974). Thermal death point of *Dermatophagoides pteronyssinus* (Trouessart 1897) (Astigmata, Pyroglyphidae), the house dust mite. *Acarologia*, 16:340–342.

Kniest F, Liebenberg B, Bischoff E (1989). Presence and transport of dust mites in clothing. In: *Science Writers' Press Conference Notes. American Academy of Allergy and Immunology 45th Meeting, San Antonio, 24 February – 1 March, 1989*. Milwaukee, WI, American Academy of Allergy and Immunology.

Korsgaard J (1979). Husstøvmider (Pyroglyphidae, Acari) i danske boliger [House dust mites (Pyroglyphidae, Acari) in Danish homes]. *Ugeskrift for Laeger*, 141:882–892 (in Danish).

Korsgaard J (1983a). Mite asthma and residency. A case-control study on the impact of exposure to house-dust mites in dwellings. *The American Review of Respiratory Disease*, 128:231–235.

Korsgaard J (1983b). House-dust mites and absolute indoor humidity. *Allergy*, 38:86–96.

Kuehr J et al. (1994). Natural variation in mite antigen density in house dust and relationship to residential factors. *Clinical and Experimental Allergy*, 24:229–237.

Larson D (1969). *The critical equilibrium activity of adult females of the house dust mite* Dermatophagoides farinae *Hughes* [PhD thesis]. Columbus, OH, Ohio State University.

Lintner TJ, Brame KA (1993). The effects of season, climate and air-conditioning on the prevalence of *Dermatophagoides* allergens in household dust. *The Journal of Allergy and Clinical Immunology*, 91:862–867.

Løvik M, Gaarder P, Mehl R (1998). The house-dust mite: its biology and role in allergy. A synopsis. *Allergy*, 53(Suppl. 48):121–135.

Lowe R (2000). Psychrometric control of dust mites in UK housing. *Building Services Engineering Research and Technology*, 21:274–276.

Mage D, Gammage R (1985). Evaluation of changes in indoor air quality occurring over the past several decades. In: Gammage RB, Kaye SV, eds. *Indoor air and human health.* Chelsea, MI, Lewis Publishers Inc.

Marsh R (1996). *Sustainable housing design: an integrated approach* [PhD dissertation]. Cambridge, United Kingdom, University of Cambridge.

Maunder J (1990). House dust mites in the work place. *Environmental Health Magazine*, Nov.:304–306.

McDonald LG, Tovey E (1992). The role of water temperature and laundry procedures in reducing house dust mite populations and allergen content of bedding. *The Journal of Allergy and Clinical Immunology*, 90:599–608.

McDonald LG, Tovey E (1993). The effectiveness of benzyl benzoate and some essential plant oils as laundry additives for killing house dust mites. *The Journal of Allergy and Clinical Immunology*, 92:771–772.

McEvoy M, Southall R (2004). Redefinition of the functions of a window to achieve improved air quality and energy performance in European housing. In: de Wit M, ed. *Built environments and environmental buildings.* PLEA [Passive and Low Energy Architecture] 2004 Conference Proceedings, 19–22 September 2004. Eindhoven, Netherlands, Technische Universiteit Eindhoven:837–842.

McIntyre DA (1980). *Indoor climate*. London, Applied Science Publishers Ltd.

McIntyre DA (1992). The control of house dust mites by ventilation: a pilot study. In: *Proceedings of 13th AIVC Conference: Ventilation for Energy Efficiency and Optimum Indoor Air Quality,* 15–18 September 1992, Nice, France. Coventry, United Kingdom, Air Filtration and Ventilation Centre:497–507.

Mitchell WF et al. (1969). House dust, mites and insects. *Annals of Allergy*, 27:93–99.

Mosbech H, Korsgaard J, Lind P (1988). Control of house dust mites by electrical heating blankets. *The Journal of Allergy and Clinical Immunology*, 81:706–710.

Mulla M et al. (1975). Some house dust control measures and abundance of *Dermatophagoides* mites in Southern California (Acari: Pyroglyphidae). *Journal of Medical Entomology*, 12:5–9.

Mumcuoglu Y (1977). House dust mites in Switzerland. III. Allergenic properties of the mites. *Acta Allergologica*, 32:333–349.

Munir AK (1998). Mite sensitization in the Scandinavian countries and factors influencing exposure levels. *Allergy*, 53(Suppl. 48):64–70.

Munir AK, Einarsson R, Dreborg SK (1993). Vacuum cleaning decreased the levels of mite allergens in house dust. *Pediatric Allergy and Immunology*, 4:136–143.

Nagakura T et al. (1996). Major *Dermatophagoides* mite allergen, Der 1, in soft toys. *Clinical and Experimental Allergy*, 26:585–589.

NHLBI (2002). *Morbidity and. mortality: 2002 chart book on cardiovascular, lung and blood diseases*. Bethesda, MD, National Heart, Lung and Blood Institute, National Institutes of Health, United States Department of Health and Human Services (http://www.nhlbi.nih.gov/resources/docs/02_chtbk.pdf, accessed 27 December 2006).

Nishioka K, Yasueda H, Saito H (1998). Preventive effect of bedding encasement with microfine fibers on mite sensitization. *The Journal of Allergy and Clinical Immunology*, 101:28–32.

Niven R et al. (1999). Attempting to control mite allergens with mechanical ventilation and dehumidification in British houses. *The Journal of Allergy and Clinical Immunology*, 103:756–762.

Palin S et al. (1993). Energy implications of domestic ventilation strategy. *Proceedings of the 14th AIVC Conference: Energy Impact of Ventilation and Air Infiltration*, 21–23 September 1993, Copenhagen, Denmark. Coventry, United Kingdom, Air Filtration and Ventilation Centre:141–148.

Pike AJ, Cunningham MJ, Lester PJ (2005). Development of *Dermatophagoides pteronyssinus* (Acari: Pyroglyphidae) at constant and simultaneously fluctuating temperature and humidity conditions. *Journal of Medical Entomology*, 42:266–269.

Platts-Mills TA, Chapman MD (1987). Dust mites: immunology, allergenic disease and environmental control. *The Journal of Allergy and Clinical Immunology*, 80:755–775.

Platts-Mills TA, de Weck A (1989). Dust mite allergens and asthma – a world-wide problem. Report of an International Workshop, Bad Kreuznach, Federal Republic of Germany, September 1987. *The Journal of Allergy and Clinical Immunology*, 83:416–427.

Platts-Mills TA et al. (1987). Seasonal variation in dust mite and grass pollen allergens in dust from the houses of patients with asthma. *The Journal of Allergy and Clinical Immunology*, 79:781–791.

Platts-Mills TA et al. (1996). Changing concepts of allergic disease: the attempt to keep up with real changes in lifestyles. *The Journal of Allergy and Clinical Immunology*, 98:S297–S306.

Prescott-Clarke P, Primatesta P, eds (1998). *Health survey for England: the health of young people '95–97*. London, Stationery Office (Department of Health Series HS No. 7; http://www.archive.official-documents.co.uk/document/doh/survey97/hs00.htm, accessed 27 December 2006).

Pretlove S et al. (2001). A combined transient hygrothermal and population model of house dust mites in beds. In: *Proceedings of Indoor Air Quality 2001 – Moisture, Microbes and Health Effects: Indoor Air Quality and Moisture in Buildings*, 4–7 November 2001, San Francisco, CA. Atlanta, GA, American Society of Heating, Refrigerating and Air Conditioning Engineers (http://www.ucl.ac.uk/bartlett-house-dustmites/Publications/Publications/ashrae.pdf, accessed 26 March 2007).

Pretlove SEC et al. (2005). A steady-state model for predicting hygrothermal conditions in beds in relation to house dust mite requirements. *Building Services Engineering Research & Technology*, 26:301–314.

Rao VR et al. (1975). A comparison of mite populations in mattress dust from hospital and from private houses in Cardiff, Wales. *Clinical Allergy*, 5:209–215.

Raw G et al. (1998). *Asthma, dust mites, ventilation and air quality: a field study of an environmental approach to reducing asthma*. Garston, United Kingdom, Building Research Establishment (BRE Internal Note N21/98).

Schober G et al. (1987). Control of house-dust mites (Pyroglyphidae) with home disinfectants. *Experimental & Applied Acarology*, 4:179–189.

Singh J (1994). The built environment and the development of fungi. In: Singh J, ed. *Building mycology management of health and decay in buildings*. London, E & FN Spon:1–21.

Solarz K (1997). Seasonal dynamics of house dust mite populations in bed/mattress dust from two dwellings in Sosnoweic (Upper Silesia, Poland): an attempt to assess exposure. *Annals of Agricultural and Environmental Medicine*, 4:253–261.

Sundell J et al. (1995). Ventilation in homes infested by house-dust mites. *Allergy*, 50:106–112.

Thompson P et al. (1991). The effect of a combined acaricide/cleaning agent on house dust mite allergen load in carpet and mattress. *Australian and New Zealand Journal of Medicine*, 21:660.

Tovey ER (1992). Allergen exposure and control. *Experimental & Applied Acarology*, 16:181–202.

Tovey ER, Mahmic A, McDonald LG (1995). Clothing – an important source of mite allergen exposure. *The Journal of Allergy and Clinical Immunology*, 96:999–1001.

Tovey ER, Marks GB (1999). Methods and effectiveness of environmental control. *The Journal of Allergy and Clinical Immunology*, 103:179–191.

Tovey ER, Woolcock AJ (1994). Direct exposure of carpets to sunlight can kill all mites. *The Journal of Allergy and Clinical Immunology*, 93:1072–1075.

Twiggs JT et al. (1991). *Dermatophagoides farinae* allergen levels from two different sources within the same home: evaluation of two different collection techniques. *Annals of Allergy*, 66:431–435.

van Asselt L (1999). Interactions between domestic mites and fungi. *Indoor and Built Environment*, 8:216–220.

van Bronswijk JEMH (1973). *Dermatophagoides pteronyssinus* (Trouessart, 1897) in mattress and floor dust in a temperate climate (Acari: Pyroglyphidae). *Journal of Medical Entomology*, 10:63–70.

van Bronswijk JEMH (1981). *House dust biology for allergists, acarologists and mycologists*. Zoelmond, Netherlands, NIB Publishers.

van Bronswijk JEMH et al. (1971). On the abundance of *Dermatophagoides pteronyssinus* (Trouessart, 1897) (Pyroglyphidae: Acarina) in house dust. *Researches on Population Ecology*, 13:67–79.

van Bronswijk JEMH et al. (1987). Effects of fungicide treatment and vacuuming on pyroglyphid mites and their allergens in mattress dust. *Experimental & Applied Acarology*, 3:271–278.

van der Heide S et al. (1997). Allergen avoidance measures in homes of dust-mite-allergic asthmatic patients: effects of acaricides and mattress encasings. *Allergy*, 52:921–927.

van Hage-Hamsten M, Johansson SG (1989). Clinical significance and allergenic cross reactivity of *Euroglyphus maynei* and other non-pyroglyphid and pyroglyphid mites. *The Journal of Allergy and Clinical Immunology*, 83:581–589.

van Strien RT et al. (1994). Mite antigen in house dust: relationship with different housing characteristics in the Netherlands. *Clinical and Experimental Allergy*, 24:843–853.

van Strien RT et al. (2004). The influence of air conditioning, humidity, temperature and other household characteristics on mite allergen concentrations in the northeastern United States. *Allergy*, 59:645–652.

Vaughan JW et al. (1999). Evaluation of materials used for bedding encasement: effect of pore size in blocking cat and dust mite allergen. *The Journal of Allergy and Clinical Immunology*, 103:227–231.

Vervloet D et al. (1982). Altitude and house dust mites. *The Journal of Allergy and Clinical Immunology*, 69:290–296.

Vervloet D et al. (1991). Epidemiologie de l'allergie aux acariens de la poussiere de maison. *Revue des Maladies Respiratoires*, 8:59–65.

Villaveces JW, Rosengren H, Evans J (1997). Use of laminar flow portable filter in asthmatic children. *Annals of Allergy*, 38:400–404.

Voorhorst R, Spieksma F, Varekamp H (1969). *House dust atopy and the house dust mite,* Dermatophagoides pteronyssinus (*Trouessart, 1897*). Leiden, Stafleu's Scientific Publishing Company.

Walshaw MJ, Evans CC (1987). The effect of seasonal and domestic factors on the distribution of *Euroglyphus maynei* in the homes of *Dermatophagoides pteronyssinus* allergic patients. *Clinical Allergy*, 17:7–14.

Wanner H et al. (1993). *Biological particles in indoor environments.* Luxemburg, Office for Official Publications of the European Communities (Indoor Air Quality and its Impact on Man, Report No. 12, EUR 14988 EN; www.inive.org/medias/ECA/ECA_Report12.pdf, accessed 27 December 2006).

Warner JA et al. (2000). Mechanical ventilation and high-efficiency vacuum cleaning: a combined strategy of mite and mite allergen reduction in the control of mite-sensitive asthma. *The Journal of Allergy and Clinical Immunology*, 105:75–82.

Wassenaar DP (1988). Effectiveness of vacuum cleaning and wet cleaning in reducing house dust mites, fungi and mite allergen in a cotton carpet: as case study. *Experimental & Applied Acarology*, 4:53–62

Watanabe M et al. (1995). Removal of mite allergens from blankets: comparison of dry cleaning and hot water washing. *The Journal of Allergy and Clinical Immunology*, 96:1010–1012.

Wharton GW (1976). House dust mites. *Journal of Medical Entomology*, 12:577–621.

Wharton GW, Duke KM, Epstein HM (1979). Water and the physiology of house dust mites. In: Rodriguez JG, ed. *Recent advances in acarology, Vol. 1.* London, Academic Press.

Wickman M et al. (1991). House dust mite sensitization in children and residential characteristics in a temperate region. *The Journal of Allergy and Clinical Immunology*, 88:89–95.

Wickman M et al. (1994). Reduced mite allergen levels in dwellings with mechanical exhaust and supply ventilation. *Clinical and Experimental Allergy*, 24:109–114.

Wilkinson T et al. (2002). Factors affecting the carrying capacity (*K*) of a mattress for the house dust mite *Dermatophagoides pteronyssinus* (Acari: Pyroglyphidae). Platform presentation at 11th International Congress of Acarology, Universidad Nacional Autonoma de Mexico, 8–13 September 2002, Merida, Mexico (http://eprints.ucl.ac.uk/archive/00002452/01/2452.pdf , accessed 26 March 2007).

Yoshikawa M (1980). Epidemic of dermatitis due to a cheyletid mite, *Chelacaropsis* sp. in tatami rooms, Part 1. *Annual Report of the Tokyo Metropolitan Research Laboratory of Public Health*, 31:253–260.

Zwemer RJ, Karibo J (1973). Use of laminar control devices as adjunct to standard environmental control measures in symptomatic asthmatic children. *Annals of Allergy*, 31:284–290.

4. Bedbugs

Harold J. Harlan, Michael K. Faulde and Gregory J. Baumann

Summary

Bedbugs have long plagued humans in their living environment. Historically, bedbugs were noted throughout the ages. After the Second World War, bedbug populations appeared to decline to a point where infestations by them were rare. In fact, just collecting specimens of bedbugs for instructing entomology became a difficult task, due to the rarity of this pest. Some people credit broadcast or wide area insecticide use with the decline in bedbug populations while others just believe it is the cyclical nature of pests that contributed to rare sightings. In the past 10 years, however, a resurgence of bedbugs has been noted. From anecdotal comments to reports of data showing multifold increases, it is clear that bedbugs are rising again. Some credit this resurgence to one or more of the following theories: loss of control products and changes in control practices for other pests that coincidentally controlled bedbug populations; increased travel; use of previously owned furniture and furnishings; and other theories. Bedbug control requires a fully integrated approach, as the bedbug is nocturnal, transient and elusive. Care must be taken to confirm proper identification, as other pests are similar in appearance and have quite different habits.

4.1. Overview of the biology, bionomics and distribution of bedbugs

4.1.1. Background

The common bedbug (*Cimex lectularius*), two tropical bedbugs (*Cimex hemipterus* and *Cimex rotundatus* and a few closely related species of blood-feeding true bugs (Hemiptera: Cimicidae) have been persistent pests to people throughout recorded history. They may have evolved as cave-dwelling ectoparasites of mammals (especially bats). As people moved from caves into tents, and then into houses, these bugs, especially the common bedbug, were probably brought along, too. With the widespread use of synthetic insecticides soon after the Second World War, bedbugs became very rare pests in many industrialized countries. By 1997, this species was so scarce in Canada, much of Europe and the United States that it was hard to find fresh specimens to use in teaching entomology classes (Snetsinger, 1997). Similar trends had been reported earlier for the United Kingdom, with a relatively constant or slightly declining level of public requests for control from 1967 to 1972 (Cornwell, 1974). Many current PMPs with 10 years of experience may never have even seen an active bedbug infestation. During the past eight years, a definite resurgence of common bedbugs has been reported in parts of Africa, Australia, Canada, some European countries and the United States. Sites infested by them have included homes, hotels, hostels and long-term care facilities (Cooper & Harlan, 2004; Doggett, Geary & Russell, 2004; Hwang et al., 2005; Johnson, 2005). The authors have also observed bedbug infestations in university dormitory housing.

4.1.2. Biology and bionomics

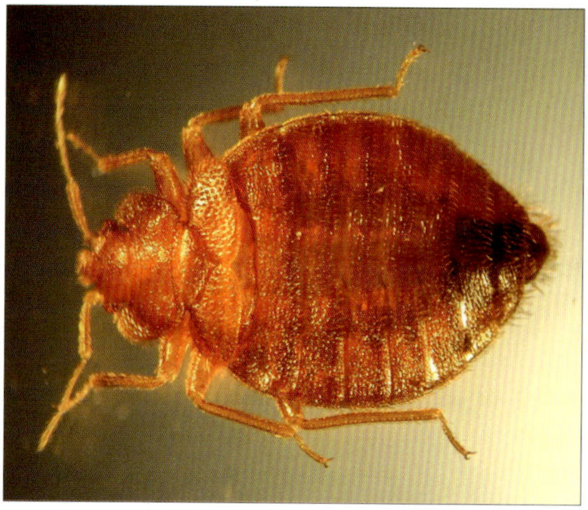

Fig. 4.1. Shape and characteristics of the bedbug

Source: Photo by H. Harlan.

Adults of the common bedbug are about 6–7 mm long, broadly oval, flat, brown to reddish-brown true bugs, with a three-segmented beak, four-segmented antennae, and vestigial wings (Fig. 4.1). They are flattened dorsoventrally and covered with short, golden hairs. The bedbug gives off a distinctive, musty, sweetish odour, consisting mainly of various aldehydes (such as *trans*-hex-2-enal, *trans*-oct-2-enal), which are produced by the defensive gland system located ventrally on the metathorax (Weatherston & Percy, 1978). They usually deposit undigested parts of earlier blood-meals in their hiding places, as a seemingly tar-like or rusty residue. Their abdomen tips are usually pointed in males and are more rounded in females (Fig. 4.2 and Fig. 4.3). Bedbugs feed only on blood, usually of mammals or birds, and mate by so-called traumatic insemination (Usinger, 1966; Stutt & Siva-Jothy, 2001). Their life-cycle, from egg to egg, may take four

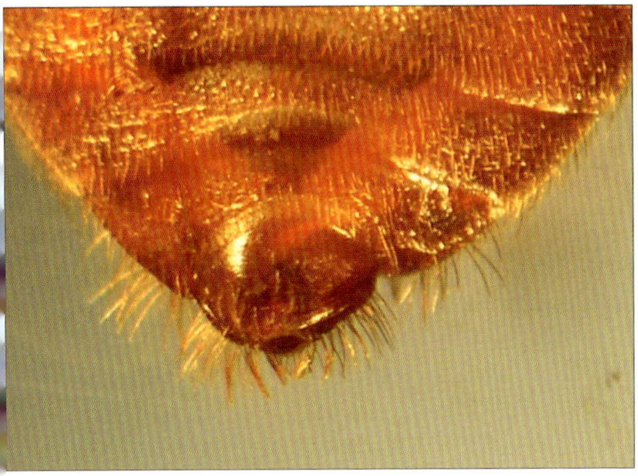

Fig. 4.2. Male bedbug abdomen, showing pointed characteristics of reproductive parts

Source: Photo by H. Harlan.

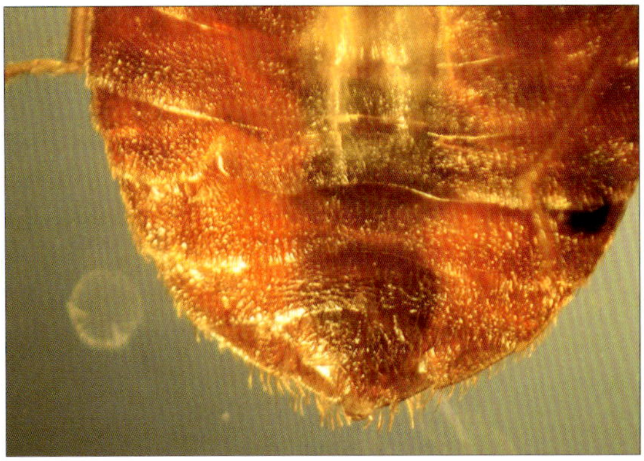

Fig. 4.3. Female bedbug, showing characteristic rounded abdomen

Source: Photo by H. Harlan.

to five weeks under good conditions – that is, 75–80% relative humidity and 28–32°C. They can survive and remain active at temperatures as low as 7°C, if they are held at an intermediate temperature for a few hours, but the temperature point of thermal death for them is 45°C (Wigglesworth, 1984). They have five nymphal developmental stages (instars), each needing at least one blood-meal to develop to the next instar. Bedbugs are nocturnal, but they will feed in full daylight when hungry. Females attach their small whitish eggs (about 1 mm long) to substrate surfaces, often in the crevices where they may hide in loose groups or clusters. Each female may lay 200–500 eggs in her lifetime, which may be 2 years or longer. Like fleas, these bugs often produce a series of bites in so-called rows or in fairly straight lines, usually along the edge beside an item of clothing or a bed sheet that was lying against their human host's skin at the time the bugs fed (Usinger, 1966; Krinsky, 2002).

4.1.3. Distribution in Europe and North America

The common bedbug can be found in all the temperate areas of the world – almost anywhere people have established dwellings and cities. They thrive in conditions of temperature and humidity that are considered comfortable for most people, and those same people usually provide them with ample blood-meals and plenty of choice harbourage nearby. The tropical bedbug, *C. hemipterus*, is distributed broadly throughout tropical and subtropical regions around the world, both north and south of the equator, but it requires a higher average temperature than does the common bedbug for its normal development and biological functions. It is seldom found in established infestations in continental Europe and is rarely found north of Mexico and Puerto Rico in the western hemisphere; however, occasional limited populations have been found in Florida. Several species of bat bugs (*Cimex* spp.) and swallow bugs (*Oeciacus* spp.) that may bite people are well established in most temperate areas of both Europe and North America (Usinger, 1966).

4.2. Resurgence of bedbug populations in Europe and North America

4.2.1. Evidence of resurgence in North America

In 1997, in the United States, reports of bedbug and specimens submitted for identification to the National Pest Management Association (NPMA) were limited to two separate infestations in two states. By September 2001, such submissions totalled 29 infestations in 18 states and the District of Columbia. By April 2004, bedbugs in samples submitted for identification had come from a total of 108 infestations in 4 Canadian provinces, 3 states of Mexico and 40 states in the United States. Also, public or media inquiries to the NPMA about bedbugs have gone from only 1 in 1997 to at least 14 in 2001and to more than 100 in 2005. Many pest management companies in the United States have seen major increases in services for bedbug infestations over the past few years (Krueger, 2000; Cooper & Harlan, 2004; Potter, 2004; Gooch, 2005). For example, one small company went from two unusual infestations in 2001 to an average of a call a day (Johnson, 2005). Also, a national company had an increase of 300% in bed bug control calls from 2000 to 2001, another 70% increase in 2002 and another 70% increase in 2003 (F. Meek, Orkin Pest Control, Atlanta, GA, personal communication, June 2005). Moreover, one small company, which formerly specialized in termite control, reported that more than 25% of its profit for 2004 came from bedbug treatments (Johnson, 2005). Resources invested by the NPMA, manufacturers, suppliers and individual pest management companies to train PMPs in bedbug management have increased greatly in the past two years (Cooper & Harlan, 2004; Potter, 2004; Gooch, 2005; Johnson, 2005). Furthermore, between the beginning and end of 2003, public health officials in Toronto, Canada, reported a 100% increase in phone complaints about bedbugs, a 100% increase in the number of commercial treatments for bedbugs in private residences and a more than 50% increased incidence of bedbugs in public shelters (Hwang et al., 2005).

4.2.2. Evidence of resurgence in Europe

In Europe, no precise data were available to document quantitatively the resurgence of the common bedbug. Nevertheless, observational reports from Germany and the United Kingdom claim a sharp increase in the frequency of infestations during the last decade. In the city of Berlin, Germany, more than a tenfold increase in the frequency of bedbug infestations has been reported, rising from five cases reported in the 1992 to 62 cases in 2002 and to 76 cases in 2004 (Bauer-Dubau, 2004). In the case of Berlin, the three major reasons for bedbug infestations are considered to be:

1. purchases of used furniture and electronic devices,
2. business travel,
3. holiday travel.

4.2.3. Resurgence in other parts of the world

Similar trends in resurgence have been occurring in several countries on other continents.

Gbakima and colleagues (2002) reported a very high prevalence of both common and tropical bedbugs (up to 98% of rooms infested) in camps for internally displaced persons in Freetown, Sierra Leone. In Australia, a government public health agency reported a 400% increase in bedbug complaints submitted during 2001–2004, compared with 1997–2000. They also reported increased interceptions of bedbugs (mainly in luggage) by national quarantine inspectors from 1986 to 2003, with 74% of those occurring from 1999 to 2003 (Doggett, Geary & Russell, 2004).

4.2.4. Other species of Cimicidae that can affect people

Besides the common bedbug, two other species of bugs in the family Cimicidae are synanthropic (ecologically associated with people) and historically well-known and significant pests of people in certain geographical regions. Those are the tropical bedbug species, *C. hemipterus*, which is distributed throughout the tropics, and *Leptocimex boueti*, which is limited to tropical western Africa.

Several other species of the Cimicidae, such as the European swallow bug (*Oeciacus hirundinis*), frequently enter human dwellings and feed on people opportunistically, especially when their normal hosts are eliminated or depart on normal natural migrations. The eastern bat bug (*Cimex adjunctus*), which feeds on several species of bats and is distributed mainly east of the Rocky Mountains in North America, frequently invades human dwellings and readily feeds on people. The cliff swallow bug (*Oeciacus vicarius*), which lives in the nests of cliff swallows (*Petrochelidon pyrrhonota*) and is distributed throughout the nearctic region (that is, the biogeographic region that includes the Arctic and temperate areas of North America and Greenland), will sometimes bite people when its hosts are removed or leave on annual migrations, or when people disturb these bugs in their habitat. The European swallow bug, which is associated with house martins (*Delicon urbica*) and is distributed throughout northern Africa, most of Europe and Turkey, will sometimes enter human dwellings and feed on people. The poultry bug (or Mexican chicken bug, *Haematosiphon inodorus*), will readily feed on people in Mexico and the south-western United States, but it is most commonly found associated with poultry nests and the nests of large raptors, such as hawks and eagles. The chimney swift bug (*Cimexopsis nyctalis*), which lives in the nests of Chimney swifts (*Chaetura pelagica*), is distributed throughout most of the eastern United States and has occasionally bitten people.

There are at least four species of bat bugs in the species group with *Cimex pipistrelli*. They collectively occur throughout most of Europe and feed mainly on local species of bats. These and other species in the *C. pipistrelli* group occasionally invade human dwellings, and one or more species have been reported to occasionally feed on people. The taxonomic status of this species group is presently somewhat unclear.

The western bat bug, *Cimex pilosellus*, which feeds on several species of bats and is found almost exclusively west of the Rocky Mountains in North America, has been repeatedly reported to invade human living spaces, but there are no confirmed reports of this species actually biting a person. Because multiple species of the Cimicidae can potentially occur in human dwellings, precise and careful identification is critical for proper implementation of pest control measures. Although it is rare that they attack people, these species are noted because infestation of human dwellings may occur.

4.2.5. Future prospects

Under prevailing conditions, it seems inevitable that bedbugs will continue to spread and cause increased problems. Since 1970, in high- and middle-income countries, the loss of non-repellent and longer residual insecticides, through regulatory actions, has made their control much more difficult. In the United States, anecdotal reports from PMPs about currently labelled products are variable and hard to verify or compare objectively. Some have reported effective control, while others claim outright failure from the same products. Also, screening for resistance to such products is rare. There are few published data or reports of screening for resistance on field populations of bedbugs, and no government or private agency is currently providing routine or even periodic screening of field populations of bedbugs in North America. A few laboratories in Europe and North America have only recently begun planning more extensive, controlled susceptibility (resistance) testing on bedbugs (E. Snell, Snell Scientifics, Barnesville, GA, personal communication, May 2005; O. Kilpinen, Danish Pest Infestation Laboratory, Lyngby, Denmark, personal communication, August 2005). Fletcher & Axtell (1993) published one of the most recent articles on the susceptibility of a population of lab-reared common bedbugs to several specific insecticide formulations, including some being used commercially at that time. At least five of the nine chemicals they tested, which include the active ingredients bendiocarb, carbaryl, dichlorvos, malathion and tetrachlorvinphos, have had their product labels changed, following the United States Food Quality Protection Act of 1996, and those products are no longer available for use in human dwellings. In addition to these factors, significant changes in application technologies against other household pests, such as cockroaches and ants, may also have unintentionally eased incidental control of bedbugs (Potter, 2004; Gooch, 2005).

4.3. Implications for public health

4.3.1. Obligate blood feeders

All species of bugs in the family Cimicidae are obligate blood feeders – that is, they survive only by feeding on blood – and many have limited host specificity. They only consume blood from a host, usually a mammal (such as human beings and bats) or bird. They must take at least one blood-meal of adequate volume in each active life stage (instar) to develop to the next instar, or to reproduce. There are five immature instars, and each one may feed multiple times if hosts are readily available. Adults of the common bedbug may feed every 3–5 days throughout their estimated typical 6–12-month lifespan. The acts involved in feeding to acquire each blood-meal – that is, biting a host – can cause both physical and psychological discomfort, as well as local allergic skin reactions to the salivary proteins injected (Feingold, Benjamini & Michaeli, 1968).

4.3.2. Potential as vectors of human pathogens

Common bedbugs have been found to naturally contain 28 human pathogens, but they have never been proven to transmit biologically or mechanically even one human pathogen (Usinger, 1966; Burton, 1968) – specifically hepatitis C (Silverman et al., 2001) and HIV (Webb et al., 1989). Nevertheless, shedding of viral DNA fragments in faecal matter and transstadial (across life stage) transmission of hepatitis B virus seem to support the possibility of mechanical transmission by contaminated faeces, or when bugs are crushed during feeding onto abraded skin by a susceptible person (Jupp et al., 1991; Blow et al., 2001). The study by Jupp and colleagues (1991) clearly indicated that common bedbugs do not biologically transmit hepatitis B. Also, it is still unclear whether or not the reported induction of skin papillomas in the European rabbit (*Oryctolagus cuniculus*) subsequent to continuous exposure to bites of common bedbugs irradiated by gamma rays (el-Mofty, Sakr & Younis, 1989) is caused by a viral pathogen.

One species of Cimicidae, the cliff swallow bug (so far, found occurring naturally infected, only in western North America), has been proven to transmit at least two identifiable virus entities in the western equine encephalitis complex: these are the Fort Morgan virus (FMV) and a distinct strain of FMV, called Buggy Creek virus (Hayes et al., 1977; Calisher et al., 1980; Brown & Brown, 2005). Transmission has thus far only been reported to occur from infected swallow bugs, mainly adult bugs that usually live in or on the swallows' nests over the winter until the next year's susceptible hatchlings of cliff swallows. So far, no risk to people from this epizootic virus cycle has been established. The bugs are not very mobile and tend to stay at, on or in the swallow nests and do not readily move to a different nest that is more than a short distance away (Foster & Olkowski, 1968). A few individual bedbugs of this species have been found very rarely in local small rodent nests and occasionally in a nearby nest of a European (or barn) swallow (*Hirundo rustica*). This species has been reported to have fed on a person, and it readily feeds on mice in laboratories (Usinger, 1966). Although some species of the Cimicidae that seek human beings over other animals have not been investigated very thoroughly, it is unlikely that any of them pose a threat as a human pathogen vector.

In Europe, infestations of human dwellings by the European swallow bug often occur, especially during the wintertime, when their normal hosts have migrated to Africa. Abandoned swallow nests may harbour hundreds of overwintering swallow bugs that search for blood hosts during warmer winter weather conditions. Within a radius of up to 5 m, swallow bugs may infest buildings through any opening, and they feed readily on people. In an extensive study in Colorado, cliff swallows and house sparrows (*Passer domesticus*), which live within the swallow nesting colonies, were the main vertebrate hosts for maintenance and amplification of FMV. However, the presence of fairly large populations of these bugs, and their transmission of FMV, reportedly had no significant impact on the health or reproduction of local populations of cliff swallows, house sparrows or barn swallows in co-located or adjacent breeding sites (Scott, Bowen & Monath, 1984).

4.3.3. Bites and health effects

Although their bite is often nearly undetectable, the saliva of bedbugs contains biologically and enzymatically active proteins that may cause a progressive immunogenic and allergenic reaction to repeated biting. Depending on the combined biting intensity and frequency, there are typically five stages of symptoms, including no reaction, delayed reaction, delayed plus immediate reaction, immediate reaction only and, finally, no reaction. A common hypersensitivity response that follows bedbug bites is papular urticaria, a disorder manifested by chronic or recurrent papules. Rarely, a disseminated bullous eruption with systemic reaction may occur (Liebold, Schliemann-Willers & Wollina, 2003). Typical symptoms include a raised, inflamed, reddish wheal at each feeding site. Such wheals may itch very intensely for several successive days.

Immediate immune reactions may appear from 1 to 24 hours after a given bite and may last 1–2 days, but delayed immune reactions usually first appear 1–3 days after a bite and may last 2–5 days (Feingold, Benjamini & Michaeli, 1968). In the cases of delayed reactions, it is often difficult to pinpoint the source, as the victim may have visited several different rooms in several locations over that period. Consequently, upon manifestation of symptoms, the previous night's lodging may be suspected; however, the bites may have occurred several nights prior. People bitten frequently by these bugs may develop a so-called sensitivity syndrome, which may include nervousness, nearly constant agitation (jumpiness), and sleeplessness. In such cases, either removing the bedbugs (either physically or by the use of an insecticide) or relocating the person has caused the syndrome to disappear over time. Several different species of Cimicidae may bite people, including tropical bedbugs, some bat bugs and some swallow bugs, as mentioned previously in section 4.2 (Ryckman & Bentley, 1979). Also, a social stigma may be associated with having an infestation of bedbugs (Usinger, 1966; Krinsky, 2002).

Currently, there is no requirement to report bedbug infestations to any public health or other government agency at any level. Medical clinicians, however, have reported the following significant symptoms as due to common bedbug bites:

- serious local redness and intense itching, both immediately and after several days delay (Sansom, Reynolds & Peachey, 1992);
- disseminated bullous eruption with systemic reaction (Liebold, Schliemann-Willers & Wollina, 2003); and
- true anaphylaxis, which has been misinterpreted as coronary occlusion (Parsons, 1955).

Besides the effects of direct bites, airborne common bedbug allergens that are always released during infestations may produce bronchial asthma. Within a group of 54 asthmatic Egyptian patients, 37.1% reacted positively to a common bedbug head and thorax extract, and 50.1% reacted positively to an abdominal common bedbug extract (Abou Gamra et al., 1991). Numerous routine bedbug bites can contribute to anaemia and may even make a person more susceptible to common diseases (Usinger, 1966; Snetsinger,

1997). Some people can develop a general malaise from numerous bedbug bites; that, along with the loss of sleep and extreme itching of bug bites, can lower a person's vitality and make individuals listless and almost constantly uncomfortable.

Because of the media attention given to bedbugs in recent years, people may more readily suspect bedbug bites as a most likely cause of a rash or skin irritation. Physicians commonly will not rule this out, as it is not possible to positively identify a skin irritation as a bedbug bite. It is common for those experienced with bedbugs to look at an irritation and eliminate bedbugs as a possible cause; however, just examining an irritated area will not determine conclusively the source of the bite. Building inspectors that search for bedbugs should make sure that an infestation is confirmed via inspection prior to any action or, at least, determine that the victim is not suffering from delusory parasitosis, where he or she believes that they have been bitten by an invisible pest. Bedbug bites can be confused with numerous dermatological conditions, including those of psychological origin (such as delusory parasitosis). If the person affected is complaining about bites, but an entomological source cannot be found, that person should seek medical attention.

4.3.4. Importance as pests

Because they are very small, nocturnal, seek cryptic harbourages, and can detect and avoid many chemicals (including cleaning agents), common bedbugs are often hard to control. Complete elimination of an established bedbug population is nearly impossible to accomplish in a single visit by most PMPs. They are easily transported on or in luggage, furniture, boxes, or even on clothes. They are very thin, except just after a blood-meal, and can fit through or hide in very narrow cracks. In the absence of hosts, adults can live without feeding for several months (to more than a year) and nymphs for three months or longer. This pest has substantially increased in importance as a result of the public's fear of bedbugs, their characteristic odour and their feeding behaviour. Because bedbug feeding can cause serious physical distress (mainly persistent, intense itching, which varies with each individual), demands for prompt, effective control are increasing significantly, too.

4.3.5. Economic impact

There are neither precise nor detailed records of the costs of bedbug control efforts nor of any of the related costs to people bitten in residential, commercial or institutional housing. Costs for the hospitality industry include:

- increased laundry expenses
- replacement of bedding and furniture
- structural cleaning and physical modifications
- lost revenue from negative publicity
- insurance claims and lawsuits.

As an example of one of the simplest cases, a reasonable cost estimate for the inspection and effective treatment of a small infestation, involving a bed and two nearby floor–wall baseboards in only one bedroom of a separate home, would be about US$ 300 in most areas of Canada, most western European nations and the United States. This is based on the following: three hours of detailed inspection time, concise customer education and a very limited insecticide application by a well-trained PMP on a single visit to the site.

Costs, however, will vary. In hotels and other multi-unit structures, for example, infestations are almost never that limited and usually require significantly more inspection, control time and effort. Because infestations may go unnoticed in these facilities, there is a much greater chance of so-called callbacks – that is, return services. Because the general public lacks knowledge of bedbug feeding activity, victims nearly always seek medical attention, usually incurring additional costs for diagnosis, or at least costs for symptomatic medical treatments. At least 17 of 65 homeless shelters in Toronto, Canada, spent an average of Can$ 3085 each to address bedbug problems in 2004 (Hwang et al., 2005).

Litigation costs can be more variable and even less well documented. Examples of judgments range from one award of US$ 382 000 down to awards in the range of US$ 20 000 plus expenses (Doggett, Geary & Russell, 2004; Gooch, 2005; Johnson, 2005). In early 2006, a lawsuit was filed against a hotel, seeking US$ 20 million in damages. It is impossible to estimate the cost of any related subsequent actions or the anxieties of bitten individuals.

4.4. Poverty, housing and bedbug infestations

Bedbugs will infest human dwellings in all social and economic groups. A number of factors more commonly associated with poverty, however, are ideal for bedbugs. Buildings and other dwellings that are crowded, cluttered and in need of repair offer bedbugs many places to hide very near their food source. Under such conditions, it is very hard to eliminate a population of bedbugs from any room or building. Also, deteriorating structures with warped woodwork or floors, loose tiles or wallpaper and large cracks around doors, windows or ductwork may be nearly impossible to seal or to treat effectively by any physical or insecticide control technique.

One key type of housing where there are widespread reports of infestations is migrant housing. Various media sources have reported that the crowded, cluttered conditions of poorly housed migrant workers have led to bedbug infestations. The cost of controlling these infestations professionally is usually more than such building owners or occupants can afford. As a result, they often try to control the infestations themselves and are seldom very effective, which further depletes already limited financial and physical resources. Apparently, because these infestations can become an endless, cyclical problem for those living in poverty, some authors have claimed that bedbugs may help cause poor living conditions – not just being typical of them.

Dispersal of bedbugs from one dwelling to another is usually passive, with the bugs or their eggs being carried in or on pieces of furniture, bedding, luggage, clothing, electronic devices or cardboard boxes. Because leasing furniture and purchasing used items are much more common in poorer communities, these practices probably help to rapidly and repeatedly spread bedbugs to new sites and to redistribute them to places from which they may have been eliminated earlier.

Large multi-unit buildings common to poor areas can be very hard to rid of bedbugs. Once bedbugs become established, any control effort that does not include checking the whole building at nearly the same time, along with a coordinated occupant education and treatment effort (as needed), will usually fail, because the bugs will frequently move away from any partially treated and potentially repellent active sites into adjacent rooms. Their movements are generally unencumbered, because they readily move through wall voids and along utility lines, heating ducts, elevator shafts, and laundry and mail chutes.

4.5. An integrated approach to bedbug management

With current pest management practices, bedbugs can only be effectively managed by combining as many different strategies, techniques and products as possible for any particular set of local conditions or infestations. This makes the bedbug an ideal candidate for IPM practices, which include:

- inspection
- identification
- establishment of threshold levels – generally, no bedbugs are acceptable
- incorporation of two or more control measures
- monitoring the effectiveness of controls.

Particular steps may differ for given pests or unique local conditions and may take into account the need to quickly reduce a given life stage of pests that pose an impending serious health risk.

4.5.1. Conducive environmental conditions

Conditions under which the common bedbug thrives include:

- an adequate supply of food (available blood-meal hosts)
- plenty of small cracks or narrow harbourage spaces within about 1.5m of a host
- ambient temperatures within a few degrees of 28–32°C
- average relative humidity of 75-80%.

Currently occupied, cluttered bedrooms with little air movement are ideal. Sanitation alone will not eliminate a population of bedbugs; however, eliminating clutter, removing all accumulated dirt and debris and sealing cracks and crevices will reduce available harbourages, will make it easier to detect remaining live bedbug populations and will increase the probability that any further treatment may succeed.

4.5.2. Inspection, detection and education

4.5.2.1. Inspection

Detailed inspection by a qualified person is the most essential basic element at the start of any effective effort to control bedbugs. The bugs must be detected promptly, correctly identified and at least a rough estimate of the extent of harbourage sites must be determined as rapidly as possible. No currently known device or technique is available to effectively attract or trap bedbugs, so a thorough visual inspection must be done. Certain pyrethrin-based flushing agents can be used to help stimulate the bugs to move around and make them much easier to detect in limited populations. Once detected, correct identification of the pest bugs is important, to focus further inspections and facilitate the application of control techniques or products that are precise and limited in scope – for example, for treatment of cimicid bugs that feed mainly on certain species of bats or birds.

Fig. 4.4. Bedbug found on a mattress dust cover

Source: Photo by G. Baumann.

4.5.2.2. Detection

Typical actions and signs that can accurately detect a bedbug infestation include:

- seeing or collecting live bugs (Fig 4.4);
- smelling the characteristic odour, finding their eggs or so-called cast skins in harbourages or near feeding sites;
- finding dark faecal deposits or lighter rust-like spots on bed linens or in traditional harbourages (Fig. 4.5); and
- noting and recording where and when bite victims know (or think) they were bitten.

The use of both sticky traps and insecticidal aerosols that flush out or excite the bugs can potentially augment monitoring. Any combination of two or more of these signs can help verify the infestation and determine the distribution and prevalence of the bugs. For species that feed mainly on bats or birds, detecting and locating the nests of their local hosts is important. The presence of typical hosts may be an early indication of a developing infestation.

Fig. 4.5. Bedbug staining is typical of a dropped partially digested blood-meal

Source: Photo by H. Harlan.

4.5.2.3. Education

Educating dwelling occupants affected by bedbug infestations is essential to ensure that they actively and voluntarily cooperate in any control programme or effort needed. These occupants are usually the ones who must improve and maintain sanitation, reduce and minimize clutter, and (perhaps) seal harbourages, to exclude or restrict the movements of the bug population. It can be helpful if dwelling occupants affected by bedbug infestations understand as fully as possible the bugs – their biology and behaviour – and the strategies and techniques proposed or being used to control them. To foster such understanding, educational efforts may include verbal explanations, answering questions, information posted on an Internet web site, or at least a concise printed handout in a language that people can read and understand. Good communications with homeowners, housing managers and any relevant government agencies should be maintained throughout a bedbug control programme.

4.5.3. Physical removal and exclusion

A number of possibilities exist for physically removing or excluding bedbugs.

4.5.3.1. Physical removal

Bedbugs can be physically removed from exposed harbourages or resting sites, such as edges of a box spring or mattress seams, by sucking them up with a vacuum cleaner. Using a HEPA-filtered vacuum, which removes more than 99% of all particles greater than 0.3μm in diameter, would ensure that allergens associated with bedbugs or their debris were being removed concurrently. Vacuuming will usually kill a large portion of those bugs and can be done at the same time as an inspection, eliminating immediately a significant portion of the pest population. Bedbugs might also be lifted from exposed resting sites with commercially available tape or by hand, or just brushed off directly into a container of rubbing alcohol or soapy water (Potter, 2004; Gooch, 2005).

4.5.3.2. Exclusion

Sealing access to harbourages can effectively isolate bedbug populations. Bedbugs have specially adapted piercing–sucking mouthparts, and three-segmented, structurally primitive tarsi (the terminal segments of the leg) with claws. That makes them incapable of chewing or clawing through even a very thin layer of sealant or an unbroken layer of paper or cloth. Sealing a layer of almost any material in place, so that it completely covers the opening of any harbourage, can stop bedbugs from passing through. If any bedbug is thus effectively sealed inside a void or harbourage, it could be permanently removed from the pest population. Even if such a bedbug were to live for another year, or two years (or longer), it must die there if that space is never unsealed while the bug is still alive. Just sealing most of the known openings between a harbourage and the bugs' usual host access site(s) will restrict the bugs' movements and help temporarily reduce the intensity of their feeding. Enclosing clothes and other items in plastic bags and similarly tightly sealed containers can greatly reduce the availability of harbourage sites.

4.5.3.3. Mattress covers

Commercially available plastic covers, at least 0.8 mm thick and usually having a zippered edge, can completely encase a mattress or box spring and stop any bedbugs harbouring in either of them from further access to bite a host using that bed. Such covers were first developed and marketed as a measure to help reduce human exposure to HDM allergens that emanated from mattresses, but they can work well to isolate bedbugs within (or keep them out of) such items. If no such covers are readily available, any plastic of similar thickness and strength can be used to completely cover a mattress or box spring, and it can be sealed tightly shut with any durable, flexible tape, such as filament tape or duct tape (Cooper & Harlan, 2004).

4.5.4. Physical elimination techniques

Heat, cold and steam are used in physical techniques for eliminating bedbugs.

4.5.4.1. Heat

Heating infested rooms or whole buildings to temperatures of at least 45°C, the thermal death point of common bedbugs, has been used to try to control bedbugs since the early 1900s. For a heat treatment to be effective, it is critical to attain a high enough temperature, low enough relative humidity and minimum length of time at those combined conditions. Some species of stored product pest beetles, which are considered to be very hard to kill, have been shown to be eliminated by exposure to a combination of 49–52°C and 20–30% relative humidity for 20–30 minutes (Dosland, 2001). Heat treatments, however, do not prevent reinfestations, and bedbugs can reoccupy any site so treated immediately after temperatures return to ambient levels. Of concern in particular situations, when using this technique for physically eliminating bedbugs, is the potential physical distortion of structures or their contents, as well as flammability risks for some kinds of heat sources (Usinger, 1966).

4.5.4.2. Cold

If bedbugs are kept cold enough long enough, exposure to cold temperatures can kill them. Bedbugs can tolerate –15°C for short periods and, if acclimated, they can survive at or below 0°C continuously for several days (Usinger, 1966). Cold treatments of rooms or buildings to control bedbugs have not been well studied or used often, but freezing of furniture or other items within containers or chambers may be a practical alternative for limited infestations or to augment other control measures.

4.5.4.3. Controlled atmospheres

In a small series of very preliminary laboratory tests conducted by the German Federal Environmental Agency, all life stages of common bedbugs were reportedly killed within 24 hours or less by constant exposure to very high concentrations of carbon dioxide gas at ambient atmospheric pressure, but they were not affected very much by high concentrations of nitrogen gas under those same conditions (Herrmann et al., 1999). Further precise testing of such a strategy may be warranted.

4.5.4.4. Steam

Some pest managers have effectively used steam treatments to quickly eliminate live bugs and their eggs from the seams of mattresses and other cloth items. Effective use of this technique requires practice and care. Manufacturer's instructions about the steam generating devices' operation, maintenance and safety precautions must be followed carefully. To be effective, the steam emission tip must be about 2.5–3.8 cm from the surface being steamed. If the tip is too far away, the steam (water vapour) may not be hot enough to kill all the bedbugs and eggs on such a surface. If the tip is too close, excess moisture may be injected into the treated material, and that can potentially lead to other problems – for example, facilitating the survival and increase of dust mite populations and creating an environment for the growth of surface molds.

4.5.4.5. Sticky monitors

Insect monitors with an adhesive layer on a flat cardboard backing are a simple means to detect many types of crawling insects. They have also been recommended to augment other techniques for increased control of some wandering spiders. Although bedbugs often get caught on such monitors, many recent reports from pest control technicians in North America have indicated these are not very effective at detecting (much less increasing the level of control of) small to moderate populations of bedbugs in rooms where other signs are obvious, where bugs are easily found by direct observation and where people are being bitten routinely. Based on this evidence, both the impact of such devices on the control of bedbugs and their reliability as a surveillance tool for the bugs are poor.

4.5.5. Pesticide applications

Currently, the exclusive use of non-insecticide control products and techniques is not effective or efficient enough to be practical. Even their use as a primary means for controlling or eliminating an established bedbug population is ineffective. Still one of the most effective, practical and quickest ways to reduce the size of an established bedbug infestation is the use of a precisely placed (but thorough) application of a properly labelled, registered and adequately formulated residual insecticide. Effective control usually consists of applying interior sprays or dusts to surfaces that the bedbugs crawl over to reach the host, as well as applying them to cracks and crevices where they rest and hide. Microencapsulated formulations and dust formulations have a longer residual effect than other formulations. Also, both synergized and natural pyrethrins are used. Synergized pyrethrins not only show high lethal activity against the bugs, but they also show the ability to flush them out, allowing quicker analysis of the infested area. Moreover, the addition of natural pyrethrins at 0.1–0.2% v/v to organophosphate, carbamate or microencapsulated insecticide formulations will increase efficacy by irritating the bedbugs and initiating an excitatory effect that causes them to leave their hiding places, thereby increasing exposure to the fresh insecticide layer.

Modified diatomaceous earths with hydrophobic surfaces can also be used to treat cracks and crevices. Retreatment, however, is essential and should be carried out at not less than two-week intervals until the population has been eradicated.

Products listed in this section are for illustrative purposes only. The choice of insecticide products and specific application techniques can depend on many factors, which include the following:

- the physical locations and structural details of the bugs' harbourages
- the availability of the product
- the products' own labels, which can vary greatly from country to country
- exact local conditions
- any applicable national and local laws or regulations
- many different, related official recommendations.

True fumigation of small volumes or limited amounts of furniture, clothing, or other personal items can kill all stages of bedbugs present, possibly including bedbug eggs. Such a treatment, however, would not prevent reinfestation immediately after the fumigant had been sufficiently removed for the items treated to be used again. True fumigation of a whole building should be equally effective at killing all mobile stages of bedbugs present, but would not prevent a reinfestation and would seldom be needed, practical or affordable (WHO, 1982; Snetsinger, 1997; Gooch, 2005).

4.5.5.1. Surface treatments

Limited, precisely placed treatments of selected surfaces, using properly formulated residual insecticides, can provide at least a temporary lethal (or perhaps repellent) barrier and thus can help reduce biting of nearby hosts.

4.5.5.2. Impregnated fabric and bednets

Fabrics and bednets that are factory- or self-impregnated with licensed and commercially available formulations of residual insecticides can help deny bedbugs access to hosts and may even kill some of the bugs that crawl across them. This can be economical, since some spray, dipping or coating formulations of products containing permethrin will remain on clothing at their full applied strength through six or more launderings, or even for the life of the fabric (Lindsay et al., 1989; Faulde, Uedelhoven & Robbins, 2003). This treatment is done during the manufacture of the fabric and not by a pest control company when the fabric is in place. There is a recent report of pyrethroid resistance to the use of treated bednets in a population of tropical bedbugs (Myamba et al., 2002).

4.5.5.3. Ultra-low-volume aerosols or foggers

Insecticide products currently labelled as ultra-low-volume (ULV) aerosols or foggers have little or no residual effects and act as a flushing agent. Most of them can stimulate bedbugs hidden in harbourages to become active and move out into the open, allowing them to be noticed more easily. Some of those bugs may be killed by prolonged or repeated exposures, but control would be successful for only a small percentage of the bugs flushed from harbourages.

4.5.5.4. Crack-and-crevice treatments

Because bedbugs habitually hide clustered together in cracks and narrow harbourages, precisely applied crack-and-crevice treatments may often be among the most effective control techniques to use against them. Several different active ingredients and formulations have been licensed and are currently used against bedbugs, and a variety of insecticide formulations and devices must be used to treat infested harbourages. As per label directions, applications of dust formulations should be used in electrical outlet boxes and in other places where it is desirable to use a minimum-risk, long-lasting insecticide.

4.5.5.5. Use of pest management products

The use of a properly labelled or licensed insecticide formulation in combination with the IPM principle recommended for control of bedbugs will often be the quickest, most practical and (in some cases) possibly the only affordable or viable control option. Most residual insecticides that are labelled (or licensed) or recommended for the control of bedbugs are only intended for use as a crack-and-crevice treatment, according to directions on their product labels. The vast majority of such currently available products are either natural or synthetic pyrethroids; some of these products include synergistic chemicals as well.

Recommendations in Canada, Europe and the United States for the control of bedbugs cover pyrethrins (such as 6 ml/m^2 synergized natural pyrethrum), pyrethroids (such as permethrin), organophosphates (such as 50 ml/m^2 fenthion, 4–8% v/v) and carbamates (such as 200 ml/m^2 propoxur, 1% w/v). For any class of insecticide product active ingredient included in products currently labelled for use to control bedbugs, type I pyrethroids are among the least toxic. The application concentration of pyrethroid active ingredients registered as bedbug control products is seldom more than 1% v/v. Also, most non-residual pyrethroids are rapidly detoxified by normal exposure to sunlight, oxygen in air, common household cleaning agents and microbes routinely found on nearly all exposed surfaces. For monitoring and for improving control results, certain aerosol formulations of pyrethroids are sometimes used to stimulate (flush) hidden bugs, so they will move out of cryptic harbourages, making their presence more readily detectable; however, such aerosol products seldom kill many of the bedbugs that emerge.

Dust formulations of insecticidal products currently labelled for use in controlling bedbugs are of low toxicity and pose little risk to people when used in accordance with label directions.

IGRs are used to alter bedbug reproductive processes. Such regulators have essentially no effect on vertebrate systems (as applied), because of their modes of action and because of the very small amount of active ingredient applied when used according to label directions. They can, however, have a significant impact on bedbug fertility and the success of egg hatching (Takahashi & Ohtaki, 1975).

Currently, the vast majority of pesticide products labelled for use on mattresses contain very low concentrations of very low-toxicity and low-volatility pyrethroid formulations or else their active ingredients are such non-insecticide products as silica aerogel or diatomaceous earth.

Any such licensed pest management products for use on mattresses, as well as nearly all of the pyrethroids labelled for control of bedbugs at any indoor site, pose no significant risk to human health, if they are used in accordance with their individual product label directions (Cooper & Harlan, 2004; WHO, 2005).

4.6. Benchmarks for success in bedbug management

Some benchmarks that may help in evaluating the success of a bedbug control effort include:

- the lack of live bedbugs, cast skins (after those present earlier have been removed), faecal spots on bed linens or harbourage sites, and unhatched eggs;

- the lack of new feeding activity, as evidenced by occupants of previously infested sites having no new bites (such as no evidence of bites at new sites) appear more than 10 days after the most recent control effort;

- no new complaints of bites (with symptoms resembling typical bedbug bites) from occupants of rooms or apartments (dwelling units) adjacent to or near, but not included among, the most recently treated similar sites;

- new knowledge received and retained by residents, from local direct inquiry or from a survey that shows that occupants of recently treated sites (and their neighbours and servicing pest management technicians) have a good general knowledge of the bugs, their biology and their signs and that they have a good understanding of preventive techniques and effective control strategies against bedbugs;

- careful and thorough follow-up surveillance at five weeks or longer after the most recent treatment still shows no signs of a presence of live bedbugs; and

- an absence of recent bedbug infestations or bites being reported to public health or other government agencies.

Note: The absence of new bites assumes that the victims were not suffering from delusory parasitosis. Continued so-called bites in the absence of bugs and after careful inspection may imply delusory parasitosis, a condition that can be only addressed by medical professionals.

4.7. Conclusions

To improve prevention and control of bedbug infestation, the following is suggested.

- Steps should be taken to make accurate and practical current information readily and widely available to PMPs, health professionals and the general public. This should include information about the biology and behaviour of bedbugs and about effective control and prevention strategies against them.

- For populations of all three bedbug species that routinely feed on people, research should be encouraged and carried out to determine the susceptibility of bugs collected in field studies to the insecticides most frequently used at present to control them.

- Research should be encouraged and carried out to determine whether or not bedbugs can successfully transmit human pathogens, especially those that cause new or emerging diseases.

- Research should be encouraged and carried out to further characterize the nature and effective treatment of the effects on people of unusual, extreme or very persistent bedbug bites.

- Research should be encouraged and carried out to clarify more specific aspects of the physiology and behaviour of bedbugs, with a secondary goal of developing effective techniques or devices that can efficiently and quickly survey for the presence of even small populations of bedbugs.

- Research should be encouraged and carried out to determine the effectiveness and practical use of extreme temperatures (especially heat) to eliminate or control bedbugs in human habitats.

- Efforts should be undertaken (or at least planned) by appropriate government agencies to address locally evident problems that relate to the difficulties encountered by poor and low-income people in dealing with bedbugs and their control and with housing or building quality. Community-wide or citywide programmes may be needed and possible, if properly supported and well coordinated.

- Research should be encouraged and carried out to discover and make available new insecticide active ingredients, products, and devices and techniques that will be effective in controlling bedbugs.

References[1]

Abou Gamra EM et al. (1991). The relation between *Cimex lectularius* antigen and bronchial asthma in Egypt. *Journal of the Egyptian Society of Parasitology*, 21:735–746.

Bauer-Dubau K (2004). Invasion in deutschen Betten: Bettwanzen. *Ärzte Zeitung*, 176:9.

Blow J (2001). Stercorarial shedding and transstadial transmission of hepatitis B virus by common bed bugs (Hemiptera: Cimicidae). *Journal of Medical Entomology*, 38:694–700.

Brown C, Brown M (2005). Between-group transmission dynamics of the swallow bug, *Oeciacus vicarious*. *Journal of Vector Ecology*, 30:137–143.

Burton G (1968). Bedbugs in relation to transmission of human diseases. Review of the literature. *Public Health Reports*, 78:513–524.

Calisher C (1980). Characterization of Fort Morgan virus, an alphavirus of the western equine encephalitis virus complex in an unusual ecosystem. *The American Journal of Tropical Medicine and Hygiene*, 29:1428–1440.

Cooper R, Harlan H (2004). Ectoparasites. Part three: bed bugs & kissing bugs. In: Hedges S, ed. *Mallis' handbook of pest control*, 9th ed. Cleveland, OH, GIE Publishing: 494–529.

Cornwell PB (1974). The incidence of fleas and bedbugs in Britain. *International Pest Control*, 16:17–20.

Doggett S, Geary M, Russell R (2004). The resurgence of bed bugs in Australia: with notes on their ecology and control. *Environmental Health*, 4:30–38.

Dosland O (2001). The Heat is on. *IPM Extra*, Brookfield, WI, Copesan, 2nd Quarter: 1–2.

el-Mofty MM, Sakr SA, Younis MW (1989). Induction of skin papillomas in the rabbit, *Oryctologus cuniculus*, by bites of a blood-sucking insect, *Cimex lectularius*, irradiated by gamma rays. *The Journal of Investigative Dermatology*, 93:630–632.

Faulde M, Uedelhoven W, Robbins R (2003). Contact toxicity and residual activity of different permethrin-based fabric impregnation methods for *Aedes aegypti* (Diptera: Culicidae), *Ixodes ricinus* (Acari: Ixodidae), and *Lepisma saccharina* (Thysanura: Lepismatidae). *Journal of Medical Entomology*, 40:935–941.

[1] The citations quoted are from scientific literature, anecdotal observations and trade journals. The authors have carefully scrutinized the references and have judged the references to be credible. Little scientific literature is available about recent observations and research on bedbugs; however, the sources quoted are, to the authors' best knowledge, credible and may be used with confidence.

Feingold B, Benjamini E, Michaeli D (1968). The allergic responses to insect bites. *Annual Review of Entomology*, 13:137–158.

Fletcher M, Axtell R (1993). Susceptibility of the bedbug, *Cimex lectularius*, to selected insecticides and various treated surfaces. *Medical and Veterinary Entomology*, 7:69–72.

Foster W, Olkowski W (1968). The natural invasion of artificial cliff swallow nests by *Oeciacus vicarious* (Hemiptera: Cimicidae) and *Ceratophyllus petrochelidoni* (Siphonaptera: Ceratophyllidae). *Journal of Medical Entomology*, 5:488–491.

Gbakima AA et al. (2002). High prevalence of bedbugs, *Cimex hemipterus* and *Cimex lectularius*, in camps for internally displaced persons in Freetown, Sierra Leone: a pilot humanitarian investigation. *West African Journal of Medicine*, 21:268–271.

Gooch H (2005). Hidden profits: there's money to be made from bed bugs – if you know where to look. *Pest Control*, 73:26–32.

Hayes R et al. (1977). Role of the cliff swallow bug (*Oeciacus vicarious*) in the natural cycle of a western equine encephalitis-related alphavirus. *Journal of Medical Entomology*, 14:257–262.

Herrmann J et al. (1999). Efficacy of controlled atmospheres on *Cimex lectularius* (L.) (Heteroptera: Cimicidae) and *Argus reflexus* Fab. (Acari: Argasidae). In: Robinson WH, Rettich F, Rambo GW, eds. *Proceedings of the Third International Conference on Urban Pests*, 19–22 July 1999, Prague, Czech Republic. Hronov, Czech Republic, Grafické Závody:637.

Hwang SW et al. (2005). Bed bug infestations in an urban environment. *Emerging Infectious Diseases*, 11:533–538.

Johnson A (2005). The hotel industry is beginning to wake up to bedbug problem. *The Wall Street Journal*, April 21 2005, 245(78): Section A-1, Column 4; Section A-12, columns 5–6.

Jupp PG et al. (1991). Attempts to transmit hepatitis B virus to chimpanzees by arthropods. *South African Medical Journal*, 79:320–322.

Krinsky W (2002). True bugs. In: Mullen G, Durden L, eds. *Medical and veterinary entomology*. Orlando, FL, Academic Press: 67–86.

Krueger L (2000). Don't get bitten by the resurgence of bed bugs. *Pest Control*, 68(3):58–64.

Liebold K, Schliemann-Willers S, Wollina U (2003). Disseminated bullous eruption with systemic reaction caused by *Cimex lectularius*. *Journal of the European Academy of Dermatology and Venereology*, 17:461–463.

Lindsay SW et al. (1989). Permethrin-impregnated bednets reduce nuisance arthropods in Gambian houses. *Medical and Veterinary Entomology*, 3:377–383.

Myamba J et al. (2002). Pyrethroid resistance in tropical bedbugs, *Cimex hemipterus*, associated with use of treated bednets. *Medical and Veterinary Entomology*, 16:448–451.

Parsons DJ (1955). Bedbug bite anaphylaxis misinterpreted as coronary occlusion. *Ohio Medicine*, 51:669.

Potter M (2004). Your guide to bed bugs. *Pest Control Technology Magazine*, 32(8): pull-out supplement between pages 12 & 13.

Ryckman RE, Bentley DG (1979). Host reactions to bug bites (Hemiptera, Homoptera): a literature review and annotated bibliography, Parts I and II. *California Vector Views*, 26(1/2):1–49.

Sansom JE, Reynolds NJ, Peachey RD (1992). Delayed reaction to bed bug bites. *Archives of Dermatology*, 128(2):272–273.

Scott TW, Bowen GS, Monath TP (1984). A field study on the effects of Fort Morgan virus, an arbovirus transmitted by swallow bugs, on the reproductive success on cliff swallows and symbiotic house sparrows in Morgan County, Colorado, 1976. *The American Journal of Tropical Medicine and Hygiene*, 33:981–991.

Silverman AL et al. (2001). Assessment of hepatitis B virus DNA and hepatitis C virus RNA in the common bedbug (*Cimex lectularius* L.) and kissing bug (*Rhodnius prolixus*). *The American Journal of Gastroenterology*, 96:2194–2198.

Snetsinger R (1997). Bed bugs & other bugs. In: Hedges S, ed. *Mallis' handbook of pest control,* 8th ed. Cleveland, OH, GIE Publishing, Inc.:392–424.

Stutt AD, Siva-Jothy MT (2001). Traumatic insemination and sexual conflict in the bed bug *Cimex lectularius*. *Proceedings of the National Academy of Science of the United States of America*, 98:5683–5687.

Takahashi M, Ohtaki T (1975). Ovicidal effects of two juvenile hormone analogs, methoprene and hydroprene, on the human body louse and the bed bug. *Japanese Journal of Sanitary Zoology*, 26:237–239.

Usinger RL (1966). *Monograph of Cimicidae (Hemiptera: Heteroptera)*. Thomas Say Foundation, Vol. VII. College Park, MD, Entomological Society of America.

Weatherston J, Percy JE (1978). Venoms of Rhyncota (Hemiptera). In: Bettini S, ed. *Arthropod venoms. Handbook of experimental pharmacology*, Vol. 48. Berlin, Springer-Verlag:489–509.

Webb P et al. (1989). Potential for insect transmission of HIV: experimental exposure of *Cimex hemipterus* and *Toxorhynchites amboinensis* to human immunodeficiency virus. *The Journal of Infectious Diseases*, 160:970–977.

WHO (1985). *Bed bugs*. Geneva, World Health Organization (Vector Control Series, Training and Information Guide; document WHO/VBC/85.2; http://whqlibdoc.who.int/hq/1985-86/VBCTS85.2.pdf, accessed 6 November 2006).

WHO (2005). *Safety of pyrethroids for public health use*. Geneva, World Health Organization (document WHO/PCS/RA/2005.1; http://whqlibdoc.who.int/hq/2005/WHOCDSWHOPESGCDPP2005.10.pdf, accessed 6 November 2006):*iv*, 69.

Wigglesworth V (1984). *Insect physiology*, 8th ed. New York, Chapman and Hall.

5. Fleas

Nancy C. Hinkle

Summary

The two groups of fleas most significant to human health are rodent fleas and fleas found on domestic animals and non-rodent urban wildlife (cat fleas, *Ctenocephalides felis*).

All adult fleas require frequent blood-meals, making them a suitable vehicle to spread blood-borne disease agents among hosts. While cat fleas play no significant role in transmitting human disease, their bites may produce substantial irritation and itching. Unlike cat fleas, rodent fleas transmit the causative agent of one of humanity's most important diseases, bubonic plague, as well as the pathogen that produces murine typhus.

Exclusion or elimination of wild flea hosts, coupled with flea control on pets and in the home, provides the best options for protecting people from exposure to cat fleas. The options for non-chemical flea suppression are limited, so most pet owners still rely on pesticides as part of their flea management strategy. Also, products are available for treating outdoor flea infestations and fleas in the home. Products applied to the flea host are particularly efficacious, because they use the animal itself as the flea lure, assuring that all fleas are exposed to the toxicant as they attempt to feed.

Rodent fleas may occur on both wild and peridomestic rodents. Human behaviour is the main predisposing factor in exposure to plague-infected fleas on wild rodents. Government agencies inspect campgrounds, parks and other locations where wild rodents may be encountered by people, closing these venues when plague is detected and then initiating flea control measures. When plague is found in urban rodents, concomitant flea and rodent eradication programmes are mobilized, to eliminate the vertebrate reservoir and ensure that residual fleas do not remain to feed on people.

Flea control is intimately tied to host control. In particular, wild and feral hosts should be excluded from residential neighbourhoods, to prevent their interaction with domestic animals. Buildings should be rodent-proofed and maintained in good repair, and human activities (such as garbage disposal) should be directed towards avoiding the creation of conditions conducive to rodents.

5.1. Introduction

While over 2200 species of fleas are known worldwide, most of them are of no public health significance and do not have an impact on people or their companion animals, because they are found only on specific wild hosts. Of primary health and veterinary concern are rodent fleas, especially *Xenopsylla* spp., and fleas found on companion animals, *Ctenocephalides* spp.. While other flea species, such as *Leptopsylla segnis* (the European mouse flea), are found on peridomestic animals, they are considered unlikely vectors (Pratt & Wiseman, 1962). Both *Xenopsylla* spp. and *Ctenocephalides* spp. have worldwide distribution.

5.2. Flea biology

Fleas have developed two host-maintenance strategies. *Host fleas* remain on the host, exhibiting strong fidelity to the vertebrate, once a host is acquired, and abandoning their vertebrate host only when it dies. The host provides food (as blood), warmth and shelter, giving the flea little incentive to leave the host, especially since there is no guarantee of it acquiring another host. The flea's need to frequently feed increases its inclination to remain near the host (Fig. 5.1). In contrast, nidicolous fleas remain in the nest, moving to the host only to feed (Krasnov, Khokhlova & Shenbrot, 2004). *Nest fleas* parasitize animals such as rodents that return to the nest daily, ensuring that the flea will be able to obtain regular blood-meals. Typically, nest fleas are poor jumpers, moving primarily by crawling, while host fleas have well-developed jumping legs.

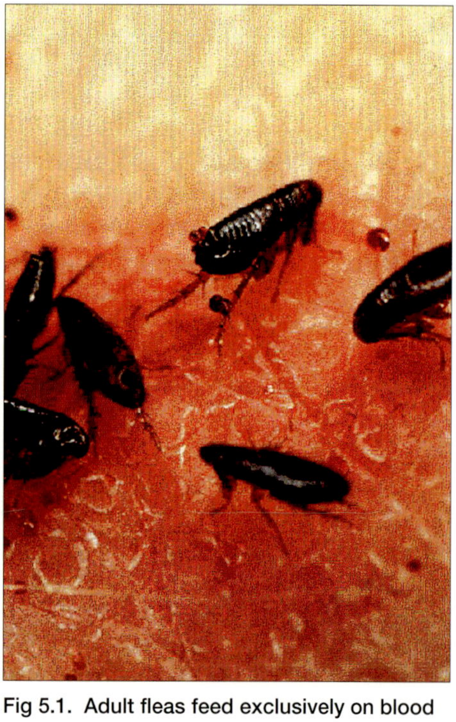

Fig 5.1. Adult fleas feed exclusively on blood

Source: Photo by N.C. Hinkle.

While the adult flea is dependent on the vertebrate host, flea larvae live off the host. Nevertheless, they also depend on host blood, in the form of adult flea faeces, for nutrition. Adult fleas ingest much more host blood than they need or can utilize for their own nutrition; thus, adult flea faeces are often described as being partially digested or undigested host blood (Rust & Dryden, 1997). Rodent and cat flea species have similar holometabolous life cycles, with free-living larval stages undergoing complete metamorphosis, resulting in the parasitic host-dependent adult stage.

5.2.1. Rodent fleas

Much of what is known about fleas is based on the biology of rodent fleas, especially *Xenopsylla* spp. on domestic rodents and *Oropsylla montana* and other flea species on ground squirrels (Metzger & Rust, 2002). Rodent fleas have been the subject of intense biological investigation (Gage & Kosoy, 2005) because they are the vectors of the causative agents of two significant human diseases, bubonic plague and murine typhus. Most rodent fleas are nest fleas, moving onto the host only to feed.

5.2.2. Cat fleas

The cat flea (*Ctenocephalides felis*) is not considered a major threat to human health because, as it is rarely found on rodents, it usually has little chance to transmit disease agents from rodent reservoirs to humans. Also, it has been shown to be an inefficient vector of plague (Pollitzer, 1954). It does, however, produce significant discomfort, due to a pruritic reaction to salivary secretions, both in human beings and in other animals.

The cat flea is a serious urban pest, infesting pets, such as dogs and cats, as well as urban wildlife, such as the northern raccoon (*Procyon lotor*), the Virginia opossum (*Didelphis virginiana*), the striped skunk (*Mephitis mephitis*), foxes (*Vulpes* spp.) and coyotes (*Canis latrans*) (Rust & Dryden, 1997). In urban settings, these wild animals, as well as feral cats and dogs, maintain cat flea populations and build a natural reservoir for reinfestation of domestic pets. Even homes without a pet can experience severe cat flea problems, if wild or feral animals nest in the crawl space or attic and share their fleas. Typically, migrant animals den under the structure in the spring; once the young leave the nest, it is abandoned, and the fleas left behind climb up through subflooring, avidly seeking a bloodmeal from any warm-blooded host.

5.2.3. Flea development

Because rodent fleas and cat fleas share many similarities in their life-cycles, the cat flea will be used to provide an overview of flea development. Differences between them are provided to distinguish the two.

The eggs of cat fleas are oval, measuring 0.5 mm in length, with tiny openings called aeropyles in one end of their white shells. A female cat flea produces about one egg an hour while she remains on the host. The smooth eggs are not sticky, so they easily sift through the host's pelage to collect in the surrounding environment, concentrating in areas where the host spends the most time. Cat flea eggs hatch in 1–10 days, depending on temperature and humidity (Dryden & Rust, 1994). The majority of eggs hatch within 36 hours at 70% relative humidity and 35 °C, while at 13 °C most of the eggs hatch within six days (Silverman, Rust & Reierson, 1981).

Flea larvae are white, legless, eyeless maggot-like creatures, covered sparsely with hairs. A neonate is scarcely larger than the egg from which it emerged, while full-grown larvae are about 5 mm long. Larvae live off the host, feeding on organic debris and adult flea faeces in their environment. Being negatively phototactic (moving away from light sources) and positively geotropic (burrowing), flea larvae avoid sunlight and actively move deep into carpet or under organic debris. Larvae typically require 5–11 days to complete their three instars, but the larval developmental period may be extended up to three weeks, depending on food availability and climatic conditions (Silverman, Rust & Reierson, 1981).

Because flea larvae are highly susceptible to desiccation, the larval environment is defined by relative humidities over 50%; larvae maintained in soil with low moisture fail to develop (Silverman, Rust & Reierson, 1981). Due to their susceptibility to heat and desiccation, flea larvae cannot survive outdoors in areas exposed to the sun. Flea hosts prefer shaded areas, so flea eggs are more likely to be deposited in shade, with the resulting flea larvae developing where the ground is shaded and moist. Similarly, flea larvae are protected under the carpet canopy in indoor habitats, where air movement is minimized and humidity is highest.

The third instar larva secretes silk and spins a cocoon within which it pupates. Because the silk is sticky, debris from the environment adheres to it, camouflaging the cocoon as a lint ball or dirt clod. The cocoon is ovoid and about 3 mm long. Inside its cocoon, the larva molts to a pupa and the pupa then molts to an adult.

The length of residence within the cocoon varies. Under conducive conditions (around 27°C and 80% relative humidity), the adult flea may emerge five days after the cocoon is formed, while under adverse environmental conditions or absence of a host the adult fleas may not emerge for many months (perhaps over a year). Typically, fleas emerge within two weeks following cocoon formation. After the adult develops inside the cocoon, a stimulus is required to cause the flea to emerge. Such stimuli as movement, heat and carbon dioxide signal the flea that a potential host is nearby, triggering emergence from the cocoon. The pre-emerged adult flea inside the cocoon is more resistant to desiccation than are eggs and larvae (Rust & Dryden, 1997).

Using its rudimentary eyes, the newly emerged flea orients itself towards a potential host by cueing in on movement. Extending its tarsal hooks, the flea leaps towards the host. If it lands on the host, the hooks permit it to cling. Once an adult cat flea acquires a host, it typically remains on that animal for the duration of its life, with host grooming being the most common mortality factor.

Unlike cat fleas, rodent fleas spend the majority of their lives in their host's nest. Because most rodents are nocturnal, rodent fleas tend to feed in the daytime, after their rodent hosts have returned to the nest and are resting. So rodent flea eggs are found in the host nest, as are most of the developmental life stages.

5.3. Health risk and exposure assessment

Plague, caused by the Gram-negative bacterium *Yersinia pestis*, is the most significant zoonosis that involves fleas. The fleas transfer the bacterial infection among their rodent hosts and to people (Gage & Kosoy, 2005). A plague infection is characterized by fever, chills, headache and malaise that lead to extreme physical weakness; the symptoms vary, depending on the form of plague. The principal clinical forms include bubonic plague, septicaemic plague and pneumonic plague. Untreated bubonic plague has a case-fatality rate of about 50–60%, but current therapies, including treatment with an appropriate course of antibiotics, markedly reduce fatality from it. Untreated primary septicaemic plague and pneumonic plague are invariably fatal. Septicaemic plague and pneumonic plague also respond to early diagnosis and treatment; however, patients who do not receive appropriate therapy for primary pneumonic plague within 18 hours after the onset of symptoms are unlikely to survive.

Historically, plague has resulted in considerable human mortality, producing significant population declines along with sociological changes. In the Middle Ages, bubonic plague (the so-called Black Death) killed between a quarter and a third of Europe's population within just a few decades (Gage & Kosoy, 2005). During the last pandemic, between 1896 and 1911, more than seven million people died of plague in India.

During the last half of the 20th century, plague outbreaks were reported in Africa, Asia, and North and South America. Nowadays, the number of annual cases is about 2000–3000 worldwide (with more than 90% of them in Africa), with a case fatality rate of about 7% (WHO, 2004). The infection obviously remains entrenched in sylvatic (rural) rodent–flea ecosystems throughout the world, and international travel and transportation make reintroduction and re-emergence likely. Plague rarely occurs in the European Region, but it does occur regularly in North America, with chronic zoonotic maintenance of the plague pathogen in the south-western United States provoking ongoing concern. Plague in non-commensal rodents and lagomorphs is dealt with in greater depth in Chapter 13 of this report.

The oriental rat flea (*Xenopsylla cheopis*) and *Xenopsylla brasiliensis* are the most important vectors of plague bacilli, from rat to rat and from rat to human. In the laboratory, other flea species, such as the northern rat flea (*Nosopsyllus fasciatus*), cat and dog fleas (*Ctenocephalides felis* and *Ctenocephalides canis*, respectively), and the human flea (*Pulex irritans*), have been shown to be capable of transmitting the plague organism, so they may play some role in disease maintenance (Traub, 1983). In the American Southwest, animals such as ground squirrels (*Spermophilus* spp.) and chipmunks (*Tamias* spp.) are reservoirs of sylvatic plague (plague in wild rodents), and rodent fleas play an important role in its maintenance and transmission (Davis et al., 2002). The risk of people becoming infected is greatest in urban–rural interfaces, parks and recreation areas, because these are areas where people are most likely to encounter infected hosts and their fleas.

The other flea-borne disease of significance is murine typhus, an endemic zoonosis caused by an obligate intracellular bacterium, *Rickettsia typhi* (Azad, 1990). Murine typhus is primarily a disease of domestic rats and mice, but is transmitted to people when the crushed bodies of infected fleas or their faeces are rubbed into an open sore or onto mucous membranes. The disease occurs worldwide, particularly in warm climates with large reservoir populations (such as rats or opossums) and flea vectors (Traub, Wisseman & Farhang Azad, 1978). Dramatic declines in reported cases of murine typhus began in the 1940s, and the current prevalence of the disease globally is estimated at fewer than 100 cases a year (Boostrom et al., 2002), although cases may be considerably underestimated due to its nonspecific symptomatology (Jensenius, Fournier & Raoult, 2004). Murine typhus often goes unrecognized and is perceived as a clinically mild disease, with a case fatality rate less than 1%. Texas and regions of Southern California have the highest prevalence in the United States, but epidemiological studies (Azad, 1990; Boostrom et al., 2002) have stimulated concern that typhus reservoirs and vectors are spreading. Although most patients are adults, children constitute up to 75% of infections in some outbreaks. Systemic involvement is evident from the frequent occurrence of abnormal laboratory findings that involve multiple organ systems, including the liver, kidney, blood and central nervous system.

In the future, ecotourism and increased international travel are likely to result in more imported cases of rickettsioses in Europe and elsewhere (Jensenius, Fournier & Raoult, 2004), indicating that medical communities in nonendemic regions must expand their differential diagnoses to include these importations.

The cat flea is also the recognized vector of *Bartonella henselae*, *Bartonella clarridgeiae* and *Rickettsia felis* (Shaw et al., 2004). Many of the infections related to these bacteria are subclinical in both pets and humans, but the increasing immunocompromised population puts more individuals at risk of clinical illness. In addition, *Bartonella quintana* and *Bartonella koehlerae* have been detected in cat fleas, indicating that there may be emerging unknown diseases caused by as yet unidentified microbes in cat fleas (Kelly, 2004; Lappin et al., 2006).

In addition to transmitting pathogens, fleabites produce pruritic lesions in both humans and animals. Flea allergy dermatitis (FAD) is a serious atopic hypersensitivity reaction to flea salivary secretions that commonly afflicts cats and, more prevalently, dogs. The itching and discomfort produced by fleabites are the main reason cat fleas are considered human and pet pests (Hinkle, 2003).

5.4. Notification and reporting

In Europe, both plague and murine typhus are notifiable diseases. In the United States, however, only plague is reportable, while local jurisdictions may maintain murine typhus under reportable status (Boostrom et al., 2002). Despite it being notifiable, WHO (2004) considers plague underreported due to:

- reluctance of some endemic countries to disclose cases
- diagnostic failures because of nonspecific clinical presentations
- inadequate facilities for laboratory confirmation.

In North America, plague has been reported primarily in the West, ranging from southwestern Canada to Mexico. Human cases average fewer than 10 a year in North America, a rate that has remained relatively constant since plague was introduced into the Americas in 1899 (WHO, 2004).

Worldwide, the number of murine typhus cases is low, but because symptoms are nonspecific and frequently misdiagnosed, infection is presumed to be much more common than reported (Gratz, 2004). In Europe, murine typhus has been reported from Bosnia and Herzegovina, Croatia, the Czech Republic, France, Greece, Italy, Montenegro, Portugal, the Russian Federation, Serbia, Slovakia, Slovenia and Spain, and it is considered likely to be present in most other countries, as well (Gratz, 2004).

5.5. Economic burden of flea infestations

As urbanization increases and pets become more important to their owners, fleas have become more and more economically significant. Few other single species of urban pests cost consumers more than the cat flea, not only in terms of costs of over-the-counter control products and pest control services, but also in terms of the costs of veterinary bills (both for flea control and for treatment of flea-caused conditions, such as FAD and tapeworm infestations).

5.5.1. Cost for control and management

In North America, costs of domiciliary flea control are borne typically by the homeowner or property owner. The average pet owner in the United States spends an estimated US$ 38 on over-the-counter (non-veterinary) flea control products annually (Hinkle, 1997). Also, in the United States, over US$ 175 million is spent every year for flea control services provided by pest control firms (G. Curl, private pest management consultant, Mendham, NJ, personal communication, 2005). In a survey of ten pest control companies in southern states and areas along the coast where environmental conditions are conducive and fleas thrive, the so-called United States flea belt, charges for a single flea control treatment averaged US$ 196 (range: $85–314) (N.C. Hinkle, unpublished data, 2006).

With new on-host veterinary products available for flea suppression, annual cat flea control costs typically average about US$ 90. Luechtefeld (2005) indicated that 63% of pet owners purchase flea control products annually. The number of households that own pets is increasing in both Europe and North America, with 30 cats and 25 dogs per 100 Americans (APPMA, 2006), compared with about 13 cats and 9 dogs per 100 Europeans (Statistics Belgium, 2003). As more homes acquire pets, the incidence of fleas is likely to increase, exposing more people to cat fleas and increasing the market for flea control products.

In plague outbreaks, public health personnel are mobilized and taxes fund rodent- and flea-suppression efforts. For instance, the Vector-Borne Disease Section of the Division of Communicable Disease Control of the California Department of Health Services monitors campgrounds where plague is enzootic – that is, infecting animals in a particular geographic area – and provides flea and rodent inspections and suppression (Gerry et al., 2005). Annual surveillance costs for a typical California campground are estimated at US$ 7920 (Kimsey et al., 1985; all estimates are adjusted for inflation). In years with epizootic episodes, control costs are estimated at US$ 63 890, while the total cost – including lost revenue, assuming the park must be closed for two months – is US$ 241 540 (Kimsey et al., 1985). Obviously this not only affects the economy of the park, but also affects the economy of the local area. These cost estimates do not take into account either the direct or indirect impacts of a human plague case and its public relations implications on regional tourism.

As with most urban pests, construction practices that deny pests entry and establishment not only are more cost effective, but are also more dependable and long lasting. Also, retrofitting buildings and post-construction remediation are more expensive and are seldom completely satisfactory.

5.5.2. Cost of health-related conditions

While fleabites cause discomfort, they are not treated as medical conditions and people typically initiate palliative care using over-the-counter medications. When flea-allergic animals suffer unremitting dermatitis, however, veterinary intervention is necessary. This usually entails anti-pruritic medications and desensitization therapy, costs of which can be substantial.

In some parts of the United States, over half the annual income of veterinary clinics is attributable to flea and tick product sales (Scheidt, 1988), amounting to about US$ 1.7billion a decade ago (Hinkle, 1997). The monthly cost of host-targeted products, such as pills or spot-ons (topical ectoparasiticides applied in small volumes to the skin), averages US$ 10 per animal, considering only charges for the product. The worldwide market for ectoparasite control products (ectoparasiticides) was estimated at about €1.1 billion, with the United States accounting for 67% of this market and western Europe 20% (Krämer & Mencke, 2001). Pet owners spend substantial sums (about US$1.12 billion in 1997) for FAD treatments (about US$ 938 million), tapeworm prophylaxis and treatment (about US$ 184 million) and other flea-related problems (Hinkle, 1997).

The cost of a non-fatal human plague case was estimated to be about US$ 19761 (adjusted for inflation; Kimsey et al., 1985), but increases in health care costs in the United States in the past two decades probably make this estimate low.

As the vast majority of murine typhus cases are mild and go undiagnosed, there are no published estimates of their treatment costs. Individuals in whom disease is more severe often must be hospitalized for about a week, with attendant charges for such items as supportive care, laboratory tests, X-rays, antibiotic therapy and pharmaceuticals (D.H. Walker, University of Texas Medical Branch, Galveston, TX, personal communication, 2006).

5.6. Impact of poverty and lifestyle on risk of infestation

While today's standard of living minimizes flea-vectored diseases in most European and North American communities, some population subsets experience chronic or sporadic disease outbreaks, requiring the involvement of public health personnel. Substandard housing, lack of appropriate veterinary care for domestic and feral dogs and cats, and other neighbourhood attributes predispose such communities as Native American reservations, inner-city apartments and homeless populations to flea exposure and the resultant transmission of disease. For such populations, pest control may not be economically feasible.

5.7. Flea management

As with most pests, the management of flea problems involves sanitation, source reduction and exclusion. There are few biological control options for fleas, so pesticides play a major role in flea suppression (Hinkle, Rust & Reierson, 1997). Habitat manipulation focuses on creating an environment that is inhospitable to flea hosts, thus eliminating the vertebrate and its ectoparasitic load from proximity to people.

The first priority is to identify the pest, because suppression strategies differ among flea species – for example, developing a control strategy for the cat flea is vastly different from dealing with discovery of a chigoe flea (*Tunga penetrans*). Most cat flea control calls to pest control firms are from residences, not businesses, although an occasional infestation may occur in a warehouse or other structure where wildlife or feral animals reside.

Flea management in and around man-made structures lends itself well to the concept of urban IPM (Rust, 2005). Developing an integrated flea control programme necessitates understanding flea biology, population assessment techniques, mechanical control systems, biological control, IGRs and traditional insecticide treatments (Hinkle 2003). Once flea numbers have been reduced, environmental modification to exclude wildlife and feral animals from the area prevents flea reinfestation. Similarly, removing pet food at night avoids luring raccoons, opossums and other flea hosts onto properties where they share their fleas with pets, and covering openings to crawl spaces excludes animals from denning beneath the structure.

New pesticidal products formulated to be used on the host have increased pet-owner compliance, due to enhanced efficacy, long-term activity and ease of application (Rust, 2005). Therefore, cat flea control has shifted to reliance on these host-directed products, in conjunction with ancillary environmental treatments.

Similarly, developing strategies to eliminate rodents from neighbourhoods will simultaneously eliminate their flea populations, along with the pathogens they transmit. Again, the fleas should be identified, to ascertain the most likely hosts as well as their potential risk to human health. Flea suppression that targets hosts should be initiated before attempts to eliminate rodents, to prevent flea problems rebounding after host removal.

The control of ectoparasites must be integrated with rodent control, in combination with ongoing and sustained habitat modification.

In urban areas, government agencies bear responsibility for managing wildlife and feral animals, to prevent their ectoparasites from infesting domestic pets. Likewise, municipalities have codes that regulate sanitation, debris accumulation and other conditions that affect rodent populations. Budgets for code enforcement, related maintenance activities and municipal pest control services vary widely by jurisdiction, as does emphasis on such expenditures.

5.7.1. Flea inspections and detection

Because all fleas require warm-blooded hosts, frequently the most efficient monitoring tactic centres on the host. Typically, visual inspections are used for cat and dog fleas, with demonstration of so-called flea dirt (adult flea faeces) used in clinical settings to confirm the presence of fleas. Homeowners typically have less defined action thresholds, initiating flea control efforts when they experience fleabites (Hinkle, 2003).

Plague surveillance focuses on rodent burrows. This surveillance uses artificial hosts inserted into the burrows to attract fleas, which then are retrieved, identified and tested for plague infection. Alternatively, rodents can be trapped and combed to determine the so-called flea index – that is, the number of fleas per rodent (Pratt & Wiseman, 1962; Krasnov, Khokhlova & Shenbrot, 2004).

5.7.2. Conducive environmental conditions

Building construction has less impact on fleas and their hosts than does maintenance. Crumbling foundations, debris accumulation and other situations that provide suitable habitats for rodents are also predictive of flea infestations. Thus, man-made habitats allow wildlife to thrive in urban and suburban areas, with small mammals living in culverts and feeding from trash cans, dumpsters and pet bowls left outside.

In some urban or suburban areas, opossums have been found at population densities of more than three animals/ha (Boostrom et al., 2002). Research has shown that not only do the wildlife host species thrive in urban and suburban areas, but also that greater host population densities in urban areas lead to elevated parasite loads, thus increasing parasite sharing with domestic animals (Dryden et al., 1995). Moreover, urban wildlife interact more closely with one another and with people, maximizing the likelihood of sharing fleas.

Both cat flea incidence and prevalence are greater in urban wildlife populations than in their rural counterparts. For instance, cat fleas were recovered from 50% to 100% of raccoons trapped in urban areas, while fewer than 17% of rural raccoons had fleas. Urban raccoons had up to 50 fleas per animal, while no rural raccoon ever had more than 6 fleas. Similar results were found in opossums, with 62% of the urban opossum population infested with fleas and only 10% of rural opossums carrying fleas (Dryden et al., 1995).

Rodent-proofing buildings and landscape modifications should be implemented to ensure the habitat is unsuitable for rodents and their fleas. Exclusion measures and sanitation practices prevent rats and other small mammals from dwelling near or entering buildings. By denying them food, water and harbourage, rodents and their ectoparasites can be eliminated from a neighbourhood.

5.7.3. Flea exclusion and physical removal

Host exclusion from the property, especially preventing access to crawl spaces beneath homes, is the most important means of excluding the fleas they carry. Pest-proofing homes also requires conscientious inspection of access points, to identify openings through which arthropods can gain access to the building, so that they can be sealed as well.

Flea traps can be effective in monitoring flea populations and in confirming infestations (Dryden & Broce, 1993). However, trapping rarely reduces population size enough to serve as a significant component of an IPM programme.

Fleas can be physically removed from a cat or dog by using a flea comb. These devices are designed with closely spaced teeth so that, as the comb is drawn through the coat, fleas become entrapped and can be removed. The successful use of flea combs depends on animal tractability, coat thickness and length, and user patience and persistence.

5.7.4. Pesticide applications for flea control

In California, plague control operations are conducted by public health agencies and typically involve the use of insecticides to reduce rodent flea populations in localized areas. Usually, insecticidal dust is applied in and around rodent burrows so that, as rodents enter or exit these treated burrows, adherent residues are transferred to nest material where the majority of adult and immature fleas are found (Gerry et al., 2005). This technique ensures that small quantities of insecticide are optimally targeted to reduce flea numbers over a large geographic area.

Control measures for plague and murine typhus should first focus on fleas and subsequently on their rodent hosts. Indiscriminate elimination of rodents creates greater risk, since their ectoparasites (the pathogen vectors) immediately seek other hosts – among them people – and transmit infections to them, as may happen potentially with plague and rickettsioses. Flea suppression should therefore precede rodent baiting by at least a week (WHO, 1998). Insecticides, IGRs, or both should be applied to rat pathways, nests and holes (Mian et al., 2004). Rats should then be eliminated using traps or rodenticides after the flea population is decreased. If separate treatments for rodents and ectoparasites are not practical, an alternate method is to dust runways and apply poison baits at the same time. This method works well only with slow-acting poisons (such as anticoagulants) so that ectoparasites are killed before rodents succumb to the rodenticide.

Avoiding secondary poisoning of predators is critical to prevent rodent populations from resurging. Following any intervention, it is important to continue monitoring, to evaluate programme effectiveness and to modify tactics for greater efficacy.

Indoor flea infestations require thorough inspections. Any location where a flea-infested animal spends time will accumulate flea eggs, so examinations should include discussions with the pet owner about the animal's habits, to ascertain likely infestation foci. Flea larvae can survive in uncarpeted areas by moving to protected sites under furniture or baseboards, so effective treatments must also target these areas.

Increasing public concern about exposure to insecticides has caused the emphasis in urban entomology to shift towards non-chemical or reduced chemical pest control methods (Hinkle, Rust & Reierson, 1997). Strategies that focus on the host have superseded the historic environmental focus (Rust, 2005).

5.7.4.1. On-animal products

Therapies that target the flea hosts essentially involve flea baits, using the host animal as a lure for adult fleas. Because all fleas feed exclusively on blood and must find a host to survive, treating potential hosts with toxicants ensures that fleas either fail to feed or attempt to feed and are killed. Rust (2005) provided an excellent overview of commercially available products that are applied to the host to suppress adult fleas. Currently, registered on-host products include avermectins, fipronil, imidacloprid, pyrethroids, pyrethrins and IGRs (including insect developmental inhibitors and juvenile hormone analogues). In addition to suppressing adult fleas, on-host products provide an ancillary benefit by suppressing off-host life stages. Shed hair, dander and other debris from treated animals carry sufficient insecticide residues to suppress flea larvae (McTier et al., 2000; Mehlhorn, Hansen & Mencke, 2001). Products administered orally include nitenpyram and lufenuron (Rust, 2005).

On-animal flea control products are prepared in a variety of forms, as topically applied spot-ons, pills, injectables, sprays, dusts, shampoos, mousses, dips and collars. Some active ingredients can be formulated in several ways, while some are delivered in only one way.

Spot-on products are made to be lipophilic, so they passively distribute in the skin oils following application. These products have gained pet-owner acceptance, due to their dependable efficacy, sustained effectiveness, low risk and ease of use (Schenker et al., 2003).

Products administered internally (such as pills, liquids and injectables) vary in their mode of action, including both flea adulticides and larvicides. Because fleas must feed on host blood to acquire these compounds, they should not be relied on as the exclusive means of flea control for animals allergic to fleas.

Topical applications of sprays, dusts, shampoos, mousses and dips likewise vary in their effectiveness, depending on the active ingredient and its persistence in the animal's hair coat. Most of these kill fleas initially, but may not offer prolonged control.

Sustained release devices, such as flea collars, may provide long-term on-animal flea suppression. However, the active ingredient determines efficacy, and flea populations have been shown to be resistant to several active ingredients, including organophosphates, carbamates and pyrethroids (Bossard, Hinkle & Rust, 1998).

Products already available for cat flea control may also be useful in rodent flea control, with modifications of how the products are applied (Davis, 1999; Metzger & Rust, 2002). Because shed residues do persist in nesting materials, long-term efficacy could be obtained in sylvatic rodents, as well as in peridomestic species (McTier et al., 2000; Mehlhorn, Hansen & Mencke, 2001).

5.7.4.2. Foggers

Aerosol foggers disperse pesticides on all horizontal surfaces, making them poorly adapted for flea suppression. Foggers cause unnecessary environmental contamination and do not distribute the pesticide under furnishings where flea larvae and pupae may be concentrated. Instead of using foggers, pet owners preferentially use other application methods (Davis, Brownson & Garcia, 1992).

5.7.4.3. General surface treatments

Household environmental treatments with flea adulticides have fallen out of favour, due to deposition of toxic chemicals on broad expanses of the living area and due to concern about airborne and dislodgeable residues (Koehler & Moye, 1995), so there are few products available for killing adult fleas. Instead, flea suppression depends primarily on larvicides. IGRs constitute critical components of effective flea IPM programmes, but disodium octaborate tetrahydrate and other borate products have also been shown to be effective in eliminating larval fleas (Rust & Dryden, 1997). Borate products are broadcast in sites where larvae develop, brushed into the carpet and then vacuumed to remove loose residue; this results in adequate material being left behind to effect larval mortality as flea larvae graze on organic debris at the base of the carpet fibres (Klotz et al., 1994).

5.7.4.4. IGRs

The use of IGRs is the preferred method of suppressing fleas, because it eliminates larval fleas before they reach the bloodsucking adult stage (Fig. 5.2). Targeting larval flea habitats is critical to effective management, so care should be taken to direct applications to areas where flea hosts spend most of their time – for example, by distributing IGRs around dog beds (Rust & Dryden, 1997). IGRs, such as methoprene and pyriproxyfen, are ineffective against adult fleas, so treatment must be initiated before the flea problem becomes severe.

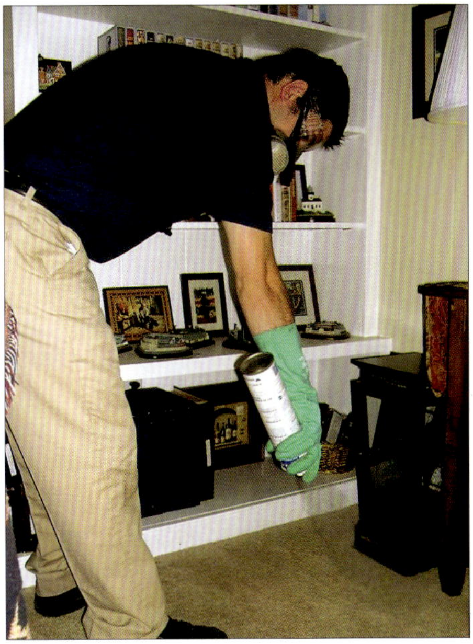

Fig. 5.2. Indoor application of an IGR forestalls development of immatures into biting adult fleas

Source: Photo by N.C. Hinkle.

5.7.4.5. Outdoor residential applications

To control fleas outdoors, insecticide applications must be targeted at areas where hosts (either pets, or feral or wild animals) spend the most time. Pyrethroids constitute the majority of products available, but products tested in outdoor habitats typically exhibit short-lived efficacy (Rust & Dryden, 1997). Such products are generally applied as sprays, but granular preparations are available that are activated by moisture and then release toxicant over a longer interval. Also, because some IGRs lack photostability, they are not useful for outdoor applications. Moreover, borates exhibit significant phytotoxicity, which makes them also inappropriate for outdoor use.

5.8. Pesticide exposure and risk

Because flea larvae can occur in carpeted areas of buildings, general carpet treatment is frequently undertaken to eliminate them. This tactic involves treating large indoor areas with pesticides, which may result in human exposure (Koehler & Moye, 1995). To minimize this risk, reduced toxicity compounds, such as IGRs, are promoted.

Because pets share human living quarters, they bring people into contact with the animal's ectoparasites and with residual insecticides used to treat them for fleas. While elimination of fleas from the pet is desirable, the products used on the animal must have low mammalian toxicity to minimize risks to pets and people. As with all pesticidal treatments, the applicator is likely to contact the most concentrated material, so manufacturers have developed packaging for host-targeted products (individual vial applicators), to reduce human exposure and protect consumers.

5.9. Benchmarks

Larvicides (especially IGRs) may be employed prophylactically to suppress fleas, but generally environmental flea control is undertaken when adult fleas are detected by a home's occupants (Hinkle, 2003). In warm regions, pets may be maintained on flea preventatives year-round, but veterinarians in cooler climates typically recommend on-host products be used during the months when flea exposure is likely.

While the action threshold for cat flea control is based on human annoyance, in situations of potential zoonotic disease outbreaks, rodent fleas must be actively monitored and the risk assessed to trigger control interventions. In cases involving imminent disease transmission, immediate and comprehensive flea suppression should be initiated. Due to the risk of fleas moving from dying hosts to people, rodent control must be coordinated with flea control.

5.10. Conclusions

5.10.1. Surveillance and environmental design

Areas with significant rodent populations at risk of flea-borne zoonoses must be regularly monitored, so that rodents and fleas can be recovered, identified, and tested for infection with *Y. pestis* or *R. typhi*. Also, conducive conditions must be eliminated or rectified, including structural features that encourage wildlife in and around people's homes. Likewise, sealing cracks and crevices is important, to prevent host-seeking fleas from entering homes. Adverse human actions, such as exposing garbage and pet food residues to wildlife, must be modified to discourage wild flea hosts from frequenting neighbourhoods. Furthermore, education is of highest importance with regard to exclusion of rodents and their fleas, to flea management and to possible health risks posed by fleas.

5.10.2. Deploy suppression strategies

When the potential for zoonoses exists, concerted efforts must be undertaken to simultaneously reduce rodents and their fleas, and responsibilities between parties involved must be clear. Rodent control must not be initiated without concomitant flea control, to avoid fleas abandoning dying hosts and attacking people.

5.10.3. Monitor and evaluate

As with any IPM programme, monitoring and ongoing evaluation constitute critical components. Monitoring, via trapping, provides an assessment of the effectiveness of suppression efforts, indicating when programme modifications are necessary. Also, epidemiological data are needed to predict the possible spread or resurgence of flea-borne diseases.

References[1]

APPMA (2006). Industry statistics & trends [web site]. Greenwich, CT, American Pet Products Manufacturers Association, Inc. (http://www.appma.org/press_industry-trends.asp, accessed 22 November 2006).

Azad AF (1990). Epidemiology of murine typhus. *Annual Review of Entomology*, 35:553–569.

Boostrom A et al. (2002). Geographic association of *Rickettsia felis*-infected opossums with human murine typhus, Texas. *Emerging Infectious Diseases*, 8:549–554.

Bossard RL, Hinkle NC, Rust MK (1998). Review of insecticide resistance in cat fleas (Siphonaptera: Pulicidae). *Journal of Medical Entomology*, 35:415–422.

Davis JR, Brownson RC, Garcia R (1992). Family pesticide use in the home, garden, orchard, and yard. *Archives of Environmental Contamination and Toxicology*, 22:260–266.

Davis RM (1999). Use of orally administered chitin inhibitor (lufenuron) to control flea vectors of plague on ground squirrels in California. *Journal of Medical Entomology*, 36:562–567.

Davis RM et al. (2002). Flea, rodent, and plague ecology at Chuchupate Campground, Ventura County, California. *Journal of Vector Ecology*, 27:107–127.

Dryden MW, Broce AB (1993). Development of a trap for collecting newly emerged *Ctenocephalides felis* (Siphonaptera: Pulicidae) in homes. *Journal of Medical Entomology*, 30:901–906.

Dryden MW, Rust MK (1994). The cat flea: biology, ecology and control. *Veterinary Parasitology*, 52:1–19.

Dryden MW et al. (1995). Urban wildlife as reservoirs of cat fleas, *Ctenocephalides felis*. In: *Proceedings of the American Association of Veterinary Parasitologists, 40th Annual Meeting*, Pittsburgh, Pennsylvania, USA, 6–10 July 1995:35–39.

Gage KL, Kosoy MY (2005). Natural history of plague: perspectives from more than a century of research. *Annual Review of Entomology*, 50:505–528.

Gerry AC et al. (2005). Worker exposure to diazinon during flea control operations in response to a plague epizootic. *Bulletin of Environmental Contamination and Toxicology*, 74:391–398.

[1] The references provided include the most pertinent research reports and reviews. Peer-reviewed journals were used when available, but non-refereed documents were cited when information, such as pet populations and expenditure data, was not available from other sources.

Gratz NG (2004). *The vector-borne human infections of Europe: their distribution and burden on public health*. Copenhagen, WHO Regional Office for Europe (document number: EUR/04/5046114; www.euro.who.int/document/E82481.pdf, accessed 30 August 2006).

Hinkle NC (1997). Economics of pet ectoparasites. In: *Proceedings of the 4th International Symposium on Ectoparasites of Pets*, Riverside, California, USA, 6–8 April 1997:107–109.

Hinkle NC (2003). Companion animal ectoparasites, associated pathogens, and diseases. In: Barker KR, ed. *Integrated pest management*. Ames, IA, Council for Agricultural Science and Technology:180–187.

Hinkle NC, Rust MK, Reierson DA (1997). Biorational approaches to flea (Siphonaptera: Pulicidae) suppression: present and future. *Journal of Agricultural Entomology*, 14:309–321.

Jensenius M, Fournier PE, Raoult D (2004). Rickettsioses and the international traveler. *Clinical Infectious Diseases*, 39:1493–1499.

Kelly PJ (2004). A review of bacterial pathogens in *Ctenocephalides felis* in New Zealand. *New Zealand Veterinary Journal*, 52:352–357.

Kimsey SW et al. (1985). Benefit-cost analysis of bubonic plague surveillance and control at two campgrounds in California, USA. *Journal of Medical Entomology*, 22:499–506.

Klotz JH et al. (1994). Oral toxicity of boric acid and other boron compounds to immature cat fleas (Siphonaptera: Pulicidae). *Journal of Economic Entomology*, 87:1534–1536.

Koehler PG, Moye HA (1995). Airborne insecticide residues after broadcast application for cat flea (Siphonaptera: Pulicidae) control. *Journal of Economic Entomology*, 88:1684–1689.

Krämer F, Mencke N (2001). *Flea biology and control: the biology of the cat flea – control and prevention with imidacloprid in small animals*. Berlin, Springer-Verlag.

Krasnov BR, Khokhlova IS, Shenbrot GI (2004). Sampling fleas: the reliability of host infestation data. *Medical and Veterinary Entomology*, 18:232–240.

Lappin MR et al. (2006). Prevalence of *Bartonella* species, haemoplasma species, *Ehrlichia* species, *Anaplasma phagocytophilum*, and *Neorickettsia risticii* DNA in the blood of cats and their fleas in the United States. *Journal of Feline Medicine and Surgery*, 8:85–90.

Luechtefeld L (2005). Veterinarians' flea-treatment market share on the rise. *Veterinary Practice News*, 17:8.

McTier TL et al. (2000). Evaluation of the effects of selamectin against adult and immature stages of fleas (*Ctenocephalides felis felis*) on dogs and cats. *Veterinary Parasitology*, 91:201–212.

Mehlhorn H, Hansen O, Mencke N (2001). Comparative study on the effects of three insecticides (fipronil, imidacloprid, selamectin) on developmental stages of the cat flea (*Ctenocephalides felis* Bouché 1835): a light and electron microscopic analysis of *in vivo* and *in vitro* experiments. *Parasitology Research*, 87:198–207.

Metzger ME, Rust MK (2002). Laboratory evaluation of fipronil and imidacloprid topical insecticides for control of the plague vector *Oropsylla montana* (Siphonaptera: Ceratophyllidae) on California ground squirrels (Rodentia: Sciuridae). *Journal of Medical Entomology*, 39:152–161.

Mian LS et al. (2004). Field efficacy of deltamethrin for rodent flea control in San Bernardino County, California, USA. *Journal of Vector Ecology*, 29:212–217.

Pollitzer R (1954). *Plague*. Geneva, World Health Organization (WHO Monograph Series No. 22; http://whqlibdoc.who.int/monograph/WHO_MONO_22.pdf, accessed 1 September 2006).

Pratt HD, Wiseman JS (1962). *Fleas of public health importance and their control.* Washington, DC, United States Department of Health, Education, and Welfare (Insect Control Series, Part VIII. PHS Publication No. 772).

Rust MK (2005). Advances in the control of *Ctenocephalides felis* (cat flea) on cats and dogs. *Trends in Parasitology*, 21:232–236.

Rust MK, Dryden MW (1997). The biology, ecology and management of the cat flea. *Annual Review of Entomology*, 42:451–473.

Scheidt VJ (1988). Flea allergy dermatitis. In: White SD, ed. *The veterinary clinics of North America: small animal practice.* Philadelphia, Saunders:1023–1042.

Schenker R et al. (2003). Comparative speed of kill between nitenpyram, fipronil, imidacloprid, selamectin and cythioate against adult *Ctenocephalides felis* (Bouché) on cats and dogs. *Veterinary Parasitology*, 112:249–254.

Shaw SE et al. (2004). Pathogen carriage by the cat flea *Ctenocephalides felis* (Bouché) in the United Kingdom. *Veterinary Microbiology*, 102:183–188.

Silverman J, Rust MK, Reierson DA (1981). Influence of temperature and humidity on survival and development of the cat flea, *Ctenocephalides felis* (Siphonaptera: Pulicidae). *Journal of Medical Entomology*, 18:78–83.

Statistics Belgium (2003). Cats and dogs in Belgium [web site]. Brussels, Statistics Belgium (News Flash No. 37; http://www.statbel.fgov.be/press/fl037_en.asp, accessed 30 August 2006).

Traub R (1983). Medical importance of the Ceratophyllidae. In: Traub R, Rothschild M, Haddow JF, eds. *The Rothschild collection of fleas. The Ceratophyllidae: key to the genera and host relationships, with notes on their evolution, zoogeography and medical importance.* London, Cambridge University Press/Academic Press:202–228.

Traub R, Wisseman CL, Farhang Azad A (1978). The ecology of murine typhus – a critical review. *Tropical Diseases Bulletin*, 75:237–317.

WHO (1998). Murine typhus, Portugal. *Weekly Epidemiological Record*, 73:262–263.

WHO (2004). Human plague in 2002 and 2003. *Weekly Epidemiological Record*, 79:301–308 (http://whqlibdoc.who.int/wer/WHO_WER_2004/79_301-308(no33).pdf, accessed 30 August 2006).

6. Pharaoh ants and fire ants

David H. Oi

Summary

Pharaoh ants are cosmopolitan pests that inhabit residential and commercial buildings. While they do not sting, they have the potential to mechanically transmit diseases and thus are of special concern in health care facilities. Their propensity to breach sterile packaging, feed on wounds, and extensively infest large buildings makes them a public health risk. However, published economic impact data on these ants is minimal. Control measures for Pharaoh ants are effective when implemented properly. Baits that contain IGRs or metabolic inhibitors can eliminate infestations within a few weeks. Faster control with applications of non-repellent residual insecticides to building perimeters has been reported. Because Pharaoh ants can be easily transported, monitoring and treatment will be an ongoing process to maintain acceptable control.

Fire ants are stinging, invasive ants from South America that have infested the southern United States since the 1930s. They now seem to be invading other parts of the world, as evidenced by recent infestations in Australia and South-East Asia. Most of Europe is too cold for the proliferation and spread of fire ants. However, countries along the Mediterranean and Black seas have suitable climates for fire ants to become established. The economic cost of fire ants in the United States is an estimated US$ 6.5 billion annually, with the majority of the losses in the urban sector. In infested areas of the United States, 30–60% of the population is stung annually, of which anaphylactic shock was conservatively estimated to occur in 1% of the victims. Litigation settlements of over US$ 1 million have been awarded for deaths related to fire ant stings. The significant impact of fire ants confirms the importance of preventing their establishment in new regions. Countries at risk for infestations should have a centralized coordinated response plan that includes regulatory clearance and manufacturing source(s) for insecticide treatments. To combat these ants, surveillance methods that can detect low levels of fire ant populations are direly needed. Baits can effectively control fire ants and non-repellent residual insecticides can provide extended control. Selection of treatment regimes for controlling fire ants should consider fire ant tolerance and the liability of a treatment regime for each land-use pattern.

In this chapter the biology of each species is briefly reviewed and is followed by discussions of the health hazard they present, their public health and economic impacts, methods of monitoring and control, and a discussion on emerging issues and needs.

6.1. Overview of biology and distribution in Europe and North America

6.1.1. Pharaoh ant biology

Pharaoh ants (*Monomorium pharaonis*) are a worldwide pest associated with human habitats. They are small (2mm long), with color variations that range from yellow to yellowish-brown to even a light red. Pharaoh ants do not sting, but are a nuisance to building occupants and an important contamination concern in medical and food preparation and processing facilities. They can nest in a variety of easily transported items, such as boxes and packaging, sheets of stationery, linen and clothes, and other items that offer harbourage (Smith, 1965). In addition, colonies can be initiated from small groups of worker caste ants and immature ants, or brood (Peacock, Sudd & Baxter, 1955a). Vail & Williams (1994) reported founding colonies from just 5 adult workers, 30 eggs, 19 larvae, and 3 pupae. Since colonies can be easily transported through commerce, the distribution of Pharaoh ants is worldwide, with these ants being present throughout Africa, Australia, Europe, the Hawaiian islands, Japan, North America (Canada and the United States), the Russian Federation and South America (Edwards, 1986; Reimer, Beardsley & Jahn, 1990).

Pharaoh ants can survive in a wide range of environmental conditions, but they thrive in warm, humid conditions of about 27–30°C and 70–80% relative humidity (Peacock & Baxter, 1949; Samšiňák, Vobrázková & Vaňková, 1984). Edwards (1986) indicated that Pharaoh ants could probably survive temperatures up to 45°C, if water was available. Since these types of conditions can occur in microenvironments within buildings, Pharaoh ant colonies typically occur indoors (Sudd, 1962; Smith, 1965). However, they are known to nest outdoors in subtropical climates (Vail, 1996) and even in temperate areas where warmth is maintained (Kohn & Vlček, 1986). The lower limit for colony survival is a sustained temperature of about 18°C. At 6–11°C, colony death can occur within 7 days, yet colonies can survive a few days of such cold and recover if conditions become favourable (Peacock, Waterhouse & Baxter, 1955; Edwards, 1986).

Pharaoh ants do not have mating flights that are typical of other ants. Instead, they mate within or near the nest, and new colonies form when a group of adult workers and brood move, or bud, from the original colony. It is not necessary for a queen(s) to be part of the budding colony (Peacock, Sudd & Baxter, 1955a). Colonies of Pharaoh ants can be comprised of several nests, with free movement among nests. Colony sizes vary tremendously, with 35 adult workers, 35 pupae, larvae or eggs, and a queen being reported as one the smallest natural colonies, while laboratory colonies of 400 queens and 50 000 workers have been reared (Peacock, Sudd & Baxter, 1955a,b; Williams & Vail, 1993) (Fig. 6.1). Egg-to-adult development time for Pharaoh ant workers is 22–54 days (Alvares, Bueno & Fowler, 1993), and adult worker longevity is 9–10 weeks, with queens living up to 39 weeks (Peacock & Baxter, 1950). Edwards (1986), however, observed queens living beyond 52 weeks in the laboratory.

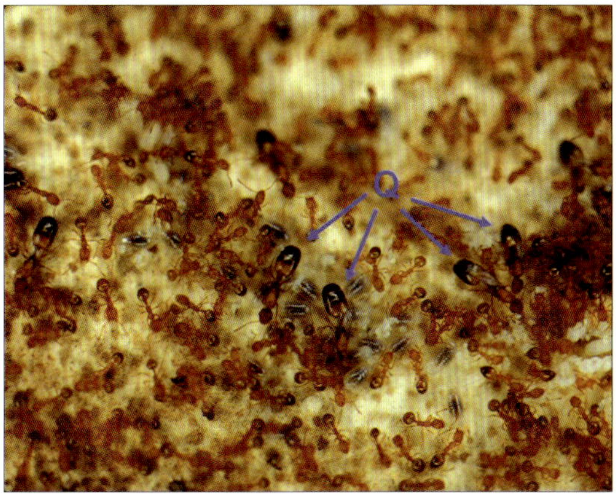

Fig. 6.1. Pharaoh ant colony with several queens (Q)

Source: Photo by S.D. Porter.

6.1.2. Fire ant biology

Fire ants are stinging ants whose name most commonly refers to the aggressive and invasive species *Solenopsis invicta*, which has an official common name of the red imported fire ant. They are a reddish brown to black ant, 3.2–6.4 mm in length. In addition to red imported fire ants, the names *fire ant* and *imported fire ant* also refer to another ant species, *Solenopsis richteri*, the black imported fire ant. Both of these species were accidentally introduced separately from South America into the United States before 1935 (Tschinkel, 2006) and have since spread throughout the southern United States. Of the two species, the red imported fire ant is more widespread, with the black imported fire ant restricted to northern pockets in the states of Mississippi and Alabama, and portions of Tennessee. In areas where the species overlap, they have hybridized. The majority of research and control efforts are for red imported fire ants, and most recommendations are applicable to both species. Since 1998, the red imported fire ant has significantly extended its geographic distribution – most likely through commerce. Well-established infestations were reported in California in 1998, in Australia in 2000, in Taiwan, China, in 2004, and in China, Hong Kong Special Administrative Region (Hong Kong SAR), China and in Mexico in 2005. Prior to the entry of black imported fire ants and red imported fire ants into the United States, four other species of fire ants were established in North America: the tropical fire ant (*Solenopsis geminata*), the southern fire ant (*Solenopsis xyloni*), and two species that inhabit the desert (*Solenopsis aurea* and *Solenopsis amblychila*). In areas where they overlapped, the red imported fire ant has displaced the southern fire ant.

Given adequate warmth and moisture, their tremendous reproductive capacity, mobility and stinging ability have allowed the red imported fire ant to become a dominant arthropod in the areas it invaded. Colonies of red imported fire ants can survive and reproduce over a temperature range of 20–35°C, with optimal temperatures of 27–32°C (Porter, 1988; Williams, 1990a). Moisture is critical for the survival of red imported fire ants, where a minimum 510 mm of annual precipitation has been estimated to be a reasonable threshold for sustaining a colony (Korzukhin et al., 2001). Because colonies are very mobile and can easily relocate within a day, red imported fire ants can occupy seemingly inhospitable habitats by moving to more favourable niches as environmental conditions change.

Fire ants develop from eggs, through four stages of larvae, to pupae and finally to adults. Depending on temperature, red imported fire ant development time, from egg to adult, ranges from 20 to 45 days. Adult workers can live as long as 97 weeks; however, depending on size and temperature, life spans normally range from 10 to 70 weeks (Hölldobler

& Wilson, 1990). Queens can live as long as 5–7 years (Tschinkel, 1987), with maximum egg-laying rates of over 2000 eggs a day (Williams, 1990a).

Fire ant colonies also contain a reproductive caste of non-stinging, winged males and females (alates) that initiate new colonies. Alates usually fly from the nest and mate during flight in late spring and early summer (although flights have been reported for all months). These mated females, or newly mated queens, have been reported to fly as far as 19.3 km from the nest (or even farther when aided by wind) during these mating flights, but most land within 1.6 km of their nest (Markin et al., 1971). After landing, newly mated queens move to a protected, moist harbourage – for example, in soil, under debris or in crevices – that can serve as an initial nesting site. Males die after mating. After landing, a queen sheds her wings and lays a clutch of eggs and tends them until adult workers develop from pupae. Worker ants will then tend the queen and additional eggs that are laid, and eventually a colony can grow exponentially. After six weeks, a new nest may be barely noticeable; after six months, however, nests 5–13 cm in diameter can be detected more easily. Very large, mature colonies, a few years old, can construct nests over a 0.9m in basal diameter and 0.9m high. Nest sizes and shapes can vary with habitat and soil type (Fig. 6.2). Over 4500 alates can be produced annually in large colonies (Tschinkel, 1986), but it is speculated that less than 0.1% of the newly mated queens will successfully found a colony (Taber, 2000).

Source: Photo by J.L. Castner. *Source:* Photo by D.H. Oi.

Source: Photo by B.M. Drees. *Source:* Photo by D.H. Oi.

Fig. 6.2. Fire ant nests

Note. (a) Inside a building; (b) under pavement stones; (c) in a pasture; (d) large, old nest in clay soil.

Colonies of red imported fire ants consist of two types:

1. colonies with only a single, fertile queen, or *monogyne* colonies
2. colonies with multiple fertile queens, or *polygyne* colonies.

Monogyne colonies are territorial, and thus fight with other colonies of red imported fire ants. As a result of this antagonistic behaviour, nests are farther apart, with densities of 99–370 nests/ha and with 100 000–240 000 ants per colony. In contrast, polygyne colonies are not antagonistic to other polygyne colonies, and thus queens, workers and immature ants (brood) can move between nests. The visible mound structure of polygyne nests are usually smaller in size and closer together than monogyne mounds, with densities of 494–1976 mounds/ha and with 100 000–500 000 ants per mature colony. Discriminating between individual nests or colonies in polygyne populations is uncertain, but in general, polygyne populations contain nearly twice the number of worker ants (35 million/ha versus 18 million/ha) and biomass per unit area than monogyne populations (Macom & Porter, 1996). Distinguishing between monogyne and polygyne colonies without locating fertile queens can now be accomplished through molecular markers (Valles & Porter, 2003).

6.2. Health hazards

6.2.1. Pharaoh ant infestations: pathogen transmission and contamination

For most residential situations, Pharaoh ants are a nuisance pest – they do not sting and their bite does not pierce human skin. However, their ability to establish colonies without constructing a separate nest structure and their large worker populations can make infestations in large buildings widespread and potentially disruptive to occupants, resulting in less productivity (Eichler, 1990). Of more serious concern are infestations in hospitals, because of the documented potential of Pharaoh ants to carry pathogens. Beatson (1972) isolated pathogenic bacteria of the genera *Pseudomonas, Salmonella, Staphylococcus, Streptococcus, Klebsiella* and *Clostridium* from Pharaoh ants collected in nine hospitals. Beatson also reported on cross-infection of a pneumonia pathogen in piglets by Pharaoh ants, despite the animals being held in an isolation unit. Mechanical transmission of a plague organism from Pharaoh ants that fed on infected animal carcasses demonstrates how their foraging behaviour can lead to transmission of disease (Alekseev et al., 1972). The propensity of Pharaoh ants to forage on wounds (Cartwright & Clifford, 1973; Eichler, 1990) and to infest institutional kitchens, thereby contaminating food, may all provide opportunities for transmitting pathogens. It has also been hypothesized that pathogens carried back to the nest may proliferate in the environs of the warm, humid nest and possibly be passed on to other colony members, increasing the probability of spread (Beatson, 1972; Edwards, 1986). Contamination of sterile instruments and supplies by Pharaoh ants chewing through packaging is a common problem (Beatson, 1972, 1973). Specific documentation of Pharaoh ant contamination affecting patients has not been reported, however.

6.2.2. Fire ant exposure: hazards related to stings and allergic reactions

The painful, burning sensation that is inflicted by the sting of a fire ant is easily the most recognizable hazard to people. While one sting is painful, it is not uncommon for a person to receive numerous stings simultaneously when ants swarm out of their nest to attack an intruder. This greatly intensifies the pain and can cause panic; thus, fear or apprehension of these ants can be present in heavily infested or newly infested areas. Stings are caused by adult worker ants injecting venom that contains mostly alkaloids. Such stings result in the immediate burning sensation at the sting site. Typically, this is followed by a wheal-and-flare response within 20 minutes, and then a sterile pustule forms within 24 hours (Kemp et al., 2000), which is accompanied by intense itching. Itchiness may persist for several days and infection may occur if pustules are broken. Large local reactions may occur in some stung individuals (17–56%), where an itching, hardened, reddish swelling develops several hours after the sting and persists 24–72 hours. This response resembles reactions caused by stings of other insects and may affect an entire extremity. Although no known treatment effectively prevents pustule formation or hastens healing (Kemp et al., 2000), topical insect bite treatments may help reduce itchiness (deShazo, Butcher & Banks, 1990).

Allergic or systemic reactions can vary from generalized itching, swelling and redness to anaphylaxis (a sudden, severe, potentially fatal, allergic systemic reaction). Surveys have reported anaphylactic reactions in 0.6–16% of individuals stung. Anaphylaxis may occur hours after a sting, with the formation of the sterile pustule(s), which distinguishes fire ant stings from other insect stings as the cause of the reaction (Kemp et al., 2000). Systemic reactions to fire ant stings usually occur in individuals sensitized by previous fire ant stings (Freeman et al., 1992). Sensitization rates of 16% and 17% from fire ant stings have been reported by Tracy and colleagues (1995) and Caplan and colleagues (2003), respectively. Thus, potentially 13 million people may be at risk for allergic reactions in fire ant infested areas in the United States (Caplan et al., 2003). Rapid sensitization may also occur. A three-week exposure to a fire ant endemic location by 107 non-sensitized individuals resulted in a sting rate of 51% and the development of fire-ant-specific antibodies in 16% of the subjects (Tracy et al., 1995).

There is also evidence of cross-reactivity with yellow-jacket wasp (*Vespula germanica*) venom. Unlike bee, hornet and wasp venoms that are mostly aqueous solutions containing proteins, red imported fire ant venom is a 95% water-insoluble alkaloid, with the remaining portion being an aqueous solution that contains four major allergenic proteins. It is a portion of these proteins that cross-reacts with *Vespula* venoms. Black imported fire ant venom has three of the four major allergenic proteins found in red imported fire ant venom (Kemp et al., 2000). Immunotherapy with injections of whole body fire ant extracts has been used to treat fire ant allergy (Freeman et al., 1992).

6.3. Exposure and risk assessment, with risk based on geographical location

6.3.1. Pharaoh ant distribution and population monitoring

Colony proliferation by budding, with and without the presence of mature queens and the suitability of small harbourages as nest sites, has contributed greatly to the worldwide spread of Pharaoh ants via commerce. This spread probably occurred before and after the original description by Linnaeus in 1758 of a Pharaoh ant specimen collected in Egypt. Pharaoh ant infestations are documented throughout Europe and North America (Edwards, 1986). In tropical and subtropical climates, infestations can extend outdoors; in temperate areas, heated buildings and man-made heat sinks permit winter survival and even colony growth (Kohn & Vlček, 1986; Vail, 1996). Colonies can have interconnected nest sites, and movement to suitable habitats, as environments change, permits the ants to become established in new sites. The movement of infested articles, packaging and luggage can also initiate infestations in buildings (Smith, 1965). Buildings with a high turnover or exchange of occupants, or shared services (such as laundry and equipment), such as hospitals and hotels, may have a higher risk of infestation (Edwards & Baker, 1981).

Monitoring populations of Pharaoh ants includes visual counts of trailing ants or counts of the number of trails present. Because Pharaoh ants are omnivorous, a variety of food lures (such as raw liver, jelly, peanut butter, honey and sugar solutions) have been used to locate and quantify their presence, generally for research studies (Edwards & Clarke, 1978; Haack, 1991; Oi et al., 1994). In laboratory testing, Williams (1990b) reported lard and several types of honey as being most accepted by Pharaoh ants. In general, a food lure is placed at various intervals on the interior and exterior of a building and near a suspected harbourage, and near food and water sources. After 2–24 hours (depending on foraging activity), lures are examined for Pharaoh ants, with the lure location and number of ants recorded.

6.3.2. Fire ant geographic range and potential expansion

The red imported fire ant is thought to originate from the Pantanal, a flood plain of the Paraguay River in south-western Brazil and parts of Bolivia and Paraguay, where it is adapted to the seasonal flooding and seems to have a confined distribution along the Paraguay and Paraná rivers (Allen et al., 1974; Buren et al., 1974; Vander Meer & Lofgren, 1990). The biotic and abiotic constraints on red imported fire ants found in South America are not present in North America (Buren et al., 1974; Porter, Fowler & Mackay, 1992), and thus its geographic distribution currently covers over 129.5 million ha in 13 states in the United States and in Puerto Rico. It also infests several islands in the West Indies (Davis, Vander Meer & Porter, 2001). Fig. 6.3 shows the counties and states in the United States under the imported fire ant quarantine, where movement of materials that potentially harbour fire ants are regulated. The red imported fire ant has been expanding its geographic range in the United States since its arrival there (Callcott & Collins, 1996). Based on climatic temperature and precipitation data, Korzukhin and colleagues

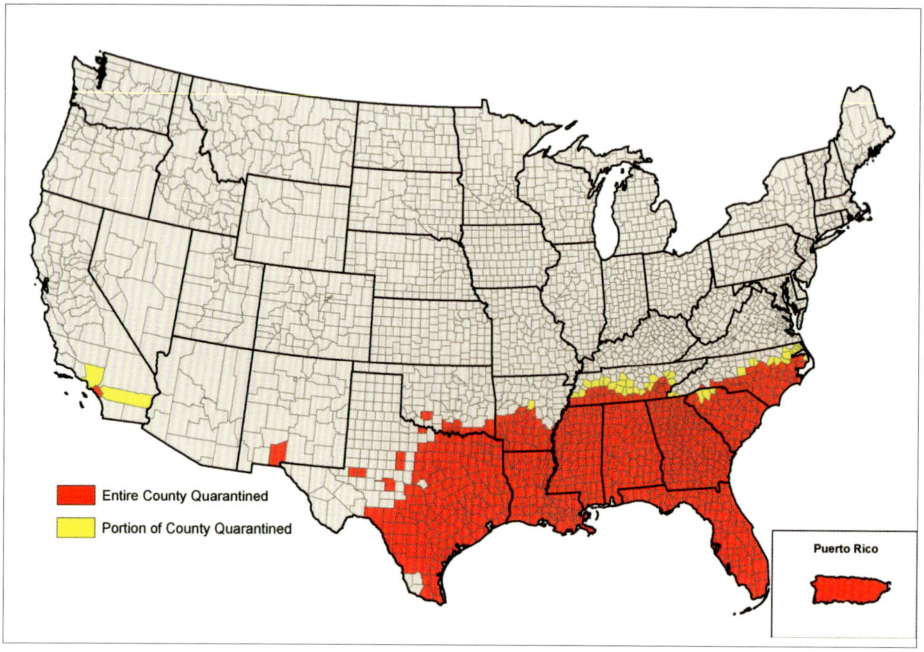

Fig. 6.3. Counties in the United States under imported fire ant quarantine

Source: Map from USDA-APHIS (2006).

(2001) predicted that its range certainly could be extended to include: the more northerly areas of Oklahoma, Arkansas and Tennessee; maritime Virginia; western Texas; and substantial portions of New Mexico, Arizona, California and Oregon. Infestations also could become established in Washington, Utah, Nevada, Delaware and Maryland (Fig. 6.4). Note that these predictions were based on interpolated weather data and do not account for natural and man-made microhabitats that may permit red imported fire ants to survive and become established.

Morrison and colleagues (2004) used worldwide temperature and rainfall patterns to predict the potential global distribution of red imported fire ants. For Europe, the areas surrounding the Mediterranean and Black seas are suitable for establishing fire ant colonies. These include, but are not limited to, the countries of Portugal, Spain, (southern) France, Italy, Slovenia, Croatia, Bosnia and Herzegovina, Montenegro, Serbia, Albania, Greece and Turkey. Temperature patterns also suggest possible infestations on the south-western coast of France and southern England. For the majority of Europe, temperatures are too cold; however, urbanized areas with artificial heat can provide suitable habitats. These predictions were based on interpolated weather station data and do not account for natural and man-made microhabitats that may permit red imported fire ants to survive and establish colonies. If fire ant incursions do become established in Europe, cold climates will most likely slow and restrict the range of geographic expansion.

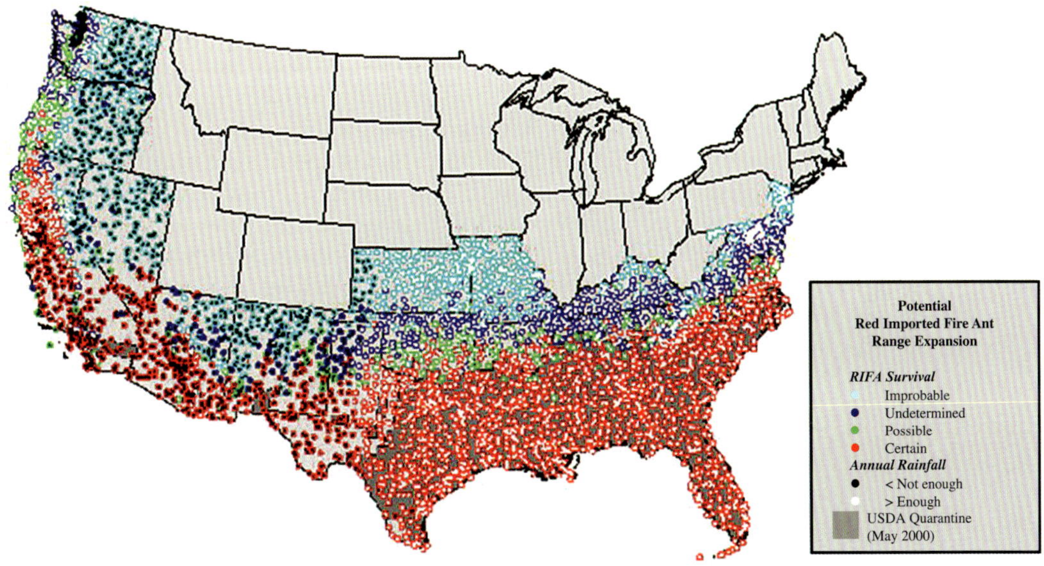

Fig. 6.4. Predicted invasion by red imported fire ants in continental United States

Source: Map from Korzukhin et al. (2001).

6.3.2.1 Fire ant population assessment and monitoring methods

Within infested areas, fire ant populations have generally been quantified by using nest density in 0.05–0.2 ha plots or along measured paths (called transects) established within study sites. In the first method of quantifying fire ant populations, the number of active nests is counted within each plot, and each individual nest is categorized as being inactive, low, moderately active or active. Category assignments are generally based on the number and rate of ants exiting a nest minimally disturbed by the insertion of a thin probe; criteria for activity categories, however, vary among researchers. Another population assessment system, used extensively by United States Department of Agriculture (USDA) researchers, assigns population indices to active nests, according to visual estimates of adult worker numbers and the presence or absence of worker caste brood after the nest is partially opened with a shovel. The absence of worker brood is an indication of a declining colony (Lofgren & Williams, 1982). Dimensions of individual nests also have been measured to calculate nest volume (Tschinkel, 1993). A more traditional method of sampling ants and other arthropods involves setting pitfall traps: arthropods fall and drown in the trap, which consists of a vial or container partially filled with a liquid (such as soapy water or automotive anti-freeze solution) inserted flush to the ground. Traps are collected after a few days to weeks and their contents identified. Pitfall traps are often used to sample both fire ants and other ants, to determine ant diversity and abundance. All the above methods are used for research, are labour intensive and require training.

Monitoring fire ant populations can be simplified by surveying for the presence or absence of fire ants with an attractive food source or lure. Vegetable oils and fatty foods have been used to detect the presence of fire ants, because they mainly attract ants that will feed on lipids. This is in contrast to liquid carbohydrates or foods sweetened with sugar, which will attract all types of ants and thus are not as selective as lipid-based foods. Examples of foods used to survey for the presence of fire ants include potato chips, cookies with high fat content, peanut butter, ground beef, canned fish packed in oil and processed meat products, such as sausages, hot dogs, wieners and canned luncheon meat. If such items are not available, any conveniently handled food with a high fat or lipid content could be substituted. Once the food lure is chosen, it is set on the ground along transects or grid patterns, for 30–60 minutes, and then checked for the presence of fire ants. Combining the use of bait stations with recording actual nest locations, while servicing the stations, can provide an efficient method of monitoring fire ant populations. However, the location of baits with fire ants does not always coincide with individual nest positions (Oi, Watson & Williams, 2004).

Long-range, pheromone-based surveillance traps are currently not available for fire ants. To help prevent fire ant incursions, agricultural border or port inspections of potential nest material or harbourages, such as nursery stock and earth-moving equipment, are necessary. The United States Federal Imported Fire Ant Quarantine, which is enforced by the United States Department of Agriculture, Animal and Plant Health Inspection Service (USDA-APHIS), lists items whose movement from quarantine areas are regulated by state and federal quarantine officials (USDA-APHIS, 2005) and could serve as a basis for targeted inspections in Europe. Identifying commodities from fire ant infested countries can further narrow inspections to high-risk importations. Making transporters and recipients of regulated items aware of signs of fire ants should also aid in early detection.

6.4. Public health impact

6.4.1. Pharaoh ants: prevalence in hospitals

A survey of half the hospitals in England indicated that Pharaoh ant infestations occurred in over 10% of the hospitals (Edwards & Baker, 1981). In South America, Pharaoh ants were reported to be common in hospitals and health care centres and were thought to be associated with hospital infections (Fowler et al., 1990). Besides the environmental conditions found in hospitals that are conducive to the establishment and growth of Pharaoh ant colonies, Burrus (2004) reported that dextrose solutions and dietary supplements commonly used in hospitals were potential food sources, as drips or spills made them accessible to ants. Pharaoh ants have been reported in giving sets (supplies for intravenous fluids) (Beatson, 1973). Whether patients lost significant amounts of medication or sustenance, either by direct ant feeding from dispensers or by equipment malfunction, was not documented.

6.4.2. Fire ants: stinging incidents

Extreme fire ant sting incidents are not systematically reported nationally or regionally. In urban areas infested with fire ants, an estimated 30–60% of the population are stung annually (deShazo, Butcher & Banks, 1990). A survey of suburban New Orleans conducted in 1973 indicated that 55% of stings occurred in children less than 10 years old (Clemmer & Serfling, 1975). Of individuals that are stung, surveys have indicated that 0.6–16% had anaphylactic reactions (Stafford et al., 1989; Kemp et al., 2000). From a survey of 29 300 physicians, Rhoades, Stafford & James (1989) reported 83 cases of fatal anaphylaxis related to fire ant stings. Of these cases, there was sufficient information to confirm 32 deaths. The fatalities ranged in age from infancy to 65 years, with the majority being healthy individuals. Dependent or immobile residents, such as the disabled, the elderly, young children and infants, are at greater risk of suffering from severe stinging incidents, where hundreds of stings may cause anaphylactic reactions, death or both. In a review of the literature (from 1966 to March 2003) and interviews, deShazo and colleagues (2004) found six cases of massive numbers of fire ant stings on elderly residents within health care facilities, of which four died within a week of being stung. Most extreme or fatal fire ant attacks, however, go to litigation, and many are settled with the stipulation that the details of these incidents are not to be disclosed (R.D. deShazo, University of Mississippi Medical Center, personal communication, April 2005). Similarly, departments of health from individual states may have records of severe stinging incidents, but the release of such information may be limited due to the potential for litigation (J. Goddard, Mississippi State Department of Health, personal communication, April 2005).

6.4.3. Pharaoh or fire ant infestations in health care facilities

Because these ants pose medical risks to patients, of stings and transmitting disease, elimination of infestations can be a major concern to health care managers and staff. Severe infestations of Pharaoh ants inside housing can be maddening, and such infestations have caused homeowners to consider selling their homes (Smith, 1965). Where maintaining sanitary and safe environments is imperative for hospitals and nursing homes, resources must be allocated to eliminate pest infestations. Ant infestations of either or both species can occur within a single building (personal observation). While control strategies within structures are similar for both species, extensive infestations can be exceedingly difficult and costly to control (Wilson & Booth, 1981). As a result, pest control companies often exclude these ant species from their contracts or require a separate contract to secure their services for Pharaoh or fire ant control. Goddard, Jarratt & deShazo (2002) recommend specifying monthly inspections and, if required, treatment and emergency service in contracts for pest control service for fire ants.

6.5. Public cost of infestation

6.5.1. Pharaoh ants: cost of control

Published documentation on costs to the public of infestation by Pharaoh ants was not found; however, the cost to treat a Pharaoh ant infestation can provide a partial indication of the potential economic impact. For example, fees in the United States in 2006 may conservatively range from US$ 35 to US$ 80 to treat a residential house with a structural perimeter of 61 linear meters. Larger buildings would cost more to service, but fees are negotiated, with the price per linear meter decreasing dramatically even though more labour and material would be required.

6.5.2. Fire ants: cost of health-related issues, control and management

The economic impact of fire ant infestations in the United States has been reported from surveys of various sectors in individual states and has been extrapolated across the infested areas of the United States. An extrapolation by Pereira and colleagues (2002) of a Texas survey reported the annual economic impact of fire ants in the United States to be more than US$ 6.5 billion across both urban and agricultural sectors. This impact was composed of costs for repair and replacement (66.1%), treatment (27.3%), medical expenditures (1.1%), and livestock and crop losses (5.5%). Based on a survey of households in Arkansas, Thompson and colleagues (1995) reported an extrapolation of annual losses due to fire ants of US$ 2.77 billion among nine heavily infested states in the United States. In the Arkansas survey, households that owned less than 0.4 ha (1 acre) of land were classified as being urban; they had estimated annual urban losses of US$ 1.2 billion across the infested states of the United States. Annual losses per urban household surveyed in Arkansas were US$ 87.10. Lard, Hall & Salin (2001) reported much higher annual losses of US$ 150.79 per household in five metropolitan areas of Texas (Austin, Dallas, Ft. Worth, Houston and San Antonio). Mean fire-ant-related expenditures per annum adjusted for a typical (or average) household in South Carolina were US$ 118 (Miller et al., 2000). Despite the variation in fire-ant-related losses or expenditures, the studies all concluded that the economic impact of fire ants could be substantial.

When urban household expenditures due to fire ants were categorized according to type of cost, treatments accounted for 53–55% of the expenditures, followed by repair and replacement costs (38–43%), and finally medical costs (2–9%). The medical costs were generally for retail medicines used to alleviate discomfort from fire ant stings (Thompson et al., 1995; Lard, Hall & Salin, 2001). However, a potentially large economic burden due to fire ants is from lawsuits that arise from severe incidents of stinging, especially at health care facilities (deShazo, Williams & Moak, 1999). Examples of awards for lawsuits related to fire ant stings include a 2005 settlement in Florida of US$ 1.875 million for the death of a bedridden patient and a jury award of US$ 1.2 million in the same state to a nursing home resident severely stung in 2002.

Households are a convenient unit for conducting economic impact surveys about fire ants. However, fire ants affect other sectors economically, particularly in the urban setting. For the five Texas cities mentioned earlier in this subsection, Lard and colleagues (2002) reported per city expenditures on fire-ant-related damages and treatments of US$ 53 628 a year. These costs were associated with controlling fire ants and replacing and repairing equipment in parks, landscapes, airports and cemeteries. It was also noted that fire ant damage to electrical and communications equipment had a total annual cost to the five cities of US$ 111 million.

6.5.3. Fire ants: cost of eradication

When the red imported fire ant was detected and identification confirmed in California, Australia and China, infestations were already quite extensive, thus making eradication more difficult and expensive. In California, the most recent outbreak was first detected in almond orchards in the Central Valley in 1997, and eradication efforts have been ongoing since then. In 1998, several more infestations, one of which covered at least 12 950 ha, were confirmed in the more urbanized areas of southern California (Klotz et al., 2003). By early 1999, additional surveys extended the infestation to 204 350 ha among several locations (California Department of Food and Agriculture, 1999). The value of a planned 10-year eradication programme in California was US$ 65.4 million (Jetter, Hamilton & Klotz, 2002). Funding for the first five years of this programme was approved in 2000, but due to budget limitations the funding and effort for eradication were curtailed in 2003. Thus, infestations still persist in California.

Projected damage estimates for the scenario where fire ants become established in both urban households and agriculture throughout California ranged from US$ 387 million to US$ 987 million a year. Relative to the climatic suitability for the establishment of fire ant colonies and to the similarity of crops produced in southern Europe to those produced in California, projected potential additional annual costs to treat fire ants in California citrus groves ranged from US$ 1.49 million to US$ 5.95 million, and for vineyards they ranged from US$ 4.11 million to US$ 16.44 million. This corresponded to a maximum 0.52% of annual farm receipts (Jetter, Hamilton & Klotz, 2002). Eradication of isolated infestations, quarantine programmes and non-irrigated desert environments are most likely to help limit the rapid spread of fire ants in California.

Environmental contamination by pesticides used in these programmes is a potential cost of eradication or control programmes. Insecticides used to eradicate fire ants in California (bifenthrin, fenoxycarb, hydramethylnon, pyriproxyfen, chlorpyrifos and diazinon) were not detected in well water. Fenoxycarb, hydramethylnon and pyriproxyfen, which are active ingredients of fire ant baits, and bifenthrin, a contact pyrethroid, were detected in surface water, mainly from nursery sites. These active ingredients are often used in quarantine treatments of nursery premises and nursery stock. Toxicity testing, using the water flea *Ceriodaphnia dubia*, revealed toxicity that could be directly linked to fire ant insecticide concentrations found in the water at nurseries (Levine et al., 2005). Kabashima and colleagues (2003) have reported practices that reduced bifenthrin run-off at a commercial nursery.

An infestation in Brisbane, Australia occupied more than 40000 ha. Treatments were initiated within two months of discovery, and a centralized eradication programme, coordinated by the Department of Primary Industries and Fisheries (DPIF, Queensland), was implemented within 18 months (Drees & Davis, 2002). After 3.5 years of a 6-year, US$133.5 million eradication programme, 99.4% of infestations were free of fire ants, and intensive surveillance continues for new or undiscovered incursions (DPIF, 2004). While the final outcome of the eradication effort in Australia is yet to be determined, the significant reductions in fire ants in an urbanized environment provides an example of the tremendous commitment, effort and organization needed to even attempt eradication. For a meaningful response to the detection of fire ant infestations, countries at risk for infestation should have regulatory clearance and a manufacturing source(s) for treatments, and a centralized coordinated response plan.

6.6. Control measures for Pharaoh and fire ants

6.6.1. Overview of general ant control tactics

Ants are one of the most diverse families of insects, with over 11800 species described worldwide (Agosti & Johnson, 2005) and with just a tiny percentage being considered pests. For example, Thompson (1990) considered about 35 of over 600 ant species as pests in the continental United States. The first step in implementing a control programme is to confirm the identification of the organism causing damage or the problem. Differentiation of Pharaoh ants or fire ants from other ants can be difficult, especially to untrained personnel or when these ants are a recent introduction. Due to regional differences in ant fauna, having a specialist verify the identity of Pharaoh or fire ants is recommended. Detailed descriptions, taxonomic keys and images of many ant species are available on several university Internet web sites and specialized web sites, such as AntWeb (California Academy of Sciences, 2006).

Pharaoh and fire ants have been the targets of innumerable methods of control. Eliminating extensive infestations from buildings permanently can be difficult to achieve; as with other pests; however, acceptable levels of suppression can be achieved with properly implemented control methods. Three general methods are currently being used to control Pharaoh and fire ants:

1. physical exclusion
2. application of residual contact insecticides
3. distribution of insecticidal baits.

The method (or combination of methods) used depends on the ant species, the extent or nature of the infestation, and the desired level of control, in terms of ant population reduction and speed of reduction.

6.6.1.1. Physical exclusion

Preventing ants from entering a building or structure is the objective of physical exclusion. This approach attempts to eliminate potential points of entry that can be used by ants to gain access to a building. Examples of eliminating entry points include sealing cracks and crevices on building exteriors and maintaining door sweeps and weather-stripping around windows. Removing access could entail pruning back tree branches that are in contact with a structure or relocating favourable harbourages, such as wood and debris piles, away from a building. Of course, making a building completely impervious to ant entry is unrealistic, given that some areas are inaccessible or cannot be made excludable, such as ventilation openings. Thus, it is more practical to focus on sealing areas where ants are observed entering and likely entry areas that are close to ant nests or harbourages. Identifying ants and possessing knowledge of their biology is essential to efficiently targeting exclusion efforts.

6.6.1.2. Residual contact insecticides

A common method of controlling ants has been to apply insecticides to building perimeters – around door frames, windows and other entryways – and also along interior baseboards and to actual ant trails and nests. Depending on the active ingredient and application rate, insecticide applications could result in a temporary barrier that immediately kills or repels ants from the treated area. However, except when nests are directly and thoroughly treated, contact insecticides affect only the non-reproducing worker caste that contact treated surfaces, generally leaving the colony intact. When the residual activity of the insecticide degrades, ants from the unaffected colonies are free to reinvade. If nests are inaccessible or not thoroughly treated, or both, insecticide applications may cause a colony to split into two or more colonies and disperse to other locations, resulting in a more widespread infestation.

Contact insecticides that are not repellent, that have a residual activity of over six months and that do not cause immediate insecticidal effects provide effective control of ants. This combination of characteristics permits extensive insecticide contact with trailing ants, because it circumvents the typical ant behaviour of avoiding deleterious substances. Also, possible insecticide transfer to the colony by trailing ants may have an impact on the colony (Soeprono & Rust, 2004a).

6.6.1.3. Insecticidal baits

Ant baits incorporate a toxicant into a food attractant that is carried to the nest by foraging ants and fed to the colony. Most bait products contain slow-acting toxicants dissolved or suspended in a vegetable oil or a sweetened aqueous liquid or syrup. The oil and sweet solution serve as lipid and carbohydrate food sources, respectively, for the ants. The toxicant-laden food is then absorbed into corn grits or some other carrier that makes the bait easier to handle and apply, as well as more available to the ants. Some baits are formulated as liquids and must be used in a bait dispenser; others are formulated as gels and dispensed through a syringe; still others are formulated as solid baits and delivered in receptacles called stations. Ants either carry the bait back to the colony and extract the toxicant-laden food from the carrier within the nest or extract it from the carrier immediately and carry it back to the colony internally. The slow action of the toxicants allows the foraging ants to feed the toxic bait to the other members of the colony before the foragers themselves die. When the toxicant is fed to the queen(s), she either dies or no longer produces new workers, and the colony will eventually die.

Utilization of the foraging and food sharing (by trophallaxis or regurgitation of food) behaviour of ants to distribute toxicant permits the treatment of inaccessible and undetected colonies. Also, the amount of active ingredient used in baits is lower than that of most formulations of residual contact insecticides. Because these bait toxicants entail delayed action, reducing populations is also delayed. Depending on the active ingredient, baits usually take 1–8 weeks to kill a colony. However, some baits will significantly reduce colony populations within three days. Ingestion of baits by the colony is necessary for effective treatment, and although ingested baits are slower than insecticides that kill immediately on contact, obtaining insecticide contact with most of the ants in a colony can be difficult with fast-acting insecticides. Factors that can interfere with bait foraging and effectiveness include seasonal food preferences of a species, competing food sources, bait spoilage and bait degradation by rain.

Overall, to be successful, the three general methods of ant control require knowledge of the ant species, skill in implementing control measures and diligence. Details on control methods specific to either Pharaoh or fire ants are provided below, in sections 6.6.2 and 6.6.3.

6.6.2. Management practices for Pharaoh ants

The following subsections cover the efficacy of management practices and the implementation of Pharaoh ant control programmes.

6.6.2.1. Efficacy of management practices

The control of Pharaoh ants has evolved and changed over the years, from trapping with raw liver to the application of residual insecticides and the utilization of ant baits (Edwards, 1986). Because Pharaoh ant nests are often difficult to observe (cryptic), inaccessible or both, the application of contact insecticides has generally been discouraged for controlling these ants. Also, the incomplete treatment of nests or the application of repellent, residual insecticides to building perimeters or foraging areas may cause colony fragmentation or migration (Edwards, 1986; Oi, Vail & Williams, 1996; Buczkowski et al., 2005). Moreover, the application of residual insecticides indoors in sensitive areas, such as hospitals and schools, may be perceived as being potentially hazardous. As a safer, more efficient alternative, ant baits have been successfully utilized to control indoor infestations of Pharaoh ants (Edwards & Clarke, 1978; Williams & Vail, 1994; Vail & Williams, 1995). Pharaoh ants can forage on the exterior of buildings, and foraging trails as long as 45m have been recorded (Vail & Williams, 1994). As a result, baits applied only to the exterior of buildings have effectively reduced indoor infestations (Oi et al., 1994; Vail, Williams & Oi, 1996). Recently, the application of non-repellent, residual insecticides to the perimeter of buildings has demonstrated effectiveness (Oi, 2005).

Table 6.1. Some Pharaoh ant baits, 2006

Mode of action	Active ingredient	Comments
Neural disrupter	Fipronil	—
Metabolic inhibitor	Hydramethylnon	—
Mid gut poison	Orthoboric acid	—
Neural disrupter	Sulfluramid [a]	Sulfluramid not sold after 2006
Neural disrupter	Indoxacarb	Label lists ants [b]
IGR	S-methoprene	—
IGR	Abamectin	Can kill adult ants
IGR	Pyriproxyfen [c]	Label lists ants [b], also contains orthoboric acid

[a] Sulfluramid = N-ethyl perfluorooctanesulfonamide.

[b] Label lists "ants" only; Pharaoh ants are not listed specifically.

[c] Pyriproxyfen = Nylar = 2-(1-Methyl-2-(4-phenoxyphenoxy)ethoxy)pyridine.

6.6.2.1.1. Active ingredients in Pharaoh ant baits

Commercially available ant baits that include Pharaoh ants on their label can have active ingredients with different modes of action (Table 6.1). The active ingredients in IGRs, such as methoprene and pyriproxyfen, can cause various deleterious effects, including death of larvae and pupae, deformities in queens and cessation of egg laying by queens. Adult workers are generally unaffected and die naturally (Edwards, 1975; Vail & Williams, 1995; Lim & Lee, 2005). Boric (or orthoboric) acid is a slow-acting mid gut toxicant that will kill Pharaoh ant adults and brood. A continuous supply of a 1% boric acid in a sucrose solution caused the demise of small, laboratory Pharaoh ant colonies in four weeks (Klotz et al., 1996). Also, metabolic inhibitors can kill small laboratory Pharaoh ant colonies within two weeks (Klotz et al., 1996).

6.6.2.1.2. Non-repellent residual insecticides

Pharaoh ants that follow the traces or scent of other colony members (trailing) on the exterior of buildings may warrant the application of insecticides to the exterior perimeter of these buildings. Non-repellent, slow-acting insecticides permit increased ant contact with treated surfaces (Soeprono & Rust, 2004b), and there is evidence that sprays for building perimeters that contain fipronil can be transferred to colonies by trailing ants (Soeprono & Rust, 2004a). In this manner, an insecticide treatment on the exterior of buildings may eliminate indoor colony infestations (Oi, 2005).

6.6.2.2. Implementation of Pharaoh ant control programmes

To implement a Pharaoh ant control programme, the following four steps are suggested.

1. Confirm the identification of the problem pest as being the Pharaoh ant (see section 6.6.1).

2. Map the extent of the infestation; it may be useful in determining where treatments should be placed or applied. Food lures that are easily handled yet attractive to Pharaoh ants, such as peanut butter on index cards, can be placed at set intervals or on potential activity sites to locate trails and harbourages. Good sanitation practices should be followed during surveys, so that alternative sources of food will not compete with lures or baits.

3. Select bait that Pharaoh ants will feed on (if bait is to be used), given that food preferences may change relative to the nutritional needs of the colony. Also, consider the type of active ingredient in the bait and its potential speed of efficient treatment relative to how soon control must be obtained. If Pharaoh ants are trailing extensively on the exterior of the building, treatment with a non-repellent, slow-acting insecticide may be useful, if exterior baiting is not effective or population declines are too slow.

4. Evaluate treatment efficacy periodically, through visual surveys, staff interviews or monitoring with food lures. Evaluations should be used to adjust future treatments and determine sources of reinfestation.

The control of Pharaoh ant infestations in both large building and small residential dwellings is feasible with the proper use of currently available materials. Reductions of 75–84% have been reported within 2–10 weeks; reductions of 100% have been reported after 16 weeks, using IGR baits (Edwards & Clarke, 1978; Vail, Williams & Oi, 1996; Lee et al., 2003); and reductions of 99% have been reported in 1 week, with metabolic inhibiting baits (Oi et al., 1994; Oi, Vail & Williams, 1996). However, when colonies located near baits are killed too quickly, rapid reinfestation by other colonies can occur. In contrast, slow-acting IGRs that do not affect workers can result in more thorough bait distribution among several colonies and a longer suppression of populations (Williams & Vail, 1994; Vail, Williams & Oi, 1996; Oi, Vail & Williams, 2000). Also, the application of non-repellent residual insecticides has provided 100% control in one week (Oi, 2005). Controlling Pharaoh ant infestations, especially in large facilities, will most likely be an ongoing process, requiring constant monitoring and treatment, as new colonies may enter with new occupants, merchandise or both.

6.6.3. Management practices for fire ants

The efficacy of management practices, home remedies and control devices, natural enemies and biological control agents, and implementation of fire ant control programmes are covered in the following subsections.

6.6.3.1. Efficacy of management practices

Fire ant control methods are based primarily on research on red imported fire ants, but are applicable to black imported fire ants and their hybrid. Because nests are often visible as mounds of excavated soil, fire ant control can be directed at individual colonies. Depending on the number of nests that need to be treated and on their accessibility, two approaches to applying treatments can be utilized. If only a few nests are in a limited area (say, less than 50 nests/ha), locating and treating individual nests would be feasible. If nest densities are high or the nests cannot be located because the management area is too large or difficult to survey (for example, nests are concealed by vegetation), or for all the preceding reasons, broadcasting a treatment over an infested area without locating individual nests is more practical.

6.6.3.1.1. Broadcasting fire ant bait

Fire ant baits that can be broadcast over an area are usually granular formulations comprised of slow-acting toxicants dissolved in vegetable oil (such as soybean oil), which is absorbed into corn grits. Most fire ant baits have a very low broadcast application rate of 1.1–2.2 kg/ha and are dispensed with manual seeders or larger seeders mounted on a tractor or all-terrain vehicle. For rough terrain, blowers have also been used. Aerial application is another option for area-wide and whole-community treatment programmes. Because individual nests do not have to be located and treated, broadcasting bait is a very efficient treatment method, both in terms of control and labour (Barr, Summerlin & Drees, 1999). Calibrating seeders accurately and dispensing bait evenly can be difficult. However, since foraging ants move to where baits are distributed, exact precision in application is not an absolute requirement. Also, broadcast application rates may not be effective for very small areas. For example, for 30 m^2, only 5g of bait should be applied at a recommended broadcast application rate of 1.65 kg/ha, which is well below label recommendations of 10.0–56.7 g/nest.

Fire ant baits do not have any residual activity. They usually must be collected by foraging fire ants within 1–2 days, before they become non-palatable, usually because of exposure to moisture prolonged heat, air and sunlight. Also, some active ingredients in baits are susceptible to photolysis (Vander Meer, Williams & Lofgren, 1982). To facilitate timely foraging, bait applications should be made when environmental conditions are conducive to foraging (air temperatures between 25°C and 32°C and no rain or irrigation 12–24 hours after broadcasting (Ferguson, Hosmer & Green, 1996)). Recalling that the oil in baits serve as a food source that colony members must ingest, baits must be fresh (oil should not be rancid) and, if possible, applied near nests to improve accessibility and competition with natural food sources. Seasonal food preferences may also affect bait acceptance. During the early summer, fire ants actively forage on oils to replenish depleted lipid reserves (Tschinkel, 1993); at this time, fire ant baits that contain oils are readily foraged, and alternative lipid sources, such as seeds and insects, are less available. However, weather permitting, fire ants will feed on baits throughout the year.

6.6.3.1.2. Active ingredients in fire ant baits

When broadcast properly, commercially available fire ant baits can reduce fire ant populations by over 90%. The mode of action of the active ingredients used in the baits will dictate the speed of its efficient action. Baits that contain metabolic inhibitors and neural disrupters can cause colony death as fast as one day to two weeks. Baits also can have active ingredients that interfere with reproduction; these are often referred to as IGRs. IGRs can prevent queens from laying eggs, and they cause a caste shift from workers to reproductive ants. As workers die off naturally, they are not replaced. Thus, colonies treated with IGRs will eventually succumb because workers will not be available to tend the queen(s), and she (they) will die. IGR baits may take 5–10 weeks to eliminate colonies, because IGRs do not affect adult workers.

While IGR baits require many weeks to kill colonies, in large treated areas (>0.4 ha), control can last for as long as a year. During the slow colony decline, remnant colonies will execute newly mated queens that try to reinfest treated areas. The duration of control in smaller areas is shorter, because they are more easily reinfested from adjacent areas. In contrast, the faster-acting metabolic and neural disruptive baits create a colony void that can be quickly reinfested within two months. Table 6.2 lists characteristics of fire ant baits that contain various active ingredients. For quick and extended suppression, one bait product is a mixture of metabolic inhibitor and IGR active ingredients.

6.6.3.1.3. Broadcasting residual insecticides

Broadcasting residual insecticides over an infested area attempts to eliminate fire ant populations and prevent reinfestation of the treated area. The most effective materials have been non-repellent, slow-acting contact insecticides, with residual activity for over six months. The absence of both repellency and the immediate death of ants facilitate insecticide contact with foraging ants and colonies located in treated areas. On the other hand, immediate death or repellency due to irritation often elicits avoidance of treated areas by fire ants (Oi & Williams, 1996) and reduces control to a short-lived suppression of fire ants. Reductions in fire ant nests of over 90% for over a year have been documented for the broadcast application of a granular insecticide containing fipronil (Barr & Best, 2002; Barr et al., 2005). Products with other active ingredients are available, but the level and duration of control they provide has not been as good. Product cost can be prohibitive for large areas and the application of contact residual insecticides must be more evenly distributed than baits.

Table 6.2. Some fire ant baits, 2005

Mode of action	Active ingredient	Speed of efficacy	Comments
Neural disrupter	Fipronil	4–6 weeks [a]	—
Metabolic inhibitor	Hydramethylnon	1–4 weeks [b]	—
Neural disrupter	Spinosad	3–14 days [c]	Organic certification
Neural disrupter	Indoxacarb	2–3 days [d]	—
IGR	Abamectin	6–8 weeks [e]	Some adult worker death
IGR	Fenoxycarb	4–8 weeks [f]	—
IGR	S-methoprene	8–10 weeks [g]	Registered for "croplands"
IGR	Pyriproxyfen	4–8 weeks [h]	Registered for various crops
Metabolic inhibitor + IGR	Hydramethylnon + S-methoprene	1–3 weeks [i]	IGR extends control

[a] From product label: EPA Registration Number 432-1219; Collins & Callcott (1998).

[b] From product label: EPA Registration No. 241-322; Barr (2004).

[c] From product label: EPA Registration No. 62719-304-239

[d] From Barr (2004).

[e] From Lofgren & Williams (1982); Williams (1985).

[f] From product label: EPA Registration No. 100-722; Collins et al. (1992).

[g] From product label: EPA Registration No. 2724-475.

[h] From product label: EPA Registration No. 1021-1728-59639.

[i] From product label: EPA Registration No. 2724-496; Barr et al. (2001).

6.6.3.1.4. Treating nests individually

Fire ant nests are often visible mounds of excavated soil that contain brood, adults and one queen or more. The visibility and accessibility of fire ant nests makes direct treatment of colonies feasible, where the objective is to eliminate the colony by killing the queen and most of the stinging adult workers. If the queen is not killed or functionally sterilized, she will continue to lay eggs and the colony will recover. In the case of multi-queen colonies, all the queens must be killed, thus making effective treatments especially difficult. Individual nest treatments are time consuming and labour intensive, because each mound must be located and treated (Barr, Summerlin & Drees, 1999). However, colonies treated properly with fast-acting insecticides can be eliminated more quickly than colonies treated with baits and residual insecticides with slow modes of action. Individual nest treatments, however, may cause the fire ants to relocate and create a new nest. Even if the queen is killed, surviving ants may still inhabit the treated nest or make a new nest until they die naturally, which may take over a month. Thus, it may be necessary to re-treat remaining nests that contain large numbers of stinging workers.

Mounds can be treated individually by chemical and non-chemical methods. Chemical methods include insecticides that are most commonly formulated as baits, liquid drenches, granules or dusts. Products formulated as drenches, granules or dusts generally contain active ingredients that are contact insecticides that will immediately affect treated ants. Because fire ant colonies move to occupy optimal temperature strata within a nest throughout the day, treatments should be applied when the colony is concentrated near the nest surface. Thus, optimal treatment times are generally limited to when air temperatures are cool (about 20–25°C) and the sun warms the nest surface. When properly treated, colonies may be eliminated within a few hours to a few days after treatment.

Bait products used for broadcast bait applications can be applied to individual nests. Because ants will distribute the bait to the colony, the emphasis with bait applications is to ensure baits are available when and where fire ants are foraging. Bait application to individual nests is relatively simple, where the recommended amount of bait, usually 15–75 ml (1–5 tablespoons), is sprinkled around the base of the nest. As with broadcast bait applications, the use of baits for individual nest treatments usually takes one to several weeks to eliminate colonies. However, there are now bait products that will kill colonies within three days.

Non-chemical treatment methods include pouring hot water onto the nests or physically excavating them. Scalding or boiling water (88–100°C) has been used to eliminate 20–60% of colonies, where about 11 litres of hot water were poured onto nests (Tschinkel & Howard, 1980). One variation of this technique uses steam generators that inject scalding water. The other non-chemical method, excavating colonies, is inefficient and impractical.

6.6.3.1.5. Combinations of baiting, residual insecticides and individual nest treatments

Each type of method used for fire ant control has advantages and disadvantages relative to speed of efficacy, residual activity and ease of application. Because fire ant stings represent a hazard, quick inactivation of colonies is often a priority. Treating nests individually with contact insecticides is potentially the fastest method to eliminate colonies; however, successful treatment can be difficult and inefficient. The combination of broadcasting bait followed by treating hazardous nests individually permits the efficient treatment of many colonies and the rapid suppression or elimination of the most dangerous colonies. It is generally recommended to bait first and then treat selected nests individually with a contact insecticide at least a day later. Baiting first allows colonies to forage and distribute baits without impediment from contact insecticides. In addition, colonies not successfully controlled by individual nest treatments may eventually succumb to ingested bait.

An alternative strategy is to combine the individual treatment of hazardous nests with a broadcast, non-repellent, residual contact insecticide. Non-repellent, contact insecticides may not suppress colonies immediately, thus the additional application of faster-acting insecticides to individual nests compensates for the delayed activity. If both types of treatments are contact insecticides, the sequence in which they are used is not critical.

The recent introduction of fire ant baits that suppress or kill colonies in 1–3 days may provide an acceptable time frame for effective treatment and may reduce the need for treatment combinations. Also, the application of a non-repellent, contact insecticide at least a day after baiting could retard reinfestation.

6.6.3.2. Home remedies and control devices

There are many home-made remedies and mechanical control devices that have not been scientifically proven to consistently eliminate fire ant colonies. Often, these so-called cures, which are usually directed at an individual nest, will kill many ants and the colony will abandon the nest. This gives the false impression that the colony was killed. Some home remedies also are dangerous to apply and contaminate the environment. These remedies include the use of gasoline or other petroleum products, battery acids, bleaches, and ammonia and other cleaning products. Such so-called remedies should never be used.

6.6.3.3. Natural enemies and biological control agents

Numerous organisms can prey on individual fire ants. Newly mated queens are attacked by other fire ants and other ant species and by dragonflies, spiders, lizards, birds and other general predators, but this predation does not reduce established fire ant populations. Direct applications of parasites and pathogens, which include mites, nematodes and fungi, to fire ant nests have not resulted in long-term control under field conditions (Williams et al., 2003). In general, these organisms required direct contact with individual ants and may be described as biological pesticides rather than self-propagating biological control agents that can spread naturally to other fire ant colonies.

In contrast, some parasites and pathogens have infiltrated the life-cycle of fire ants and are self-sustaining. These include two species of phorid flies (*Pseudacteon tricuspis* and *Pseudacteon curvatus*) and the microsporidium (protozoan) *Thelohania solenopsae*, which are well established in the United States (Williams et al., 2003). The phorid flies have been shown to reduce the short-term foraging activity of fire ants (Morrison & Porter, 2005) and reductions in field populations infected with *T. solenopsae* have been documented (Oi & Williams, 2002). While biological control agents are detrimental to fire ants, their impact at the population level may only be apparent with the establishment of several types of agents and after several years (Morrison & Porter, 2005). As such, the usefulness of these agents for immediately reducing the risk of fire ant stings in the urban environment is limited.

6.6.3.4. Implementation of fire ant control programmes

To implement a fire ant control programme, the following four steps are recommended.

1. **Confirmation.** Confirm that fire ants are the species causing the problem (see section 6.6.1).

2. **Determine where control is needed.** Fire ant population densities and distribution will vary with the degree to which habitats are conducive to colony growth. Determining whether control is needed should be based on the extent fire ants can be tolerated for specific land use patterns. Locating and mapping areas where fire ant control is required will help limit potential treatment areas.

3. **Design monitoring and treatment regimes.** Assess fire ant population levels when conditions are conducive to the type of monitoring method used. For example, if the number of fire ant nests in an area will be used to estimate ant populations, soil should be moist and vegetation low so that nests are easily seen. If food lures are used, set lures when weather conditions are conducive to fire ant foraging (see subsection 6.3.2.1 on monitoring fire ant populations). Consider deadlines for achieving control when scheduling sites for monitoring and treatment. Many fire ant treatments take at least a few weeks to obtain population reductions.

   The intensity of the control effort should reflect the potential hazard fire ants present, which is a function of the probability of a fire ant sting and the consequences if a sting occurs (such as a lawsuit). Treatment regimes relative to fire ant tolerance and liability will vary among land-use patterns. For example:

a. no treatment is needed for freeway median strips;

b. an annual broadcast application of an IGR bait is needed for an infrequently used park; and

c. a broadcast application of baits in the spring, summer and fall, plus an individual treatment of hazardous nests with fast-acting contact insecticides or baits and weekly monitoring for new nests or the presence of fire ants are all needed for a toddler playground.

   Thresholds have been used to initiate bait applications in cattle pastures. For example, to maintain a fire ant population below five nests per 1000 m^2 (50 nests/ha), fire ants on more than 40% of monitoring food lures would trigger bait applications (Pereira, 2003; R.M. Pereira, unpublished data, 2005). Thresholds based on the percentage of fire ants on monitoring lures would have to be adjusted for site layout (such as landscaping and land-use pattern), fire ant tolerance and monitoring scheme.

4. **Evaluation and adjustment of control programme.** Mapping areas where fire ant infestations are located and recording pre- and post-treatment population levels allow the control programme to be evaluated. Monitoring population levels at times when the potential for stings is high (such as the outdoor recreational season) will provide an indication of a treatment's efficacy and timeliness. Population levels can be based on nest densities, the percentage of lures with fire ants, the number of sting incidents or complaints, or a combination of these indicators. Maintaining site-specific historical records of treatment regimes, dates and weather conditions during treatment applications and population levels will allow for more precise adjustments to control programmes.

The examples of treatment regimes listed above are simple illustrations of possible control programmes. More complex programmes have been proposed that are tailored to more specific environments, such as health care facilities (Goddard, Jarratt & deShazo, 2002). These published programmes are only models and generally must be modified to suit site-specific needs.

6.7. Emerging problems and policy options

Early detection and a rapid response to eliminate infestations are vital to prevent the establishment of both fire ant and Pharaoh ant colonies in new areas. In the long run, preventing establishment is also more cost effective than eradication of established populations.

6.7.1. Fire ants

With regard to fire ants and countries at risk for fire ants becoming established, surveillance mechanisms, clearance for treatments, a manufacturing source(s) for treatments and centralized coordinated response plans should be kept in place. As discussed in section 6.5.3, the eradication programme in Australia has shown the best potential of being a model for fire ant eradication in a relatively large urban area. Moreover, initiatives to prevent the incursion of invasive ants, such as the Pacific Ant Prevention Plan (Invasive Species Specialist Group, 2004), are being developed. The Plan was generated for the Pacific islands and countries. The Pacific Invasive Ant Group provides guidelines that address legislation or policy to prevent entry of invasive ants, with a focus on red imported fire ants and outlines surveillance and response procedures for preventing ants from becoming established. These guidelines are broad in scope and adaptable to other at-risk regions, such as southern Europe.

The need to efficiently improve the ability to detect low levels of fire ant populations is dire. Current monitoring methods that utilize food lures are time sensitive and lack species specificity; also, the placement of lures must be within the foraging range of a colony, which can be small for small colonies. Alternatively, visual surveys for nests made by teams of inspectors are very labour intensive, and quality assurance must be maintained. Thus, research to develop more sensitive and efficient surveillance tools is needed to support plans and programmes to prevent and eradicate fire ant incursions.

While preparation for fire ant eradication is prudent for at-risk areas, improvements in fire ant IPM, including pesticide application, are needed for well-established fire ant infestations. Products with active ingredients that have long residual activity can provide control of fire ants for over nine months. These products are contact insecticides that require thorough coverage of an area to maintain control (reductions of more than 90%). Inadequate coverage and the proliferation of similar products (albeit less expensive) that may not have the same efficacy can lead to reapplications and greater exposure to pesticides. Fire ant baits, however, typically result in the application of less active ingredient per unit area than residual contact treatments (Drees, 2003). While the residual activity of fire ant baits is limited, as a tool for controlling these ants these baits are efficient and environmentally compatible. Integrating the efficient use of both baits and long-lasting residual contact insecticides, relative to the risk management of fire ant stings and exposure to pesticides, would improve fire ant IPM.

6.7.2. Pharaoh ants

With regard to Pharaoh ants, the first step towards control is physical exclusion. Access to a building, especially to such sensitive facilities as hospitals, must be reduced to a minimum by internal and external structural measures (such as sealing of cracks and crevices and moving possible garden harbourages away from a building). Also, regular monitoring and targeted insecticide application by specialists is fundamental and should be compulsory.

6.7.3. Research

Research on the dynamics of the reinfestation of treated areas and how to significantly delay reinfestation of sensitive areas (such as hospitals and preschool playgrounds) by fire ants and Pharaoh ants, while minimizing pesticide use, is needed to develop improved control strategies.

References[1]

Agosti D, Johnson NF, eds (2005). Antbase.org [web site]. New York, American Museum of Natural History (http://antbase.org, version 05/2005, accessed 2 August 2006).

Alekseev AN et al. (1972). [On the preservation of plague microbes on the surface and in the intestinal tract of the little yellow ant, *Monomorium pharaonis* L.: experimental study]. *Medicinskaja Parazitologija i Parazitarnye Bolezni*, 41:237–239 (in Russian).

Allen GE et al. (1974). Red imported fire ant, *Solenopsis invicta*: distribution and habitat in Mato Grosso, Brazil. *Annals of the Entomological Society of America*, 67:43–46.

Alvares LE, Bueno OC, Fowler HG (1993). Larval instars and immature development of a Brazilian population of Pharaoh's ant, *Monomorium pharaonis* (L.) (Hym., Formicidae). *Journal of Applied Entomology*, 116:90–93.

Barr CL (2004). How fast is fast? Indoxacarb broadcast bait. In: Pollet D, ed. *Proceedings of the 2004 Imported Fire Ant Conference*, Baton Rouge, LA, 21–23 March 2004. Baton Rouge, Louisiana State University:46–49.

Barr CL, Best RL (2002). Product evaluations, field research and new products resulting from applied research. *Southwestern Entomologist Supplement*, 25:47–52.

Barr CL, Summerlin W, Drees BM (1999). A cost/efficacy comparison of individual mound treatments and broadcast baits. In: *Proceedings of the 1999 Imported Fire Ant Conference, Charleston, South Carolina, 3–5 March 1999.* Clemson, SC, Clemson University (http://cphst.aphis.usda.gov/sections/SIPS/Conf_Proceedings/1999_%20IFA_Conference.pdf, accessed 26 January 2007):31–36.

Barr CL et al. (2001). Different ratios of s-methoprene and hydramethylnon baits as hopper blends for the suppression of red imported fire ants. In: *Red imported fire ant research and management program result demonstration handbook, 1999–2003*. College Station, TX, Texas A&M University (http://fireant.tamu.edu/research/arr/year/99-03/res_dem_9903/pdf/9_different_ratios.pdf, accessed 2 August 2006):28–31.

Barr CL et al. (2005). *Broadcast baits for fire ant control.* College Station, Texas, Texas A&M University (System B-6099).

Beatson SH (1972). Pharaoh's ants as pathogen vectors in hospitals. *Lancet*, 1:425–427.

Beatson SH (1973). Pharaoh's ants enter giving-sets. *Lancet*, 1:606.

[1] Information was gleaned mostly from pertinent research and review journal articles, with an emphasis on recent publications. However, many older articles provided the only relevant information available and thus were included. Trade publications were cited if they provided data in support of their findings.

Buczkowski G et al. (2005). Efficacy of simulated barrier treatments against laboratory colonies of Pharaoh ant. *Journal of Economic Entomology*, 98:485–492.

Buren WF et al. (1974). Zoogeography of the imported fire ants. *Journal of the New York Entomological Society*, 82:113–124

Burrus RG (2004). *Pharaoh ant consumption of fluids used in hospital environments* [MSc thesis]. Gainesville, FL, University of Florida.

California Academy of Sciences (2006). AntWeb [web site]. San Francisco, CA, California Academy of Sciences (http://www.antweb.org/index.jsp, accessed 12 November 2006)

California Department of Food and Agriculture (1999). Red imported fire ants: be on the lookout [web site]. Sacramento, CA, California Department of Food and Agriculture (http://www.cdfa.ca.gov/phpps/pdep/rifa/html/english/mediaroom/RIFAactionplan.html, accessed 12 November 2006).

Callcott AA, Collins HL (1996). Invasion and range expansion of imported fire ants (Hymenoptera: Fomicidae) in North America from 1918–1995. *Florida Entomologist*, 79:240–251.

Caplan EL et al. (2003). Fire ants represent an important risk for anaphylaxis among residents of an endemic region. *The Journal of Allergy and Clinical Immunology*, 111:1274–1277.

Cartwright RY, Clifford CM (1973). Letter: Pharaoh's ants. *Lancet*, 2:1455–1456.

Clemmer DI, Serfling RE (1975). The imported fire ant: dimensions of the urban problem. *Southern Medical Journal*, 68:1133–1138.

Collins HL, Callcott AMA (1998). Fipronil: an ultra-low-dose bait toxicant for control of red imported fire ants (Hymenoptera: Formicidae). *Florida Entomologist*, 81:407–415.

Collins HL et al. (1992). Seasonal trends in effectiveness of hydramethylnon (AMDRO) and fenoxycarb (LOGIC) for control of red imported fire ants (Hymenoptera: Formicidae). *Journal of Economic Entomology*, 85:2131–2137.

Davis LR Jr, Vander Meer RK, Porter SD (2001). Red imported fire ants expand their range across the West Indies. *Florida. Entomologist*, 84:735–736.

deShazo RD, Butcher BT, Banks WA (1990). Reactions to the stings of the imported fire ant. *The New England Journal of Medicine*, 323:462–466.

deShazo RD, Williams DF, Moak ES (1999). Fire ant attacks on residents in health care facilities: a report of two cases. *Annals of Internal Medicine*, 131:424–429.

deShazo RD et al. (2004). Fire ant attacks on patients in nursing homes: an increasing problem. *The American Journal of Medicine*, 116:843–846.

DPIF (2004). *National fire ant eradication program, November 2004 – progress report*. Brisbane, Australia, The State of Queensland, Department of Primary Industries and Fisheries (http://www2.dpi.qld.gov.au/extra/pdf/fireants/progressreport2004.pdf, accessed 2 January 2007).

Drees BM (2003). *Estimated amounts of insecticide ingredients used for imported fire ant control using various treatment approaches*. College Station, TX, Texas A&M University (Fire Ant Plan Fact Sheet No. 042; http://fireant.tamu.edu/materials/factsheets_pubs/pdf/042_jun03.pdf, accessed 2 August 2006).

Drees BM, Davis PR (2002). *Scientific review of the red imported fire ant eradication program – report to the red imported fire ant national consultative committee (October 2002)*. Brisbane, Australia, The State of Queensland, Department of Primary Industries and Fisheries (http://www2.dpi.qld.gov.au/extra/pdf/fireants/sapreport.pdf, accessed 2 January 2007).

Edwards JP (1975). The effects of a juvenile hormone analogue on laboratory colonies of Pharaoh's ant *Monomorium pharaonis* (L.) (Hymenoptera: Formicidae). *Bulletin of Entomological Research*, 65:75–80.

Edwards JP (1986). The biology, economic importance, and control of the Pharaoh's ant, *Monomorium pharaonis* (L.). In: Vinson SB, ed. *Economic impact and control of social insects*. New York, Praeger: 257–271.

Edwards JP, Baker LF (1981). Distribution and importance of the Pharaoh's ant *Monomorium pharaonis* (L.) in National Health Service Hospitals in England. *The Journal of Hospital Infection*, 2:249–254.

Edwards JP, Clarke B (1978). Eradication of Pharaoh's ants with baits containing the insect juvenile hormone analogue methoprene. *International Pest Control*, 20:5–10.

Eichler W (1990). Health aspects and control of *Monomorium pharaonis*. In: Vander Meer RK. Jaffe K, Cedeno A, eds. *Applied myrmecology: a world perspective*. Boulder, CO, Westview Press:671–675.

Ferguson JS, Hosmer AJ, Green ME (1996). Rate of removal of fenoxycarb (Logic®) fire ant bait by red imported fire ants (Hymenoptera: Formicidae) from treated pastures. *Journal of Entomological Science*, 31:20–32.

Fowler HG et al. (1990). Major ant problems of South America. In: Vander Meer RK, Jaffe K, Cedeno A, eds. *Applied myrmecology: a world perspective*. Boulder, CO, Westview Press:3–14.

Freeman TM et al. (1992). Imported fire ant immunotherapy: effectiveness of whole body extracts. *The Journal of Allergy and Clinical Immunology*, 90:210–215.

Goddard J, Jarratt J, deShazo RD (2002). Recommendations for prevention and management of fire ant infestation of health care facilities. *Southern Medical Journal*, 95: 627–633.

Haack KD (1991). Elimination of Pharaoh ants: an analysis of field trials with Pro-control and Maxforce ant baits. *Pest Control Technology*, 19:32–33, 36, 38, 42.

Hölldobler B, Wilson EO (1990). *The ants.* Cambridge, MA, Belknap Press of Harvard University Press.

Invasive Species Specialist Group (2004). *Pacific ant prevention plan.* Auckland, New Zealand, World Conservation Union (http://www.issg.org/database/species/reference_files/papp.pdf, accessed 12 November 2006).

Jetter KM, Hamilton J, Klotz JH (2002). Red imported fire ants threaten agriculture, wildlife and homes. *California Agriculture*, 56:26–34.

Kabashima JN et al. (2003). Pesticide runoff and mitigation at a commercial nursery site. In: Gan J et al., eds. *Deactivation and detoxification of biocides and pesticides.* ACS Symposium Series. Washington, DC, American Chemical Society:213–230.

Kemp SF et al. (2000). Expanding habitat of the imported fire ant (*Solenopsis invicta*): a public health concern. *The Journal of Allergy and Clinical Immunology*, 105:683–691.

Klotz JH et al. (1996). Laboratory evaluation of a boric acid liquid bait on colonies of *Tapinoma melanocephalum*, Argentine ants, and Pharaoh ants (Hymenoptera: Formicidae). *Journal of Economic Entomology*, 89:673–677.

Klotz JH et al. (2003). An insect pest of agricultural, urban, and wildlife areas: the red imported fire ant. In: Sumner DA, ed. *Exotic pests and diseases: biology and economics for biosecurity*. Ames, IA, Iowa State Press:151–166.

Kohn M, Vlček M (1986). Outdoor persistence throughout the year of *Monomorium pharaonis* (Hymenoptera: Formicidae). *Entomologia Generalis*, 11:213–215.

Korzukhin MD et al. (2001). Modeling temperature-dependent range limits for the fire ant *Solenopsis invicta* (Hymenoptera: Formicidae) in the United States. *Environmental Entomology*, 30:645–655.

Lard CF, Hall C, Salin V (2001). *The economic impact of the red imported fire ant on the homescape, landscape, and the urbanscape of selected metroplexes of Texas.* . College Station, TX, Department of Agricultural Economics, Texas A&M University (Faculty Series

Paper FP01–3; http://agecon.tamu.edu/publications/facultyPapers/2001/fp01-3.pdf, accessed 2 January 2007).

Lard CF et al. (2002). Economics assessments of red imported fire ant on Texas' urban and agricultural sectors. *Southwestern Entomologist Supplement*, 25:123–137.

Lee CY et al. (2003). Evaluation of methoprene granular baits against foraging Pharaoh ants, *Monomorium pharaonis* (Hymenoptera: Formicidae). *Sociobiology*, 41:717–723.

Levine J et al. (2005). *Surface and ground water monitoring of pesticides used in the red imported fire ant control program*. Sacramento, CA, California Department of Pesticide Regulation, Environmental Monitoring Branch (Report EH05-02; http://www.cdpr.ca.gov/docs/empm/pubs/ehapreps/EH0502.pdf, accessed 3 August 2006).

Lim SP, Lee CY (2005). Effects of juvenile hormone analogs on new reproductives and colony growth of Pharaoh ant (Hymenoptera: Formicidae). *Journal of Economic Entomology*, 98:2169–2175.

Lofgren CS, Williams DF (1982). Avermectin B_1a: highly potent inhibitor of reproduction by queens of the red imported fire ant (Hymenoptera: Formicidae). *Journal of Economic Entomology*, 75:798–803.

Macom TE, Porter SD (1996). Comparison of polygyne and monogyne red imported fire ant (Hymenoptera: Formicidae) population densities. *Annals of the Entomological Society of America*, 89:535–543.

Markin GP et al. (1971). Nuptial flight and flight ranges of the imported fire ant, *Solenopsis saevissima richteri* (Hymenoptera: Formicidae). *Journal of the Georgia Entomological Society*, 6:145–156.

Miller S et al. (2000). Contingent valuation of South Carolina households' willingness to pay for imported fire ant control. In: Croker JL, Vail, KM, Pereira RM, eds. *Proceedings of the 2000 Imported Fire Ant Conference*. Chattanooga, Tennessee, 5–7 April 2000. Chattanooga, Tennessee, The University of Tennessee (http://cphst.aphis.usda.gov/sections/sips/conf_proceedings/2000_%20ifa_conference.pdf, accessed 26 January 2007).

Morrison LW, Porter SD (2005). Testing for population-level impacts of introduced *Pseudacteon tricuspis* flies, phorid parasitoids of *Solenopsis invicta* fire ants. *Biological Control*, 33:9–19.

Morrison LW et al. (2004). Potential global range expansion of the invasive fire ant, *Solenopsis invicta*. *Biological Invasions*, 6:183–191.

Oi DH, Vail KM, Williams DF (1996). Field evaluation of perimeter treatments for Pharaoh ant (Hymenoptera: Formicidae) control. *Florida Entomologist*, 79:252–263.

Oi DH, Vail KM, Williams DF (2000). Bait distribution among multiple colonies of Pharaoh ants (Hymenoptera: Formicidae). *Journal of Economic Entomology*, 93:1247–1255.

Oi DH, Watson CA, Williams DF (2004). Monitoring and management of red imported fire ants in a tropical fish farm. *Florida Entomologist*, 87:522–527.

Oi DH, Williams DF (1996). Toxicity and repellency of potting soil treated with bifenthrin and tefluthrin to red imported fire ants (Hymenoptera: Formicidae). *Journal of Economic Entomology*, 89:1526–1530.

Oi DH, Williams DF (2002). Impact of *Thelohania solenopsae* (Microsporidia: Thelohaniidae) on polygyne colonies of red imported fire ants (Hymenoptera: Formicidae). *Journal of Economic Entomology*, 95:558–562.

Oi DH et al. (1994). Indoor and outdoor foraging locations of Pharaoh ants (Hymenoptera: Formicidae), with implications for control strategies using bait stations. *Florida Entomologist*, 77:85–91.

Oi FM (2005). Household ants still #1 pest. *Florida Pest Pro Magazine*, 1:12, 14, 16.

Peacock AD, Baxter AT (1949). Studies in Pharaoh's ant, *Monomorium pharaonis* (L.). 1. The rearing of artificial colonies. *Entomologist's Monthly Magazine*, 85:256–260.

Peacock AD, Baxter AT (1950). Studies in Pharaoh's ant, *Monomorium pharaonis* (L.). 3. Life history. *Entomologist's Monthly Magazine*, 86:171–178.

Peacock AD, Sudd JH, Baxter AT (1955a). Studies in Pharaoh's ant, *Monomorium pharaonis* (L.). 11. Colony foundation. *Entomologist's Monthly Magazine*, 91:125–129.

Peacock AD, Sudd JH, Baxter AT (1955b). Studies in Pharaoh's ant, *Monomorium pharaonis* (L.). 12. Dissemination. *Entomologist's Monthly Magazine*, 91:130–133.

Peacock AD, Waterhouse FL, Baxter AT (1955). Studies in Pharaoh's ant, *Monomorium pharaonis* (L.). 10. Viability in regard to temperature and humidity. *Entomologist's Monthly Magazine*, 91:37–42.

Pereira RM (2003). Areawide suppression of fire ant populations in pastures: project update. *Journal of Agricultural and Urban Entomology*, 20:123–130.

Pereira RM et al. (2002). Yellow head disease caused by a newly discovered *Mattesia* sp. in populations of the red imported fire ant, *Solenopsis invicta*. *Journal of Invertebrate Pathology*, 81:45–48.

Porter SD (1988). Impact of temperature on colony growth and developmental rates of the ant, *Solenopsis invicta*. *Journal of Insect Physiology*, 34:1127–1133.

Porter SD, Fowler HG, Mackay WP (1992). Fire ant mound densities in the United States and Brazil (Hymenoptera: Formicidae). *Journal of Economic Entomology*, 85:1154–1161.

Reimer N, Beardsley JW, Jahn G (1990). Pest ants in the Hawaiian islands. In: Vander Meer RK, Jaffe K, Cedeno A, eds. *Applied myrmecology: a world perspective*. Boulder, CO., Westview Press:40–50.

Rhoades RB, Stafford CT, James FK Jr (1989). Survey of fatal anaphylactic reactions to imported fire ants stings. Report of the Fire Ant Subcommittee of the American Academy of Allergy and Immunology. *The Journal of Allergy and Clinical Immunology*, 84:159–162.

Samšiňák K, Vobrázková E, Vaňková J (1984). A laboratory-reared stock of *Monomorium pharaonis* L. (Hymenoptera: Formicidae). *Zeitschrift für Angewandte Entomologie*, 97:399–402.

Smith MR (1965). *House-infesting ants of the eastern United States: their recognition, biology, and economic importance*. USDA Technical Bulletin 1326. Washington, DC, United States Department of Agriculture.

Soeprono AM, Rust MK (2004a). Effect of horizontal transfer of barrier insecticides to control Argentine ants (Hymenoptera: Formicidae). *Journal of Economic Entomology*, 97:1675–1681.

Soeprono AM, Rust MK (2004b). Effect of delayed toxicity of chemical barriers to control Argentine ants (Hymenoptera: Formicidae). *Journal of Economic Entomology*, 97:2021–2028.

Stafford CT et al. (1989). Imported fire ant as a health hazard. *Southern Medical Journal*, 82:1515–1519.

Sudd JH (1962). The natural history of *Monomorium pharaonis* (L.) (Hym., Formicidae) infesting houses in Nigeria. *Entomologist's Monthly Magazine*, 98:164–166.

Taber SW (2000). *Fire ants*. College Station, TX, Texas A&M University Press.

Thompson CR (1990). Ants that have pest status in the United States. In: Vander Meer RK, Jaffe K, Cedeno A, eds. *Applied myrmecology: a world perspective*. Boulder, CO, Westview Press:51–67.

Thompson LC et al. (1995). Fire ant economic impact: extending Arkansas' survey results over the south. In: Vinson SB, Drees BM, eds. *Proceedings of the Fifth International Pest Ant Symposia and 1995 Annual Imported Fire Ant Research Conference*, San Antonio, Texas, 2–4 May 1995:155–156.

Tracy JM et al. (1995). The natural history of exposure to the imported fire ant (*Solenopsis invicta*). *The Journal of Allergy and Clinical Immunology*, 95:824–828.

Tschinkel WR (1986). The ecological nature of the fire ant: some aspects of colony function and some unanswered questions. In: Lofgren CS, Vander Meer R, eds. *Fire ants and leaf cutting ants: biology and management.* Boulder, CO, Westview Press: 72–87.

Tschinkel WR (1987). Fire ant queen longevity and age: estimation by sperm depletion. *Annals of the Entomological Society of America*, 80:263–266.

Tschinkel WR (1993). Sociometry and sociogenesis of colonies of the fire ant *Solenopsis invicta* during one annual cycle. *Ecological Monographs*, 64:425–457.

Tschinkel WR (2006). *The fire ants*. Cambridge, MA, Belknap Press of Harvard University Press.

Tschinkel WR, Howard DF (1980). A simple, non toxic home remedy against fire ants. *Journal of the Georgia Entomological Society*, 15:102–105.

USDA-APHIS (2005). *Imported fire ant 2005: quarantine treatments for nursery stock and other regulated articles.* Washington, DC, United States Department of Agriculture, Animal and Plant Health Inspection Service (APHIS Program Aid No. 1822; www.aphis.usda.gov/lpa/pubs/ifapub.pdf, accessed 2 January 2007).

USDA-APHIS (2006). Imported fire ant quarantine [web site]. Washington, DC, United States Department of Agriculture, Animal and Plant Health Inspection Service (http://www.aphis.usda.gov/ppq/maps/fireant.pdf, accessed 26 January 2007).

Vail KM (1996). *Foraging, spatial distribution, and control of the Pharaoh ant,* Monomorium pharaonis *(L.)* [Ph.D. dissertation]. Gainesville, FL, University of Florida.

Vail KM, Williams DF (1994). Foraging of the Pharaoh ant, *Monomorium pharaonis*: an exotic in the urban environment. In: Williams DF, ed. *Exotic ants: biology, impact, and control of introduced species*. Boulder, CO, Westview Press:228–239.

Vail KM, Williams DF (1995). Pharaoh ant (Hymenoptera: Formicidae) colony development after consumption of pyriproxyfen baits. *Journal of Economic Entomology*, 88:1695–1702.

Vail KM, Williams DF, Oi DH (1996). Perimeter treatments with two bait formulations of pyriproxyfen for control of Pharaoh ants (Hymenoptera: Formicidae). *Journal of Economic Entomology*, 89:1501–1507.

Valles SM, Porter SD (2003). Identification of polygyne and monogyne fire ant colonies (*Solenopsis invicta*) by multiplex PCR of *Gp-9* alleles. *Insectes Sociaux*, 50:199–200.

Vander Meer RK, Lofgren CS (1990). Chemotaxonomy applied to fire ant systematics in the United States and South America. In: Vander Meer RK, Jaffe K, Cedeno A, eds. *Applied myrmecology: a world perspective*. Boulder, CO, Westview Press:75–84.

Vander Meer RK, Williams DF, Lofgren CS (1982). Degradation of the toxicant AC 217,300 in Amdro imported fire ant bait under field conditions. *Journal of Agricultural and Food Chemistry*, 30:1045–1048.

Williams DF (1985). Laboratory and field evaluation of avermectin against the imported fire ant. *Southwestern Entomologist Supplement*, 7:27–33.

Williams DF (1990a). Oviposition and growth of the fire ant *Solenopsis invicta*. In: Vander Meer RK, Jaffe K, Cedeno A, eds. *Applied myrmecology: a world perspective*. Boulder, CO, Westview Press:150–157

Williams DF (1990b). Effects of fenoxycarb baits on laboratory colonies of the Pharaoh's ant, *Monomorium pharaonis*. In: Vander Meer RK, Jaffe K, Cedeno A, eds. *Applied myrmecology: a world perspective*. Boulder, CO, Westview Press:676–683.

Williams DF, Vail KM (1993). Pharaoh ant (Hymenoptera: Formicidae): fenoxycarb baits affect colony development. *Journal of Economic Entomology*, 86:1136–1143.

Williams DF, Vail KM (1994). Control of a natural infestation of the Pharaoh ant (Hymenoptera: Formicidae) with a corn grit bait of fenoxycarb. *Journal of Economic Entomology*, 87:108–115.

Williams DF et al. (2003). Biological control of imported fire ants (Hymenoptera: Formicidae). *American Entomologist*, 49:150–163.

Wilson GR, Booth MJ (1981). Pharaoh ant control with IGR in hospitals. *Pest Control*, 49(3):14–19, 74.

7. Flies

Jerome R. Hogsette and Jens Amendt

Summary

Flies constitute a major group of nuisance species in rural and urban environments worldwide. Many species are collectively called filth flies, because of their association with potentially contaminated substrates, such as food wastes, faeces, animal manures and carrion. Through this association, they can quite easily and accidentally become disease vectors, by transmitting pathogens, especially those that cause enteric infections (such as *Salmonella* and *Campylobacter*), from contaminated to uncontaminated substrates. The epidemiological association of flies with various diseases is well documented and it has been established that certain flies are capable of contaminating food with pathogens. Nevertheless, there is still much discussion about the role filth flies play in actually transmitting pathogens to people and, more importantly, about the extent to which this transmission leads to disease. In fact, in urban areas of the northern hemisphere, the main complaint at present is about the annoying presence of flies, but rising temperatures due to changes in the climate may lead in the future to an increase in fly populations and a concomitant increase in fly-borne diseases.

A number of management practices or techniques can be used in urban areas to combat flies, and they are presented here. Among these practices is trapping. Outdoors and indoors, it is a good way to manage fly populations around homes, apartments and stores. Many fly traps do not involve the use of pesticides and are safe to use around people and their companion animals. Indoors and outdoors, sanitation is the key to effective fly control. Elimination of food leftovers, breeding sites and shelter will minimize fly populations. It is therefore important to inform the public and health care officials about fly biology and management. Benchmarks for a good fly-management programme include further research on fly biology, the development of perimeter control techniques, the restriction of pesticide use to outbreak scenarios, the use of regular monitoring and the improvement of control devices.

7.1. Introduction

Flies, from the insect order Diptera, constitute a major group of nuisance species in rural and urban environments worldwide. Some 120 000 different species of flies have been described and they inhabit almost all marine and non-marine ecosystems. Flies can be a prevalent and important pest, as determined by the LARES pan-European housing survey made in eight European cities (WHO Regional Office for Europe, 2006) and their presence alone can be an indication of unsanitary conditions. Many flies bear the name filth flies because of their association with potentially contaminated substrates, such as food wastes, faeces, animal manures and carrion (Ebling, 1975). Through this association, flying from contaminated to uncontaminated substrates and transmitting pathogens, filth flies can quite easily and accidentally become disease vectors (Greenberg, 1971, 1973; Olsen, 1998; Hogsette & Farkas, 2000; Graczyk et al., 2001). The present chapter provides an overview of possible pathogens carried by filth flies and discusses the potential of flies being disease vectors in the northern hemisphere.

7.2. Biology and bionomics of filth flies in Europe and North America

Most filth flies have the ability to feed on and reproduce rapidly in a variety of organic substrates, such as carrion, faeces and food wastes, and to mechanically transmit pathogens (Fig. 7.1; Greenberg, 1973). The hair-like structures on the bodies of filth flies, their deeply channelled mouthparts and the hairs and sticky pads on their feet become easily contaminated as they walk and feed on contaminated substrates. In addition, some flies, such as houseflies, frequently regurgitate digestive juices while feeding and also defecate on surfaces where they feed or rest. Whether or not they have been contaminated with pathogens, some flies can be annoying simply because they are present in large

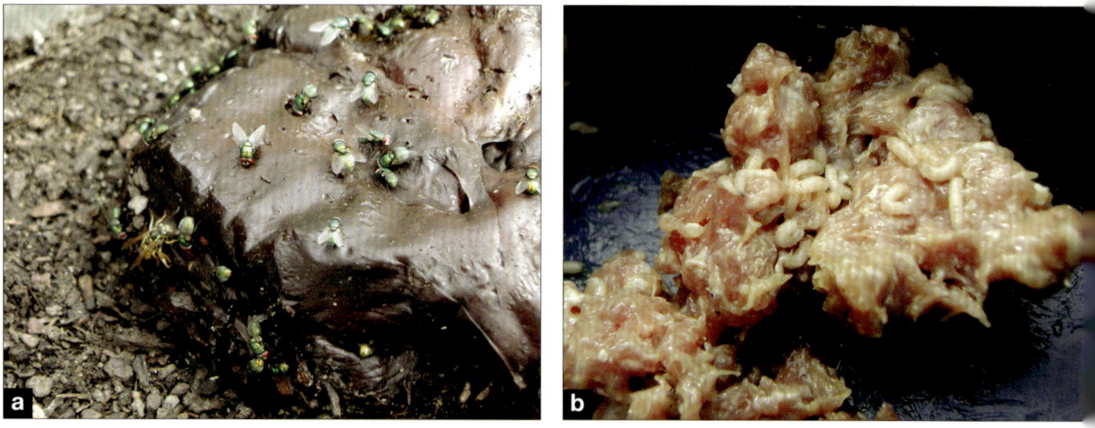

Fig. 7.1. Filth flies can transmit pathogens mechanically from contaminated substrates

a. Blow flies feeding on rotting liver; b. Fly larvae developing in poultry meat.

Source: Photos by J. Amendt.

Fig. 7.1c. Filth fly adults feeding on bovine manure

Source: Photo by J. Hogsette.

numbers. In fact, the mere presence of flies, particularly houseflies, in sensitive locations (such as food preparation areas), is considered an indication of poor hygiene. Where flies routinely come into contact with contaminated substrates, they can be noteworthy vectors of disease (Olsen, 1998; Graczyk et al., 2001; Clavel et al., 2002; De Jesus et al., 2004). They can aggravate the situation after the occurrence of natural disasters, such as the tsunami in South-East Asia in 2004 and Hurricane Katrina in New Orleans in 2005. The millions of flies produced by the sudden availability of corpses, raw sewage, hospital waste and animal carcasses presented a serious threat to public health and could have played an important role in disease transmission had control methods not been implemented (Srinivasan et al., 2006).

From a medical perspective, the most important pest species of flies belong to the superfamilies Muscoidea and Oestroidea – most notably the houseflies (Muscidae), lesser houseflies (Fanniidae), blow flies (Calliphoridae) and flesh flies (Sarcophagidae) (Greenberg, 1973; Olsen, 1998; Graczyk et al., 2001).

Broadly speaking, flies that are important as pests in urban environments can be divided into two groups: biting flies and non-biting flies. The non-biting flies constitute a larger group and will be discussed first.

The major urban and agricultural species of pest fly in the world is the housefly (*Musca domestica*). The housefly is important because it is ubiquitous and utilizes many proteinaceous materials, including garbage and human and animal faeces (James, 1947; Demény, 1989; Farkas & Papp, 1989). Houseflies are very prolific and large populations can develop very quickly (Hogsette, 1981). Adults are 6–9 mm in length, although size depends on the nutrients, crowding and moisture present in the habitat where larva develop. Adult females can be maintained for 30–40 days in the laboratory; field populations, however, probably do not exceed 10–14 days of age (Hogsette & Farkas, 2000). For males, mortality increases abruptly after they have copulated. Females can produce 1000 or more eggs in their lifetime (LaBrecque, Meifert & Weidhaas, 1972) and lay these in clutches of 100–150 eggs in suitable substrates (James, 1947). The life-cycle from egg to adult can be as short as 6.5 days at about 33.2°C (Larsen & Thompsen, 1940) and up to a month or more when substrate temperatures are much lower. Under optimum conditions, eggs hatch in 12–18 hours. There are three larval instars, which can complete their development in 3–5 days. Subsequent pupation occurs inside the integument of the third-instar larva, and adults can emerge after another 4–5 days of pupal development (Lysyk & Axtell, 1987).

Houseflies tend to disperse randomly and may move from contaminated to clean substrates several times in the course of a day. Their flight speed, without wind, is 8 km an hour and their known daily flight range is between 3 km (Winpisinger et al., 2005) and 30 km (Bishopp & Laake, 1921; Murvosh & Thaggard, 1966), but flies can also be distributed by wind, animals, and vehicles.

As adults, houseflies overwinter in a quiescent state and become active intermittently when microclimate temperatures exceed about 15°C. Adults remain active year-round in protected environments, such as animal housing (Sømme, 1961). Populations can grow to large numbers over the winter in animal housing and the adults disperse to nearby urban areas in the spring, when the housing is opened and cleaned out. At times, houseflies are a nuisance simply because of their sheer numbers and fly populations can limit outdoor recreational activities, especially if food is involved.

The face fly (*Musca autumnalis*), the false stable fly (*Muscina stabulans*) and the lesser housefly (*Fannia canicularis*) behave like houseflies, being mainly a nuisance to people, but they have also been associated with a number of pathogenic organisms (James, 1947; Treese, 1960; Greenberg, 1971; Skidmore, 1985). The lesser housefly, especially, is one of the most abundant flies found in human dwellings in many parts of the world. The black dump fly (*Hydrotaea aenescens*) has been shown to transmit pathogens under certain circumstances and is considered to be a pest species in some countries (Greenberg, 1991).

Calliphorids, such as the green blowfly (*Lucilia sericata*), the blue blowfly (*Calliphora vicina*) and Chrysomya spp., are ubiquitous; species composition, however, may vary from one location to another and from one season to another. These flies are easy to identify in nature because of their body colours: shiny metallic green, blue or bronze. Calliphorids are usually associated with animal carcasses, garbage and faecal material; they will, however, enter structures and land on food. This movement between contaminated and clean substrates makes them a potential pathogen vector. These flies can be very pestiferous at the outdoor or open meat markets still found in various parts of Europe. Females can land on unrefrigerated meat and quickly conceal large numbers of eggs in folds and openings in the meat. Other foods sold by outdoor vendors are also subject to attack, unless foods are properly wrapped and maintained at standard temperatures (James, 1947; Harwood & James, 1979; Kettle, 1995).

A number of species of flesh flies, in the family Sarcophagidae, can be present in urban areas, but rarely develop in pestiferous numbers. These flies are attracted to animal carcasses and decaying meat, and many deposit living larvae instead of eggs. Flesh flies, notably the spotted flesh fly (*Wohlfahrtia magnifica*), are known to cause myiasis (a disease that results from infestation of living tissue by fly larvae) in humans and animals (James, 1947; Hall & Wall, 1995).

Other flies that can be pestiferous in urban areas on occasion include *Hydrotaea dentipes*, *Hydrotaea ignava*, *Fannia manicata*, the latrine fly (*Fannia scalaris*), and some phorids and piophilids that could cause myiasis.

The major biting fly in urban areas is the stable fly (*Stomoxys calcitrans*), which has a long, bayonet-style mouthpart designed for sucking blood. Stable flies are a cosmopolitan pest (Zumpt, 1973; Skidmore, 1985; Soós & Papp, 1986), mainly of livestock, but also of people in villages, in the suburbs of larger cities and in recreational areas near shorelines of lakes, rivers and larger bodies of water (Newson, 1977; Betke, Schultka & Ribbeck, 1986; Hogsette, Ruff & Jones, 1987; Steinbrink, 1989). Both sexes require blood to reproduce, although nectar can be substituted for survival (not reproduction) when blood is unavailable (Jones et al., 1985). Stable flies are persistent in their feeding activities and continue to feed intermittently until replete (Hogsette & Farkas, 2000). Their bite is very painful, because they inject no anaesthetic when feeding. Some people have allergic reactions to the stable fly bite, some of which can be life threatening. In the laboratory, stable flies have been shown to mechanically transmit pathogenic organisms, such as those causing cutaneous leishmaniasis, anthrax, brucellosis, equine infectious anaemia and bovine diarrhoea virus, through their intermittent feeding behaviour; in the field, however, successful transmission is rare (Greenberg, 1971; Zumpt, 1973; Tarry, Bernal & Edwards, 1991).

The life-cycle of the stable fly is slightly longer than that of the housefly and is about 12–13 days in length at 27°C (Larsen & Thomsen, 1940). Its life-cycle is longer at cooler temperatures, and adults overwinter in a quiescent state, like houseflies (Hogsette & Farkas, 2000). Stable flies are not as prolific as houseflies, laying 60–800 eggs during their lifespan (Killough & McKinstry, 1965).

In urban areas, the preferred hosts of stable flies are dogs and people (Foil & Hogsette, 1994). Dogs are viciously attacked and their ears can be bloodied and notched by repeated feeding activity (Hogsette, Ruff & Jones, 1987). Preferred feeding areas on people are the lower legs and elbows (Hansens, 1951). When populations are large, flies will attempt to feed wherever they land and landing rates on people can exceed 100 stable flies a minute (Hogsette, Ruff & Jones, 1989). Stable flies can be a frustrating nuisance at zoos (Rugg, 1982), where they viciously attack the animals. Stable flies also disperse at about 8 km an hour (Hogsette, Ruff & Jones, 1987) and have been shown to move long distances with synoptic (large-scale) weather systems, such as cold fronts. The known flight range is 225 km (Hogsette & Ruff, 1985).

7.3. Health hazards

7.3.1. Diseases

Filth flies can become contaminated with more than a hundred different pathogens (Table 7.1) that cause human disease (Olsen, 1998). They develop in (and feed on) animal manures, human excrement, garbage and many types of decaying organic matter. As they disperse randomly, possibly several times a day, between contaminated and clean substrates, it is reasonable to expect that these flies transfer pathogens. Greenberg (1964) demonstrated that houseflies transmit *Salmonella typhimurium* to people; there is still, however, much discussion about the role played by filth flies in actually transmitting

Table 7.1. Pathogens known to be carried by flies in central Europe

Prions	Viruses	Bacteria
Scrapie prion (experimental)	Keratoconjunctivitis epidemica adenovirus	*Aeromonas hydrophila*
	Poliovirus	*Bacillus* spp.
	Rotavirus (experimental)	*Campylobacter jejuni*
		Chlamydia trachomatis
		Clostridium spp.
		Corynebacterium spp.
		Enterococcus faecalis
		Enterococcus spp.
		Erysipelothrix rhusiopathiae
		Escherichia coli
		Enterotoxic *Escherichia coli*
		Enterohaemorrhagic *E. coli* 0157:H7
		Flavobacterium spp.
		Helicobacter pylori
		Klebsiella pneumoniae
		Klebsiella spp.
		Micrococcus indolicus
		Micrococcus spp.
		Moraxella bovis
		Moraxella spp.
		Neisseria spp.
		Proteus vulgaris
		Proteus spp.
		Pseudomonas aeruginosa
		Salmonella enteritidis
		Salmonella paratyphi A
		Salmonella paratyphi B
		Salmonella typhimurium
		Salmonella typhosa
		Salmonella spp.
		Shigella dysenteriae
		Shigella sonnei
		Shigella spp.
		Staphylococcus aureus
		Staphylococcus epidermidis
		Staphylococcus pyogenes
		Streptococcus spp.
		Treponema pallidum pertenue subsp.
		Vibrio cholerae
		Vibrio fluvialis
		Yersinia enterocolitica

Table 7.1. (contd.)

Protozoa	Helminths	Fungi
Balantidium coli	*Ascaris lumbricoides*	*Absidia* spp.
Cryptosporidium parvum	*Capillaria hepatica*	*Alternaria* spp.
Entamoeba histolytica	*Hymenolepis* spp.	*Aspergillus niger*
Giardia intestinalis	*Necator americanus*	*Aspergillus* spp.
Trichomonas spp.	*Taenia* spp.	*Candida albicans*
	Toxocara spp.	*Candida* spp.
	Trichuris trichiura	*Fusarium* spp.
		Geotrichum spp.
		Microsporum gypseum
		Microsporum spp.
		Penicillium spp.
		Scopulariopsis spp.
		Trichophyton terrestre
		Trichothecium spp.

Source: Faulde (2002).

pathogens to people and more importantly, about the extent to which this transmission leads to the development of disease. Nevertheless, there is strong evidence that flies play an important role in certain human enteric bacterial infections; for example, flies can mechanically transfer pathogenic organisms, such as those that cause salmonellosis, shigellosis, and cholera (Levine & Levine, 1991; Khalil et al., 1994; Olsen & Hammack, 2000), all of which are severe diarrhoeal diseases. Adults can also carry enterohaemorrhagic *E. coli* serotype O157:H7, a virulent enteric pathogen (Sasaki, Kobayashi & Agui, 2000), and *Campylobacter* spp. (Rosef & Kapperud, 1983; Szalanski et al., 2004). Also, *Helicobacter pylori* might be transmitted by contaminated houseflies (Grübel et al., 1997), but both the minimum infectious dose of the bacterium for people and the quantity of *H. pylori* that can be carried by houseflies are unknown (Grübel et al., 1998). Moreover, the role of flies as a vector (or even as a reservoir for *H. pylori*) is disputed (Osato et al., 1998; Allen et al., 2004).

Synanthropic flies – that is, flies ecologically associated with humans – may carry bacteria resistant to a number of antibiotics (Fotedar et al., 1992; Rady et al., 1992; Rahuma et al., 2005), possibly playing an epidemiological role in health facilities (Sramova et al., 1992). They also have been identified as vectors of protozoan parasites, such as *Toxoplasma gondii* (Wallace, 1972; Graczyk, Knight & Tamang, 2005) and *Cryptosporidium parvum* (Graczyk et al., 1999, 2000, 2001). According to Clavel and colleagues (2002), the housefly acts as a transport vector of human cryptosporidiosis. Flies harbour the oocysts of *C. parvum* in their digestive tracts and on their external surfaces and can deposit them with their excreta in substantial quantities (about 100 oocysts/cm^2). Flies were found to be capable of carrying up to 131 oocysts each for at least three weeks. Because the natural hosts of *Cryptosporidium* are cattle and sheep, this problem may be restricted to rural areas.

Synanthropic flies have also been incriminated in the transmission of viral pathogens, including poliovirus, coxsackievirus and enteroviruses (Gregorio et al., 1972; Greenberg, 1973; Graczyk et al., 2001). Food exposed to flies in homes of patients with poliomyelitis in areas where epidemics occurred acquired enough poliovirus to produce a non-paralytic infection or asymptomatic carrier state when consumed by chimpanzees (Ward, Melnick & Horstmann, 1945). Melnick (1951) stated that sampled flies frequently tested positive for poliovirus, which was the only seasonal factor that could be correlated with summer epidemics of poliomyelitis.

Flies, moreover, are capable of transferring the eggs and cysts of various cestodes and nematodes (Olsen, 1998), particularly hookworms and ascarids. Furthermore, hamsters have been experimentally infected with scrapie, a disease classified as a transmissible spongiform encephalopathy, after eating extracts of the larvae and pupae of the flesh fly *Sarcophaga carnaria* that fed on scrapie-infected hamster brains (Post et al., 1999). Field transmission of the scrapie agent, however, has not been verified.

As stated in section 7.2, on "Biology and bionomics of filth flies in Europe and North America", the stable fly is quite inefficient at transmitting disease under field conditions (Greenberg, 1971; Zumpt, 1973; Tarry, Bernal & Edwards, 1991) and the major concern is the pain associated with its bites and the few people who are allergic to the proteins injected by the fly. Nevertheless, there are presumptions that biting flies are involved in the transmission of Lyme disease (Luger, 1990).

Most of the diseases caused by flies in urban areas are intestinal in nature, and victims may suffer a series of flu-like symptoms, including elevated temperature, diarrhoea and vomiting. Treatment varies, depending on the causative agent and finding a physician who is knowledgeable in this area of disease management is of utmost importance. Some bacteria, such as *E. coli* serotype 0157:H7, are extremely pathogenic and may cause death.

7.3.2. Myiasis

Myiasis, the infestation of living human or animal tissue with fly larvae (Hall & Wall, 1995), has the potential for tremendous human morbidity and mortality (Sherman, 2000). The classification of myiasis may be based on the separation of myiasis-producing Diptera into the groups that produce *accidental*, *facultative*, or *obligatory* myiasis. Accidental myiasis is most often the result of ingestion of food contaminated with maggots, the worm-like larva of any of various flies. Most ingested fly larvae are unable to complete their life cycles in the human digestive system; however, enteric myiasis can cause malaise, vomiting, pain and bloody diarrhoea. The larvae of more than 50 fly species are known to cause enteric myiasis, the most common of them are the housefly, the lesser housefly, the latrine fly and the false stable fly (James, 1947).

Facultative myiasis occurs when fly species that normally develop in faeces or dead animals lay their eggs or deposit their larvae in the tissues of living humans or animals. Maggots of these flies can develop in a living host, by feeding on dead tissue, but they sometimes invade living tissue as well. Urogenital and traumatic (open wound) facultative myiases occur most frequently. The former is associated mainly with the housefly, the lesser housefly, the latrine fly and the false stable fly; the latter is commonly caused by blow flies (Calliphoridae) and flesh flies (Sarcophagidae). Normally attracted to the rotting tissue of carrion, the maggots feed primarily on necrotic tissue, but they may also invade living tissue (James, 1947; Harwood & James, 1979).

In the third type of myiasis, called obligatory, the species is incapable of reproducing without a living host for larvae to feed upon. In Europe, obligatory myiasis is caused by blow flies, flesh flies and bot flies (family Oestridae). Their feeding can result in dermal creeping myiasis (where the path of the larvae beneath the skin can be traced), pain and inflammation (James, 1947).

7.3.3. Nuisance

During its lifetime, the female housefly is capable of producing up to 1000 eggs, and the resulting larvae will develop into adults in about 7–10 days (Larsen & Thomsen, 1940). The potential for a population explosion under the proper conditions is obvious. It is difficult to quantify the emotional effects of large numbers of flies on people already living under stress and expecting a fly-free environment. But people have been known to vacate their homes and apartments simply because of huge fly populations. In many parts of the world, urban development has extended into farming areas, resulting in significant increases in housefly populations in communities adjacent to farms, even though the source of flies may be up to 6.4 km away (Winpisinger et al., 2005). Flies can cause tremendous problems in these situations, by restricting outdoor recreational activities, particularly those that involve cooking or consumption of food (Thomas & Skoda, 1993; Winpisinger et al., 2005).

In a number of studies in the Middle East (J. Hogsette et al., unpublished observations, 1997), large numbers of flies were produced in expected and unexpected circumstances. In the Gaza Strip, large numbers of houseflies were produced in the towns and cities from lack of adequate storage, collection and disposal systems for garbage. Although public health did not seem to be adversely affected by the fly populations, reduction in fly populations would constitute an improvement in the quality of life. In many areas, flies had access to raw sewage, so the potential for contamination and transmission of disease was high. Farmers in Israel and Jordan significantly contributed to the increase in the numbers of flies at different times of the year by their farming practices. Housefly populations were at times unbearable in both countries as a result. To reduce the populations, one village in Israel used a large barrier of box traps. Although thousands of flies were trapped daily, results were mainly psychological. The nuisance factor was the main complaint, and diseases were probably not transmitted, because the flies were not contaminated in the harsh terrain.

In Europe and the United States, flies have long been considered to be a public health threat. This is based mainly on their past notoriety as a disease vector, as well as on their habit of developing in (and feeding on) manures and other undesirable organic wastes. In the late 19th century, a few flies were considered to be a normal part of every house-

Fig. 7.2. The presence of flies is considered an indication of unsanitary conditions *Source:* Photo by J. Hogsette.

hold life. But by the early 20th century, flies had been incriminated as disease vectors and public health laws were enacted in many countries. Typhoid was a major problem in Europe and the United States, and the housefly became known as the typhoid fly. Cities with poor sanitation had large numbers of typhoid cases while cities with good sanitation did not (Greenberg, 1973). After the Second World War, sewer lines were closed and sanitation was vastly improved. As a result, flies were denied access to contamination points and fly-borne disease essentially disappeared. Despite this, flies are still considered to be an indication of unsanitary conditions in homes, hospitals and restaurants (Fig. 7.2). Also, in times of disaster, proper sanitation levels are compromised, increasing the opportunity for flies to become contaminated and transmit disease. At present, a few flies in a home or restaurant may not constitute a serious health risk. But this may not be the case in the future.

7.4. Exposure and risk assessment

As already indicated in subsection 7.3.1, relatively little has been done to assess the risk of flies, especially in urban areas of the northern hemisphere. Some studies show a relationship between the source of an infection and the potential for people living nearby to become infected (for example, Greenberg, 1964), but the relative risk associated with flies transmitting foodborne pathogens has not been quantified. In fact, it has been questioned whether there are scientific studies that demonstrate that an organism as small as a fly can deliver an infective dose of a pathogen to exposed food.

Published research has shown that the incidence of enteric disease in people decreased with the distance from the source of infection, which was a large dairy farm with confined cows (Kobayashi et al., 1999). In a Japanese day-care centre, people who lived closest to the dairy had the greatest chance of coming in contact with the coliform *E. coli* serotype 0157:H7, carried by the cows and probably transmitted by houseflies. A similar incident was observed in southern Chile, where enteric diseases increased in a village after houseflies began developing in astronomical numbers in several tons of a mixture of turkey manure and straw stored at a nearby vineyard (J. Hogsette, unpublished observation, 1999). Houseflies can maintain *E. coli* serotype 0157:H7 internally for up to 30 days (Keen et al., 2003), and a single fly could potentially deliver a lethal dose of this highly pathogenic bacterium under certain circumstances – for example, by falling into a child's glass of water. The results from small-scale studies indicate strongly that flies play a role in transmitting pathogenic organisms, but exposure and risk assessment have not been quantified.

As one might expect, flies of certain species remain active year-round in areas between (and not too far beyond) the tropics of Cancer and Capricorn. Having year-round populations of flies increases the chance of a long-term fly nuisance and of the accompanying risk of fly-borne disease transmission.

As one approaches the poles, flies become seasonal, depending on their temperature tolerance. In this region, nuisance-level populations are of short duration, thus limiting the risk of disease transmission to a few weeks or months.

Over the past 100 years, the global average temperature has increased by about 0.6°C and this trend may show a fast rise in the future (Houghton et al., 2001; Root et al., 2003). This warming can affect the world's biota and the functioning of ecosystems in many indirect ways (Stenseth et al., 2002; Parmesan & Yohe, 2003). It is also possible that warmer conditions will promote the transmission of diseases by allowing a broader geographical distribution and an increase in the abundance of local disease-vector populations (Peterson & Shaw, 2003; Brownstein, Holford & Fish, 2005; Ogden et al., 2006; Poulin, 2006). When the relationship between fly numbers and weather conditions was examined, results showed that fly population changes are driven more by climatic conditions than by biotic factors (Goulson et al., 2005). With a simulated model of climate change, using recently predicted values for warmer temperatures, Goulson and colleagues (2005) predicted a potential increase in fly populations of 244% by 2080, compared with current levels. If this were to occur, concomitant increases in fly-borne diseases are expected.

7.5. Public health impact

The epidemiological association of flies with various diseases is well documented (Olsen, 1998; Graczyk et al., 2001; Nichols, 2005). The documentation shows that certain flies are capable of contaminating food with more than one foodborne pathogen and that natural populations of flies harbour these pathogens. It also shows that *E. coli* serotype 0157:H7 actively proliferates in the minute spaces of the housefly mouthparts and that this proliferation leads to persistence of the bacteria in fly faeces (Kobayashi et al., 1999; Sasaki, Kobayashi & Agui, 2000). The authors cited used DNA techniques to implicate houseflies as the source of *E. coli* in an outbreak in a day-care centre in a Japanese village (see section 7.4.). Also, a reduction in the transmission rate of shigellosis has been positively correlated with improved fly control (Watt & Lindsay, 1948; Cohen et al., 1991). The relationships between houseflies and a number of other pathogens that cause gastroenteritis can also be found in the literature (Nayduch, Noblet & Stutzenberger, 2002; Nichols, 2005), but the importance of this insect in causing illness in people through field transmission of these pathogens has not been verified, particularly in densely populated urban areas. Studies that show correlations between the suppression of flies and the concomitant reduction of enteric diseases can be found in the literature, but these studies were performed in rural settings.

For a number of reasons, cases of fly-borne illness or disease are at times difficult to verify. For example, when examining sick patients, most physicians, despite their medical training, have little practical knowledge of entomology. Also, flies may cause low-grade infections in large numbers of people, but these people may not be associated collectively unless they all attended a related event or unless the infections cause fatalities.

Flies must be sampled and subjected to microbiological culturing to determine their status as a carrier of pathogens. Few communities can justify such sampling programmes unless large numbers of residents are affected.

To aid in disease surveillance, the European Centre for Disease Prevention and Control (ECDC), counterpart of the Centers for Disease Control and Prevention (CDC) in the United States, was founded in 2004. Its mission is to help strengthen the European Union's defences against such infectious diseases as influenza, severe acute respiratory syndrome (SARS) and human immunodeficiency virus/acquired immunodeficiency syndrome (HIV/AIDS). To achieve this goal, its small core staff works with an extended network of partners across the European Union and in the European Economic Area/European Free Trade Association Member States (Iceland, Liechtenstein, Norway and Switzerland). The ECDC works in partnership with national health protection bodies to strengthen and develop continentwide disease surveillance and early warning systems and to develop authoritative scientific opinions about the risks posed by new and emerging infectious diseases. Weekly and monthly releases about incidences of communicable disease in the European Community, plus archives of releases back to 1995, can be found on the ECDC web site (ECDC, 2006).

The mission of the CDC, which was founded in 1946 to help control malaria, is to promote health and quality of life by preventing and controlling disease, injury, and disability. The CDC applies research and findings to improve people's daily lives and responds to health emergencies. More can be learned about the CDC mission and activities by visiting its web site (CDC, 2006).

7.6. Cost to the public of fly infestation

7.6.1. Cost of health-related conditions

The burden of disease caused by flies is difficult to estimate, since there are generally no large-scale epidemics known to be caused by fly-transmitted pathogens and since disease levels are low in the northern hemisphere (Prüss et al., 2002). In the United States, diseases caused by major pathogens alone were estimated to cost up to US$ 35 billion annually in 1997 (WHO, 2002), but the contribution, if any, of filth flies to the transmission of the pathogens responsible for these diseases is not traceable. Costs for illnesses that may have been caused by flies cannot be determined.

A large number of food-related illnesses and deaths are reported annually in many countries, suggesting that transmission by flies is a possibility. In 2003, there were 63 044 cases of salmonellosis reported in Germany. In the United States, about 40 000 cases of salmonellosis are reported annually. Because many milder cases are not diagnosed by (or reported to) health care professionals, the actual number of infections may be thirty or more times greater. Each year, an estimated 600 people die from acute salmonellosis. Second to salmonellosis, *Campylobacter* enteritis is the most common illness in Germany. In the United States, *Campylobacter* spp. are one of the most common causes of bacterial diarrhoea, causing an estimated 2.45 million illnesses and 124 deaths annually. These enormous numbers of infections suggest the tremendous cost of diseases potentially transmitted by flies. The percentage of cases that may have been caused by fly-transmitted pathogens is unknown, however, thus preventing the calculation of any meaningful estimates.

7.6.2. Cost of control and management

The cost to the public of fly infestation in the United States is measured by the cost of pest control contracts for killing flies and other pests in and around homes, supermarkets, restaurants, hospitals, hotels, warehouses and chains of discount (food) stores. Contracts for home protection can cost between US$ 30 and US$ 50 a month. Commercial contracts can cost between US$ 240 and US$ 300 a month for chains of discount stores, between US$ 160 and US$ 225 a month for supermarkets, and between US$ 90 and US$150 a month for fast food restaurants. Thus, a supermarket company with 20 stores in a city would pay between US$ 3200 and US$ 4500 a month or between US$ 38400 and US$ 54000 a year for management of flies and other pests. A comparable cost calculation may reveal similar pricing in some parts of Europe, but pricing data are rarely published. Costs for managing fly infestations are also incurred by local, state or national government agencies that on occasion must control flies in public recreational areas. State-funded aerial spray programmes, such as the one for stable fly management in western Florida, may operate on budgets of between US$ 45000 and US$ 50000 a year.

7.7. Impact of poverty

Because people at lower income levels may live under conditions that are more attractive to flies, poverty may play a role in attracting flies, fostering their development and transmitting disease. For the poor, money is limited and none is available for fly control.

Marginal and unkempt housing can be attractive to flies and fly entry can be difficult to prevent. Unsanitary conditions inside and outside of houses attract flies and fly breeding may occur on the premises. Also, traditional methods of sanitation and food storage, preparation and disposal may be conducive to attracting flies and may aid their proliferation.

Education is required to teach people affected by poverty the need for proper sanitation and the principles of fly management. This will build awareness among them and allow people to help themselves. A quantitative measure of the impact of poverty might be difficult to estimate, however.

7.8. Fly management

A number of IPM practices or techniques can be used to combat flies in urban areas. Some cities and countries encourage composting of garden waste, which is a good method of recycling. But if compost temperatures are not high enough (≥ 50°C), flies, particularly stable flies, can use compost piles to develop their immature stages. The same is true of straw and other stall bedding used for horses in urban fringes. Stall bedding and litter should be stacked to reduce the surface area and to promote internal composting. If small enough, stacks should be covered with plastic film to preclude flies and promote composting.

Trapping flies outdoors is a good way to manage the fly populations around homes, apartments and stores. Many flytraps do not involve the use of pesticides and are safe to use around people and companion animals. Traps are available locally and through the Internet.

There is a good selection of jar or bag traps that employ a foul-smelling attractant in water, to attract and capture flies. These traps are very effective for treating the perimeter of buildings, but not for general use close to entry doors or at outdoor lounging or eating areas. Sticky traps that attract and capture flies on a sticky surface can also be effective. No pesticides are associated with these traps, and all of these traps work best if placed in sunny areas.

For a number of reasons, insect traps that use ultraviolet light as an attractant are most effective when used indoors and are not recommended for outdoor use. First, flies are active in daylight hours, and ultraviolet light traps mounted outdoors work best after dark. Second, if used outdoors, ultraviolet light traps should be switched off after dark, because they will become clogged by night-flying insects, particularly moths. Finally, an ultraviolet light trap outdoors will attract insects into areas they never intended to go. The trap will attract these insects, but they will not necessarily enter the trap after they are in the vicinity.

To capture flies indoors without pesticides, sticky tapes, ribbons, strips or cylinders hung at or near ceiling level can be used.

Residents of houses and apartments can make a major contribution to curbing the proliferation of flies by properly managing their organic wastes, particularly garbage. The following measure should help reduce fly populations. Garbage cans should have tight-fitting lids, and the garbage inside should be contained in closed plastic bags. Garbage cans should be sanitized periodically to remove associated odours and should be kept indoors, or outside away from entry doors. The area around the garbage cans should be clean, and any spilled garbage should be removed. This is particularly important for restaurants, where large quantities of garbage can accumulate. If possible, garbage cans and dumpsters at restaurants should be sanitized regularly and kept with lids closed. Spillage, particularly of grease and oils, should be avoided, because thorough cleaning can be difficult to impossible. This is especially true if spillage occurs on asphalt or cement.

At commercial establishments, such as restaurants, fly exclusion fans or air curtains can be used at outside doors to prevent fly entry. These can be effective if airflow is properly maintained (8 m/s). However, fly exclusion fans are unpopular with customers and staff and are seldom seen in operation at the proper speed.

Ceiling fans can be used to keep flies out of food-preparation and -serving areas. Flies tend to prefer places where the airflow is negligible, and they also avoid airflows much less than the 8 m/s required to prevent them from passing though an entry door. Ceiling fans mounted over counters in bakeries and delicatessens can be very effective, and the airflow is low enough to prevent discomfort to staff and customers. Traps can also be placed in dead air locations near fans (but at floor level) to capture the displaced flies in a push–pull system (J. Hogsette, unpublished observation, 2001).

Temperature is commonly used indoors to combat flies. Meat cutting and processing rooms kept at 15°C will be free of flies.

The key to good fly control, both indoors and outdoors, is sanitation. Elimination of food, possible breeding sites and shelter will minimize fly populations.

7.8.1. Conducive environmental conditions

To minimize resting sites for flies in residential settings – as they wait for an opportunity to get into food handling areas and health care facilities – landscape plantings, trash receptacles and vending machines should be away from entry doors. If not managed properly, garbage and grease stored outdoors can be highly attractive to flies. Both garbage and grease must be kept in clean, closed containers that are emptied regularly, and garbage and grease containers must not be kept near entrance or service doors. Also, food odours must be vented away from entry doors and proper lighting must be used to minimize attracting flies to buildings. In residential settings, outside pet areas must be properly maintained, with daily removal of faecal materials and excess food.

A major factor in fly management in urban areas is odour management. If odours attract flies to a site, flies can then create a nuisance or contamination problem.

7.8.2. Fly inspection, detection and surveillance

Surveys of filth flies help determine the effectiveness of sanitation practices, identify filth fly breeding sites and determine the need for control measures, such as improved exclusion in a restaurant or application of a pesticide. Surveys of filth flies are also necessary to determine baseline fly populations, track population trends and evaluate the effectiveness of control measures. Sanitation is the cornerstone of a sound filth fly control programme. Unfortunately, breeding areas may lie outside of surveillance areas, placing sanitation beyond the charge of local authorities and making it necessary to concentrate on adult surveillance and control. The Armed Forces Pest Management Board of the United States lists five elements of effective fly surveillance in Technical Guide No. 30 (Armed Forces Pest Management Board, 2006):

1. surveillance to identify their presence, species population size and conditions that favour breeding;

2. sustained monitoring of fly populations and conditions that favour breeding;

3. evaluation of survey results;

4. initiation of control measures when established thresholds have been passed and notification of appropriate units responsible for conducting control measures; and

5. continued surveillance, to determine the success of control measures.

An effective surveillance programme must have a mechanism for determining the need for control measures. Surveillance usually, but not always, means trapping. Such trapping should be conducted at a standardized time and at the same locations. Locations must be accurately identified, so the trap will be placed in the same location for each subsequent survey. Traps must be selected carefully to determine the fly species responsible for the problem. Most flies in urban areas will be in the adult stage; larval populations, however, might be encountered if suitable habitats (such as garbage and sewage) are available. An index of adult flies may be obtained most efficiently by so-called fly grills (Scudder, 1947, 1949), but sticky traps, fly baits, spot counts and many other methods can be used. Larval or pupal specimens, or both, can be found by checking the breeding habitats and, if necessary, they can be identified with appropriate taxonomic keys. Because proper identification in the field can be critical, it is important that inspectors be familiar with all life stages.

The presence of flies does not automatically initiate a recommendation for control. To help predict when control measures are needed, thresholds are established. The threshold value itself is an index calculated from surveillance data. Continuous surveillance over an extended period of time may be required to establish reliable threshold values. Long-term surveillance data may also reveal identifiable trends that can be used to protect the people involved, by allowing control measures to be initiated just before a serious fly problem occurs. Threshold values will vary at different geographical locations, depending on such factors as species, area involved, habitat, collection technique, number of complaints and disease potential. After gaining some experience, a nuisance threshold may be established. In residential areas, for example, if fly complaints are numerous when the average grill index is 25 flies a week (or whatever the sampling interval may be), then this may be at or near the nuisance threshold.

7.8.3. Fly exclusion practices

There are good mechanical devices that can be used around homes and businesses to control flies, and windows and doors should have good screens. Businesses should have doors that open outwards, to minimize flies being drawn into a building, and doors should be equipped with self-closing devices. Businesses that are air-conditioned should have a positive air pressure inside, so that air blows out of the doors when customers enter. Also, exhaust vents and air intakes should be screened. Moreover, if properly installed and maintained, mechanical air curtains or blowers mounted over and around outside doors can be beneficial in preventing flies from entering.

7.8.4. Pesticide applications for fly control

Pesticides are not widely used for controlling flies in urban areas. Most urban fly populations are adults, and it is difficult to treat local populations effectively with pesticides. The exception is during times of large-scale disasters, when fly populations can reach catastrophic levels very quickly. During these situations, aerial applications of labelled pesticides may result in exposing some people to the pesticides. However, the expected public exposure in urban areas should be minimal. Around structures, the only pesticide applications made, are those applied topically to outdoor walls where flies rest. Pest control operators make these applications, and the general public receives little or no exposure. Topically applied residual pesticides are available, but not highly recommended. Unless applied to known fly resting sites, and reapplied regularly, these chemicals may increase the chances of developing pesticide resistance in local fly populations (Hemingway & Ranson, 2000).

In private dwellings, commercial aerosol pesticides should give adequate protection against small numbers of flies that enter. The only other pesticides in use around buildings are granular baits, and these are localized in bait stations and are out of the reach of people and animals. There still may be some indoor or outdoor spray systems in operation, with pyrethrins or synthetic pyrethroids most likely used as toxicants. Whenever possible, pesticide use should be minimized in urban areas.

For a review of fly-management techniques, including pesticides, see Farkas & Hogsette (2000).

7.8.5 Granular baits

These baits are in granular form and can be placed in bait stations or scattered on the ground, depending on label restrictions and local regulations. Baits contain stomach poisons and are best used for housefly management. Most of the older granular baits on the market contain methomyl as the active ingredient, but there are at least two new baits on the market with new active ingredients (spinosad and imidacloprid) that should prove useful in urban areas, if applied properly. Unless flies are present in large numbers on patios or in garden areas, these baits would not normally be recommended for home owners or apartment dwellers. Granular baits have been used successfully around commercial establishments, such as restaurants and supermarkets, to manage moderate fly populations in the outer rear areas of stores. Bait stations are available, and they consist of granular baits affixed to cardboard strips and encased in a protective mesh. These can be hung or attached to walls to attract and kill flies. Granular baits can be used to quickly reduce housefly populations, but seldom can the use of these baits alone control houseflies.

7.8.6. Traps

Many types of trap are used in urban areas for managing flies inside and outside of commercial and private dwellings. These include: ultraviolet light traps; sticky traps; jar or bag traps; window traps; and sticky tubes, tapes and ribbons.

7.8.6.1. Ultraviolet light traps

These traps attract flies with long-wavelength ultraviolet light and then either kill them by electrocution or trap them on boards covered with glue. These traps are the main method used by many commercial pest management companies to control flies in commercial establishments, particularly supermarkets and restaurants. They are not recommended for individual houses or apartments. There are many trap models to choose from; however, to maximize the number of flies captured, the location of their placement is critical. Whenever possible, traps should be mounted within a metre of the floor, and models having an open front with a direct display of the lights are generally most effective. Traps that kill by electrocution should not be used indoors, especially in food preparation or consumption areas. When electrocution occurs, very small particles of the insects and any associated microorganisms are projected into the air where they may contaminate food or be inhaled by people (Urban & Broce, 2000).

7.8.6.2. Sticky traps

Several sticky flytraps are commercially available. These can be used in commercial locations or in backyard garden areas to capture flies outside without pesticides. These traps are also suitable for surveillance work.

7.8.6.3. Jar or bag traps

These traps are charged with an attractant mixed with water to draw flies inside the trap, where they die. These traps can be used by commercial establishments, around trash compactors and in garbage collection areas. They can also be used outdoors by private individuals – for example, in garden areas. No pesticides are used these traps.

7.8.6.4. Window traps

These traps can be used in commercial establishments and in homes, to capture flies that enter buildings and eventually come to the closed, glass windows. These are passive traps and flies attracted to windows eventually fumble down into the traps and become captured on a sticky strip. Some of the traps fit in the corners of the windows, and others merely fit along the sill. Window traps capture the flies on glue boards out of view and without pesticides.

7.8.6.5. Sticky tubes, tapes and ribbons

These traps, hung at or near ceiling level, can be used to capture flies in commercial kitchens, away from food preparation areas. If fly populations are large, these traps should be changed frequently, as they become filled or nearly filled with flies. When selecting sites in which to hang these traps, it should be kept in mind that warm temperatures in food preparation areas can cause the adhesives to liquefy and drip onto surfaces below.

7.8.7. Perimeter treatments

Perimeter treatments that use vaporized or liquid repellents or pesticides are just now being developed for flies, although various products on the market also claim effectiveness against mosquitoes. Jar traps and sticky traps have been used successfully to prevent or minimize the passage of flies, but the former would not be recommended for use in urban areas.

Some pest management companies apply various pesticides and growth regulators to outside walls of commercial buildings that might be used by flies as resting sites. Some of these sites are near rear entry doors and treatments are thought to prevent fly entry into the structure. The effectiveness of such treatments has not been evaluated.

7.8.8. Crack-and-crevice treatments

As the name implies, crack-and-crevice treatments are usually treatments with pesticides placed in cracks and crevices of walls. Generally, however, these treatments do not apply to flies, because in most instances flies are not found in such locations. The exception might be overwintering flies. If this is the case, standard liquid-formulation pesticides could be used to kill the flies. If flies are entering a structure, the entryways should be closed, as suggested in subsection 7.8.3. If cracks in walls, around windows and under doors are large enough to allow fly entry, these should be repaired by the party responsible for maintenance.

7.8.9. Biological control

Biological control is not recommended at present for urban settings. This is because most of the available options work best against the immature stages of the fly and essentially, it is the adults that are causing the problems. If immature stages are found in temporary urban habitats, these can usually be eliminated quickly by nonbiological means. The exception might be their elimination in compost piles.

The most promising biological organisms being investigated for urban use against adult flies are entomopathogenic fungi. A number of these have been identified and tested, but spore preservation techniques need to be improved, so spores remain viable in the field for long periods. This will allow the use of feeding or infection stations, where flies are attracted to bait and become infected with a fungus.

7.8.10. Attractants

Attractants are desperately needed for fly management in urban areas. Several housefly attractants are available for agricultural use, but these are in many cases too odoriferous for use indoors. Attractants for use indoors must compete well with common indoor odours, such as those from pastries in bakeries and meats in delicatessens. Commercial buildings in some locations are air-conditioned and the air inside can be completely replaced up to five times an hour. Thus, any attractants designed to be used with indoor traps must be dispensed in appropriate quantities. Various pest control companies have added housefly sex pheromones to the glue boards in their light traps as attractants; however, increases in efficacy have not been substantiated.

7.8.11. Air flow

To prevent or minimize flies from entering structures, airflow from air curtains can be used around doorways (Mathis, Smith & Schoof, 1970). In some places, air curtains are required on service entrances where doors remain open for long periods. Air curtains can also be used over entrances used by customers, although customer acceptance is usually low. New research is revealing improved methods for air curtain use that will maximize results and improve customer satisfaction. Also, ceiling fans can be used indoors to prevent flies from entering and resting in selected areas, such as on counters or in food preparation areas in commercial establishments. Airflow does not need to be extremely high to keep flies away. Fans can be used in conjunction with ultraviolet light traps in a push–pull system, by placing fans over areas where fly exclusion is desired and placing light traps in nearby dead air spaces, to capture the flies.

7.9. Guidelines for fly control

The following guidelines are used to control flies.

- Monitor regularly the urban fly populations, with a special focus on hospitals, retirement facilities, kindergartens and the like. Evaluate their pathogen load and determine the percentage of infected flies.

- Devise a system to better estimate when disease outbreaks are related to pathogens transmitted by flies.

- Restrict pesticide usage to outbreak scenarios. Determine the peak season of various pest species, to predict potential outbreaks. These flies could then be managed with the focused use of pesticides during a small window of time.

- Develop improved attractants to use in traps and baits.

- Improve trap designs.

- Improve fly-exclusion and trapping devices for use around entrances.

- Develop effective perimeter barriers or perimeter control techniques that do not involve the use of pesticides.

- Educate the public about the biology, ecology and control of nuisance flies, the role of flies as transmission agents and the need for proper sanitation to prevent flies from becoming contaminated.

- Refine our knowledge of fly biology and of their habits and behaviour in urban areas.

- Refine our knowledge of the infestation rate of natural fly populations in urban areas.

7.10 Conclusions

For dealing with fly infestations, the following items are suggested.

1. Proper sanitation is the key to fly control. Deny flies access to food, shelter and a place to lay their eggs.

2. Do not allow flies to come in contact with contaminated substances and thus contaminate themselves.

3. Although management of adult flies can provide temporary relief, the location and elimination of development sites for immature stages is the best method for long-term control. Although people make the distinction between urban and rural flies, flies do not, so fly management in urban areas may involve surveillance at (and management of) potential fly-producing sites outside the urban perimeter.

4. Prevent flies from entering buildings, by keeping doors closed and window screens in proper repair.

5. If flies do enter structures, eliminate them with traps or other suitable methods as quickly as possible.

6. If people experience fly problems, particularly if such problems are associated with illness, health authorities should be contacted immediately.

7. Health authorities with entomological expertise should have properly trained personnel to identify flies and assess the extent of fly outbreaks with or without associated pathogenic organisms. Should their assistance be needed, health authorities should have contacts with outside entomologists and medical personnel.

8. There is a need to improve education in entomology at the biological branches of universities. Such education will establish and produce expertise and knowledge, which is currently being eroded dramatically, because of the lack of financial support.

9. Public awareness and educational programmes are essential to minimize the transmission of pathogens by flies, especially in times of disaster. It is particularly important to teach the benefits of exclusion of flies from foods and from food-preparation and dining areas.

10. Communities should develop fly-management guidelines that indicate action thresholds for adult populations and that suggest corrective measures to be taken when

thresholds have been exceeded. Corrective measures may include legal action to be taken against individuals or companies that fail to control flies when a nuisance situation has been identified.

References[1]

Allen SJ et al. (2004). Flies and *Helicobacter pylori* infection. *Archives of Disease in Childhood*, 89:1037–1038.

Armed Forces Pest Management Board (2006). *Filth flies: significance, surveillance and control in contingency operations.* Washington, DC, Armed Forces Pest Management Board, Defense Pest Management Information Analysis Center (Technical Guide No. 30; http://www.afpmb.org/pubs/tims/TG30/TG30.htm, accessed 14 August 2006).

Betke P, Schultka H, Ribbeck R (1986). *Stomoxys calcitrans:* Plage in einer Milchviehanlage. *Angewandte Parasitologie*, 27:39–44.

Bishopp FC, Laake EW (1921). Dispersion of flies by flight. *Journal of Agricultural Research*, 21:729–766.

Brownstein JS, Holford TR, Fish D (2005). Effect of climate change on Lyme disease risk in North America. *EcoHealth*, 2:38–46.

CDC [web site] (2006). Atlanta, GA, Centers for Disease Control and Prevention (http://www.cdc.gov/about/default.htm, accessed 26 November 2006).

Clavel A et al. (2002). House fly (*Musca domestica*) as a transport vector of *Cryptosporidium parvum*. *Folia Parasitologica*, 49:163–164.

Cohen D et al. (1991). Reduction of transmission of shigellosis by control of houseflies (*Musca domestica*). *Lancet*, 337:993–997.

De Jesus AJ et al. (2004). Quantitative contamination and transfer of *Escherichia coli* from foods by houseflies, *Musca domestica* L. (Diptera: Muscidae). *International Journal of Food Microbiology*, 93:259–262.

Deményny A (1989). [Filth fly breeding sites on a Hungarian large-scale cattle farm]. *Parasitologica Hungarica*, 22:99–107.

Ebeling W (1975). *Urban entomology.* Los Angeles, CA, University of California, Division of Agricultural Sciences.

ECDC (2006). Eurosurveillance [web site]. Stockholm, European Centre for Disease Prevention and Control (http://www.eurosurveillance.org/index-02.asp, accessed 26 November 2006).

[1] The literature cited is a representative sample of what is relevant and important to the topic, but is not intended to constitute a comprehensive systematic review. The references are considered to be highly reputable and will give the reader an overview of the state-of-the-art knowledge in this field.

Farkas R, Hogsette JA (2000). Control possibilities of filth-breeding flies in livestock and poultry production. In: Papp L, Darvas B, eds. *Contributions to a manual of palaearctic Diptera. Vol. 1. General and applied dipterology.* Budapest, Science Herald:889–904.

Farkas R, Papp L (1989). [Species composition and breeding sites of fly communities (Diptera) in caged-layer houses in Hungary]. *Parasitologica Hungarica*, 22:93–98.

Faulde M (2002). *Vorkommen und Epidemiologie vektorassoziierter Infektionserkrankungen in Mitteleuropa*. Augsburg, Germany, U-Books Verlag.

Foil LD, Hogsette JA (1994). Biology and control of tabanids, stable flies, and horn flies. *Revues Scientifique et Technique de l'Office Internationale des Epizooties*, 13:1125–1158.

Fotedar R et al. (1992). The housefly (*Musca domestica*) as a carrier of pathogenic microorganisms in a hospital environment. *The Journal of Hospital Infection*, 20:209–215.

Goulson D et al. (2005). Predicting calyptrate fly populations from the weather, and probable consequences of climate change. *Journal of Applied Ecology*, 42:795–804.

Graczyk TK, Knight R, Tamang L (2005). Mechanical transmission of human protozoan parasites by insects. *Clinical Microbiology Reviews*, 18:128–132.

Graczyk TK et al. (1999). House flies (*Musca domestica*) as transport hosts of *Cryptosporidium parvum. The American Journal of Tropical Medicine and Hygiene*, 61:500–504.

Graczyk TK et al. (2000). Mechanical transport and transmission of *Cryptosporidium parvum* oocysts by wild filth flies. *The American Journal of Tropical Medicine and Hygiene*, 63:178–183.

Graczyk TK et al. (2001). The role of non-biting flies in the epidemiology of human infectious diseases. *Microbes and Infection*, 3:231–235.

Greenberg B (1964). Experimental transmission of *Salmonella typhimurium* by house flies to man. *American Journal of Hygiene*, 80:149–156.

Greenberg B (1971). *Flies and disease. Vol I. Ecology, classification, and biotic associations*. Princeton, Princeton University Press.

Greenberg B (1973). *Flies and disease. Vol II. Biology and disease transmission*. Princeton, Princeton University Press.

Greenberg B (1991). Flies as forensic indicators. *Journal of Medical Entomology*, 28:565–577.

Gregorio SB, Nakao JC, Beran GW (1972). Human enteroviruses in animals and arthropods in the central Philippines. *The Southeast Asian Journal of Tropical Medicine and Public Health*, 3:45–51.

Grübel P et al. (1997). Vector potential of houseflies (*Musca domestica*) for *Helicobacter pylori*. *Journal of Clinical Microbiology*, 35:1300–1303.

Grübel P et al. (1998). Detection of *Helicobacter pylori* DNA in houseflies (*Musca domestica*) on three continents. *Lancet*, 352:788–789.

Hall MJR, Wall R (1995). Myiasis in humans and domestic animals. *Advances in Parasitology*, 35:258–334.

Hansens EJ (1951). The stable fly and its effect on seashore recreational areas in New Jersey. *Journal of Economic Entomology*, 44:482–487.

Harwood RF, James MT (1979). *Entomology in human and animal health*, 7th ed. New York, Macmillan Publishing Company, Inc.

Hemingway J, Ranson H (2000). Insecticide resistance in insect vectors of human disease. *Annual Review of Entomology*, 45:371–391.

Hogsette JA (1981). Fly control by composting manure at a south Florida equine facility. In: Patterson RS et al., eds. *Status of biological control of filth flies: proceedings of a workshop*. Gainesville, Florida, United States Department of Agriculture Science and Education Administration:105–113.

Hogsette JA, Farkas R (2000). Secretophagous and haematophagous higher Diptera. In: Papp L, Darvas B, eds. *Contributions to a manual of palaearctic Diptera. Vol. 1. General and applied dipterology*. Budapest, Science Herald: 769–792.

Hogsette JA, Ruff JP (1985). Stable fly (Diptera: Muscidae) migration in Northwest Florida. *Environmental Entomology*, 14:206–211.

Hogsette JA, Ruff JP, Jones CJ (1987). Stable fly biology and control in northwest Florida. *Journal of Agricultural Entomology*, 4:1–11.

Hogsette JA, Ruff JP, Jones CJ (1989). Dispersal behaviour of stable flies (Diptera: Muscidae). In: Petersen JJ, Greene GL, eds. Current status of stable fly (Diptera: Muscidae) research. *Miscellaneous Publications of the Entomological Society of America*, 74:23–32.

Houghton JT et al., eds. (2001). *Climate change 2001: the scientific basis*. Cambridge, United Kingdom, Cambridge University Press.

James MT (1947). *The flies that cause myiasis in man*. Washington, DC, United States Government Printing Office (United States Department of Agriculture Miscellaneous Publications, No. 631).

Jones CJ et al. (1985). Effects of natural saccharide and pollen extract feeding on stable fly (Diptera: Muscidae) longevity. *Environmental Entomology*, 14:223–227.

Keen JE et al. (2003). Isolation of Shiga-toxigenic *Escherichia coli* (STEC) from livestock pest flies. In: *Proceedings of the 84th Annual Meeting of the Conference of Research Workers in Animal Diseases*, Chicago, IL, 9–11November 2003 (https://www.ars.usda.gov/research/publications/publications.htm?seq_no_115=138624, accessed 2 March 2007).

Kettle DS (1995). *Medical and veterinary entomology*, 2nd ed. Cambridge, United Kingdom, CAB International.

Khalil K et al. (1994). Flies and water as reservoirs for bacterial enteropathogens in urban and rural areas in and around Lahore, Pakistan. *Epidemiology and Infection*, 113:435–444.

Killough RA, McKinstry DM (1965). Mating and oviposition studies of the stable fly. *Journal of Economic Entomology*, 58:489–491.

Kobayashi M et al. (1999). Houseflies: not simple mechanical vectors of enterohemorrhagic *Escherichia coli* O157:H7. *The American Journal of Tropical Medicine and Hygiene*, 61:625–629.

LaBrecque GC, Meifert DW, Weidhaas DE (1972). Dynamics of house fly and stable fly populations. *Florida Entomologist*, 55:101–106.

Larsen EB, Thomsen M (1940). The influence of temperature on the development of some species of Diptera. *Videnskabelige meddelelser fra Dansk Naturhistorisk Forening*, 104:1–75.

Levine OS, Levine MM (1991). Houseflies (*Musca domestica*) as mechanical vectors of shigellosis. *Reviews of Infectious Diseases*, 13:688–696.

Luger SW (1990). Lyme disease transmitted by a biting fly. *The New England Journal of Medicine*, 322:1752.

Lysyk TJ, Axtell RC (1987). A simulation model of house fly (Diptera: Muscidae) development in poultry manure. *Canadian Entomologist*, 119:427–437.

Mathis W, Smith EA, Schoof HF (1970). Use of air barriers to prevent entrance of house flies. *Journal of Economic Entomology*, 63:29–31.

Melnick JL (1951). Poliomyelitis and poliomyelitis-like viruses of man and animals. *Annual Review of Microbiology*, 5:309–332.

Murvosh CM, Thaggard CW (1966). Ecological studies of the house fly. *Annals of the Entomological Society of America*, 59:533–547.

Nayduch D, Noblet GP, Stutzenberger FJ (2002). Vector potential of house flies for the bacterium *Aeromonas caviae*. *Medical and Veterinary Entomology*, 16:193–198.

Newson HD (1977). Arthropod problems in recreation areas. *Annual Review of Entomology*, 22:333–353.

Nichols GL (2005). Fly transmission of *Campylobacter*. *Emerging Infectious Diseases*, 11:361–364.

Ogden NH et al. (2006). Climate change and the potential for range expansion of the Lyme disease vector *Ixodes scapularis* in Canada. *International Journal for Parasitology*, 36:63–70.

Olsen AR (1998). Regulatory action criteria for filth and other extraneous materials. III. Review of flies and foodborne enteric disease. *Regulatory Toxicology and Pharmacology*, 28:199–211.

Olsen AR, Hammack TS (2000). Isolation of *Salmonella* spp. from the housefly, *Musca domestica* L., and the dump fly, *Hydrotaea aenescens* (Wiedemann) (Diptera: Muscidae), at caged-layer houses. *Journal of Food Protection*, 63:958–960.

Osato MS et al. (1998). Houseflies are an unlikely reservoir or vector for *Helicobacter pylori*. *Journal of Clinical Microbiology*, 36:2786–2788.

Parmesan C, Yohe G (2003). A globally coherent fingerprint of climate change impacts across natural systems. *Nature*, 421:37–42.

Peterson AT, Shaw J (2003). *Lutzomyia* vectors for cutaneous leishmaniasis in Southern Brazil: ecological niche models, predicted geographic distributions, and climate change effects. *International Journal of Parasitology*, 33:919–931.

Post K et al. (1999). Fly larvae and pupae as vectors for scrapie. *Lancet*, 354:1969–1970.

Poulin R (2006). Global warming and temperature-mediated increases in cercarial emergence in trematode parasites. *Parasitology*, 132:143–151.

Prüss A et al. (2002). Estimating the burden of disease from water, sanitation, and hygiene at a global level. *Environmental Health Perspectives*, 110:537–542.

Rady MH et al. (1992). Bacterial contamination of the housefly *Musca domestica*, collected from 4 hospitals at Cairo. *Journal of the Egyptian Society of Parasitology*, 22:279–288.

Rahuma N et al. (2005). Carriage by the housefly (*Musca domestica*) of multiple-antibiotic-resistant bacteria that are potentially pathogenic to humans, in hospital and other urban environments in Misurata, Libya. *Annals of Tropical Medicine and Parasitology*, 99:795–802.

Root TL et al. (2003). Fingerprints of global warming on wild animals and plants. *Nature*, 421:57–60.

Rosef O, Kapperud G (1983). House flies (*Musca domestica*) as possible vectors of *Campylobacter fetus* subsp. *jejuni*. *Applied and Environmental Microbiology*, 45:381–383.

Rugg D (1982). Effectiveness of Williams traps in reducing the numbers of stable flies (Diptera: Muscidae). *Journal of Economic Entomology*, 75:857–859.

Sasaki T, Kobayashi M, Agui N (2000). Epidemiological potential of excretion and regurgitation by *Musca domestica* (Diptera: Muscidae) in the dissemination of *Escherichia coli* 0157:H7 to food. *Journal of Medical Entomology*, 37:945–949.

Scudder HL (1947). A new technique for sampling the density of house fly (*Musca domestica*) populations. *Public Health Reports*, 62:681–686.

Scudder HL (1949). Some principles of fly control for the sanitarian. *The American Journal of Tropical Medicine and Hygiene*, 29:609–623.

Sherman RA (2000). Wound myiasis in urban and suburban United States. *Archives of International Medicine*, 160:2004–2014.

Skidmore P (1985). *The biology of the Muscidae of the world.* Dordrecht, Netherlands, Dr. W. Junk Publishers.

Sømme L (1961). On the overwintering of house flies (*Musca domestica* L.) and stable flies (*Stomoxys calcitrans* (L.)) in Norway. *Norsk Entomologisk Tidsskrift*, 11:191–223.

Soós Á, Papp L, ed. (1986). *Catalogue of palaearctic Diptera. Vol. 11. Scathophagidae-Hypodermatidae.* Budapest, Akadémiai Kiadó,

Sramova H et al. (1992). Epidemiological role of arthropods detectable in health facilities. *The Journal of Hospital Infection*, 20:281–292.

Srinivasan R et al. (2006). Muscoid fly populations in tsunami-devastated villages of southern India. *Journal of Medical Entomology*, 43:631–633.

Steinbrink H (1989). Zur Verbreitung von *Stomoxys calcitrans* (Diptera: Muscidae) in Ställen. *Angewandte Parasitologie*, 30:57–61.

Stenseth NC et al. (2002). Ecological effects of climate fluctuations. *Science*, 297:1292–1296.

Szalanski AL et al. (2004). Detection of *Campylobacter* and *Escherichia coli* O157:H7 from filth flies by polymerase chain reaction. *Medical and Veterinary Entomology*, 18:241–246.

Tarry DW, Bernal L, Edwards S (1991). Transmission of bovine virus diarrhoea virus by blood feeding flies. *The Veterinary Record*, 128:82–84.

Thomas GD, Skoda SR, eds (1993). *Rural flies in the urban environment?* Lincoln, University of Nebraska (North Central Regional Research Publication No. 335).

Treese RE (1960). Distribution, life history, and control of the face fly in Ohio. In: *Proceedings of the North-Central Branch of the Entomological Society of America*, 15:107–108; Milwaukee, 23–25 March 1960.

Urban JE, Broce AB (2000). Killing of flies in electrocuting insect traps releases bacteria and viruses. *Current Microbiology*, 41:267–270.

Wallace GD (1971). Experimental transmission of *Toxoplasma gondii* by filth-flies. *The American Journal of Tropical Medicine and Hygiene*, 20:411–413.

Ward R, Melnick JL, Horstmann DM (1945). Poliomyelitis virus in fly-contaminated food collected at an epidemic. *Science*, 101:491–493.

Watt J, Lindsay DR (1948). Diarrheal disease control studies: I. Effect of fly control in a high morbidity area. *Public Health Reports*, 63:1319–1334.

WHO (2002). *Food safety and food-borne illnesses.* Geneva, World Health Organization (WHO Fact Sheet No. 237; http://www.who.int/mediacentre/factsheets/fs237/en/print.html, accessed 7 December 2006)

WHO Regional Office for Europe (2006). *LARES: Large Analysis and Review of European housing and health Status.* Copenhagen, WHO Regional Office for Europe (http://www.euro.who.int/Document/HOH/LARES_results.pdf, accessed 9 August 2006).

Winpisinger KA et al. (2005). Spread of *Musca domestica* (Diptera: Muscidae), from two caged layer facilities to neighboring residences in rural Ohio. *Journal of Medical Entomology*, 42:732–738.

Zumpt F (1973). *The Stomoxyine biting flies of the world.* Stuttgart, Gustav Fischer Verlag.

8. Birds

Zdenek Hubálek

Summary

Beside the harm some wild urban bird species (mostly feral pigeons) cause to buildings by their activity and droppings, their nesting sites can be the source of abundant ectoparasites (such as argasid ticks, mites, bugs and fleas) that produce allergic reactions in people. Also, certain microorganisms pathogenic to people have been found to be associated with wild urban birds:

- some arboviruses (the agent of diseases such as St. Louis encephalitis virus and West Nile virus);
- *Chlamydophila psittaci* (the etiological agent of ornithosis);
- *Borrelia burgdorferi* sensu lato (the agent of Lyme disease);
- *Campylobacter jejuni* (the agent of campylobacterosis);
- *Salmonella enterica* serovars Enteritidis and Typhimurium (the agents of salmonellosis);
- *Histoplasma capsulatum* (the agent of histoplasmosis); and
- *Cryptococcus neoformans* (the agent of cryptococcosis).

Cases of human disease acquired directly from urban birds or from their habitats have been reported for ornithosis, histoplasmosis, salmonellosis, campylobacterosis, mycobacteriosis, cryptococcosis, and toxoplasmosis. Monitoring zoonotic and sapronotic diseases associated with birds in urban areas is the first essential step in controlling these diseases. In circumstances of established risk, managing urban bird populations includes:

- restricting their feeding at public sites;
- controlling scavenging birds on landfill sites and at harbours (if bird numbers create hygienic problems); and
- controlling and sanitizing large communal roosts of birds in city parks and suburban habitats.

Also, proactive and reactive control measures can be implemented, such as:

- dispersing birds – for example, by acoustic or light signal methods, predation by trained raptors or water-mist sprayers;
- modifying habitats – for example, by thinning or clearing vegetation;
- inhibiting birds from breeding on buildings by blocking the loft orifices and perch sites in, on, and below the roofs, using netting, spike systems, repellent gels or electroshock deterrents;

- collecting and inactivating eggs;
- trapping and euthanizing the captured birds, if permitted; and
- sterilizing birds chemically with a treated feed bait, if permitted.

This integrated management approach should also involve educational and legal components.

8.1. Introduction

This chapter aims to review and assess the pertinent literature on wild birds in urban ecosystems – that is, birds that live in both urban and suburban habitats – and it includes the feral pigeon (*Columba livia* f. *domestica*) as a potential health hazard for people. The survey does not include wild birds kept in captivity (such as in zoological parks and aviaries), reared game, poultry and other domestic birds as a source of human infection, nor does it include pet birds in urban areas. The term *synanthropic birds* is sometimes used alternatively and means wild birds that live in association with people (in human habitats).

8.1.1. Free-living birds as so-called pests in the urban ecosystem

Urban free-living birds could be called companion animals, especially for children, the elderly and lonely people: they are often watched or fed (or both) with pleasure by many city dwellers around the world who are deprived of the countryside (Fig. 8.1 and 8.2). These birds are usually considered to be an agreeable, pleasant component of the urban environment. In that sense, urban birds are not formidable *pests*.

Fig. 8.1. Contact between birds (house sparrows and feral pigeons) and people, Paris

Source: Photo by Z. Hubálek.

Fig. 8.2. Contact between birds (feral pigeons) and people, Venice

Source: Photo by Z. Hubálek.

Fig. 8.3. Adjustment of historical buildings against feral pigeon plagues[a], Salzburg

a. The word "plagues" refers to feral pigeon excreta and activity during roosting and breeding.

Source: Photo by Z. Hubálek.

Fig. 8.4. Massive droppings from a communal roost of rooks at Brno

Source: Photo by Z. Hubálek.

However, under some circumstances, some urban bird species (especially, pigeons, doves, gulls, house sparrows, starlings, blackbirds, grackles and corvids) that congregate at too high population densities produce droppings that harm such human artefacts as historical monuments, buildings, statues, fountains and cars (Fig. 8.3). Also, certain synanthropic avian species can be extremely noisy at feeding sites, breeding colonies and communal roosts (Fig. 8.4). Moreover, some of them can be harmful to urban vegetation (trees and such fruits as cherries) in gardens (blackbirds and starlings, for example) or cause additional pollution problems with their droppings, which foul yards, sidewalks (creating the risk of slipping and physical injury for pedestrians) and roads, and also produce foul odours. In so doing, such species as pigeons, gulls, starlings, grackles, blackbirds and corvids become a nuisance (Lesaffre, 1997; Odermatt et al., 1998). In addition, some medium-sized gregarious birds, such as gulls, rooks and lapwings, can cause dan gerous aircraft accidents, during take-off and landing at suburban airfields.

8.1.2. Urban birds as a source of ectoparasites harmful to people

In urban areas, nests of feral pigeons in the lofts of houses can result in invasions of soft ticks (such as the pigeon tick, *Argas reflexus*) into closely situated flats and apartments. Such invasions can result in infestations of their occupants (Dusbábek & Rosický, 1976) that often lead to allergic reactions (Glünder, 1989; Veraldi et al., 1998; Rolla et al., 2004). Other pigeon ectoparasites can occasionally attack humans as well. Examples of these are: mites, such as the chicken mite (*Dermanyssus gallinae*), which may produce allergic reactions, especially in children and susceptible adults; bugs, such as the pigeon bug (*Cimex columbarius*); and fleas, such as the pigeon flea (*Ceratophyllus columbae*) (Sixl, 1975; Glünder, 1989; Haag-Wackernagel, 2005).

8.1.3. Wild birds and human pathogenic microorganisms in general

Yet another (and more serious) problem could arise: some urban birds are associated with microorganisms pathogenic to people; thus, these birds present a potential human health hazard. It is well established that wild birds, including those living in the urban ecosystem, can harbour pathogenic microorganisms and also spread them to people (Keymer, 1958; McDiarmid, 1969; Davis et al., 1971; Sixl, 1975; Lvov & Ilyichev, 1979; Glünder, 1989; Cooper, 1990; Glünder et al., 1991; Hubálek, 1994; Lesaffre, 1997; Wobeser, 1997; Haag-Wackernagel & Moch, 2004; Hubálek, 2004). In general, birds may be involved in the circulation of pathogenic microorganisms in a number of diverse ways:

1. as biological hosts (the pathogen multiplies in or on the avian body) with an acute infection (such as ornithosis), chronic infection (such as tuberculosis) or latent infection, and as carriers shedding the causative agent for a prolonged period (such as salmonellosis);

2. as mechanical hosts or carriers (the pathogen only survives and does not multiply in or on the avian body): the carriage can be either external (on the surface of the bird's body) or internal (when the causative agent passes through the digestive tract and is excreted in a viable state);

3. as carriers of infected haematophagous ectoparasites, especially ixodid (hard) ticks (Hoogstraal, 1967); birds may also serve as non-infectious blood donors, just supporting the life-cycle and increasing the population density of potential arthropod vectors of transmissible diseases; and

4. as so-called lessors, providing a substrate (such as droppings, pellets and nest linings) suitable for reproduction or survival (or both) of pathogenic agents (such as *Histoplasma capsulatum* and *Cryptococcus neoformans*); birds are lessors when the pathogen multiplies in avian excreta or nests – that is, outside the bird's body (Hubálek, Juřicová & Halouzka, 1995).

From an epidemiological perspective, these infections can be regarded as either zoonoses (diseases transmitted from animals to people, as in cases 1–3) or sapronoses (diseases caused by organisms of the environment, as in case 4, where the causative agent requires a non-animal reservoir or site to replicate or complete the development) (Hubálek, 2003).

8.2. Human microbial pathogens associated with wild birds

This section concisely reviews human microbial pathogens – viruses, bacteria, fungi and protozoa – that have been isolated from free-living birds in urban and suburban areas and from their haematophagous ectoparasites, excreta and nests. Omitted from consideration here are agents of avian diseases that usually do not cause clinical symptoms in people (such as Newcastle disease virus and other paramyxoviruses, avian circoviruses, avian coronaviruses, avian poxviruses, avian herpesviruses, *Pasteurella multocida*, avian trypanosomes, avian plasmodia, other avian haematozoa, and coccidiae), as well as microorganisms in whose circulation birds play a negligible role, compared with mammals (such as Venezuelan equine encephalitis virus and leptospirae). This section also assesses infections of people reported as attributable to, or directly associated with, urban birds.

8.2.1. Viruses

8.2.1.1. Togaviridae: genus Alphavirus

8.2.1.1.1. Eastern and Western equine encephalitis viruses

In North America, birds are the principal hosts in the natural transmission of mosquito-borne Eastern and Western equine encephalitis viruses (EEEV and WEEV, respectively). These mosquito-borne viruses have also been isolated from synanthropic birds (mostly passerines and feral pigeons), but mainly during epidemics and epizootic episodes (Holden, 1955; Dardiri et al., 1957; Holden et al., 1973; Beran, 1981; Soler, Portales & Del Barrio, 1985). Some wild birds develop a relatively high-level, long-term viraemia after inoculation with WEEV and EEEV and can even maintain persistent virus infection, with viraemia lasting up to 10 months for WEEV (Hammon, Reeves & Sather, 1951; Kissling et al., 1954, 1957a,b; Reeves et al., 1958). Birds serve as a source of infectious blood to mosquito vectors and latently infected birds can be regarded as a virus reservoir. WEEV affects a wide range of wild birds, but it occurs less often than EEEV in epizootic episodes. For instance, mortality in *Agelaius* spp. blackbirds was observed after experimental infection with WEEV (Hardy, 1987). However, no EEEV or WEEV infections of people have been reported as attributable to, or directly associated with, urban birds.

8.2.1.1.2. Sindbis virus

Birds are the principal hosts and disseminators of Sindbis virus (SINV; Scandinavian subtype: Ockelbo), while the vectors are largely ornithophilic mosquitoes. SINV was isolated from synanthropic species: carrion crows (*Corvus corone*) in Egypt (Taylor et al., 1955) and European starlings (*Sturnus vulgaris*) in Slovakia (Ernek et al., 1977). It causes encephalitis and occasional death in experimentally infected chickens and pigeons. The chronic course of the infection in pigeons has been ascertained experimentally (Semenov et al., 1973). However, no SINV infections of people have been reported as attributable to, or directly associated with, urban birds.

8.2.1.2. Flaviviridae: genus Flavivirus

8.2.1.2.1. St. Louis encephalitis virus

Birds are principal amplifying hosts in endemic foci of St. Louis encephalitis virus (SLEV) in North and Central America (Theiler & Downs, 1973; Beran, 1981). House sparrows, feral pigeons and other synanthropic birds have often been used as sentinels for monitoring and predicting the epidemiological situation in urban areas (Holden et al., 1973; Lord et al., 1974; Lord, Calisher & Doughty, 1974; McLean et al., 1983). The house sparrow develops a relatively high-level, long-term viraemia (up to one month) and can serve as a source of infectious blood to mosquito vectors (Hammon, Reeves & Sather, 1951; Chamberlain et al., 1957). However, no SLEV infections of people have been reported as attributable to, or directly associated with, urban birds.

8.2.1.2.2. West Nile virus

Birds are principal vertebrate hosts of this mosquito-borne agent. West Nile virus (WNV) was isolated from a number of synanthropic species – for example, feral pigeons and carrion crows in Egypt (Work, Hurlbut & Taylor, 1953; Hurlbut et al., 1956), black-headed gulls (*Larus ridibundus*) in Slovakia (Grešíková, Sekeyová & Prazniaková, 1975; Ernek et al., 1977), rooks (*Corvus frugilegus*) in Ukraine, carrion crows in the southern part of the European part of Russia, American crows (*Corvus brachyrhynchos*) in North America, and the common blackbird (*Turdus merula*) in Azerbaijan (Gaidamovich & Sokhey, 1973; Lvov & Ilyichev, 1979; Vinograd et al., 1982; Eidson et al., 2001). WNV sometimes causes a clinically manifest disease in feral pigeons and other free-living birds, with an occasional death (Hurlbut et al., 1956). Nevertheless, mass mortality of corvids (mostly American crows) and other birds has been observed during recent outbreaks of West Nile fever in North America, due to a high virulence of the virus strain for birds; the dying crows have therefore been used to monitor the epidemiological situation in urban areas of the United States (Eidson et al., 2001; Komar et al., 2003; Caffrey, Smith & Weston, 2005), and they probably contributed to the rapid east–west spread of WNV across North America since 2001. However, no WNV infections of people have been reported as attributable to, or directly associated with, urban birds.

8.2.1.2.3. Flaviviruses of the tick-borne encephalitis complex

The principal hosts of tick-borne encephalitis (TBE) complex viruses are forest rodents, while birds are occasional hosts and disseminators of preimaginal (larval and nymphal stages of) ixodid ticks infected with TBE viruses. One of these viruses, the central European encephalitis virus (CEEV), was isolated occasionally from a few synanthropic birds – for example, the common blackbird and the song thrush (*Turdus philomelos*) in Finland (Brummer-Korvenkontio et al., 1973) and from common blackbirds and the great tit (*Parus major*) in the eastern Baltic region (Lvov & Ilyichev, 1979). Two CEEV strains were recovered from nymphal *Ixodes ricinus* (castor-bean) ticks collected on common blackbirds in Slovakia (Ernek et al., 1968). Most bird species are resistant to CEEV, including synanthropic house sparrows and great tits (Grešíková & Ernek, 1965); certain species, however, have been observed to develop viraemia that lasted several days after experimental infection with CEEV: common starling, song thrush, common blackbird (Saikku, 1973; Lvov & Ilyichev, 1979). Morbidity, mortality, and shedding of TBE viruses have been observed occasionally in experiments with the house sparrow (Lvov & Ilyichev, 1979). No human cases of TBE, however, have been reported as attributable to, or directly associated with, urban birds.

8.2.1.3. Orthomyxoviridae: genus Orthomyxovirus

Influenza A virus has often been isolated from wild birds worldwide, mainly from ducks, gulls, terns, shearwaters and shorebirds, and less often from passerines (Lvov & Ilyichev, 1979; Wobeser, 1997; Lipkind, Shihmanter & Shoham, 1982; Hinshaw et al., 1985; Stallknecht & Shane, 1988; Webster et al., 1992; Munster et al., 2005), most strains being so-called low pathogenic avian influenza (LPAI) viruses. Wild aquatic birds are the primordial source and reservoir of all influenza viruses for other vertebrates (Hinshaw et al., 1981), and these birds perpetuate all known antigenic subtypes of influenza A viruses (H1–H16 and N1–N9). In wild ducks, influenza viruses replicate in the cells of the intestinal tract and are excreted in high concentrations in the faeces; the virus shedding can continue for 2–4 weeks (Slemons & Easterday, 1977; Hinshaw et al., 1980). Experimental infection of adult mallard ducks with the H5N2, H5N3 and H5N9 influenza A virus subtypes resulted in histologically demonstrated mild pneumonia, although clinical signs were not evident (Cooley et al., 1989). On the other hand, experimental infection of common starlings and house sparrows with an H7N7 influenza virus isolate from Australian chickens caused high mortality (100% in starlings and 30% in house sparrows) and transmission to contact birds, although it caused no mortality in ducks (Nestorowicz et al., 1987). In Italy, highly pathogenic avian influenza A (HPAI) strain H7N1 was isolated from two synanthropic birds (a house sparrow and a collared dove, *Streptopelia decaocto*) found dead, close to a chicken farm with an HPAI epizootic (Capua et al., 2000). In addition to HPAI viruses H5 and H7, LPAI strains circulate at a much higher frequency in such wild waterfowl as mallards (*Anas platyrhynchos*), and could convert to HPAI after being transmitted to poultry (Munster et al., 2005).

South-east Asia is regarded as a region where influenza viruses co-circulate in urban and rural environments among people, domestic pigs and ducks, which provides the opportunity for interspecies transmission and genetic exchange among viruses (Webster et al., 1992). An extensive HPAI epornitic (an outbreak of disease in a bird population), caused by an H5N1 strain, started in domestic poultry in South-East Asia in 1997 and has raised special concern, because of its very rapid spread across Eurasia in 2005 and 2006 (Perdue & Swayne, 2005; WHO Global Influenza Program Surveillance Network, 2005). Some wild urban and suburban birds have also been involved – for example, feral pigeons and tree sparrows (*Passer montanus*) died from HPAI in China, Hong Kong SAR, Thailand and the Russian Federation (Ellis et al., 2004; Brown et al., 2005; FAO Avian Influenza Task Force, 2005; Morris & Jackson, 2005; Nguyen et al., 2005).

HPAI is mainly a veterinary (and economic) problem, and bird-to-human transmissions have been only sporadically reported: 271 human cases (165 of them fatal) have been confirmed in some countries in Asia and Africa but, as of 3 February 2007, no cases have been confirmed in Europe or North America (WHO, 2007). Obviously, people can become infected with HPAI only under very special circumstances: when they inhale or ingest a massive dose of virus, resulting from very intimate contact with infected domestic birds. Moreover, no human HPAI infection acquired from a wild bird has been described before April 2006. In brief, free-living urban birds do not present a risk of infecting people with HPAI.

8.2.2. Bacteria

8.2.2.1. Chlamydiaceae

Chlamydophila psittaci causes ornithosis (chlamydiosis, also known as psittacosis) in gulls, pigeons, passerines and other birds, especially young ones. Adult birds are more resistant to the disease, but they can carry and intermittently shed the infectious agent for months (Roberts & Grimes, 1978). Clinical symptoms in birds are extremely variable, ranging from infection without symptoms to death caused by septicaemia. Some avian species may remain serologically negative despite active chlamydiosis. The importance of chlamydiosis to the population dynamics of wild birds is probably underrated. Epizootic episodes of disease – that is, the rapid spread among animals – with high mortality have been described; such episodes have occurred in gulls in North America (Brand, 1989).

Also, *C. psittaci* was isolated from synanthropic species such as mallards, herring gulls (*Larus argentatus*), black-headed gulls, feral pigeons, wood pigeons (*Columba palumbus*), collared doves, common starlings and house sparrows (Strauss, Bednář & Šerý, 1957; Šerý & Strauss, 1960; Illner, 1961; Terskikh, 1964; Davis et al., 1971; Wobeser, 1997; Odermatt et al., 1998; Gautsch et al., 2000; Dovc et al., 2004; Haag-Wackernagel & Moch, 2004; Tanaka et al., 2005). Synanthropic doves, grackles and starlings may transmit *C. psittaci* to domestic birds (Grimes, Owens & Singer, 1979), and wild ducks, gulls and grackles may disperse the infectious agent (Page, 1976; Lvov & Ilyichev, 1979). The mean worldwide prevalence of *C. psittaci* in feral pigeons is 13% by isolation and 30% by serology (Davis et al., 1971), but as many as 38% and 56% of pigeons were found to be seropositive, for instance, in Slovakia and Switzerland, respectively (Řeháček & Brezina, 1976; Haag-Wackernagel & Moch, 2004). The feral pigeon plays a major role in the epidemiology of human ornithosis (Meyer, 1959; Terskikh, 1964; Sixl & Sixl, 1971; Sixl, 1975). At least 500 cases have been reported as acquired by people from feral pigeons or other synanthropic avian species in many parts of the world, often in urban conditions (Meyer, 1959; Henry & Crossley, 1986; Pospíšil et al., 1988; Wobeser & Brand, 1982; Haag-Wackernagel & Moch, 2004).

8.2.2.2. Coxiellaceae

Coxiella burnetii, the causative agent of Q fever (also known as coxiellosis), was isolated from more than 40 species of wild birds, including synanthropic barn swallows (*Hirundo rustica*) and white wagtails (*Motacilla alba*) in the Czech Republic (Raška & Syrůček, 1956), feral pigeons in Italy (Babudieri & Moscovici, 1952), house sparrows in central Asia and crows in India (Lvov & Ilyichev, 1979; Řeháček & Tarasevich, 1988). Pigeons and house sparrows are susceptible to experimental infection: they excrete *C. burnetii* in faeces, and the infectious agent is recoverable from the lungs, spleen and kidneys for 48–58 days post inoculation, while the birds do not develop any symptoms of the infection and are seronegative (Babudieri & Moscovici, 1952; Schmatz et al., 1977). Five members of a family living in southern France became ill with acute Q fever in May 1996, and an epidemiological investigation suggested that this outbreak resulted from exposure to contaminated pigeon faeces and their argasid ticks (Stein & Raoult, 1999). Also, the collared dove, a synanthropic species of dove with an expanding breeding range, has been suggested to be responsible for the introduction of the Q fever to Ireland (Řeháček & Tarasevich, 1988).

8.2.2.3. Anaplasmataceae

Anaplasma phagocytophilum, the causative agent of human granulocytic anaplasmosis (previously known as human granulocytic ehrlichiosis), can be carried in immature deer ticks (*Ixodes scapularis*) and castor-bean ticks attached to such synanthropic birds as some turdids, as it was detected in North America (Daniels et al., 2002), European parts of the Russian Federation (Alekseev et al., 2001) and Sweden (Bjoersdorf et al., 2001). However, no *Anaplasma* infections of people have been reported as attributable to, or directly associated with, urban birds.

8.2.2.4. Spirochaetaceae

Borrelia burgdorferi s.l., the causative agent of Lyme borreliosis, was repeatedly detected in preimaginal deer and castor-bean vector ticks collected from birds, including synanthropic species of the American robin (*Turdus migratorius*), the common blackbird, and the song thrush in North America and Europe (Anderson et al., 1986; Magnarelli, Stafford & Bladen, 1992; Humair et al., 1993; Hubálek, Juřicová & Halouzka, 1995; Hubálek et al., 1996; Hanincová et al., 2003, Stern et al., 2006). Tick larvae and nymphs have parasitized some species of birds as frequently as they have parasitized the white-footed mouse (*Peromyscus leucopus*), the principal reservoir host of *B. burgdorferi* s.s. in North America. Bird-feeding larval and nymphal deer ticks were removed from their hosts, left to molt, and shown to transmit *B. burgdorferi* trans-stadially (that is, from one stage of the life-cycle to the next) into the adult stage (Anderson, Magnarelli & Stafford, 1990). In Switzerland, 16% of larval and 22% of nymphal castor-bean ticks collected from passerine birds were infected with borreliae; the highest infection rate was observed in ticks removed from thrushes (Turdidae) (Humair et al., 1993). In the Czech Republic, borreliae were detected in 7% of larval and 12% of nymphal castor-bean ticks collected from wild birds; positive ticks were collected from synanthropic species: the common blackbird, the European robin (*Erithacus rubecula*) and the great tit. Three massively infected ticks with hundreds of borreliae were collected from robins and blackbirds, and *Borrelia garinii* was isolated from a nymphal tick, collected from a young blackbird (Hubálek et al., 1996). A high rate of infection with borreliae (mainly *B. garinii* and *Borrelia valaisiana*) was also found in preimaginal ticks, collected from Turdidae in Slovakia (Hanincová et al., 2003). Moreover, some synanthropic bird species are competent amplifying hosts for *B. burgdorferi* s.l. – for instance, the common blackbird (Humair et al., 1998) and the American robin (Richter et al., 2000; Ginsberg et al., 2005). However, no *B. burgdorferi* s.l. infections of people have been reported as attributable to, or directly associated with, urban birds.

8.2.2.5. Campylobacteraceae

Campylobacter jejuni is the most frequently isolated *Campylobacter* spp. from a wide variety of aquatic and terrestrial wild birds (including synanthropic anatids, gulls, pigeons, corvids, starlings and sparrows), followed by *Campylobacter coli* and *C. laridis* (Rosef, 1981; Kapperud & Rosef, 1983; Kinjo et al., 1983; Kaneuchi et al., 1987; Mégraud, 1987; Whelan et al., 1988; Glünder & Petermann, 1989; Maruyama et al., 1990; Glünder et al., 1991; Quessy & Messier, 1992; Sixl et al., 1997; Odermatt et al., 1998; Gautsch et al., 2000; Haag-Wackernagel & Moch, 2004). *Campylobacter* spp. cause either asymptomatic carriership in adult birds or signs of anorexia, diarrhoea and (occasionally) mortality in fledglings. Among birds found in the city of Oslo, very high isolation rates of *Campylobacter* spp. were detected (Kapperud & Rosef, 1983) in omnivorous and scavenging species: carrion crows (90%), herring gulls (50%) and black-headed gulls (43%). *C. jejuni* was also recovered from 3% and 14% of feral pigeons in Norway and Croatia, respectively (Lillehaug et al., 2005; Vlahovič et al., 2004). As many as 46% of Japanese crow (also known as jungle crow, *Corvus levaillantii*) and carrion crow dropping samples contained *C. jejuni*, mostly biovar I serogroup 2, reflecting a close epidemiological association between the crow and human isolates (Maruyama et al., 1990). *Campylobacter* spp. are transmissible to other birds through droppings and aerosols. The carrier state of young black-headed gulls for *C. jejuni* has been shown to last for about 3–4 weeks (Glünder, Neumann & Braune, 1992).

Campylobacter jejuni and *C. coli* cause human campylobacterosis (food-borne enterocolitis) worldwide, but some avian biotypes are distinct from human isolates. In *C. jejuni* isolated from 53% of pigeon faecal samples in the Bordeaux region of France; the main biotype was I – that is, the most common biotype in human disease (Mégraud, 1987). Seagulls and common grackles (*Quiscalus quiscula*) were implicated in the indirect spread of *Campylobacter* spp. to domestic animals and people, via feedstuffs and water, respectively (Sacks et al. 1986). Jackdaws (*Corvus monedula*) and European magpies (*Pica pica*) caused a milk-borne *Campylobacter* epidemic (with 59 human cases) by pecking milk bottles in England (Hudson et al., 1991).

8.2.2.6. Enterobacteriaceae

8.2.2.6.1. Escherichia

Escherichia coli enteropathogenic strains, such as serotype O157:H7 (which produces verotoxin or Shiga-like toxin), have quite often been isolated from healthy and ill wild synanthropic birds, such as the feral pigeon (Haag-Wackernagel & Moch, 2004; Grossmann et al., 2005), the wood pigeon (McDiarmid, 1969), the Canada goose (*Branta canadensis*) (Hussong et al., 1979), the common starling (Nielsen et al., 2004) and seagulls (Makino et al. 2000) and from nestlings, dead embryos and eggs of *Passer* spp. (Pinowski, Kavanagh & Górski, 1991). In addition, wild urban birds may become the carriers of *E. coli* strains resistant to antibiotics and may be responsible for the spread of R plasmids (the extra-chromosomal self-replicating structures sometimes found in bacterial cells that carry the genetic information for antibiotic resistance) in a wide area (Kanai, Hashimoto & Mitsuhashi, 1981). However, no enteropathogenic *E. coli* infections of people have been reported as attributable to, or directly associated with, urban birds.

8.2.2.6.2. Salmonella

Numerous *Salmonella enterica* serovars, particularly Typhimurium and Enteritidis were isolated from many species of synanthropic birds, such as gulls, pigeons, sparrows, starlings and corvids, and some may present a potential threat to the health of domestic animals and the health of people (Keymer, 1958; Nielsen, 1960; Snoeyenbos, Morin & Wetherbee, 1967;Wilson & Macdonald, 1967; Pohl & Thomas, 1968; Cornelius, 1969; McDiarmid, 1969; Tizard, Fish & Harmerson, 1979; Niida et al., 1983; Simitzis-le Flohic et al., 1983; Odermatt et al., 1998; Toro et al., 1999; Gautsch et al., 2000; Dovc et al., 2004; Haag-Wackernagel & Moch, 2004). Many authors reported the isolation of Typhimurium, Enteritidis and other *Salmonella* serovars from synanthropic gulls – such as black-headed gulls, herring gulls/Caspian gulls (*Larus cachinnans*) and California and ring-billed gulls (*Larus californicus* and *Larus delawarensis*, respectively) – in Europe or North America (Strauss, Bednář & Šerý, 1957; Šerý & Strauss, 1960; Nielsen, 1960; Petzelt & Steiniger, 1961; Snoeyenbos, Morin & Wetherbee, 1967; Müller, 1970; Berg & Anderson, 1972; Pannwitz & Pulst, 1972; Wuthe, 1972, 1973; Fennel, James & Morris, 1974; Macdonald & Brown, 1974; Pagon, Sonnabend & Krech, 1974; Williams, Richards & Lewis, 1976; Hall et al., 1977; Plant, 1978; Fenlon, 1981, 1983; Benton et al., 1983; Butterfield et al., 1983; Kapperud & Rosef ,1983; Fricker, 1984; Girdwood et al., 1985; Literák & Kraml, 1985; Monaghan et al., 1985; Glünder, Siegmann & Köhler, 1991; Glünder et al., 1991; Selbitz et al., 1991; Literák et al., 1992; Quessy & Messier, 1992;

Lévesque et al., 1993; Hubálek, Juřicová & Halouzka, 1995). Gulls could thus play a significant role in dispersing pathogenic salmonellae, even in urban areas.

Salmonellae (Typhimurium, Enteritidis, Paratyphi B serovars) have been recovered repeatedly from feral pigeons or synanthropic sparrows (Petzelt & Steiniger, 1961; Dózsa, 1964; Wobeser & Finlayson, 1969; Tizard, Fish & Harmerson, 1979; Tanaka et al., 2005). In pigeons, the serovar Typhimurium var. *copenhagen* is especially common, and its cells are usually firmly attached to the duodenal epithelium of the birds, causing long-term or persistent infection (Grund & Stolpe, 1992). A number of house sparrows from 36 sites in Poland were examined: the prevalence of salmonellae (serovars Typhimurium, Dublin and Paratyphi B) varied from 0% to 40% (Pinowska, Chylinski & Gondek, 1976). Certain *Salmonella* serovars can cause lethal enteritis and hepatitis of nestlings, especially gulls and other colonial water birds; the serovars Gallinarum and Typhimurium were encountered among the causes of death in British sparrows (Keymer, 1958; Baker, 1977). Experimental infection of the fledgling house sparrow with the serovar Typhimurium resulted in a rather severe disease, while some birds remained relatively asymptomatic, but excreted the salmonellae (Stepanyan et al., 1964). Epizootic episodes of salmonellosis with high mortality caused by the serovar Typhimurium have repeatedly been described among sparrows, greenfinches and other urban passerines at bird feeders in Europe and North America, particularly during the winter and spring (Englert et al., 1967; Wilson & Macdonald, 1967; Bouvier, 1969; Cornelius, 1969; Schneider & Bulling, 1969; Greuel & Arnold, 1971; Locke, Shillinger & Jareed, 1973; Nesbitt & White, 1974; Macdonald, 1977; Brittingham & Temple, 1986; Bowes, 1990; Tizard, 2004). Garden bird feeders can become heavily contaminated with salmonellae (serovar Typhimurium), and the infected birds may transmit the infection to people (directly or via cats feeding on them: Tizard, 2004). Also, an epornitic of salmonellosis occurred in common starlings overwintering in Israel (Reitler, 1955). Osteomyelitic and arthritic salmonellosis was described in carrion crows (Daoust, 1978) and pigeons, and cutaneous salmonellosis was described in house sparrows (Macdonald, 1977). Human cases of salmonellosis were reported after the handling of injured gulls (Macdonald & Brown, 1974) or drinking water contaminated by gulls (Benton et al., 1983), even in urban areas.

8.2.2.6.3. Yersinia

Yersinia enterocolitica infects wild anatids, gulls, pigeons, and some passerines (Mair, 1973; Hacking & Sileo, 1974; Kapperud & Olsvik, 1982; Simitzis-le Flohic et al., 1983; Shayegani et al., 1986; Kaneuchi et al., 1989; Odermatt et al., 1998; Gautsch et al., 2000). In the United States, 3% of wild avian faecal samples were positive for *Y. enterocolitica*, but no serogroup O:8 or O:3 human pathogenic isolates were recovered (Shayegani et al., 1986). The isolation rate in Japan for *Y. enterocolitica* was higher, at 7% (Kato et al. 1985). The avian hosts are usually asymptomatic, but sometimes the clinical signs include anorexia, diarrhoea and weight loss. Unlike *Yersinia pseudotuberculosis*, *Y. enterocolitica* is less frequent in birds than in humans, and there is no evidence that birds are significant reservoirs for it. No *Y. enterocolitica* infections of people have been reported as attributable to, or directly associated with, urban birds.

Yersinia pseudotuberculosis can cause mortality in wild birds, including such synanthropic species as the wood pigeon, *Passer* spp., the barn swallow, the common starling, the common grackle, the common blackbird, the jackdaw, the rook and the American crow (Jennings, 1955–1957; Keymer, 1958; Jennings, 1959, 1961; Clark & Locke, 1962; McDiarmid, 1969; Lipaev et al., 1970; Davis et al., 1971; Mair, 1973; Hacking & Sileo, 1974). Avian pseudotuberculosis sometimes occurs in epizootic episodes (especially during severe winter conditions), and its manifestations are varied: ruffled feathers, anorexia, diarrhoea, lack of coordination and sudden death. Some wild avian species, however, are known to be refractory to natural infection. Isolations of *Y. pseudotuberculosis* were reported from healthy synanthropic birds: common starlings in France and Switzerland (Simitzis-le Flohic et al., 1983; Odermatt et al., 1998; Gautsch et al., 2000), and pied wagtails in Japan (Fukushima & Gomyoda, 1991). Gulls were also found to be infected in the Far East (Lvov & Ilyichev, 1979; Kaneuchi et al., 1989). For people, free-living birds that carry and shed the causative agent via faeces may represent a source of infection. In Japan, *Y. pseudotuberculosis* isolates from wild ducks were of serotypes 1b and 4b, which represent the most frequent serovars in local human strains, and they contained the same plasmid types (3 and 1, respectively) as human isolates (Hamasaki et al., 1989; Fukushima & Gomyoda, 1991).

8.2.2.7. Gram-positive cocci

8.2.2.7.1. Staphylococcus

Staphylococcus aureus (Micrococcaceae) was isolated from the faeces of gulls (Cragg & Clayton, 1971), corvids (Golebiowski, 1975; Hájek & Balusek, 1988) and other synanthropic birds (Keymer, 1958). Staphylococcosis associated with trauma caused the death of a house sparrow (Keymer, 1958), and a mixed fungal cutaneous infection with *S. aureus* was described in the same species (Hubálek, 1994). Staphylococcal arthritis in gulls and necrotic arthritic lesions on the feet of European robins were observed in England (Macdonald, 1965; Blackmore & Keymer, 1969). *Staphylococcus aureus* was also isolated from 14% of dead wild birds (such as passerines) in England (Harry, 1967), dead embryos and eggs of sparrows in Poland (Pinowski, Kavanagh & Górski, 1991: together with *S. epidermidis*). Staphylococcal foot dermatitis (an inflammation of the ball of the foot of fowls, called bumblefoot) has been reproduced experimentally in common starlings infected with *S. aureus* (Cooper & Needham, 1981), and from that species the causative agent was also isolated (Odermatt et al., 1998; Gautsch et al., 2000). A disseminated *S. aureus* infection caused the death of four mallards overwintering in Saskatchewan (Wobeser & Kost, 1992). Also, a number of coagulase-positive staphylococcal strains from synanthropic birds (such as pigeons, rooks, gulls and ducks) have been shown to represent *Staphylococcus intermedius* (Hájek & Balusek, 1988; Hájek et al., 1991). However, no infections of people with *Staphylococcus* spp. have been reported as attributable to, or directly associated with, urban birds.

8.2.2.7.2. Streptococcus

Streptococcus bovis has been described as an important cause of septicaemia in pigeons of all ages (Devriese et al., 1990). In Belgium, the causative agent was frequently isolated from both healthy pigeons (carriers) and those dead from septicaemia. *Streptococcus bovis* strains from pigeons were of biotypes similar to ovine and bovine strains, but differed from those isolated from human beings (De Herdt et al., 1995). Another streptococcus, of group C, was isolated from great tits and common starlings (Jennings, 1955–1957, 1959). However, no *S. bovis* infections of people have been reported as attributable to, or directly associated with, urban birds.

8.2.2.8. Regular nonsporing Gram-positive rods

8.2.2.8.1. Listeria

Listeria monocytogenes, the etiological agent of listeriosis, can cause sporadic septic cases of the disease even in wild birds, such as common starlings and European robins (Macdonald, 1968; McDiarmid, 1969). Faecal samples from synanthropic gulls that fed at Scottish sewage works had a higher rate of carriage of the infectious agent (15%) than those species that fed elsewhere (4%); the infection rate of rooks was generally lower (6%) (Fenlon, 1985). Also, gulls may play a significant role in contaminating silage with members of the genus *Listeria*. *Listeria monocytogenes* was isolated from healthy synanthropic collared doves and house sparrows in the Czech Republic (Treml et al., 1993) and from roosting starlings in Switzerland (Odermatt et al., 1998; Gautsch et al., 2000). However, no infections of people have been reported as attributable to, or directly associated with, urban birds.

8.2.2.8.2. Erysipelothrix

Erysipelothrix rhusiopathiae, the etiological agent of erysipeloid in man, which is a localized, self-limited cutaneous lesion, can cause epornitics and even mass mortality in waterbirds. Other cases have included gulls in Denmark, pigeons, wild ducks and thrushes in Germany, and starlings in the United States (Faddoul, Fellows & Baird, 1968; Macdonald, 1968; McDiarmid, 1969). However, no *E. rhusiopathiae* infections of people have been reported as attributable to, or directly associated with, urban birds.

8.2.2.9. Mycobacteriaceae

Mycobacterium avium causes tuberculosis (mycobacteriosis) in many wild avian species, including synanthropic columbiforms, sparrows, corvids, gulls and anatids (Plum, 1942; Mitchell & Duthie, 1950; Jennings, 1955–1957, 1959; Keymer, 1958; Jennings, 1961; Brickford, Ellis & Moses, 1966; Švrček et al., 1966; Blackmore & Keymer, 1969; McDiarmid, 1969; Davis, et al., 1971; Schaefer et al., 1973; Baker, 1977; Wobeser, 1997; Smit et al., 1987; Hejlíček & Treml, 1993a–c). It is one of the most widespread wild avian infections, often resulting in a marked weight loss, severe muscle atrophy and death. Some synanthropic birds (such as wood pigeons in England, house sparrows and rooks elsewhere) could be carriers of *M. avium* and could play a role in the spread of avian tuberculosis to poultry and domestic animals (Kubín & Matějka, 1967). Plum (1942) examined 1000 house and tree sparrows from 40 farms in Denmark and found *M. avium* in 9% of them; tuberculous lesions were present in 2% of the birds. Several strains of *M. avium* were isolated from *Passer* spp. in the Czech Republic; the birds were considered a possible source of tuberculosis infection in cattle (Matějka & Kubín, 1967; Rossi & Dokoupil, 1967; Hejlíček & Treml, 1993a, c). In the Czech Republic, infected chickens are the main source of mycobacteria for house sparrows (Hejlíček & Treml, 1993a, c). *Mycobacterium avium* was also isolated from a collared dove (Volner, 1978); in experiments, this dove species was found to be much less susceptible than the chickens, but could be a potential carrier of the infectious agent (Rossi, 1969; Hejlíček & Treml, 1993b). Mycobacteria, including *M. avium-intracellulare* complex, were detected in 19% of 153 faecal samples of feral pigeons collected in Japan (Tanaka et al., 2005). However, biological and molecular typing (sometimes even nucleotide sequencing) of *M. avium* strains from free-living birds seems to be necessary for a proper epidemiological evaluation of the birds as sources of human infections – that is, to compare the avian and clinical human isolates closely and to detect whether they are identical or not (Schaeffer et al., 1973; McFadden et al., 1992). Also, there are a few unconfirmed, anecdotal cases of *M. avium* clinical infections of people (ornithologists) who collected and examined urban owl and raptor pellets.

Mycobacterium xenopi was occasionally found to cause human disease – for example, a nosocomial outbreak in Le Havre, France, where 558 cases were diagnosed for the period 1965–1967; free-living birds were the probable source of the outbreak, and the causative agent was isolated from droppings of local tree sparrows and common blackbirds (Joubert, Desbordes-Lize & Viallier, 1971). Other mycobacteria potentially pathogenic to people (*M. avium-intracellulare, M. aquae, M. flavescens* and *M. fortuitum*) were isolated from rooks (Kubín & Matějka, 1967), anatids (Schaefer et al., 1973), gulls and jackdaws with necrosis in the liver and spleen (Smit et al., 1987).

8.2.3. Fungi

8.2.3.1. Yeasts and yeast-like fungi

8.2.3.1.1. Candida

Candidosis (also called candidiasis and moniliasis) is, after aspergillosis, the second most significant mycosis of domestic and captive birds. In wild birds, however, the disease is virtually unknown, although *C. albicans* has often been isolated from the gastrointestinal tract and excretions of gulls (Kawakita & van Uden, 1965; Cragg & Clayton, 1971; Buck, 1983, 1990), pigeons and other wild birds (Littman & Schneierson, 1959; Frágner, 1962; Partridge & Winner, 1965; Brandsberg et al., 1969; Schönborn, Schütze & Pöhler, 1969; Kocan & Hasenclever, 1974; Hasenclever & Kocan, 1975; Gugnani, Sandhu & Shome, 1976; Refai et al., 1983; Pinowski, Kavanagh & Górski, 1991; Hubálek, 1994; Haag-Wackernagel & Moch, 2004), as well as from pellets of synanthropic rooks, nests of collared doves and feathers of house sparrows (Hubálek, 1994). *Candida albicans* seems to be especially frequent in fresh droppings of gulls: it was recovered from about half of the samples tested along the eastern coastline of the United States (Buck, 1983, 1990). Experimental observations of a gull fed fish containing *C. albicans* demonstrated a heavy shedding of the yeast via faeces for the next 13 days, and even 40 days post-feeding the gull excreted the yeast sporadically, even though it was treated with ketoconazole (Buck, 1986). The gulls may thus serve as carriers of *C. albicans* or as a reservoir for it. However, no *C. albicans* infections of people have been reported as attributable to, or directly associated with, urban birds.

Other *Candida* spp. pathogenic to people have been isolated from a number of synanthropic wild avian samples (from excretions, the intestinal tract and other organs): *Candida guilliermondii*, *Candida krusei*, *Candida tropicalis*, *Candida pseudotropicalis* and *Candida parapsilosis* (Frágner, 1962; Kawakita & van Uden, 1965; Schönborn, Schütze & Pöhler, 1969; Cragg & Clayton, 1971; Kocan & Hasenclever, 1972; Monga, 1972; Gugnani, Sandhu & Shome, 1976; Guiguen, Boisseau-Lebreuill & Couprie, 1986; Pinowski, Kavanagh & Górski, 1991; Hubálek, 1994; Haag-Wackernagel & Moch, 2004). An outbreak of peritonitis due to *C. parapsilosis* in 12 patients undergoing peritoneal dialysis was attributed to contaminated pigeon excreta on window sills (Greaves et al., 1992).

8.2.3.1.2. Crypotococcus

Avian (especially feral pigeon) excretions represent a significant natural source of cryptococcosis in people. The first isolation of *C. neoformans* (its teleomorph stage is *Filobasidiella neoformans*, a basidomycetous yeast) from the nests and droppings of feral pigeons (Emmons, 1955, 1960) was followed by a number of similar reports worldwide (Yamamoto, Ishida & Sato, 1957; Littman & Schneierson, 1959; Frágner, 1962; Tsubura, 1962; Bergman, 1963; Muchmore et al., 1963; Silva & Paula, 1963; Frey & Durie, 1964; Partridge & Winner, 1965; Procknow et al., 1965; Randhawa, Clayton & Riddell, 1965; Taylor & Duangmani, 1968; Hubálek, 1975; Gugnani, Sandhu & Shome, 1976; Refai et al., 1983; Ruiz, Vélez & Fromtling, 1989; Yildiran et al., 1998; Haag-Wackernagel & Moch, 2004; Chee & Lee, 2005). The association between *C. neoformans* var. *grubii* (formerly *C. neoformans* serotype A) and the feral pigeon is remarkable (Denton & Di Salvo, 1968), whereas the occurrence of the fungus in faeces of other wild bird species is surprisingly scarce. As many as a hundred thousand to a million viable *C. neoformans* cells have been detected per gram of pigeon excreta in different parts of the world (Emmons, 1960; Hubálek, 1975; Ruiz, Fromtling & Bulmer, 1981; Ruiz, Neilson & Bulmer, 1982; Ruiz, Vélez & Fromtling, 1989).

This association was shown to be conditioned nutritionally, due to the ability of the yeast-like organism to utilize all basic low-molecular-weight nitrogenous substances from avian urine – that is, uric acid, creatinine, xanthine, guanine and urea – and due to the tenacity of the causative agent, *C. neoformans* (Staib, 1962, 1963; Walter & Yee, 1968; Hubálek, 1975; Ruiz, Neilson & Bulmer, 1982). The birds (pigeons) serve therefore largely as a lessor (Hubálek, Juřicová & Halouzka, 1995) for the fungus. However, carriage of the fungus has also been proved by isolating it from the feet and bills of feral pigeons (Littman & Borok, 1968) or from their lower and upper digestive tracts (Sethi & Randhawa, 1968; Swinne-Desgain, 1976; Khan et al., 1978; Guiguen, Boisseau-Lebreuill & Couprie, 1986; Rosario et al., 2005). In an experiment, pigeons with *C. neoformans* administered into the crop excreted the fungus sporadically in faeces up to 22 days, but harboured it for at least 86 days in the crop (Swinne-Desgain, 1976); this study demonstrated that pigeons could carry the fungus in their upper digestive tract. Spontaneous (natural) cryptococcosis has only rarely been observed in feral columbiforms (Hermoso de Mendoza et al., 1984). The course of experimental avian infections is usually abortive; only intracerebral inoculation was sufficient to cause death in some pigeons; the agent persisted in the brain for up to 11 weeks (Littman, Borok & Dalton, 1965; Böhm et al., 1974). The low susceptibility of birds to cryptococcosis may be due to the poor or nil growth of the fungus at 41°C – that is, at the avian body temperature.

There are several reports on human cryptococcal meningitis or pneumonia acquired by contact with pigeon habitats contaminated with the infectious agent (Muchmore et al., 1963, Procknow et al., 1965). *Cryptococcus neoformans* has been repeatedly isolated – even from the air – in feral pigeons' breeding sites (Staib & Bethäuser, 1968; Powell et al., 1972; Ruiz & Bulmer, 1981; Ruiz, Neilson & Bulmer, 1982), and the cell size of airborne yeast-like particles (0.6–3.5µm) is compatible with alveolar deposition after inhalation. Some studies have found that antibodies to *C. neoformans* are more frequent in people who have been in contact with pigeons than in people of the control group (Walter & Atchison, 1966; Newberry et al., 1967). Nearly all isolates from pigeon excreta are of serotype A (*C. neoformans* var. *grubii*), the same as most strains isolated from human cases of cryptococcosis (Walter & Coffee, 1968). The incidence of human cryptococcosis is 5–30% in patients with AIDS (or in people who are HIV-positive); a study from Burundi indicated that many of these cases could have acquired the mycosis by contact with contaminated pigeon habitats (Swinne et al., 1991). In Australia, a man died from cryptococcal meningitis four months after removing a Welcome swallow (*Hirundo neoxena*) nest heavily contaminated with *C. neoformans* serotype A from the roof (Glasziou & McAleer, 1984). A human case of osteomyelitis caused by *C. neoformans* var. *grubii* was described in Sweden; the patient became infected by cleaning a wooden box that had been used for several consecutive years for breeding by starlings and that contained the infectious agent (Kumlin et al., 1998).

8.2.3.2. Ascomycetes

8.2.3.2.1. Histoplasma

Histoplasma capsulatum (teleomorph *Ajellomyces capsulatus*) has occasionally been isolated from pigeon excreta and feathers (Suthill & Campbell, 1965). However, the most important source of human histoplasmosis in urban ecosystems of North America are roosting sites for common starlings and blackbirds (mostly the red-winged blackbird (*Agelaius phoeniceus*) and common grackle), which are often occupied by tens of thousands of these birds. In such sites, the ground under the trees is covered with a thick layer of droppings, becoming a habitat of *H. capsulatum* for years, and the soil is a source of human infection (Furcolow et al., 1961; Murdock et al., 1962; Ajello, 1964; Dodge, Ajello & Engelke, 1965; Tosh et al., 1966, 1970; Younglove et al., 1968; Brandsberg et al., 1969; Storch et al., 1980).

Like *C. neoformans, H. capsulatum* can grow in bird droppings, by utilizing uric acid (the main and specific constituent of avian urine) (Vanbreuseghem & Eugene, 1958; Smith, 1964, 1971; Lockwood & Garrison, 1968), and also on feathers (Suthill & Campbell, 1965). However, unlike *C. neoformans,* there is no reliable information that birds can carry *H. capsulatum*, and spontaneous histoplasmosis has not been confirmed in birds. Also, experimental inoculation of birds, even intraocular, did not cause disease (Schwarz et al., 1957; Sethi & Schwarz, 1966).

Contaminated areas could pose a human health hazard for a prolonged period of time (Latham et al., 1980). For example, the endemic reactivity of children from a school in the United States to the skin-test antigen histoplasmin was ascribed to the presence of a nearby roosting site of blackbirds, the soil of which was the source of *H. capsulatum* (Dodge, Ajello & Engelke, 1965). Major outbreaks of histoplasmosis (300 human cases) occurred in Mason City, Iowa, following two repeated clearings (in 1962 and 1964) of a park area where large numbers of starlings had been roosting for years (D'Alessio et al., 1965; Tosh et al., 1966). A total of 355 students showed symptoms of histoplasmosis after the soil was rototilled in an Indiana school courtyard known as a blackbird–starling roosting site; *H. capsulatum* was then also isolated from filters in the school air-conditioning system (Chamany et al., 2004). In another case, children were infected through contact with a nesting place of common grackles, contaminated with the fungus. Also, a great number of people working on a ring-billed gull (*Larus delawarensis*) nesting colony site in winter developed acute pulmonary histoplasmosis, and *H. capsulatum* was isolated from the nesting site (Waldman et al., 1983).

8.2.4. Protozoa

8.2.4.1. Microsporidia[1]

DNA from three microsporidian species – *Enterocytozoon bieneusi*, *Encephalitozoon hellem* and *Encephalitozoon intestinalis* – was recently detected in excreta of urban feral pigeons in Spanish parks (Haro et al., 2005). Children and the elderly are among the main visitors to these parks and, at the same time, they are the population groups at risk for microsporidiosis. However, no microsporidial infections of people have yet been reported as attributable to, or directly associated with, urban birds.

8.2.4.2. Babesiidae

It is probable that the protozoan parasite *Babesia microti*, one of several infectious agents of human babesiosis, may occur on, and be dispersed by, some synanthropic birds via attached infected preimaginal deer and castor-bean vector ticks. However, no *Babesia* infections of people have yet been reported as attributable to, or directly associated with, urban birds.

8.2.4.3. Eimeriidae

The protozoan parasite *Toxoplasma gondii* has been recorded relatively often in such wild birds as corvids (Finlay & Manwell, 1956; Lvov & Ilyichev, 1979; Literák et al., 1992), ducks, gulls (Lvov & Ilyichev, 1979; Literák et al., 1992) and columbids, including feral pigeons (Jacobs, Melton & Jones, 1952; Berger, 1966; Orlandella & Coppola, 1969; Lvov & Ilyichev, 1979; Literák et al., 1992; Haag-Wackernagel & Moch, 2004); it has been recorded less frequently in sparrows (Pak, 1976; Lvov & Ilyichev, 1979; Hejlíček, Prošek & Treml, 1981; Literák et al., 1992), starlings (Pak, 1976; Haslett & Schneider, 1978; Lvov & Ilyichev, 1979; Peach, Fowler & Hay, 1989) and several other passerines (Pak, 1976; Hejlíček, Prošek & Treml, 1981; Literák et al., 1992). On a communal roost in Leicester, England, 8% of starlings examined were infected; starlings could thus play an important role in the maintenance of toxoplasmosis in urban environments (Peach, Fowler & Hay, 1989). Infected starlings, sparrows and other small passerines can be the source of *T. gondii*

[1] Microsporidia have been newly classified as organisms closer to fungi than to protozoa.

infection for domestic cats. Many synanthropic avian species have shown to be susceptible to *T. gondii* when infected with it experimentally. For instance, pigeons inoculated with *T. gondii* have shown distinct clinical abnormalities (Biancifiori et al., 1986). Infected birds, such as pigeons, are less mobile and more susceptible to predation by cats which, as a definite host, could transmit the disease to other mammals, including human beings. On several occasions, pigeons were found to be the source of human toxoplasmosis (Orlandella & Coppola, 1969; Neto & Levi, 1970).

8.2.4.4. Cryptosporidiidae

Cryptosporidium is an enteric intracellular coccidian parasite that causes gastrointestinal and respiratory tract disorders in birds, or (more often) subclinical and asymptomatic infections. *Cryptosporidium* oocysts were found in faeces and cloacal specimens of gulls (herring gulls and black-headed gulls) in Scotland (Smith et al., 1993), and *Cryptosporidium baileyi* has been detected very frequently in black-headed gulls (28–100% chicks positive) in the Czech Republic; respiratory cryptosporidiosis was also diagnosed in several young gulls (Pavlásek, 1993). However, no cryptosporidial infections of people have been reported as attributable to, or directly associated with, urban birds.

8.3. Zoonoses and sapronoses of wild birds in the urban ecosystem

Although zoonoses of avian origin remain relatively infrequent (Cooper, 1990), many of the agents considered here (see Table 8.1) have been found to cause sporadic human cases or epidemics of corresponding zoonoses and sapronoses in urban areas, while others, though already established as associated with urban birds, have not yet been reported to cause infections in people attributable to, or directly caused by, urban birds. Nonetheless, vigilance is necessary, as the incidence of disease might be significantly underreported, and there might be a number of undiagnosed cases.

Pathogens significantly associated with wild and feral birds in urban areas are, for example:

- mosquito-borne SLEV (several epidemics in North American cities);
- WNV (epidemics with hundreds of patients in Bucharest, 1996–1997; in Volgograd, 1999–2000; and in United States cities, 1999–2006);
- *Chlamydophila psittaci* (at least 500 cases of ornithosis acquired from feral pigeons have been reported in the world since 1966);
- *Borrelia burgdorferi* s.l. (blackbirds in urban parks are carriers of infected ticks and are amplifying hosts of the agent);
- *Campylobacter jejuni*;
- enterotoxigenic *Escherichia coli*;

Table 8.1. Important human microbial pathogens associated with wild birds

Agent	Arthropod vector	Avian association with the agent	Human disease (No. of cases) [a]	Geographic area	Avian disease
St. Louis encephalitis flavivirus	Mosquito	Principal hosts	NDC[b]	Americas	None
West Nile flavivirus	Mosquito	Principal hosts	NDC[b]	Worldwide	Encephalitis (some strains)
TBE flavivirus complex	*Ixodes* tick	Occasional hosts, tick carriers	NDCc	Europe, Asia	Occasional (e.g., in the grouse)
Chlamydophila psittaci	None	Principal hosts	About 500	Worldwide	Ornithosis
Coxiella burnetii	Argasid tick	Occasional hosts	5	Worldwide	None
Borrelia burgdorferi sensu lato	*Ixodes* tick	Occasional hosts, tick carriers	NDCc	N. America, Eurasia	None
Campylobacter jejuni	None	Principal hosts	About 60	Worldwide	Campylobacterosis (in juveniles)
Escherichia coli enteropathogenic	None	Occasional hosts	–	Worldwide	None
Salmonella enterica	None	Occasional hosts	Tens	Worldwide	Salmonellosis
Yersinia pseudotuberculosis	None	Occasional hosts	–	Worldwide	Pseudotuberculosis
Listeria monocytogenes	None	Occasional hosts	–	Eurasia, Asia	Listeriosis
Erysipelothrix rhusiopathiae	None	Occasional hosts	–	Worldwide	Erysipelas
Mycobacterium avium	None	Principal hosts	–	Worldwide	Tuberculosis
Mycobacterium xenopi	None	Occasional hosts	Tens	Worldwide	Mycobacteriosis
Candida albicans	None	Hosts, lessors	–	Worldwide	Candidosis
Candida parapsilosis	None	Hosts, lessors	12	Worldwide	None
Cryptococcus neoformans	None	Lessors, occasional hosts	Tens	Worldwide	None
Histoplasma capsulatum	None	Lessors	About 1000	Americas	None
Toxoplasma gondii	None	Occasional hosts	Several	Worldwide	Toxoplasmosis

[a] Reported human disease (cases or epidemics) directly associated with urban birds or their habitats.

[b] The number of cases acquired directly from urban wild birds has yet been described, but the birds can serve as amplifying hosts that contribute to the circulation of the agent in urban areas.

[c] The number of cases acquired directly from birds has been described, but wild birds can serve as carriers of infected preimaginal ixodid ticks and can thus contribute to the circulation of the agent in urban areas.

Notes. NDC: no direct cases; –: no data or no detailed reports.

Source: Prepared by the author from numerous sources, all of them present in the reference list.

- *Salmonella enterica*, particularly serovars Typhimurium and Enteritidis;
- *Cryptococcus neoformans* (pigeon droppings can be the microhabitat for the growth of the yeast-like fungus at urban gathering places of feral pigeons, such as church belfries, old buildings and squares); and
- *Histoplasma capsulatum* (hundreds of human cases of histoplasmosis acquired near communal roosts of blackbirds and starlings in North American city parks).

The ability of some fungi (less so of certain bacteria) to grow in avian excreta and nests is remarkable. These pathogens assimilate uric acid and other low-molecular-weight nitrogenous compounds contained in bird droppings, and the birds thus serve not as hosts for these microorganisms, but as *lessors* for them. This could pose a public health hazard in the case of mass communal roosts or large nesting colonies of birds at urban or suburban sites inhabited, exploited or attended by people. In these cases, airborne inhalation is the main means of infection in people, and it usually occurs during rebuilding or related work in contaminated areas.

In general, the prevailing modes of transmission of pathogens from birds to people in urban areas are either via airborne or alimentary routes. An airborne, respiratory infection is the commonest mode in, for example, ornithosis, avian tuberculosis, cryptococcosis and histoplasmosis, while infection by ingestion might occur in, for example, salmonellosis, campylobacterosis, colibacillosis, listeriosis, toxoplasmosis and cryptosporidiosis, which are food-borne or water-borne diseases. Less frequently, some zoonotic infections in urban areas are transmitted by direct contact with birds, suffering from, for example, erysipeloid, streptococcosis, or pseudotuberculosis. Finally, arthropod-borne diseases (such as Eastern and Western equine encephalomyelitis, Sindbis fever, West Nile fever, TBE and Lyme borreliosis) are caused by haematophagous invertebrates (such as mosquitoes and ixodid ticks) that acquired the infectious agent previously, by feeding on viraemic or bacteraemic urban birds.

A number of ecological factors might affect the risk of bird-borne illness in urban settings. For example, bird species that have higher population densities (such as feral pigeons); that nest in colonies (such as colonial waterbirds, gulls and rooks); that roost gregariously (such as starlings, grackles, corvids and gulls); and that congregate at water and food sources (such as terrestrial birds) or in other particular places in urban areas for mass moulting (such as waterfowl in summer), migration stops and overwintering (such as waterfowl and gulls) are more important epidemiologically (due to frequent interindividual contacts that enable effective horizontal transmission of disease agents) than species with low population densities that live solitarily or in small groups. Habitat preferences by birds also affect their epidemiological role. For example, aquatic birds, even in urban situations, attract higher numbers of haematophagous insects (such as mosquitoes) than do terrestrial birds, while woodland, ground-foraging birds are parasitized by ixodid ticks. Avian mobility and migratory habits are other crucial factors; they make the transport and spread of various agents by carrier birds more effective.

In brief, the most important factors that increase the risk of bird-borne infections are:

- overpopulation of (infected) urban birds;
- overpopulation of infected invertebrate vectors (in the case of arthropod-borne diseases); and
- enhanced, intimate contacts between people and infected urban birds or their habitats (the latter case when urban birds serve as a lessor of the agent) or both.

8.4. Management implications

8.4.1. Benchmarks

In general, the benchmarks for urban bird management should take into consideration the following (K. Sweeney, United States Environmental Protection Agency, personal communication, 2005):

- identifying the real public health threat posed by urban birds, using monitoring and surveillance methods;
- determining the level of threat that requires a response to protect the public;
- deciding the type, intensity and timing of a response;
- measuring the success of the response.

This chapter has attempted to describe the recent state of knowledge of urban birds as a public health hazard. Some synanthropic avian species can obviously serve as hosts, reservoirs, transporters or lessors of a few human pathogens. However, the extent and significance of this epidemiological hazard might vary enormously, and it should be measured according to local conditions in particular urban areas. It is therefore very difficult to propose and apply a set of general benchmarks for urban birds as a public health threat. In general, the evidence and estimation of the level of threat must be based on data on the incidence of bird-borne illnesses in particular urban settings, using standard epidemiological surveillance methods. The straightforward way of determining the epidemiological hazard is to first establish whether a human bird-borne infection does or does not occur at a particular urban setting. If it does, the second step is to establish how often it occurs (incidence of the disease). The final step then involves a decision about the level (such as one human case or more cases) at which public funds should be spent on preventive and control programmes and measures.

At least several cases of human disease directly associated with urban birds or their habitats have been reported for ornithosis, histoplasmosis, campylobacterosis, salmonellosis, mycobacteriosis, cryptococcosis, toxoplasmosis and Q fever (Table 8.1). However, compared with other communicable diseases, in general, their annual incidence is quite low, but underreporting should be taken into account. Also, although no cases acquired directly from birds have been described for a number of other infectious diseases (mostly arthropod-borne infections), certain wild urban birds can serve as amplifying hosts or carriers of infected preimaginal ixodid ticks, contributing thus to the circulation of disease agents in urban areas (Table 8.1). Moreover, allergic responses to ectoparasites of feral pigeons have not been included in Table 8.1 and are another consideration.

8.4.2. Monitoring and surveillance

A prerequisite for managing urban birds and their potential public health hazard is the monitoring of avian populations and surveillance for associated zoonoses and sapronoses. The majority of the public health problems caused by wild birds are associated with feral pigeons, gulls, blackbirds, grackles, starlings, corvids and house sparrows. For instance, at least 800 reported transmissions of a pathogen (mostly *C. psittaci*) from feral pigeons to people have been found (Sixl, 1975; Pospíšil et al., 1988; Glünder, 1989; Haag-Wackernagel & Moch, 2004), this probably being only the so-called tip of the iceberg. Similarly, hundreds of cases of histoplasmosis in people have been acquired via the air-borne route during, or after, work on communal roosts of birds in urban areas in North America (Furcolow et al., 1961; Murdock et al., 1962; Ajello, 1964; D'Alessio et al., 1965; Dodge, Ajello & Engelke, 1965; Tosh et al., 1966, 1970; Younglove et al., 1968; Latham et al., 1980; Storch et al., 1980; Waldman et al., 1983; Chamany et al., 2004).

Public health surveillance should involve both a passive and active monitoring approach, the former based mainly on reports of disease, while the latter also includes serological surveys of urban birds and city dwellers, further microbiological examination of competent haematophagous invertebrate vectors and avian hosts (their infection rate), and investigations of habitats as sources of disease. Monitoring population density of the avian hosts and invertebrate vectors, and their spatial (mapping) and temporal (seasonal) distribution, is also necessary. Management priorities should then be established and objectives defined for prevention and control of bird-related infections.

8.4.3. Control of wild and feral birds in urban areas

The control of wild bird populations (especially those of feral pigeons) in urban and sub-urban areas is difficult and sometimes ineffective. However, a few so-called public-friendly methods are available to control potentially infected urban bird populations. To prevent risks that arise from the presence of infections in birds or infectious materials in their droppings, several tasks should be performed, as soon as microbiologists and epidemiologists have demonstrated an infection (zoonosis or sapronosis) (Lesaffre, 1997; Haag-Wackernagel, 1995, 2000; Rödl, 1999). These tasks include:

- restricting feeding at public places;
- inhibiting breeding on buildings, by mechanically blocking the loft orifices and perching sites in, on and below the roofs, using netting, spikes, repellent gels and electroshock deterrent systems;
- collecting and inactivating avian (pigeon) eggs;
- controlling scavenging birds, such as gulls and corvids, on landfill sites (garbage deposits);
- controlling seagulls at harbours and airports (if their numbers create hygienic and safety problems);
- controlling and sanitizing large communal roosts of birds in city parks – for example, effective decontamination of the soil under communal roosts infected with sapronotic

fungi (*H. capsulatum*) is possible by repeated (3–4 times) use of 3% formaldehyde (Tosh et al., 1967);

- reducing the number of nest boxes available for starlings;
- trapping birds (in nets or various bird traps) and euthanizing the captured birds, if permitted; and
- sterilizing birds chemically, with a treated grain bait, if permitted.

These activities should be carried out in an integrated approach to bird management, since individual steps alone do not produce success. Furthermore, inspection and control measures must be performed by, or under the supervision of, veterinary public health agencies – and only when they are substantiated and necessary. Ornithologists, wildlife managers and citizen representatives (such as consumers) should be involved in implementing the control measures. The integrated approach to bird management also needs a public education component (media) and a legal (regulatory) component – that is, political support – as necessary parts of the process. Finally, a risk–benefit analysis should also be performed.

8.4.4. Techniques for dispersing birds in cities

Birds can be dispersed by various techniques (Frings & Jumber, 1954; Bickerton & Chapple, 1961; Schmitt, 1962; Brough, 1969; Gorenzel & Salmon, 1992, 1993). Briefly, they include the use of:

- tape strips;
- airport strobe lights and flashlights;
- fireworks, rockets and shell crackers (cartridges that contain no lead shot) against approaching flocks which, although effective, are less acceptable in cities;
- shooting, pistols, explosives and screamers;
- tape-recorded distress or scary calls of birds (such as those of owls);
- falcons and other trained raptors;
- water-mist sprayers;
- plastic netting; and

habitat modification, by thinning, clearing or even eliminating the vegetation (such as reeds, brush and trees) on communal roosts.

Birds, however, usually get accustomed to being disturbed by various acoustic or light signals. Some of the methods could be used only under certain circumstances and should be used respectfully in residential environments.

8.4.5. Economic impact of wild urban birds on human health and of controlling birds

The financial costs of bird-borne diseases that affect people in urban areas are extremely difficult to estimate at present. Some reasons for this are as follows.

1. The incidence of these diseases (and especially the proportion attributable to urban birds as the source of a particular disease) is largely unknown or underestimated, in that some of the diseases are not reportable. Table 8.1 indicates only published cases, and the incidence of allergy-related illness due to contact with ectoparasites of feral pigeons is not included (because it is underreported or unreported).

2. The public health costs of treating particular diseases can be roughly estimated (see item 1).

3. The costs of prevention – that is, proactive and reactive control measures – can vary widely, according to such factors as the disease agent and bird species, species distribution, particular country and local situation.

In any economic analysis on this issue, there will always be a very significant margin of error, in that there are no exact data available. So for now, even estimates by experts will be of limited value.

Acknowledgement

The Czech Science Foundation (grant number 206/03/0726) provided partial funding for this review.

References[2]

Ajello L (1964). Relationship of *Histoplasma capsulatum* to avian habitats. *Public Health Reports*, 79:266–270.

Alekseev AN et al. (2001). Evidence of ehrlichiosis agents found in ticks (Acari: Ixodidae) collected from migratory birds. *Journal of Medical Entomology*, 38:471–474.

Anderson JF, Magnarelli LA, Stafford KC (1990). Bird-feeding ticks transstadially transmit *Borrelia burgdorferi* that infect Syrian hamsters. *Journal of Wildlife Diseases*, 26:1–10.

Anderson JF et al. (1986). Involvement of birds in the epidemiology of the Lyme disease agent *Borrelia burgdorferi*. *Infection and Immunity*, 51:394–396.

Babudieri B, Moscovici C (1952). Experimental and natural infection of birds by *Coxiella burnetii*. *Nature*, 169:195–196.

Baker JR (1977). The results of post-mortem examination of 132 wild birds. *The British Veterinary Journal*, 133:327–333.

Benton C et al. (1983). The contamination of a major water supply by gulls (*Larus* sp.). *Water Research*, 17:789–798.

Beran GW, section ed. (1981). *Viral zoonoses, Vol. 1.* Boca Raton, CRC Press.

Berg RW, Anderson AW (1972). Salmonellae and *Edwardsiella tarda* in gull feces: a source of contamination in fish processing plants. *Applied Microbiology*, 24:501–503.

Berger J (1966). Zur Epidemiologie der Toxoplasmose. I. Die Toxoplasmose der verwilderten Stadttauben. *Zeitschrift für Medizinische Mikrobiologie und Immunologie*, 153:68–82.

Bergman F (1963). Occurrence of *Cryptococcus neoformans* in Sweden. *Acta Medica Scandinavica,* 174:651–655.

Biancifiori F et al. (1986). Avian toxoplasmosis: experimental infection of chicken and pigeon. *Comparative Immunology, Microbiology and Infectious Diseases*, 9:337–346.

Bickerton BM, Chapple W (1961). Starling roosts and their dispersal. *Agriculture*, 67:624–626.

Bjoersdorf A et al. (2001). *Ehrlichia*-infected ticks on migrating birds. *Emerging Infectious Diseases*, 7:877–879.

[2] The quoted literature in this review is largely based on articles in peer-reviewed journals and in books. In a few cases, information on the Internet published by renowned institutions or organizations has been used.

Blackmore DK, Keymer IF (1969). Cutaneous diseases of wild birds in Britain. *British Birds*, 62:316–331.

Böhm KH et al. (1974). Infektionsversuche an Tauben und Hühnern mit *Cryptococus neoformans* sowie weitere epidemiologische Untersuchungen. *Mycopathologia*, 54:317–328.

Bouvier G (1969). La salmonellose chez les oiseaux sauvages, notamment chez les petits passereaux des environs de Lausanne. *Nos Oiseaux*, 29:293–295.

Bowes V (1990). Paratyphoid infection in English sparrows. *Canadian Veterinary Journal*, 31:592.

Brand CJ (1989). Chlamydial infections in free-living birds. *Journal of the American Veterinary Medical Association*, 195:1531–1535.

Brandsberg JW et al. (1969). A study of fungi found in association with *Histoplasma capsulatum*: three birds' roosts in S.E. Missouri, U.S.A. *Mycopathologia et Mycologia Applicata*, 38:71–81.

Brickford AA, Ellis GH, Moses HE (1966). Epizootiology of tuberculosis in starlings. *Journal of the American Veterinary Medical Association*, 149:312–318.

Brittingham MC, Temple SA (1986). A survey of avian mortality at winter feeders. *Wildlife Society Bulletin*, 14:445–450.

Brough T (1969). The dispersal of starlings from woodland roosts and the use of bioacoustics. *Journal of Applied Ecology*, 6:403–410.

Brown I et al., eds (2005). Mission to Russia to assess the avian influenza situation in wildlife and the national measures being taken to minimize the risk of international spread. Paris, World Organization for Animal Health (http://www.oie.int/downld/Missions/2005/ReportRussia2005Final2.pdf, accessed 24 August 2006).

Brummer-Korvenkontio M et al. (1973). Arboviruses in Finland. I. Isolation of tick-borne encephalitis virus from arthropods, vertebrates, and patients. *The American Journal of Tropical Medicine and Hygiene*, 22:382–389.

Buck JD (1983). Occurrence of *Candida albicans* in fresh gull feces in temperate and subtropical areas. *Microbial Ecology*, 9:171–176.

Buck JD (1986). A note on the experimental uptake and clearance of *Candida albicans* in a young captive gull (*Larus* sp.). *Mycopathologia*, 94:59–61.

Buck JD (1990). Isolation of *Candida albicans* and halophilic *Vibrio* spp. from aquatic birds in Connecticut and Florida. *Applied and Environmental Microbiology*, 56:826–828.

Butterfield J et al. (1983). The herring gull *Larus argentatus* as a carrier of Salmonella. *The Journal of Hygiene*, 91:429–436.

Caffrey C, Smith SCR, Weston TJ (2005). West Nile virus devastates an American crow population. *Condor*, 107:128–132.

Capua I et al. (2000). Monitoring for highly pathogenic avian influenza in wild birds in Italy. *The Veterinary Record*, 147:640.

Chamany S et al. (2004). A large histoplasmosis outbreak among high school students in Indiana, 2001. *The Pediatric Infectious Disease Journal*, 23:909–914.

Chamberlain RW et al. (1957). Virus of St. Louis encephalitis in three species of wild birds. *American Journal of Hygiene*, 65:110–118.

Chee HY, Lee KB (2005). Isolation of *Cryptococcus neoformans* var. *grubii* (serotype A) from pigeon droppings in Seoul, Korea. *Journal of Microbiology (Seoul, Korea)*, 43:469–472.

Clark GM, Locke LN (1962). Observations on pseudotuberculosis in common grackles. *Avian Diseases*, 6:506–510.

Cooley AJ et al. (1989). Pathological lesions in the lungs of ducks infected with influenza A viruses. *Veterinary Pathology*, 26:1–5.

Cooper JE (1990). Birds and zoonoses. *Ibis*, 132:181–191.

Cooper JE, Needham JR (1981). The starling (*Sturnus vulgaris*) as an experimental model for staphylococcal infection of the avian foot. *Avian Pathology*, 10:273–280.

Cornelius LW (1969). Field notes on *Salmonella* infection in greenfinches and house sparrows. *Bulletin of the Wildlife Disease Association*, 5:142–143.

Cragg J, Clayton YM (1971). Bacterial and fungal flora of seagull droppings in Jersey. *Journal of Clinical Pathology*, 24:317–319.

D'Alessio DJ et al. (1965). A starling roost as the source of urban epidemic histoplasmosis in an area of low incidence. *The American Review of Respiratory Disease*, 92:725–731.

Daniels TJ et al. (2002). Avian reservoirs of the agent of human granulocytic ehrlichiosis? *Emerging Infectious Diseases*, 8:1524–1525.

Daoust PY (1978). Osteomyelitis and arthritis caused by *Salmonella typhimurium* in a crow. *Journal of Wildlife Diseases*, 14:483–485.

Dardiri AH et al. (1957). The isolation of eastern equine encephalomyelitis virus from brains of sparrows. *Journal of the American Veterinary Medical Association*, 130:409–410.

Davis JW et al., eds (1971). *Infectious and parasitic diseases of wild birds*. Ames, IA, Iowa State University Press.

De Herdt P et al. (1992). Biochemical and antigenic properties of *Streptococcus bovis* isolated from pigeons. *Journal of Clinical Microbiology*, 30:2432–2434.

Denton JF, Di Salvo AF (1968). The prevalence of *Cryptococcus neoformans* in various natural habitats. *Sabouraudia*, 6:213–217.

Devriese LA et al. (1990). *Streptococcus bovis* infections in pigeons. *Avian Pathology*, 19:429–434.

Dodge HJ, Ajello L, Engelke OK (1965). The association of a bird-roosting site with infection of school children by *Histoplasma capsulatum*. *American Journal of Public Health*, 55:1203–1211.

Dovc A et al. (2004). Health status of free-living pigeons (*Columba livia domestica*) in the city of Ljubljana. *Acta Veterinaria Hungarica*, 52:219–226.

Dózsa I (1964). Der Haussperling (*Passer domesticus*) als *Salmonella typhimurium* Reservoir. *Aquila*, 69–70:227–229.

Dusbábek F, Rosický B (1976). Argasid ticks (Argasidae, Ixodoidea) of Czechoslovakia. *Acta Scientiarum Naturalium Academiae Scientiarum Bohemoslovacae, Brno*, 10:1–43.

Eidson M et al. (2001). Crow deaths as a sentinel surveillance system for West Nile virus in the northeastern United States, 1999. *Emerging Infectious Diseases*, 7:615–620.

Ellis TM et al. (2004). Investigation of outbreaks of highly pathogenic H5N1 avian influenza in waterfowl and wild birds in Hong Kong in late 2002. *Avian Pathology*, 33:492–505.

Emmons CW (1955). Saprophytic sources of *Cryptococcus neoformans* associated with the pigeon (*Columba livia*). *American Journal of Hygiene*, 62:227–232.

Emmons CW (1960). Prevalence of *Cryptococcus neoformans* in pigeon habitats. *Public Health Reports*, 75:362–364.

Englert HK et al. (1967). Eine enzootisch auftretende Salmonellose unter den Singvögeln in Baden. *Berliner und Münchener Tierärztliche Wochenschrift*, 80:277–279.

Ernek E et al. (1968). The role of birds in the circulation of tick-borne encephalitis virus in the Tribeč region. *Acta Virologica*, 12:468–470.

Ernek E et al. (1977). Arboviruses in birds captured in Slovakia. *Journal of Hygiene, Epidemiology, Microbiology, and Immunology*, 21:353–359.

Faddoul GP, Fellows GW, Baird J (1968). *Erysipelothrix* infection in starlings. *Avian Diseases*, 12:61–66.

FAO Avian Influenza Task Force (2005). Update on the avian influenza situation. Potential risk of Highly Pathogenic Avian Influenza (HPAI) spreading through wild water bird migration. *FAO AIDE News*, 33:1–7 (http://www.fao.org/ag/againfo/subjects/documents/ai/AVIbull033.pdf, accessed 25 August 2006; later reports under the suffixes /AVIbull034.pdf and /AVIbull035.pdf).

Fenlon DR (1981). Seagulls (*Larus* spp.) as vectors of salmonellae: an investigation into the range of serotypes and numbers of salmonellae in gull faeces. *The Journal of Hygiene*, 86:195–202.

Fenlon DR (1983). A comparison of *Salmonella* serotypes found in the faeces of gulls feeding at a sewage works with serotypes present in the sewage. *The Journal of Hygiene*, 91:47–52.

Fenlon DR (1985). Wild birds and silage as reservoirs of Listeria in the agricultural environment. *The Journal of Applied Bacteriology*, 59:537–543.

Fennel H, James DB, Morris J (1974). Pollution of a storage reservoir by roosting gulls. *Water Treatment and Examination*, 23:5–20.

Finlay P, Manwell RD (1956). *Toxoplasma* from the crow, a new natural host. *Experimental Parasitology*, 5:149–153.

Frágner P (1962). Nálezy kryptokokû v ptačím trusu. [Findings of cryptococci in avian excreta]. *Československá Epidemiologie, Mikrobiologie, Imunologie,* 11:135–139 (in Czech).

Frey D, Durie EB (1964). The isolation of *Cryptococcus neoformans* (*Torula histolytica*) from soil in New Guinea and pigeon droppings in Sydney, New South Wales. *The Medical Journal of Australia*, 14:947–949.

Fricker CR (1984). A note on *Salmonella* excretion in the black headed gull (*Larus ridibundus*) feeding at sewage treatment works. *Journal of Applied Bacteriology*, 56:499–502.

Frings H, Jumber J (1954). Preliminary studies on the use of a specific sound to repel starlings (*Sturnus vulgaris*) from objectionable roosts. *Science*, 119:318–319.

Fukushima H, Gomyoda M (1991). Intestinal carriage of *Yersinia pseudotuberculosis* by wild birds and mammals in Japan. *Applied and Environmental Microbiology*, 57:1152–1155.

Furcolow ML et al. (1961). The emerging pattern of urban histoplasmosis. Studies on an epidemic in Mexico, Missouri. *The New England Journal of Medicine*, 264:1226–1230.

Gaidamovich SY, Sokhey J (1973). Studies on antigenic peculiarities of West Nile virus strains isolated in the USSR by three serological tests. *Acta Virologica*, 17:343–350.

Gautsch S et al. (2000). Die Rolle des Gemeinen Stars (*Sturnus vulgaris*) in der Epidemiologie bakterieller, potentiell humanpathogener Krankheitserreger. *Schweizer Archiv für Tierheilkunde*, 142:165–172.

Ginsberg HS et al. (2005). Reservoir competence of native North American birds for the Lyme disease spirochete, *Borrelia burgdorferi*. *Journal of Medical Entomology*, 42:445–449.

Girdwood RW et al. (1985). The incidence and significance of *Salmonella* carriage by gulls (*Larus* spp.) in Scotland. *The Journal of Hygiene*, 95:229–241.

Glasziou D, McAleer R (1984). Cryptococcosis: an unusual avian source. *The Medical Journal of Australia*, 140:447.

Glünder G (1989). Infektionen der Tauben als Risiko für die Gesundheit von Mensch und Tier. *Deutsche Tierärztliche Wochenschrift*, 96:112–114.

Glünder G, Neumann U, Braune S (1992). Occurrence of *Campylobacter* spp. in young gulls, duration of *Campylobacter* infection and reinfection by contact. *Zentralblatt für Veterinärmedizin. Reihe B*, 39:119–122.

Glünder G, Petermann S (1989). Vorkommen und Charakterisierung von *Campylobacter* spp. bei Silbermöwen (*Larus argentatus*), Dreizehenmöwen (*Rissa tridactyla*) und Haussperlingen (*Passer domesticus*). *Zentralblatt für Veterinärmedizin. Reihe B*, 36:123–130.

Glünder G, Siegmann O, Köhler W (1991). Krankheiten und Todesursachen bei einheimischen Wildvögeln – eine Literaturübersicht. *Zentralblatt für Veterinärmedizin. Reihe B*, 38:241–262.

Glünder G et al. (1991). Zum Vorkommen von *Campylobacter* spp. und *Salmonella* spp. bei Möwen in Norddeutschland. *Deutsche Tierärztliche Wochenschrift*, 98:152–155.

Golebiowski S (1975). Nosicielstwo bakterii chorobotwórzych wśród wolnożyjacych ptaków [The carriage of pathogenic bacteria in free-living birds]. *Medycyna Weterynarna*, 31:143–144 (in Polish).

Gorenzel WP, Salmon TP (1992). Urban crow roosts in California. In: Borrecco JE, Marsh RE, eds. *Proceedings of the Fifteenth Vertebrate Pest Conference*, 2–5 March 1992, Monterey, California, University of California at Davis:97–102.

Gorenzel WP, Salmon TP (1993). Tape-recorded calls disperse American crows from urban roosts. *Wildlife Society Bulletin*, 21:334–338.

Greaves I et al. (1992). Pigeons and peritonitis? *Nephrology, Dialysis, Transplantation*, 7:967–969.

Grešíková M, Ernek E (1965). The role of some bird species in the ecology of tick-borne encephalitis. In: Rosický B, Heyberger K, eds. *Theoretical questions of natural foci of diseases*. Prague, Publishing House of the Czechoslovak Academy of Sciences:193–198.

Grešíková M, Sekeyová M, Prazniaková E (1975). Isolation and identification of group B arboviruses from the blood of birds captured in Czechoslovakia. *Acta Virologica*, 19:162–164.

Greuel E, Arnold J (1971). Epidemiologische Studien zum Vorkommen von Salmonellen bei Singvögeln. *Berliner und Münchener Tierärztliche Wochenschrift*, 84:292–294.

Grimes JE, Owens KJ, Singer JR (1979). Experimental transmission of *Chlamydia psittaci* to turkeys from wild birds. *Avian Diseases*, 23:915–926.

Grossmann K et al. (2005). Racing, ornamental and city pigeons carry Shiga toxin producing *Escherichia coli* (STEC) with different Shiga toxin subtypes, urging further analysis of their epidemiological role in the spread of STEC. *Berliner und Münchener Tierärztliche Wochenschrift*, 118:456–463.

Grund S, Stolpe H (1992). Adhesion of *Salmonella typhimurium* var. *copenhagen* in the intestine of pigeons. *International Journal of Food Microbiology*, 15:299–306.

Gugnani HC, Sandhu RS, Shome SK (1976). Prevalence of *Cryptococcus neoformans* in avian excreta in India. *Mykosen*, 19:183–187.

Guiguen C, Boisseau-Lebreuill MT, Couprie B (1986). Flore fongique du tube digestif isolee de pigeons de ville a Bordeaux. *Bulletin de la Société Française de Mycologie Médicale*, 15:151–154.

Haag-Wackernagel D (1995). Regulation of street pigeon in Basel. *Wildlife Society Bulletin*, 23:256–260.

Haag-Wackernagel D (2000). Behavioural responses of the feral pigeon (Columbidae) to deterring systems. *Folia Zoologica*, 49:101–114.

Haag-Wackernagel D (2005). Parasites from feral pigeons as a health hazard for humans. *Annals of Applied Biology*, 147:203–210.

Haag-Wackernagel D, Moch H (2004). Health hazards posed by feral pigeons. *The Journal of Infection*, 48:307–313.

Hacking MA, Sileo L (1974). *Yersinia enterocolitica* and *Yersinia pseudotuberculosis* from wildlife in Ontario. *Journal of Wildlife Diseases*, 10:452–457.

Hájek V, Balusek J (1988). Biochemical properties and differentiation of coagulase-positive staphylococci from rooks and gulls. *Research in Veterinary Science*, 44:242–250.

Hájek V et al. (1991). Characterization of coagulase-positive staphylococci isolated from free-living birds. *Journal of Hygiene, Epidemiology, Microbiology, and Immunology*, 35:407–418.

Hall RF et al. (1977). Isolation of *Salmonella* spp. from dead gulls (*Larus californicus* and *Larus delawarensis*) from an Idaho irrigation reservoir. *Avian Diseases*, 21:452–454.

Hamasaki SI et al. (1989). A survey for *Yersinia pseudotuberculosis* in migratory birds in coastal Japan. *Journal of Wildlife Diseases*, 25:401–403.

Hammon WM, Reeves WC, Sather GE (1951). Western equine and St. Louis encephalitis viruses in the blood of experimentally infected wild birds and epidemiological implications of findings. *Journal of Immunology*, 67:357–367.

Hanincová K et al. (2003). Association of *Borrelia garinii* and *B. valaisiana* with songbirds in Slovakia. *Applied and Environmental Microbiology*, 69:2825–2830.

Hardy JL (1987). The ecology of western equine encephalomyelitis virus in the Central Valley of California, 1945–1985. *The American Journal of Tropical Medicine and Hygiene*, 37(Suppl. 3):18S–32S.

Haro M et al. (2005). First detection and genotyping of human-associated microsporidia in pigeons from urban parks. *Applied and Environmental Microbiology*, 71:3153–3157.

Harry EG (1967). Some characteristics of *Staphylococcus aureus* isolated from the skin and upper respiratory tract of domesticated and wild (feral) birds. *Research in Veterinary Science*, 8:490–499.

Hasenclever HF, Kogan RM (1975). *Candida albicans* associated with the gastrointestinal tract of the common pigeon (*Columba livia*). *Sabouraudia*, 13:116–120.

Haslett TM, Schneider WJ (1978). Occurrence and attempted transmission of *Toxoplasma gondii* in European starlings (*Sturnus vulgaris*). *Journal of Wildlife Diseases*, 14:173–175.

Hejlíček K, Prošek F, Treml F (1981). Isolation of *Toxoplasma gondii* in free-living small mammals and birds. *Acta Veterinaria*, 50:233–236.

Hejlíček K, Treml F (1993a). Výskyt aviární mykobakteriózy u volně žijícího ptactva v různých epizootologických poměrech tuberkulózy drůbeže. [The occurrence of avian mycobacteriosis in free-living birds during various epizootic conditions of tuberculosis in poultry]. *Veterinární Medicina*, 38:305–317 (in Czech).

Hejlíček K, Treml F (1993b). Epizootologie a patogeneze mykobakteriózy hrdličky (*Streptopelia* sp.). [Epizootiology and pathogenesis of avian mycobacteriosis in doves (*Streptopelia* sp.)]. *Veterinární Medicina*, 38:619–628 (in Czech).

Hejlíček K, Treml F (1993c). Epizootologie a patogeneze aviární mykobakteriózy vrabce domácího (*Passer domesticus*) a vrabce polního (*Passer montanus*). [Epizootiology and pathogenesis of avian mycobacteriosis in the house sparrow (*Passer domesticus*) and tree sparrow (*Passer montanus*)]. *Veterinární Medicina*, 38:667–685 (in Czech).

Henry K, Crossley K (1986). Wild-pigeon-related psittacosis in a family. *Chest*, 90:708–710.

Hermoso de Mendoza M. et al. (1984). Criptococosis espontánea en palomo (*Columba livia*) [Spontaneous cryptococcosis in pigeon]. *Archivos de Zootécnica (Cordoba)*, 33:27–41 (in Spanish).

Hinshaw VS et al. (1980). Genetic reassortment of influenza A viruses in the intestinal tract of ducks. *Virology*, 102:412–419.

Hinshaw VS et al. (1981). Replication of avian A influenza viruses in mammals. *Infection and Immunity*, 34:354–361.

Hinshaw VS et al. (1985). Circulation of influenza viruses and paramyxoviruses in waterfowl: comparison of different migratory flyways in North America. *Bulletin of the World Health Organization*, 63:711–719.

Holden P (1955). Recovery of western equine encephalomyelitis virus from naturally infected English sparrows of New Jersey, 1953. *Proceedings of the Society for Experimental Biology and Medicine*, 88:490–492.

Holden P et al. (1973). House sparrows, *Passer domesticus* (L.), as hosts of arboviruses in Hale County, Texas. I. Field studies, 1965–1969. *The American Journal of Tropical Medicine and Hygiene*, 22:244–262.

Hoogstraal H (1967). The potential of virus-infected and tick-infested migrating birds in the spread of disease. *Annual Review of Entomology*, 12:87–91.

Hubálek Z (1975). Distribution of *Cryptococcus neoformans* in a pigeon habitat. *Folia Parasitologica*, 22:73–79.

Hubálek Z (1994). Pathogenic microorganisms associated with free-living birds (a review). *Acta Scientiarum Naturalium Academiae Scientiarum Bohemoslovacae, Brno*, 28:1–74.

Hubálek Z (2003). Emerging human infectious diseases: anthroponoses, zoonoses, and sapronoses. *Emerging Infectious Diseases*, 9:403–404.

Hubálek Z, Juřicová Z, Halouzka J (1995). A survey of free-living birds as hosts and 'lessors' of microbial pathogens. *Folia Zoologica*, 44:1–11.

Hubálek Z et al. (1996). Borreliae in immature *Ixodes ricinus* (Acari: Ixodidae) ticks parasitizing birds in the Czech Republic. *Journal of Medical Entomology*, 33:766–771.

Hudson SJ et al. (1991). Jackdaws and magpies as vectors of milkborne human *Campylobacter* infection. *Epidemiology and Infection*, 107:363–372.

Humair PF et al. (1993). *Ixodes ricinus* immatures on birds in a focus of Lyme borreliosis. *Folia Parasitologica*, 40:237–242.

Humair PF et al. (1998). An avian reservoir (*Turdus merula*) of the Lyme borreliosis spirochetes. *Zentralblatt für Bakteriologie: International Journal of Medical Microbiology*, 287:521–538.

Hurlbut HS et al. (1956). A study of the ecology of West Nile virus in Egypt. *The American Journal of Tropical Medicine and Hygiene*, 5:579–620.

Hussong D et al. (1979). Microbial impact of Canada geese (*Branta canadensis*) and whistling swans (*Cygnus columbianus columbianus*) on aquatic ecosystems. *Applied and Environmental Microbiology*, 37:14–20.

Illner F (1961). Über das Vorkommen des *Ornithosevirus* beim Haussperling (*Passer domesticus*). *Zeitschrift für Veterinärmedizin*, 16:933–935.

Jacobs L, Melton ML, Jones EF (1952). The prevalence of toxoplasmosis in wild pigeons. *Journal of Parasitology*, 38:457–461.

Jennings AR (1955). Diseases of wild birds. *Bird Study*, 2:69–72.

Jennings AR (1956). Diseases of wild birds. *Bird Study*, 3:270–272.

Jennings AR (1957). Diseases of wild birds. *Bird Study*, 4:216–220.

Jennings AR (1959). Diseases of wild birds. *Bird Study*, 6:19–22.

Jennings AR (1961). An analysis of 1000 deaths in wild birds. *Bird Study*, 8:25–31.

Joubert L, Desbordes-Lize J, Viallier J (1971). Enquete épidémiologique vétérinaire complémentaire sur une forte endémie d'hospitalisme *Mycobacterium xenopi*. *Bulletin de la Société Scientifique Vétérinaire et Médicale Comparée (Lyon)*, 73:303–325.

Kanai H, Hashimoto H, Mitsuhashi S (1981). Drug-resistance and conjugative R plasmids in *Escherichia coli* strains isolated from wild birds (Japanese tree sparrows, green pheasants and bamboo partridges). *Japanese Poultry Science*, 18:234–239.

Kaneuchi C et al. (1987). Thermophilic campylobacters in seagulls and DNA-DNA hybridization test of isolates. *Nippon Juigaku Zasshi [The Japanese Journal of Veterinary Science]*, 49:787–794.

Kaneuchi C et al. (1989). Occurrence of *Yersinia* spp. in migratory birds, ducks, seagulls, and swallows in Japan. *Nippon Juigaku Zasshi [The Japanese Journal of Veterinary Science]*, 51:805–808.

Kapperud G, Olsvik O (1982). Isolation of enterotoxigenic *Yersinia enterocolitica* from birds in Norway. *Journal of Wildlife Diseases*, 18:247–248.

Kapperud G, Rosef O (1983). Avian wildlife reservoir of *Campylobacter fetus* subssp. *jejuni*, *Yersinia* spp. and *Salmonella* spp. in Norway. *Applied and Environmental Microbiology*, 45:375–380.

Kato Y et al. (1985). Occurrence of *Yersinia enterocolitica* in wild-living birds and Japanese serows. *Applied and Environmental Microbiology*, 49:198–200.

Kawakita S, Van Uden N (1965). Occurrence and population densities of yeast species in the digestive tracts of gulls and terns. *Journal of General Microbiology*, 39:125–129.

Keymer IF (1958). A survey and review of the causes of mortality in British birds and the significance of wild birds as disseminators of disease. *Veterinary Record*, 70:713–720, 736–740.

Khan ZU et al. (1978). Carriage of *Cryptococcus neoformans* in the crops of pigeons. *Journal of Medical Microbiology*, 11:215–218.

Kinjo T et al. (1983). Prevalence of *Campylobacter jejuni* in feral pigeons. *Nippon Juigaku Zasshi [The Japanese Journal of Veterinary Science]*, 45:833–835.

Kissling RE et al. (1954). Studies on the North American arthropod-borne encephalitis. III. Eastern equine encephalitis in wild birds. *American Journal of Hygiene*, 60:251–265.

Kissling RE et al. (1957a). Western equine encephalitis in wild birds. *American Journal of Hygiene*, 66:48–55.

Kissling RE et al. (1957b). Birds as winter hosts for eastern and western equine encephalomyelitis viruses. *American Journal of Hygiene*, 66:42–47.

Kocan RM, Hasenclever HF (1972). Normal yeast flora of the upper digestive tract of some wild columbids. *Journal of Wildlife Diseases*, 8:365–368.

Kocan RM, Hasenclever HF (1974). Seasonal variation of the upper digestive tract yeast flora of feral pigeons. *Journal of Wildlife Disease*, 10:263–266.

Komar N et al. (2003). Experimental infection of North American birds with the New York 1999 strain of West Nile virus. *Emerging Infectious Diseases*, 9:311–322.

Kubín M, Matějka M (1967). Atypická mykobakteria u havranů (*Corvus frugilegus*). [Atypical mycobacteria in the rook (*Corvus frugilegus*)]. *Veterinární Medicina*, 12:499–504 (in Czech).

Kumlin U et al. (1998). Cryptococcosis and starling nests. *Lancet*, 351:1181.

Latham RH et al. (1980). Chronic pulmonary histoplasmosis following the excavation of a bird roost. *American Journal of Medicine*, 68:504–508.

Lesaffre G (1997). *Birds in the city*. Copenhagen, WHO Regional Office for Europe (Local Authorities, Health and Environment Briefing Pamphlets Series, No. 15).

Lévesque B et al. (1993). Impact of the ring-billed gull (*Larus delawarensis*) on the microbiological quality of recreational water. *Applied and Environmental Microbiology*, 59:1228–1230.

Lillehaug A et al. (2005). Screening of feral pigeon (*Columba livia*), mallard (*Anas platyrhynchos*) and graylag goose (*Anser anser*) populations for *Campylobacter* spp., *Salmonella* spp., avian influenza virus and avian paramyxovirus. *Acta Veterinaria Scandinavica*, 46:193–202.

Lipaev VM et al. (1970). [Pseudotuberculosis of swallows in Khabarovsk]. *Zoologiceskij Zurnal*, 49:1386–1390 (in Russian).

Lipkind M, Shihmanter E, Shoham D (1982). Further characterization of H7N7 avian influenza virus isolated from migrating starlings in Israel. *Zentralblatt für Veterinärmedizin. Reihe B*, 29:566–572.

Literák I, Čížek A, Honza M (1992). Using examinations of young black-headed gulls (*Larus ridibundus*) for the detection of salmonellae in the environment. *Acta Veterinaria*, 61:141–146.

Literák I, Kraml F (1985). Výskyt salmonel u divokých ptáků (Passeriformes, Lariformes). [The occurrence of salmonellae in wild birds (Passeriformes, Lariformes)]. *Veterinární Medicina (Praha)*, 30:353–358 (in Czech).

Literák I et al. (1992). Incidence of *Toxoplasma gondii* in populations of wild birds in the Czech Republic. *Avian Pathology*, 21:659–665.

Littman ML, Borok R (1968). Relation of the pigeon to cryptococcosis: natural carrier state, heat resistance and survival of *Cryptococcus neoformans*. *Mycopathologia et Mycologia Applicata*, 35:329–345.

Littman ML, Borok R, Dalton TJ (1965). Experimental avian cryptococcosis. *American Journal of Epidemiology,* 82:197–207.

Littman ML, Schneierson SS (1959). *Cryptococcus neoformans* in pigeon excreta in New York City. *American Journal of Hygiene*, 69:49–59.

Locke LN, Shillinger RB, Jareed T (1973). Salmonellosis in passerine birds in Maryland and West Virginia. *Journal of Wildlife Diseases*, 9:144–145.

Lockwood GF, Garrison RG (1968). The possible role of uric acid in the ecology of *Histoplasma capsulatum*. *Mycopathologia et Mycologia Applicata*, 35:377–388.

Lord RD, Calisher CH, Doughty WP (1974). Assessment of bird involvement in three urban St. Louis encephalitis epidemics. *American Journal of Epidemiology*, 99:364–367.

Lord RD et al. (1974). Urban St. Louis encephalitis surveillance through wild birds. *American Journal of Epidemiology*, 99:360–363.

Lvov DK, Ilyichev VD (1979). *[Avian migrations and transport of infectious agents].* Moscow, Nauka (in Russian).

Macdonald JW (1965). Mortality in wild birds. *Bird Study*, 12:181–195.

Macdonald JW (1968). Listeriosis and erysipelas in wild birds. *Bird Study*, 15:37–38.

Macdonald JW (1977). Cutaneous salmonellosis in a house sparrow. *Bird Study*, 25:59.

Macdonald JW, Brown DD (1974). *Salmonella* infection in wild birds in Britain. *The Veterinary Record*, 94:321–322.

Magnarelli LA, Stafford KC 3rd, Bladen VC (1992). *Borrelia burgdorferi* and *Ixodes dammini* (Acari: Ixodidae) feeding on birds in Lyme, Connecticut, U.S.A. *Canadian Journal of Zoology*, 70:2322–2325.

Mair NS (1973). Yersiniosis in wildlife and its public health implications. *Journal of Wildlife Diseases*, 9:64–71.

Makino S et al. (2000). Detection and characterization of Shiga toxin-producing *Escherichia coli* from seagulls. *Epidemiology and Infection*, 125:55–61.

Maruyama S et al. (1990). Prevalence of thermophilic campylobacters in crows (*Corvus levaillantii, Corvus corone*) and serogroups of the isolates. *Nippon Juigaku Zasshi [The*

Japanese Journal of Veterinary Science], 52:1237–1244.

Matějka M, Kubín M (1967). Vrabci (*Passer domesticus*) jako zdroj infekce aviárními mykobakteriemi pro skot. [House sparrows as a source of cattle infection with avian mycobacteria]. *Veterinární Medicina*, 12:491–497 (in Czech).

McDiarmid A (1969). *Diseases in free-living wild animals*. New York, Academic Press.

McFadden J et al. (1992). Mycobacteria in Crohn's disease: DNA probes identify the wood pigeon strain of *Mycobacterium avium* and *Mycobacterium paratuberculosis* from human tissue. *Journal of Clinical Microbiology*, 30:3070–3073.

McLean RG et al. (1983). The house sparrow (*Passer domesticus*) as a sentinel for St. Louis encephalitis virus. *The American Journal of Tropical Medicine and Hygiene*, 32:1120–1129.

Mégraud F (1987). Isolation of *Campylobacter* spp. from pigeon feces by a combined enrichment-filtration technique. *Applied and Environmental Microbiology*, 53:1394–1395.

Meyer KF (1959). Some general remarks and new observations on psittacosis and ornithosis. *Bulletin of the World Health Organization*, 20:1–182.

Mitchell CA, Duthie RC (1950). Tuberculosis in the common crow. *Canadian Journal of Comparative Medicine,* 14:109.

Monaghan P et al. (1985). *Salmonella* carriage by herring gulls in the Clyde area of Scotland in relation to their feeding ecology. *Journal of Applied Ecology*, 22:669–680.

Monga DP (1972). Prevalence of pathogenic fungi in wild birds. *The Indian Journal of Medical Research*, 60:517–519.

Morris R, Jackson R (2005). *Epidemiology of H5N1 avian influenza in Asia and implications for regional control* (Massey report). Rome, Food and Agricultural Organization of the United Nations. (http://www.fao.org/ag/againfo/subjects/documents/ai/HPAI-Masseyreport.pdf, accessed 21 February 2007)

Muchmore HG et al. (1963). Occurrence of *Cryptococcus neoformans* in the environment of three geographically associated cases of cryptococcal meningitis. *The New England Journal of Medicine*, 268:1112–1114.

Müller G (1970). Möwen als Ausscheider und Verbreiter von Salmonellen. *Naturwissenschaftliche Rundschau*, 23:104–107.

Munster VJ et al. (2005). Mallards and highly pathogenic avian influenza ancestral viruses, northern Europe. *Emerging Infectious Diseases*, 11:1545–1551.

Murdock WT et al. (1962). Acute pulmonary histoplasmosis after exposure to soil contaminated by starling excreta. *The Journal of the American Medical Association*, 179:73–75.

Nesbitt SA, White FH (1974). A *Salmonella typhimurium* outbreak at a bird feeding station. *Florida Field Naturalist*, 2:46–47.

Nestorowicz A et al. (1987). Molecular analysis of the hemagglutinin genes of Australian H7N7 influenza viruses: role of passerine birds in maintenance or transmission? *Virology*, 160:411–418.

Neto VA, Levi GC (1970). Ocorrencia simultanea de casos de toxoplasmose - doenca entre moradores de um nucleo habitacional restrito da cidade de Sao Paulo. [Simultaneous occurrence of toxoplasmosis among inhabitants of Sao Paulo city]. *Revista do Instituto Medicinae Tropical de São Paulo*, 12:41–45 (in Portuguese).

Newberry WM et al. (1967). Epidemiologic study of *Cryptococcus neoformans*. *Annals of International Medicine*, 67:724–732.

Nguyen D et al. (2005). Isolation and characterization of avian influenza viruses, including highly pathogenic H5N1, from poultry in live bird markets in Hanoi, Vietnam, in 2001. *Journal of Virology*, 79:4201–4212.

Nielsen BB (1960). *Salmonella typhimurium* carriers in seagulls and mallards as a possible source of infection to domestic animals. *Nordisk Veterinaermedicin*, 12:417–424.

Nielsen EM et al. (2004). Verocytotoxin-producing *Escherichia coli* in wild birds and rodents in close proximity to farms. *Applied and Environmental Microbiology*, 70:6944–6947.

Niida M et al. (1983). Genetic and molecular characterization of conjugative R plasmids detected in *Salmonella* strains isolated from humans and feral pigeons in the same district. *Nippon Juigaku Zasshi [The Japanese Journal of Veterinary Science]*, 45:647–658.

Odermatt P et al. (1998). Starenschwärme in Basel: ein Naturphänomen, eine Belästigung oder ein Gesundheitsrisiko? *Gesundheitswesen*, 60:749–754.

Orlandella V, Coppola L (1969). Ruolo del colombo nella diffusione di alcune pericolose zoonosi. II. La toxoplasmosi. La toxoplasmosi nell'uomo. [The role of pigeons in the spread of some dangerous zoonoses. II. Human toxoplasmosis]. *Zooprofilassi*, 24:347–390, 437–480.

Page LA (1976). Observations on the involvement of wildlife in an epornitic of chlamydiosis in domestic turkeys. *Journal of the American Veterinary Medical Association*, 169:932–935.

Pagon S, Sonnabend W, Krech U (1974). Epidemiologische Zusammenhänge zwischen menschlichen und tierischen *Salmonella*-Ausscheidern und deren Umwelt im schweizerischen Bodenseeraum. *Zentralblatt für Bakteriologie, Parasitenkunde, Infektionskrankheiten und Hygiene. Erste Abteilung Originale. Reihe B: Hygiene, präventive Medizin*, 158:395–411.

Pak SM (1976). *[Toxoplasmosis in birds in Kazakhstan]*. Alma-Ata, Nauka (in Russian).

Pannwitz E, Pulst H (1972). Vorkommen von Salmonellen bei Stadttauben und Möwen. *Monatshefte für Veterinärmedizin*, 27:573–575.

Partridge BM, Winner HI (1965). *Cryptococcus neoformans* in bird droppings in London. *Lancet*, 1:1060–1061.

Pavlásek I (1993). Racek chechtavý (*Larus ridibundus* L.), nový hostitel *Cryptosporidium baileyi* (Apicomplexa: Cryptosporidiidae). [The black-headed gull (*Larus ridibundus*), a new host for *Cryptosporidium baileyi* (Apicomplexa: Cryptosporidiidae)]. *Veterinární Medicina*, 38:629–638 (in Czech, with an English summary).

Peach W, Fowler J, Hay J (1989). Incidence of *Toxoplasma* infection in a population of European starlings *Sturnus vulgaris* from Central England. *Annals of Tropical Medicine and Parasitology*, 83:173–177.

Perdue ML, Swayne DE (2005). Public health risk from avian influenza viruses. *Avian Diseases*, 49:317–327.

Petzelt K, Steiniger F (1961). Die Vögel der Kläranlage von Hannover und die von ihnen ausgeschiedenen Salmonellen. *Archiv für Hygiene und Bakteriologie*, 145:605–619.

Pinowska B, Chylinski G, Gondek B (1976). Studies on the transmitting of salmonellae by house sparrows (*Passer domesticus* L.) in the region of Žulawy. *Polish Ecological Studies*, 2:113–121.

Pinowski J, Kavanagh BP, Górski W, eds (1991). *Nestling mortality of granivorous birds due to microorganisms and toxic substances*. Warsaw, PWN-Polish Scientific Publishers.

Plant CW (1978). Salmonellosis in wild birds feeding at sewage treatment works. *The Journal of Hygiene*, 81:43–48.

Plum N (1942). Studies on the occurrence of avian tuberculosis among wild birds, especially gulls and sparrows, and rats and hares. *Skandinavisk Veterinaertidskrift*, 32:465–487 (in Swedish).

Pohl P, Thomas J (1968). *Salmonella* chez l'étourneau (*Sturnus vulgaris*). *Annales de Médecine Véterinaire*, 112:704–711.

Pospíšil R et al. (1988). Ornitóza konzervátorov v katedrále sv. Alžbiety v Košiciach. [Ornithosis in conservators of the St. Elisabeth cathedral in Ko?ice]. *Pracovní léka?ství*, 40:246–248 (in Slovak).

Powell KE et al. (1972). Airborne *Cryptococcus neoformans*: particles from pigeon excreta compatible with alveolar deposition. *The Journal of Infectious Diseases*, 125:412–415.

Procknow JJ et al. (1965). Cryptococcal hepatitis presenting as a surgical emergency. First isolation of *Cryptococcus neoformans* from point source in Chicago. *The Journal of the American Medical Association*, 191:269–274.

Quessy S, Messier S (1992). Prevalence of *Salmonella* spp., *Campylobacter* spp. and *Listeria* spp. in ring-billed gulls (*Larus delawarensis*). *Journal of Wildlife Diseases*, 28:526–531.

Randhawa HS, Clayton YM, Riddell RW (1965). Isolation of *Cryptococcus neoformans* from pigeon habitats in London. *Nature*, 208:801.

Raška K, Syrůček L (1956). Q fever in domestic and wild birds. *Bulletin of the World Health Organization,* 15:329–337.

Reeves WC et al. (1958). Chronic latent infections of birds with Western equine encephalomyelitis virus. *Proceedings of the Society for Experimental Biology and Medicine*, 97:733–736.

Refai M et al. (1983). Isolation of *Cryptococcus neoformans*, *Candida albicans* and other yeasts from pigeon droppings in Egypt. *Sabouraudia*, 21:163–166.

Řeháček J, Brezina R (1976). Ornithosis in domestic pigeons gone wild in Bratislava. *Journal of Hygiene, Epidemiology, Microbiology, and Immunology*, 20:252–253.

Řeháček J, Tarasevich IV (1988). *Acari-borne rickettsiae and rickettsioses in Eurasia*. Bratislava, Veda Publishing House.

Reitler R (1955). Salmonellosis in migratory birds (*Sturnus vulgaris*). *Acta Medica Orientalia*, 14:52–54.

Richter D et al. (2000). Competence of American robins as reservoir hosts for Lyme disease spirochetes. *Emerging Infectious Diseases*, 6:133–138.

Roberts JP, Grimes JE (1978). *Chlamydia* shedding by four species of wild birds. *Avian Diseases*, 22:698–706.

Rödl P (1999). Metodická doporučení Státního zdravotního ústavu pro ochranu budov před zdivočelými holuby a prevence šíření patogenů z holubích hnízdišť'. [Methodical recommendations of the National Public Health Institute for protection of buildings against feral pigeons and prevention of pathogen dispersal from pigeon habitats]. *Zpravodaj Sdružení DDD (Praha)*, 7:120–122 (in Czech).

Rolla G et al. (2004). Allergy to pigeon tick (*Argas reflexus*): demonstration of specific IgE-binding components. *International Archives of Allergy and Immunology*, 135:293–295.

Rosario I et al. (2005). Isolation of *Cryptococcus* species including *C. neoformans* from cloaca of pigeons. *Mycoses*, 48:424–424.

Rosef O (1981). [The occurrence of *Campylobacter fetus* subsp. *jejuni* and *Salmonella* bacteria in some wild birds]. *Nordisk Veterinaermedicin*, 33:539–543 (in Norwegian).

Rossi L (1969). Studie průběhu experimentální infekce hrdličky zahradní (*Streptopelia decaocto*) původcem ptačí tuberkulózy *Mycobacterium avium* [A study on the course of experimental infection with *Mycobacterium avium* in *Streptopelia decaocto*]. *Acta Veterinaria (Brno)*, 38:435–439 (in Czech).

Rossi L, Dokoupil S (1969). Vrabci (*Passer domesticus*) jako zdroj infekce u hospodářských zvířat [House sparrows (*Passer domesticus*) as a source of avian infection in farm animals]. *Veterinářství*, 19:356–358 (in Czech).

Ruiz A, Bulmer GS (1981). Particle size of airborne *Cryptococcus neoformans* in a tower. *Applied and Environmental Microbiology*, 41:1225–1229.

Ruiz A, Fromtling RA, Bulmer GS (1981). Distribution of *Cryptococcus neoformans* in a natural site. *Infection and Immunity*, 31:560–563.

Ruiz A, Neilson JB, Bulmer GS (1982). A one year study on the viability of *Cryptococcus neoformans* in nature. *Mycopathologia*, 77:117–122.

Ruiz A, Vélez D, Fromtling RA (1989). Isolation of saprophytic *Cryptococcus neoformans* from Puerto Rico: distribution and variety. *Mycopathologia*, 106:167–170.

Sacks JJ et al. (1986). Epidemic campylobacteriosis associated with a community water supply. *American Journal of Public Health*, 76:424–428.

Saikku P (1973). Blackbird, *Turdus merula* L., and tick-borne encephalitis virus. *Acta Virologica*, 17:442.

Schaefer WB et al. (1973). A bacteriological study of endemic tuberculosis in birds. *The Journal of Hygiene*, 71:549–557.

Schmatz HD et al. (1977). Zum Verhalten des Q-Fieber-Erregers, *Coxiella burnetii*, in Vögeln. 2. Experimentelle Infektion von Tauben. *Deutsche Tierärztliche Wochenschrift*, 84:60–63.

Schmitt N (1962). Neuere Erfahrungen bei der Starenabwehr im Lande Rheinland-Pfalz, unter besonderer Berücksichtigung der phono- und pyroakustischen Verfahren. *Annales de Épiphytie*, 13:57–68.

Schneider J, Bulling E (1969). Weitere Untersuchungen zur Salmonellose der Singvögel. *Berliner und Münchener Tierärztliche Wochenschrift*, 82:287–288.

Schönborn C, Schütze B, Pöhler H (1969). Sprosspilze im Kot von Zoovögeln, freilebenden einheimischen Vögeln und verwilderten Tauben. *Mykosen*, 12:471–490.

Schwarz J et al. (1957). Successful infection of pigeons and chickens with Histoplasma capsulatum. *Mycopathologia et Mycologia Applicata*, 8:189–193.

Selbitz HJ et al. (1991). Nachweis und Charakterisierung von Salmonella-Stämmen bei Lachmöwen (*Larus ridibundus*). *Berliner und Münchener Tierärztliche Wochenschrift*, 104:411–414.

Semenov BF et al. (1973). [Study of chronic forms of arbovirus infections in birds. 1. Experiments with West Nile, Sindbis, Bhanja and Sicilian sandfly fever viruses]. *Vestnik Akademii Medicinskich Nauk SSSR*, 2:79–83 (in Russian).

Šerý V, Strauss J (1960). Ornitóza a salmonelóza u volně žijících ptáků v přírodní rezervaci a u drůbeže i lidí v nejbližším okolí. [Ornithosis and salmonellosis in wild birds in a reserve and in domestic fowls and humans in the area]. *Veterinární Medicina*, 5:799–808 (in Czech).

Sethi KK, Randhawa HS (1968). Survival of *Cryptococcus neoformans* in the gastrointestinal tract of pigeons following ingestion of the organism. *The Journal of Infectious Diseases*, 118:135–138.

Sethi KK, Schwarz J (1966). Experimental ocular histoplasmosis in pigeons (*Columba livia domestica, Histoplasma capsulatum*). *American Journal of Ophthalmology*, 61:538–543.

Shayegani M et al. (1986). *Yersinia enterocolitica* and related species isolated from wildlife in New York State. *Applied and Environmental Microbiology*, 52:420–424.

Silva ME, Paula LA (1963). Isolamento de *Cryptococcus neoformans* de excrementos e ninhos de pombos (*Columba livia*) en Salvador, Bahia (Brasil) [Isolation of *Cryptococcus neoformans* from excrement and nests of pigeons (*Columba livia*) in Salvador, Bahia (Brazil)]. *Revisto do Instituto Medicina Tropical de São Paulo*, 5:9–11, 394–396 (in Portuguese).

Simitzis-Le Flohic AM et al. (1983). Essai d'appréciation de l'importance épidémiologique des concentrations hivernales d'étournoux sansonnets (*Sturnus vulgaris* L.) dans la région portuaire de Brest, Finist re. *Cahiers d'Office de la Recherche Scientifique et Technique Outre-Mer, Série Entomologie Médicale et Parasitaire*, 21:159–164.

Sixl W (1975). Zum Problem der verwilderten Stadttaube (Aves, Columbiformes, Columbidae). *Mitteilungen der Abteilung für Zoologie am Landesmuseums Joanneum (Graz)*, 4:87–97.

Sixl W et al. (1997). *Campylobacter* spp. and *Salmonella* spp. in black-headed gulls (*Larus ridibundus*). *Central European Journal of Public Health*, 5:24–26.

Sixl W, Sixl H (1971). Zur Frage der Ornithose verwilderter Stadttauben. *Archiv für Hygiene und Bakteriologie*, 154:612–613.

Slemons RD, Easterday BC (1977). Type A influenza viruses in the feces of migratory waterfowl. *Journal of the American Veterinary Medical Association*, 171:947–948.

Smit T et al. (1987). Avian tuberculosis in wild birds in the Netherlands. *Journal of Wildlife Diseases*, 23:485–487.

Smith CD (1964). Evidence of the presence in yeast extract of substances which stimulate the growth of *Histoplasma capsulatum* and *Blastomyces dermatitidis* similarly to that found in starling manure extract. *Mycopathologia et Mycologia Applicata*, 22:99–105.

Smith CD (1971). The role of birds in the ecology of *Histoplasma capsulatum*. In: Ajello L, Chick EW, Furcolow ML, eds. *Histoplasmosis: proceedings of the second national conference*. Springfield, IL, Charles C. Thomas Publisher:140–148.

Smith HV et al. (1993). Occurrence of oocysts of *Cryptosporidium* sp. in *Larus* spp. gulls. *Epidemiology and Infection*, 110:135–143.

Snoeyenbos GH, Morin EW, Wetherbee DK (1967). Naturally occurring *Salmonella* in 'blackbirds' and gulls. *Avian Diseases*, 11:642–646.

Soler NM, Portales JM, Del Barrio G (1985). Identificación de una cepa de encefalomielitis equina del este (EEE) aislada de una paloma *Columba livia domestica* [Identification of a strain of Eastern equine encephalomyelitis isolated from a *Columba livia domestica* pigeon]. *Revista Cubana de Medicina Tropical*, 37:12–18 (in Spanish).

Staib F (1962). Vogelkot, ein Nährsubstrat für die Gattung *Cryptococcus*. *Zentralblatt für Bakteriologie, Abteilung 1, Originale*, 186:233–247.

Staib F (1963). Zur Widerstandsfähigkeit von *Cryptococcus neoformans* gegen Austrocknung und hohe Temperaturen. *Archiv für Mikrobiologie*, 44:323–333.

Staib F, Bethäuser G (1968). Zum Nachweis von *Cryptococcus neoformans* im Staub von einem Taubenschlag. *Mykosen*, 11:619–624.

Stallknecht DE, Shane SM (1988). Host range of avian influenza virus in free-living birds. *Veterinary Research Communications*, 12:125–141.

Stein A, Raoult D (1999). Pigeon pneumonia in Provence: a bird-borne Q fever outbreak. *Clinical Infectious Diseases*, 29:617–620.

Stepanyan EG et al. (1964). Breslavskij salmonellez u vorobjev v eksperimentě. [Experimental infection of sparrows with *Salmonella typhimurium*]. *Izvestija Akademii Nauk Turkmenskoj SSR, Serija Biologicheskich Nauk*, 6:50–56 (in Russian).

Stern C et al. (2006). Die Rolle von Amsel (*Turdus merula*), Rotdrossel (*Turdus iliacus*) und Singdrossel (*Turdus philomelos*) als Blutwirte für Zecken (Acari: Ixodidae) und Reservoirwirte für vier Genospezies des *Borrelia burgdorferi*-Artenkomplexes. *Mitteilungen der Deutschen Gesellschaft für Allgemeine und Angewandte Entomologie*, 15:349–355.

Storch G et al. (1980). Acute histoplasmosis. Description of an outbreak in northern Louisiana. *Chest*, 77:38–42.

Strauss J, Bednář B, Šerý V (1957). Výskyt ornithosy a salmonellosy u racka chehtavého (*Larus ridibundus* L. II. Isolace a identifikace virusu ornithosy z racka chechtavého se současným nálezem *S. typhi murium* [Occurrence of ornithosis and salmonellosis in the black-headed gull, *Larus ridibundus* L. II. Isolation and identification of ornithosis virus in gull with simultaneous detection of *Salmonella typhimurium*]. *Československá Epidemiologie, Mikrobiologie, Imunologie*, 6:231–240 (in Czech).

Suthill LC, Campbell CC (1965). Feathers as substrate for *Histoplasma capsulatum* in its filamentous phase of growth. *Sabouraudia*, 4:1–2.

Švrček Š et al. (1966). Volne žijúce vtáctvo a domáce holuby ako zdroje aviárnej tuberkulózy. [Free-living birds and domestic pigeons as a source of avian tuberculosis]. *Rozhledy v tuberkulose a nemocech plic*, 26:659–667 (in Slovak).

Swinne-Desgain D (1976). *Cryptococcus neoformans* in the crops of pigeons following its experimental administration. *Sabouraudia*, 14:313–317.

Swinne D et al. (1991). AIDS-associated cryptococcosis in Bujumbura, Burundi: an epidemiological study. *Journal of Medical and Veterinary Mycology*, 29:25–30.

Tanaka C et al. (2005). Bacteriological survey of feces from feral pigeons in Japan. *The Journal of Veterinary Medical Science*, 67:951–953.

Taylor RL, Duangmani C (1968). Occurrence of *Cryptococcus neoformans* in Thailand. *American Journal of Epidemiology*, 87:318–322.

Taylor RM et al. (1955). Sindbis virus: a newly recognized arthropod-transmitted virus. *The American Journal of Tropical Medicine and Hygiene*, 4:844–862.

Terskikh II (1964). Epidemiologija i prirodnaja ochagovost ornitoza. [Epidemiology and natural focality of ornithosis]. *Voprosy Meditsinskoi Virusologii*, 10:163–184 (in Russian).

Theiler M, Downs WG (1973). *The arthropod-borne viruses of vertebrates*. New Haven, Yale University Press.

Tizard I (2004). Salmonellosis in wild birds. *Seminars in Avian and Exotic Pet Medicine*, 13(2):50–66.

Tizard IR, Fish NA, Harmeson J (1979). Free flying sparrows as carriers of salmonellosis. *The Canadian Veterinary Journal*, 20:143–144.

Toro H et al. (1999). Health status of free-living pigeons in the city of Santiago. *Avian Pathology*, 28:619–623.

Tosh FE et al. (1966). The second of two epidemics of histoplasmosis resulting from work on the same starling roost. *The American Review of Respiratory Disease*, 94:406–413.

Tosh FE et al. (1967). The use of formalin to kill *Histoplasma capsulatum* at an epidemic site. *American Journal of Epidemiology*, 85:259–265.

Tosh FE et al. (1970). Relationship of starling-blackbird roosts and endemic histoplasmosis. *The American Review of Respiratory Disease*, 101:283–288.

Treml F et al. (1993). Volně žijící obratlovci – možný zdroj zoopatogenních mikroorganismů pro hospodářská zvířata. [Free-living vertebrates – a possible source of zoopathogenic microorganisms for farm animals]. *Veterinářství*, 43:241–245 (in Czech).

Tsubura E (1962). Experimental studies on cryptococcosis. 1. Isolation of *Cryptococcus neoformans* from avian excreta and some considerations on the source of infection. *Japanese Journal of Medical Mycology*, 3:50–55.

Vanbreuseghem R, Eugene J (1958). Culture d'*Histoplasma capsulatum* et d'*Histoplasma duboisii* sur un milieu a bas de terre de matieres fecales provenant de divers animaux. *Comptes Rendus de Séances de la Société Biologie et de ses Filiales*, 152:1602–1605.

Veraldi S et al. (1998). Skin manifestations caused by pigeon ticks (*Argas reflexus*). *Cutis*, 61:38–40.

Vinograd IA et al. (1982). Vydělenije virusa Zapadnogo Nila na juige Ukrajiny. [Isolation of West Nile virus in the Southern Ukrainian]. *Voprosy Virusologii*, 27:55–57 (in Russian).

Vlahovič K et al. (2004). *Campylobacter*, *Salmonella* and *Chlamydia* in free-living birds of Croatia. *European Journal of Wildlife Research*, 50:127–132.

Volner Z (1978). Mycobacteria in pigs and the importance of wild birds in tuberculosis transmission. *Veterinarski Arhiv*, 48:107–110.

Waldman RJ et al. (1983). A winter outbreak of acute histoplasmosis in northern Michigan. *American Journal of Epidemiology*, 117:68–75.

Walter JE, Atchison RW (1966). Epidemiological and immunological studies of *Cryptococcus neoformans*. *Journal of Bacteriology*, 92:82–87.

Walter JE, Coffee EG (1968). Distribution and epidemic significance of the serotypes of

Cryptococcus neoformans. *American Journal of Epidemiology*, 87:167–172.

Walter JE, Yee RB (1968). Factors that determine the growth of *Cryptococcus neoformans* in avian excreta. *American Journal of Epidemiology*, 88:445–450.

Webster RG et al. (1992). Evolution and ecology of influenza A viruses. *Microbiological Reviews*, 56:152–179.

Whelan CD et al. (1988). The significance of wild birds (*Larus* sp.) in the epidemiology of *Campylobacter* infections in humans. *Epidemiology and Infection*, 101:259–267.

WHO (2006). *Cumulative number of avian influenza A/H5N1 cases reported to WHO*. Geneva, World Health Organization (http://www.who.int/csr/disease/avian_influenza/en/index.html, accessed 5 February 2007).

WHO Global Influenza Program Surveillance Network (2005). Evolution of H5N1 avian influenza viruses in Asia. *Emerging Infectious Diseases*, 11:1515–1521.

Williams BM, Richards DW, Lewis J (1976). *Salmonella* infection in the herring gull (*Larus argentatus*). *The Veterinary Record*, 98:51.

Wilson JE, MacDonald JW (1967). Salmonella infection in wild birds. *The British Veterinary Journal*, 123:212–219.

Wobeser GA (1997). *Diseases of wild waterfowl*, 2nd ed. New York, Plenum Press.

Wobeser G, Brand CJ (1982). Chlamydiosis in 2 biologists investigating disease occurrences in wild waterfowl. *Wildlife Society Bulletin*, 10:170–172.

Wobeser GA, Finlayson MC (1969). *Salmonella typhimurium* infection in house sparrows. *Archives of Environmental Health*, 19:882–884.

Wobeser G, Kost W (1992). Starvation, staphylococcosis, and vitamin A deficiency among mallards overwintering in Saskatchewan. *Journal of Wildlife Diseases*, 28:215–222.

Work TH, Hurlbut HS, Taylor RM (1953). Isolation of West Nile virus from hooded crow and rock pigeon in the Nile delta. *Proceedings of the Society for Experimental Biology and Medicine*, 84:719–722.

Wuthe HH (1972). Salmonellen in Ausscheidungen von Möwen an der Ostseeküste. *Zentralblatt für Bakteriologie, Parasitenkunde, Infektionskrankheiten und Hygiene. Erste Abteilung Originale. Reihe A: Medizinische Mikrobiologie und Parasitologie*, 221:453–457.

Wuthe HH (1973). Salmonellen in einer Brutkolonie von Lachmöwen. *Berliner und Münchener Tierärztliche Wochenschrift*, 86:255–256.

Yamamoto S, Ishida K, Sato A (1957). Isolation of *Cryptococcus neoformans* from pulmonary granuloma of a cat and from pigeon droppings. *Nippon Juigaku Zasshi [The Japanese Journal of Veterinary Science]*, 19:179–194.

Yildiran ST et al. (1998). Isolation of *Cryptococcus neoformans* var. *neoformans* from pigeon droppings collected throughout Turkey. *Medical Mycology*, 36:391–394.

Younglove RM et al. (1968). An outbreak of histoplasmosis in Illinois associated with starlings. *Illinois Medical Journal*, 134:259–263.

9. Human body lice

Ian Burgess

Summary

Human lice are the most widespread ectoparasites and cause low-level morbidity in vast numbers of people worldwide. High- and medium-income countries suffer relatively minor levels of infestation, compared with some low-income societies, yet are still unable to effectively eliminate these organisms, which can be found easily. Although human body lice historically have been vectors of devastating outbreaks of disease, they are unlikely to do so again. However, their ability to still transmit infections, more than half a century after transmission was believed to have ceased in Western society, indicates that they have a capacity for a greater impact. Simply addressing the control problem rationally would be an important development, as so many local and national policy approaches to body lice are currently inadequate.

9.1. Introduction

Human lice are host-specific, haematophagous, obligate ectoparasites that live their whole lives on their hosts. Few studies of their biology or physiology have been conducted (Burgess, 2004), and most of them were conducted in the first half of the 20th century, primarily and were focused on body or clothing lice (*Pediculus humanus*). That species constituted a major health risk as a vector of transmissible disease, especially during the First World War in Europe, the depression of the 1930s, and around 1942/1943 in parts of North Africa and southern Europe (Soper et al., 1945, 1947).

Environmental health practitioners who had responsibility for lice in most countries until the 1970s considered human lice of all three species – body lice (*P. humanus*), head lice (*Pediculus capitis*) and crab lice (*Phthirus pubis*) – as pests. Subsequently, rationalization of health and environmental services has transferred responsibility for control of lice to other services or to parent carers, as in the case head lice. Should the need arise, legislation that still exists in many countries could be used to control these insects – for example, in an arthropod-borne outbreak of disease. Thus, environmental health officers, say in the United Kingdom, can address *verminous conditions* – that is, people infested with parasites, such as body lice – by using Section 85 of the Public Health Act of 1936. However, such legislation requires a foreknowledge of the existence of an infestation, which limits its effectiveness.

Body lice, although relatively uncommon now in most high- and medium-income countries, do still constitute a health risk. Many of the services that formerly were in place to deal with these infestations no longer exist and, for the most part, the comparatively small numbers of individuals affected can obtain relief from a variety of voluntary agencies. Recently recognized changes in disease demography, however, suggest that body lice may still constitute a small risk of disease to people (Raoult & Roux, 1999).

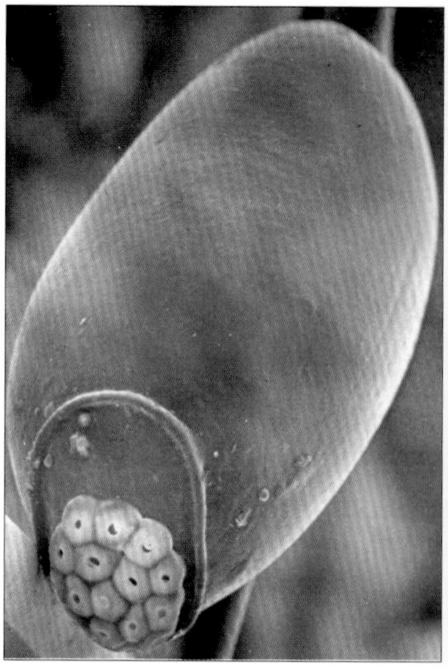

Fig. 9.1. Louse egg, showing the cap and micropyles (breathing pores)

Source: Scanning electron micrograph by J. Maunder, Medical Entomology Centre.

This chapter concentrates on body lice as an environmental health problem. Head lice and crab lice, although essentially similar in physiology and behaviour, constitute specific clinical hazards for which the remedy is found in the clinical medical context rather than the environmental health context. Therefore, these two species are outside the scope of this report and are mentioned less extensively.

9.2. Biological factors

Using glue-like secretions from glands attached to the oviduct, all human lice fix their eggs (Fig. 9.1) to hairs or fabric fibres (Nuttall, 1917). The length of time required for eggs to hatch has never satisfactorily been determined and the only study of body lice (Leeson, 1941) showed that hatching was achieved between 6 and 11 days at temperatures likely to be encountered under clothing (29–37°C).

Fig. 9.2. Young adult body lice on cloth

Source: Photo by E. Kidman, Medical Entomology Centre.

After emerging from the egg, lice go through three nymphal stages, each of which takes about three days to complete development (Nuttall, 1917; Lang, 1975). The final moult from the third nymphal stage gives rise to the adult louse (Fig. 9.2). Male lice often develop more rapidly than their female counterparts. Lang (1975) also observed that females mature for about one day after the final moult before they are able to lay eggs.

In all stages of development, lice feed at regular intervals throughout the day, with the youngest nymphs requiring more frequent blood-meals than third instar nymphs and adults (Lang, 1975), presumably due to their smaller size. All lice produce dry faecal matter that may accumulate in clothing or hair; this dry matter is the vehicle for transmitting the infective disease organisms for which lice may be vectors.

All bionomic studies of lice are inadequate. Theoretical estimates of the growth of louse populations (for example, Evans & Smith, 1952) are based on limited data and appear to overestimate the rate of increase, by not addressing juvenile mortality, migration and destruction by the host. In some cases, destruction by the host appears to account for more than 50% of juveniles, so that some people who have never knowingly had an infestation can be found with long hatched nits (empty eggshells). If populations grew as fast as estimated, the numbers of insects would soon become apparent to the host; however, experience indicates the contrary, in that infestations may go undetected for weeks or even months.

When deprived of access to food, body lice survive longer than other species of human lice, presumably due to their larger body mass and to the historical selection of insects more capable of resisting the dehydration that results from starvation induced by their host removing their clothing. Buxton (1947) reported survival for up to 10 days at 15°C, with shorter survival at higher and lower temperatures, ranging from 7 days at 10°C to 2 days at 36°C.

All three species of human lice are distributed throughout human populations, and it is likely that they affect all communities to some extent. However, the increase of affluence in Europe and North America over the past 50 years has relegated body louse infestations to vagrants, the homeless and, in recent conflicts, refugees (Gratz, 1985). Head lice have remained primarily an infestation of children. In communities where reasonably effective methods of treatment are available, adults are normally only affected by head lice through contact with infested children in their care. Crab or pubic lice mostly affect sexually active adults, although they may be passed to children and others by close nonsexual physical contact. Few studies have investigated the epidemiology of this louse. One study, however, showed that they were more common on females between the ages of 15 and 19 years and on males more than 20 years of age (Fisher & Morton, 1970).

In high- and medium-income countries, body lice are more common in urban environments; that is because the majority of people prevented from laundering or changing clothing, due to poverty, congregate in urban environments. In the past, the prevalence of head lice (Mellanby, 1941, 1943), and of infestations with lice (lousiness) in general (Lindsay, 1993), was also greatest in urban industrial slums and communities that were in close proximity. However, in the United Kingdom, increased mobility of populations during the latter half of the 20th century led to a more even spread of infestations, with only small differences between rural and urban areas and between different regional communities (Downs, Harvey & Kennedy, 1999; Smith et al., 2003).

9.3. Louse infestation in Europe and North America

As a result of improved hygiene and economic advances, body lice declined steadily throughout the 20th century in all high- and medium-income countries. In North America, some researchers believe that in homeless populations crab lice may be more prevalent that body lice (Meinking, 1999), but in Europe body lice are still common in the vulnerable homeless (Raoult & Roux, 1999; Foucault et al., 2006). Control of infestations in these groups is difficult, due to constant reinfestation through contact with others in similar circumstances.

Body lice have been shown to move freely between the clothing of individuals in close proximity in shelters, and these lice can move onto bedding and remain viable until the bed is inhabited the next night, but not beyond (Peacock, 1916; MacLeod & Craufurd-Benson, 1941). Consequently, a residue of infestation that could expand into the wider community – given appropriate circumstances – remains in most countries. In some countries, as a result of increased unemployment, homelessness and reduced social welfare programmes, the prevalence of lice is believed to have increased during the 1990s (Downs, Harvey & Kennedy, 1999; Meinking, 1999; Raoult & Roux, 1999; Willems, et al., 2005). During the late 20th century, localized increases also arose in the Balkans (Valenciano et al., 1999; Kondaj, 2002) and other regions of conflict (Raoult et al., 1998) – with the risk of infestation disseminating farther afield.

Despite a range of interventions that use synthetic insecticides, head louse infestations in children have persisted (Downs, Harvey & Kennedy, 1999; Willems et al., 2005). The limited number of studies available indicate that insecticides have had a local impact on infestation rates at various times, but that this has not been sustained. Overall levels of infestation declined through the 20th century, although there are few recorded data to confirm this (British Department of Health and Social Security, 1987; Lindsay, 1993). However, the appearance of resistance to the most commonly used insecticides is believed to have had an impact on the number of children infested in most communities in high- and middle-income countries (Burgess & Brown, 1999; Downs, Harvey & Kennedy, 1999; Meinking, 1999). The studies by Lindsay (1993) and Willems and colleagues (2005) showed that socioeconomic factors influence both the risk of infestation and the ability to cure it, indicating a broader spectrum of social and environmental health problems associated with louse infestations.

The burden of crab lice infestation is largely unknown. Anecdotal evidence suggests that infestations diminished significantly in the 1980s, but no epidemiological studies have been performed in any large population group. More recently, Meinking (1999) suggested that the levels of infestation have recovered.

9.4. Implications for public health

In general, low-grade morbidity is the principal effect of louse infestations on public health. Constant infestation affects general well-being through disturbed sleep, diminished concentration and itching that results in excoriation. Body lice have a greater impact on infested individuals whose skin is not only excoriated, but is also thickened and discoloured over time through constant exposure to louse bites. Other pathological effects seen in some people with longer-term infestations include lymphadenopathy and impetigo due to *S. aureus* and Group A *S. pyogenes* (Taplin & Meinking, 1988), with a potential reduction of their immune status rendering them more susceptible to other infections.

Body lice are primary vectors for: classical typhus, caused by *Rickettsia prowazekii*; trench fever, caused by *B. quintana*; and louse-borne relapsing fever (LBRF), caused by *Borrelia recurrentis*. They are also secondary vectors for murine typhus, caused by *Rickettsia mooseri*. Endemics of these diseases were still found throughout Europe and parts of North America in the early 20th century, resulting in considerable mortality and morbidity during and immediately following the two world wars (Soper et al., 1945, 1947; Jackson & Spach, 1996). For centuries, these diseases followed warfare, social disruptions and natural disasters. Surprisingly, no typhus outbreak was identified after the conflicts in the Balkans (the region in Europe where typhus had most recently been endemic) in the 1990s or after the 1991 Gulf War, when large numbers of prisoners of war became lousy in temporary camps.

Typhus disappeared from high- and middle-income countries with the recession of body louse infestations during the late 20th century. The current endemic zones are primarily limited to a few tropical regions, mostly in upland areas of Ethiopia, parts of Somalia and southern Sudan, Rwanda and Burundi, and to parts of the high Andes and Himalayas, where the combination of poverty and colder climate makes laundering or complete changes of clothing impractical or impossible. Small outbreaks occur in other regions, and it is suspected that some occurrences are underreported or under-diagnosed (Gratz,

1985). The distribution of LBRF is similar to that of typhus, but the number of cases is greater, particularly in East Africa (Gratz, 1985).

Recent serological investigations have shown that trench fever, thought to have died out in the 1940s, is still actively transmitted and more widespread than previously believed (Jackson & Spach, 1996; Raoult & Roux, 1999). Although the infection may produce high fever and debilitation, such as in the million or so combatants infected during the First World War, many low-grade infections may have passed unnoticed –as influenza-like fever of unknown origin (Jackson & Spach, 1996; Fournier et al, 2002). Trench fever has not figured highly in the index of suspicion of physicians investigating such episodes, especially in the absence of lice. Trench fever now appears to be disseminated in the homeless of many nations and has probably spread into the wider population, as louse faeces are dispersed from the clothing of infested individuals and come into contact with mucosae of uninfested people (Drancourt et al., 1995; Jackson & Spach, 1996; Raoult & Roux, 1999). Also, head lice may now be vectors of these infections, bringing with them the possibility that children could become infected (Sasaki et al., 2006). So with increased human mobility and more-intense and longer-term infestations, there is a hypothetical risk of an increase in disease transmission in all geographical areas of high- and middle-income countries.

9.5. Socioeconomic influences

The general increase of affluence throughout Europe and North America has resulted in an overall decline in body louse infestations, as most people now have access to facilities to launder or change their clothing regularly. Data on the distribution of lice within and between communities are sparse, as is documentation on the impact of social welfare and economics on the prevalence of lice. This is an area of study that has been neglected due to difficulties in obtaining funding for studies that are large enough to be meaningful and due to the triumph of political correctness over common sense in fostering the pretence that socioeconomic status plays no role in either catching lice or the ability to eliminate infestations. However, the study by Willems and colleagues (2005) gave the first clear indication of this link in modern times and confirms the retrospective analysis conducted by Lindsay (1993). The other evidence that exists suggests that the prevalence of all lice in many communities has increased since the beginning of the 1990s and shows no sign of abatement (Downs, Harvey & Kennedy, 1999; Meinking, 1999; Raoult & Roux, 1999).

9.6. Conducive environmental conditions

Sucking lice are host-specific, obligate ectoparasites that depend on their hosts for food, water, warmth and habitat. Lice may accidentally leave their hosts on a temporary basis, but unless they are able to re-colonize within a short period they die from cumulative dehydration. They do not infest other mammals and there is no reservoir or long-term harbourage off the body where lice can hide.

Over time, body lice have developed a greater body volume than head lice and, in turn, take larger blood-meals (Busvine, 1978). They can tolerate longer periods of starvation and in most cases, can withstand being removed from their host for about 36 hours, but do not normally survive 48 hours (McLeod & Craufurd-Benson, 1941). They live in the folds and seams of clothing and use the fibres of cloth as a substrate for laying their eggs.

All human lice attach their eggs to hairs or fibres in an area where the temperature is appropriate for incubation of the embryos and the local humidity is high enough to protect the eggs from desiccation. Body lice only visit the body surface to feed, so after removal of clothing it is rare that lice are found on the skin. While not feeding, body lice of all stages are found alongside the eggs and empty eggshells in the clothing seams. In all cases, lice require optimum surface body temperature and humidity. In febrile individuals, lice move away from the skin and, given the opportunity, will move to another person (Lloyd, 1919).

9.7. Louse management

9.7.1. Inspection and detection

Discovering the presence of living adult or nymphal lice provides the only positive diagnosis. Body lice shelter in seams of clothing and a diagnosis of their presence is primarily based on inspection of these in pruritic people at risk. Small numbers of lice may be difficult to find, but indicators of infestation include excoriation of the skin in the proximity of clothing seams and nits in the seams of clothing.

9.7.2. Physical removal

Physical methods for eliminating body lice are currently the norm for treating people and were the only method of treatment during the First World War, when most soldiers experienced infestation at some point (Nuttall, 1918). Methods employed now include replacing all clothing and at the same time, killing the insects and their eggs, by using extreme environments – that is, various forms of application of heat or cold. Studies have shown that temperatures above 50°C sustained for more than 20 minutes are sufficient to kill all lice and their eggs (Leeson, 1941; Buxton, 1947), so adequate control of infestations can be achieved with limited equipment and normal domestic hot-water washing systems, by laundering clothes or exposing them to heat – for example, in a hot-air dryer. If desired, items of clothing that cannot be washed could be dry-cleaned, although cost may make this inappropriate in most circumstances.

9.7.2. Pesticides

In most high- and medium-income countries, neurotoxic pesticides have been the mainstream method for eliminating louse infestations. In many cases, these have been used inappropriately, so that in the past lice have readily developed resistance. Resistance to pesticides is now potentially one of the main factors that govern approaches to louse control, although the actual resistance of body lice throughout most of the world is currently completely unknown.

Following the use of pyrethrin and DDT in controlling outbreaks of louse-borne typhus in North Africa and central Italy during the Second World War, neurotoxic insecticides – first DDT, then lindane, followed by malathion and permethrin – became the treatments of choice until resistance rendered most of them ineffective in mass treatment programmes (Hurlburt, Altman & Nibley, 1952; Kitaoka, 1952; Nicoli & Sautet, 1955; Miller et al., 1972; Cole et al., 1973; Sholdt et al., 1977; Gratz, 1985).

9.7.3.1. Impregnated fabric

Fabrics impregnated with insecticides are one way to prevent or restrict body louse infestations. Where deemed a viable option, treating the garments of people likely to be affected is the approach of choice. For example, among other functions, impregnation of military uniforms with permethrin may help to prevent their infestation should exposures arise – say, through contact with refugees or prisoners of war who have acquired a body louse infestation. Impregnation of clothing of so-called at-risk groups, such as vagrants in high- and middle-income communities or people living in areas where body louse infestation is endemic, is often impractical, as it is usually easier to provide alternative means of control – for example, by offering replacement clothing or by treating garments with heat. Under some circumstances, it may be possible to use impregnation as a control measure – for example, under conditions of social disruption.

Products applied to skin are inappropriate for control of body lice, as the insects colonize clothing rather than the skin.

9.8. Benchmarks for lice management

All efforts to treat louse infestations are expected to achieve 100% success. Survival of a single insect constitutes failure. However, the validation of a cure is not clear and has spawned ideas that may result in inappropriate actions, such as insistence on the removal of every louse eggshell after head louse treatment before children can return to school. When such an approach has been applied rigorously, despite other potential measures of effectiveness being available, it has had considerable social and economic impact in such countries as the United States (Williams et al., 2001).

The threat of louse infestation is perceived differently in different societies and cultures. In the high- and middle-income countries, it often results in stigmatization.

The public perception of lice as an important health topic greatly outweighs their importance in real terms. From a practical clinical viewpoint, lice do not constitute a major threat to public health, unless normal sanitation measures break down and the risk of disease transmission increases, as in a natural disaster.

Recent studies indicate that lice have been underestimated or overlooked as vectors able to transmit disease. The widespread distribution of homeless people with positive serology for trench fever indicates that this disease is probably transmitted regularly when lousy individuals come in contact with and are exposed to infected louse faeces (Jackson & Spach, 1996). This also indicates that infestation with body lice may be widespread and that, as there is no mechanism for monitoring the infestation due to the withdrawal of appropriate environmental health services, the level of lousiness in society may be considerably higher than previously thought. It is unlikely that epidemic typhus or LBRF could enter Europe or North America under current conditions, although individual cases, resulting from travel to endemic areas, could arise; in those circumstances, however, onward transmission would be highly unlikely. Unforeseen social disruptions in the future, however, could alter this position.

9.9. Conclusions

The majority of environmental health and public health workers currently consider body lice a low priority. The exception to this is a small group of investigators who have identified residual body louse populations as potentially active vectors of disease causing organisms, mainly rickettsiae – some of which are zoonotic. The survival of both body lice and the diseases they have carried for the past half century, despite intensive efforts to eradicate them in the middle of the 20th century, indicates that they are not in decline. Indeed, the limited (and often anecdotal) information about body louse infestation in populations in high- and middle-income countries suggests that levels of infestation are increasing slowly, despite various efforts to reduce poverty and deprivation.

Irrespective of the possibility of body louse prevalence increasing, the risk of disease transmission to a larger segment of the populace remains, because rickettsial organisms are spread primarily by dried faecal matter that can drift on air currents from the clothing of infested individuals to others in the vicinity that are not infested. Any services directed towards the homeless also require some matched services for the remainder of society, because body lice carry a greater stigma than other infestations and because the public facilities established long ago for dealing with this infestation have now disappeared from most countries. To reduce the risk of disease from body louse infestations, some policy options should be considered by environmental and public health services as a first stage of the process:

- make voluntary organizations that offer assistance to homeless and vagrant people more aware of body lice and their disease vector capacity;
- monitor lousiness through medical services directed towards the homeless community, especially when an infestation is associated with febrile illness;
- advise the homeless about louse-borne disease, with the aim to encourage them to seek help;
- increase the availability of treatment options for lousiness – such as disinfestation or replacement of clothing, the use of impregnated materials, and possibly group ivermectin therapy (Foucault et al., 2006) – to reduce the prevalence of lice; and

- evaluate treatment options by recording the number of cases for each district.

It is unlikely that any policy is going to eliminate body lice or any other human lice, but reducing the risk of possible disease transmission should have a recognized priority, especially as incidents of infection in the general population make diagnosis, in the absence of infestation, a relatively prolonged and costly process. Even where infections due to *Bartonella* bacilli and other rickettsial microorganisms are derived from flea, cat scratch or other zoonotic sources, it is possible that further transmission could occur via body lice in infested homeless (Jackson & Spach, 1996; Jacomo et al., 2002).

References[1]

British Department of Health and Social Security (DHSS) (1987). Department of Health and Social Security 1986 statistics. In: *School health surveillance by nurses*. London, DHSS Statistics Division: section 8m(i).

Burgess IF (2004). Human lice and their control. *Annual Review of Entomology*, 49:457–481.

Burgess IF, Brown CM (1999). Management of insecticide resistance in head lice *Pediculus capitis* (Anoplura: Pediculidae). In: Robinson WH, Rettich F, Rambo GW, eds. *Proceedings of the 3rd International Conference on Urban Pests, 19–22 July 1999, Prague, Czech Republic*. Hronov, Czech Republic, Grafické Závody:249–253.

Busvine JR (1978). Evidence from double infestations for the specific status of human head and body lice (Anoplura). *Systematic Entomology*, 3:1–8.

Buxton PA (1947). *The louse*, 2nd edn. London, Edward Arnold & Company.

Cole MM et al. (1973). Resistance to malathion in a strain of body lice from Burundi. *Journal of Economic Entomology*, 66:118–119.

Downs AMR, Harvey I, Kennedy CTC (1999). The epidemiology of head lice and scabies in the UK. *Epidemiology and Infection*, 122:471–477.

Drancourt M et al. (1995). *Bartonella* (*Rochalimaea*) *quintana* endocarditis in three homeless men. *The New England Journal of Medicine*, 332:419–423.

Evans FC, Smith FE (1952). The intrinsic rate of natural increase for the human louse *Pediculus humanus*. *The American Naturalist*, 86:299–310.

Fisher RI, Morton RS (1970). *Phthirus pubis* infestation. *The British Journal of Venereal Diseases*, 46:326–329.

Foucault C et al. (2006). Oral ivermectin in the treatment of body lice. *The Journal of Infectious Diseases*, 193:474–476.

Fournier PE et al. (2002). Human pathogens in body and head lice. *Emerging Infectious Diseases*, 8:1515–1518.

Gratz NG (1985). Epidemiology of louse infestations. In: Orkin M, Maibach HI, eds. *Cutaneous infestations and insect bites*. New York, Marcel Dekker:187–198.

[1] The literature review conducted for this chapter is not intended to be a comprehensive review of the scientific literature. The references cited are as extensive as possible for a general overview of the subject. Whenever available, data presented in peer-reviewed articles were considered. In some cases, however, it was necessary to refer to other sources of nonetheless trustworthy origin.

Hurlburt HS, Altman RM, Nibley C Jr. (1952). DDT resistance in Korean body lice. *Science*, 115:11–12.

Jacomo V, Kelly PJ, Raoult D (2002). Natural history of *Bartonella* infections (an exception to Koch's postulate). *Clinical and Diagnostics Laboratory Immunology*, 9:8–18.

Jackson LA, Spach DH (1996). Emergence of *Bartonella quintana* infection among homeless persons. *Emerging Infectious Diseases*, 2:141–144.

Kitaoka M (1952). DDT-resistant louse in Tokyo. *Japanese Journal of Medical Science & Biology*, 5:75–88.

Kondaj R (2002). Management of refugee crisis in Albania during the 1999 Kosovo conflict. *Croatian Medical Journal*, 43:190–194.

Kristensen M et al. (2005). Survey of permethrin and malathion resistance in human head lice populations from Denmark. *Journal of Medical Entomology*, 43:533–538.

Lang JD (1975). *Biology and control of the head louse,* Pediculus humanus capitis *(Anoplura: Pedciculidae) in a semi-arid urban area* [PhD thesis]. Tucson, University of Arizona.

Leeson HS (1941). The effect of temperature upon the hatching of the eggs of *Pediculus humanus corporis* De Geer (Anoplura). *Parasitology*, 33:243–249.

Lindsay SW (1993). 200 Years of lice in Glasgow: an index of social deprivation. *Parasitology Today*, 9:412–417.

Lloyd L (1919). *Lice and their menace to man.* London, Henry Frowde, Oxford University Press.

MacLeod J, Craufurd-Benson HJ (1941). Casual beds as a source of louse infestation. *Parasitology*, 33:211–213.

Meinking TL (1999). Infestations. *Current Problems in Dermatology*, 11:73–120.

Mellanby K (1941). The incidence of head lice in England. *Medical Officer*, 65:39–42.

Mellanby K (1943). The incidence of head lice in England after four years of war. *Medical Officer*, 70:205–207.

Miller RN et al. (1972). First report of resistance of human body lice to malathion. *Transactions of the Royal Society of Tropical Medicine and Hygiene*, 66:372–375.

Nicoli JL, Sautet J (1955). *Rapport sur la fréquence et la sensibilité aux insecticides de Pediculus humanus humanus L. dans le sud-est de la France.* Paris, Institut National d'Hygiene (Monographie de l'Institut National d'Hygiene (Paris), No. 8).

Nuttall GHF (1917). The biology of *Pediculus humanus*. *Parasitology*, 10:80–185.

Nuttall GHF (1918). Combating lousiness among soldiers and civilians. *Parasitology*, 10:411–586.

Peacock AD (1916). The louse problem at the Western Front. *Journal of the Royal Army Medical Corps*, 27:31–60.

Raoult D, Roux V (1999). The body louse as a vector of reemerging human diseases. *Clinical Infectious Diseases*, 29:888–911.

Raoult D et al. (1998). Outbreak of epidemic typhus associated with trench fever in Burundi. *Lancet*, 352:353–358.

Sasaki T et al. (2006). First molecular evidence of *Bartonella quintana* in *Pediculus humanus capitis* (Phthiraptera: Pediculidae), collected from Nepalese children. *Journal of Medical Entomology*, 43:110–112.

Sholdt LL et al. (1977). Resistance of human body lice to malathion in Ethiopia. *Transactions of the Royal Society of Tropical Medicine and Hygiene*, 70:532–533.

Smith S et al. (2003). Head lice diagnosed in general practice in the West Midlands between 1993 and 2000: a survey using the General Practice Research Database. *Communicable Disease and Public Health*, 6:139–143.

Soper FL et al. (1945). Louse powder studies in North Africa 1943. *Archives de l'Institute Pasteur d'Algérie*, 23:183–223.

Soper FL et al. (1947). Typhus fever in Italy 1943–1945 and its control with louse powder. *American Journal of Hygiene*, 45:305–334.

Taplin D, Meinking TL (1988). Infestations. In: Schachner LA, Hansen RC, eds. *Pediatric dermatology*, 1st ed. Vol. 2. New York, Churchill Livingstone:1465–1515.

Valenciano M et al. (1999). Surveillance of communicable diseases among the Kosovar refugees in Albania, April–June 1999. *Euro Surveillance*, 4:92–95.

Willems S et al. (2005). The importance of socio-economic status and individual characteristics on the prevalence of head lice in schoolchildren. *European Journal of Dermatology*, 15:387–392.

Williams LK et al. (2001). Lice, nits, and school policy. *Pediatrics*, 107:1011–1015.

10. Ticks

Howard S. Ginsberg and Michael K. Faulde

Summary

The most common vector-borne diseases in both Europe and North America are transmitted by ticks. Lyme borreliosis (LB), a tick-borne bacterial zoonosis, is the most highly prevalent. Other important tick-borne diseases include TBE (tick-borne encephalitis) and Crimean-Congo haemorrhagic fever in Europe, Rocky Mountain spotted fever (RMSF) in North America, and numerous less common tick-borne bacterial, viral, and protozoan diseases on both continents. The major etiological agent of LB is *Borrelia burgdorferi* in North America, while in Europe several related species of *Borrelia* can also cause human illness. These *Borrelia* genospecies differ in clinical manifestations, ecology (for example, some have primarily avian and others primarily mammalian reservoirs), and transmission cycles, so the epizootiology of LB is more complex in Europe than in North America.

Ticks dwell predominantly in woodlands and meadows, and in association with animal hosts, with only limited colonization of human dwellings by a few species. Therefore, suburbanization has contributed substantially to the increase in tick-borne disease transmission in North America by fostering increased exposure of humans to tick habitat. The current trend toward suburbanization in Europe could potentially result in similar increases in transmission of tick-borne diseases. Incidence of tick-borne diseases can be lowered by active public education campaigns, targeted at the times and places of greatest potential for encounter between humans and infected ticks. Similarly, vaccines (e.g., against TBE) are most effective when made available to people at greatest risk, and for high-prevalence diseases such as LB. Consultation with vector-borne disease experts during the planning stages of new human developments can minimize the potential for residents to encounter infected ticks (e.g., by appropriate dwelling and landscape design). Furthermore, research on tick vectors, pathogens, transmission ecology, and on geographic distribution, spread, and management of tick-borne diseases can lead to innovative and improved methods to lower the incidence of these diseases. Surveillance programs to monitor the distribution and spread of ticks, associated pathogens, and their reservoirs, can allow better-targeted management efforts, and provide data to assess effectiveness and to improve management programs.

10.1. Introduction

Ticks transmit more cases of human disease than any other arthropod vector in Europe and North America. They are also important worldwide as disease vectors to people and domestic animals, and they cause substantial economic losses, both by transmitting disease and by direct negative effects on cattle (Jongejan & Uilenberg, 2004). Lyme borreliosis (LB), in particular, is the most commonly reported vector-borne disease in both Europe and North America (Steere, Coburn & Glickstein, 2005). In Europe, Tick-Borne Encephalitis is also prevalent, especially in central and eastern Europe, while in North America, Rocky Mountain spotted fever (RMSF), caused by a rickettsial agent, is responsible for a few hundred to over a thousand cases a year. In addition to their importance as disease vectors, some hard tick species can directly cause adverse effects, such as tick paralysis, a toxicosis (systemic poisoning) due to toxic salivary proteins. Similarly, soft ticks can provoke severe allergenic bite reactions in people (IgE-mediated type-I allergy).

The response to tick-borne diseases (TBDs) in the United States has been substantial, including federally sponsored research programmes, public health programmes within individual states (partly funded by the CDC [United States Centers for Disease Control and Prevention]) and several smaller programmes funded by states, localities and non-profit-making organizations. States with a high incidence of disease have numerous public education programmes, and several novel methods of tick and disease management have been developed (Stafford & Kitron, 2002). However, coordination and evaluation of programmes is spotty, and the incidence of disease remains high in many locales and has increased nationwide (Piesman & Gern, 2004). Ecological differences in transmission dynamics from site to site mean that the approach to management needs to be tailored to conditions at each locale. Methods for developing effective IPM programmes and evaluations of efficacy remain high priorities (Ginsberg & Stafford, 2005).

The situation in Europe is different in that national reporting strategies differ among countries (Table 10.1), and little has been done to routinely implement measures that protect individuals against tick bites or TBDs. Some notable exceptions are vaccination against TBE (Nuttall & Labuda, 2005) and the use of skin repellents in some areas. Fabrics impregnated with acaricides (agents that kill ticks and mites), such as permethrin, are widely unknown and difficult to procure, even for personnel occupationally exposed to tick-infested areas of endemic TBDs. So far, few research efforts have been initiated to reduce tick populations by ecological changes, biological control or IPM.

10.2. Ticks of Europe and North America

Ticks are arachnids (the class Arachnida includes spiders, scorpions, ticks and mites) in the subclass Acari, which includes mites and ticks. There are three families of ticks (Barker & Murrell, 2004): the hard ticks, Ixodidae (713 species), which includes most ticks of medical importance to people; the soft ticks, Argasidae (185 species), which includes a few species that transmit diseases to humans; and Nuttalliellidae, which includes just one species from Africa with no known medical importance.

Table 10.1. TBDs in Europe to be notified to national health authorities, as of 2005

Country/locale	TBE/CEE	Lyme borreliosis	Other diseases
Albania	- (Endemic)	- (Endemic)	-
Austria	+	(+) Only meningoencephalitis caused by Lyme borreliosis	-
Belarus	+	+	Tularaemia, Q fever, tick-borne haemorrhagic fevers
Belgium	-	- (Endemic)	-
Bosnia and Herzegovina	-	+	CCHF
Bulgaria	- (Endemicity status unclear)	- (Endemic)	CCHF
Croatia	+	+	Tick-borne tularaemia, ehrlichiosis, human granulocytic anaplasmosis
Czech Republic	+	+	-
Denmark	- (Not endemic)	(+) Neuroborreliosis only	-
Estonia	+	+	Tick-borne tularaemia
Finland	+	+	
France	- (Endemic)	- (Endemic)	-
Germany	+	(+) Only the federal states of Brandenburg, Mecklenburg-Western Pomerania, Berlin, Lower Saxony, Saxony-Anhalt and Thuringia (about 25 % coverage of population)	-
Greece	- (Not endemic)	+	-
Hungary	+	+	-
Ireland	- (Not endemic)	- (Endemic)	Louping ill
Italy	+	+	-
Latvia	+	+	Tick-borne tularaemia
Lithuania	+	+	Tick-borne tularaemia
Luxembourg	- (Not endemic)	+	
Netherlands	-	- (Endemic)	-
Norway	+	+	-
Poland	+	+	-
Portugal	-	+	-
Republic of Moldova	- (Endemic)	- (Endemic)	-
Romania	+	+	MSF
Russian Federation	+	+	Tularaemia, Q fever, tick-borne haemorrhagic fevers
Serbia	+	+	-
Slovakia	+	+	-
Slovenia	+	+	HGE
Spain	-	- (Endemic)	-
Sweden	+	- (Endemic)	-
Switzerland	+	- (Endemic)	-
The former Yugoslav Republic of Macedonia	-	+	-
Ukraine	+	+	Tularaemia, Q fever, tick-borne haemorrhagic fevers
United Kingdom	-	(+) Scotland only	-

Note. -: not notifiable disease; +: disease notifiable by national health organisations.

Source: The information in this table has been provided by M.K. Faulde and is based on official civil and military country sources.

Table 10.2. Tick vectors of medical importance that are endemic in Europe

Species	Geographical distribution	Habitat	Host [a]	Remarks
Castor-bean tick *(Ixodes ricinus)*	North-western Europe (westwards to Baltic states)	Humid microhabitats in woodlands, rough grasslands and moorlands	Many different kinds of wild and domestic animals; readily feeds on man; three-host tick	Most common tick species in north-west Europe
Taiga tick *(Ixodes persulcatus)*	North-eastern Europe (eastwards to Baltic states)	Humid microhabitat in taiga woodlands, rough grasslands and moorlands	Many different kinds of wild and domestic animals; readily feeds on man; three-host tick	Most common tick species in north-east Europe and northern Asia
Ornate sheep tick *(Dermacentor marginatus)*	Southern Europe, southwards to 50th parallel	Scrub steppes, temperate forests, grasslands and sheep pastures	Larvae and nymphs feed on rabbits and small mammals, also on birds; adults feed on large mammals, such as deer, cattle and sheep; readily feeds on man; three-host tick	As many as 200 adult ticks can be found on one sheep
Marsh tick (also called the ornate cow tick; *Dermacentor reticulatus*)	Generally southern Europe southwards to 50th parallel; localized populations occur in north-western Europe in Belgium, south-west England and Wales	Grasslands, pastures and woodlands	Larvae and nymphs feed on small mammals, occasionally on birds; adults feed on wild and domestic mammals, such as deer, dogs, foxes, cattle and sheep; occasionally feeds on man; three-host tick	Spreading geographically in Germany
Brown dog tick *(Rhipicephalus sanguineus)*	Mediterranean Europe and Africa, with focal populations in Belgium, United Kingdom, France, Germany, Denmark, Norway	Arid microhabitat in scrub steppes, buildings and kennels	Larvae, nymphs and adults primarily feed on dogs (90%), but are also found on cattle, cats, foxes and human beings; three-host tick	Local populations may survive in kennels and other sheltered places with dogs in central and northern Europe; transported over longer distances by dogs
Bont-legged tick *(Hyalomma marginatum)*	Southern Europe (southwards to 40th parallel), southern Asia and most of Africa	Scrub steppes, temperate forests, grasslands and sheep pastures, and migrating birds	Larvae and nymphs feed on one bird host for 12–26 days; adults actively search for mammal hosts, such as cows, donkeys, dogs, foxes and human beings; two-host tick	Transported over longer distances by migrating birds and may sporadically appear in northern Europe (such as Denmark and Norway)
Coastal red tick *(Haemaphysalis punctata)*	Throughout Europe, except Ireland	Wide variety: from relatively cold and humid coasts (such as United Kingdom) to semi-desert zones of central Asia	Larvae and nymphs feed on small mammals and rarely birds; adults primarily feed on sheep and cattle, occasionally on human beings, three-host tick	Bite may cause tick paralysis

[a] Some ticks, called one-host ticks, feed on only one host throughout all three stages of life (larval, nymphal and adult). Other ticks, called two-host ticks, feed and remain on the first host during the larval and nymphal stages of life, and then drop off and attach to a different host as an adult. Finally, three-host ticks feed, drop off and reattach to progressively larger hosts subsequently to each moulting.

Source: Data presented have been collected by the authors from numerous sources.

Endemic tick species in Europe can be peridomestic or can be associated with pets and farm animals (Table 10.2). European ticks that can infest buildings in urban environments include: the ixodid brown dog tick, *Rhipicephalus sanguineus*, as far north as southern Germany; and the argasids: the European pigeon tick, *Argas reflexus* (associated with pigeons), and the fowl tick, *Argas persicus* (associated with poultry in south-eastern Europe). Long-term infestations with brown dog ticks can occur in human dwellings, if control efforts are neglected (Gothe, 1999). The only survey thus far for European pigeon ticks was performed in the city of Berlin, where more than 200 infested buildings were discovered between 1989 and 1998 (Dautel, Scheurer & Kahl, 1999). Most of the infestations were found in older buildings constructed before 1918. Control is difficult and requires professional expertise and time.

Recent studies in Germany have shown increases in urban and periurban collections of castor-bean ticks, *Ixodes ricinus* (Mehnert, 2004). According to studies conducted in north-eastern Germany, Lyme borreliosis (LB) is most often acquired in city parks and gardens near forests (Ammon, 2001; Anonymous, 2005a). Other ticks, such as the soft tick *Ornithodoros erraticus*, and the hard ticks *Dermacentor* spp., *Hyalomma* spp. and *Haemaphysalis* spp., are associated with pigs, sheep and cattle and are known vectors of both animal and human disease agents. They usually do not infest houses, but can be found in stables and in houses that incorporate stables.

The most common hard ticks that regularly bite people in North America (Table 10.3) include: the black-legged or deer tick, *Ixodes scapularis*, in eastern and central North America; the western black-legged tick, *Ixodes pacificus*, in west coastal areas; the American dog tick, *Dermacentor variabilis*, in the east and Midwest; the Rocky Mountain wood tick, *Dermacentor andersoni*, in the Rocky Mountain region; the Pacific Coast tick, *Dermacentor occidentalis*, on the Pacific coast; and the lone star tick, *Amblyomma americanum*, in eastern and central North America. The brown dog tick attaches to dogs and can be found in the home, but rarely attaches to people. The primary soft ticks that affect people are *Ornithodoros* spp. in western areas.

These ticks are found primarily in natural areas and are often encountered by recreational users of parks and woodlands (Ginsberg & Ewing, 1989). However, increasing suburbanization around major urban centres has resulted in substantial contact between people and ixodid ticks, and most disease transmission from ticks to people occurs in the peridomestic environment (Maupin et al., 1991). Some nidicolous species (including soft ticks, such as *Ornithodoros* spp.) are found in animal nests in rustic cabins and can transmit pathogens (such as relapsing fever borreliae) to recreational users of these dwellings (Barbour, 2005).

10.3. Tick-borne diseases

The epidemiology and distribution of TBDs in Europe and North America are generally similar, but differ in some important details. In Europe, 31 viral, 14 bacterial, and 5 *Babesia* species are known endemic tick-borne pathogens of people (Table 10.4). Among

Table 10.3. Tick vectors of medical importance that are endemic in North America

Species	Geographical distribution	Habitat	Host [a]	Remarks
Deer tick (*Ixodes scapularis*)	Eastern North America and northern Midwest	Closed-canopy woodlands; adults extend into open habitats	Broad range of hosts, including mammals, birds and reptiles; adults on large mammals, such as deer; three-host tick	Especially common in north-eastern United States
Western black-legged tick (*Ixodes pacificus*)	Western North America	Woodlands, scrub and open habitats	Broad range of hosts; adults on large mammals, such as deer; immatures on lizards, birds, and diverse mammals; three-host tick	Immatures more common on lizards than on rodents
American dog tick (*Dermacentor variabilis*)	Eastern and central North America, especially the Carolinas to Oklahoma	Woodlands, shrublands and grasslands, especially along animal trails	Adults on large mammals; immatures on small mammals, such as rodents; three-host tick	Can be found in urban parks as well as natural areas
Rocky Mountain wood tick (*Dermacentor andersoni*)	Rocky Mountain region	Woodlands, low shrub vegetation and grasslands	Adults on large mammals; immatures on small mammals; three-host tick	—
Pacific Coast tick (*Dermacentor occidentalis*)	Pacific region of North America	Woodlands	Adults on large mammals; immatures on small mammals; three-host tick	—
Lone star tick (*Amblyomma americanum*)	South-eastern and south-central North America, expanding northward	Woodlands, shrublands and grasslands	All three life stages attach readily to large mammals, especially deer; immatures also on birds; three-host tick	Extremely aggressive and fast-moving tick
Ornithodoros spp.	Western North America	Rodent nests	Generally rodents, but can attach to a variety of mammals	Can bite human beings who utilize rustic cabins with rodent nests

[a] Some ticks, called one-host ticks, feed on only one host throughout all three stages of life (larval, nymphal and adult). Other ticks, called two-host ticks, feed and remain on the first host during the larval and nymphal stages of life, and then drop off and attach to a different host as an adult. Finally, three-host ticks feed, drop off and reattach to progressively larger hosts subsequently to each moulting.

Note. —: no remarks.

Source: Data presented have been collected by the authors from numerous sources.

European TBDs, only TBE is a widely notifiable disease (Table 10.1), with more than 10 000 clinical cases annually. Detailed epidemiological information is not available on other TBDs, despite the fact that the most frequent TBD in Europe is LB (with possibly hundreds of thousands of clinical cases a year). Germany alone claims 20 000–60 000 cases a year (O'Connell et al., 1998; Wagner, 1999). Yearly rates of incidence in hyperendemic foci (sites where disease organisms exist in host populations at very high rates) can exceed 300 cases per 100 000 population, with average occupational seroprevalence rates of up to 48% in forest workers. Other TBDs occur, but their rates of incidence remain largely unknown.

Several isolated regional studies in Europe show that tick abundance is increasing regionally while TBDs are simultaneously emerging and spreading geographically. The changing urban landscape in Germany, specifically in the federal state of Brandenburg, where LB has been a notifiable disease since 1996, shows a steady increase in exposure to castor-bean ticks. Other studies have shown that urban parks in Berlin and Munich have growing tick populations and contribute to a growing number of cases of LB. In the Czech Republic, castor-bean tick populations spread an average of 161 meters into higher altitude sites (from about 780 m to 960 m above sea level) during the last 30 years. This resulted in exposure to ticks and TBDs in higher mountainous areas that were formerly not endemic for castor-bean ticks and diseases associated with them. Since the 1990s at least two TBDs, TBE and Mediterranean spotted fever (MSF), have been reported to be extending their geographical ranges. TBE is spreading geographically into the north-eastern parts of Germany. MSF, transmitted by the brown dog tick, is reportedly spreading northwards along the French Rhone Valley, as far north as Belgium, where the first autochthonous (locally acquired) human cases of MSF were recently reported. Data from the Baltic states show that landscape-level ecological changes (resulting from agricultural practices) have led to increases in ecotopes (the smallest ecologically distinct features in a landscape mapping and classification system) suitable for tick infestation. Finally, the reported increase in incidence of TBDs may in part result from increased awareness of TBDs, better diagnostic tools, and markedly higher leisure and sporting activities that result in increased exposure to endemic disease foci.

The most common TBD in North America, as in Europe, is Lyme borreliosis (also called Lyme disease). Other important TBDs (Sonenshine, Lane & Nicholson, 2002) include RMSF, human monocytic ehrlichiosis (HME), human granulocytic anaplasmosis (HGA), Q fever, and tularaemia (Table 10.5). All of these diseases are notifiable in the United States (Groseclose et al., 2004).

Less common or non-emerging TBDs in North America include such infections as babesiosis, which is caused by the protozoan *Babesia microti* and is transmitted by the black-legged tick, primarily in southern New England and mid-Atlantic coastal areas (Spielman, 1976; Spielman et al., 1979). Powassan encephalitis is a rarely reported viral disease related to European TBE (Ebel, Spielman & Telford, 2001). Colorado tick fever (CTF) is a viral disease transmitted by the Rocky Mountain wood tick in the Rocky Mountain region (McLean et al., 1981). The lone star tick transmits *Ehrlichia ewingi*, which causes human ehrlichiosis. Q fever, caused by *Coxiella burnetii*, is primarily a livestock disease (McQuiston & Childs, 2002). Tick-borne relapsing fever, caused by several *Borrelia* spp. and transmitted by associated *Ornithodoros* spp., is primarily contracted by people in intermittently used recreational cabins in wild areas in western North America. Important vectors include *Ornithodoros hermsi* (which transmits the spirochete *Borrelia hermsii*), *Ornithodoros parkeri* (which transmits *Borrelia parkerii*), and *Ornithodoros turicata* (which transmits *Borrelia turicatae*) (Barbour, 2005). Tularaemia, caused by the bacterium *Francisella tularensis*, is usually acquired by rabbit hunters that handle infected rabbits (especially in eastern North America), but is sometimes transmitted by ticks (especially in western states).

Table 10.4. Human pathogenic TBDs that are endemic in Europe

Tick genus	Tick species	Common name	Viral pathogens	Bacterial and parasitic pathogens
Hyalomma	*marginatum*	Bont-legged tick	SINV, WNV, TBEV, CCHFV, Dhorivirus	—
Ixodes	*ricinus*	Castor-bean tick	Louping-ill virus, TBEV, Negishi virus, Uukuniemi virus, Erve virus, Eyach virus, Tribec virus, Lipovnik virus, CCHFV, Bhanjavirus	*B. burgdorferi* s.l., *C. burnetii, A. phagocytophilum, R. slovaca, R. helvetica, B. divergens, B. bovis, B. microti, F. tularensis, E. chaffeensis, Ehrlichia equi*
	persulcatus	Taiga tick	TBEV, Negishi virus, Uukuniemi virus	*R. slovaca, E. equi, B. burgdorferi* s.l.
	gibbosus		TBEV	—
	hexagonus		TBEV, Erve virus	*B. burgdorferi* s.l.
	arboricola		TBEV	—
	uriae		Tyuleniy virus, Avalon virus	*B. burgdorferi* s.l.
	ventalloi		Erve virus, Eyach virus	—
Dermacentor	*marginatus*	Ornate sheep tick	TBEV, Bhanjavirus, Erve virus, Dhori virus, CCHFV, WNV, OHFV	*F. tularensis, B. bovis, R. slovaca, R. helvetica, C. burnetii, Ehrlichia canis*
	reticulatus	Marsh tick (also called the ornate cow tick)	TBEV, OHFV	*C. burnetii, R. slovaka, Rickettsia sibirica* strain 246
Haemaphysalis	*inermis*		TBEV	R. slovaka
	concinna		TBEV	*C. burnetii*
	punctata	Coastal red tick	TBEV, CCHFV, Bhanjavirus, Tribec virus, Lipovnik virus	*C. burnetii, B. microti, B. burgdorferi* s.l.
Rhipicephalus	*bursa*	Brown dog tick	CCHFV, Thogoto virus	—
	sanguineus		CCHFV, Lipovnik virus	*R. conori, Rickettsia massiliae* GS, *Rickettsia rhipicephali, A. phagocytophilum, C. burnetii, B. burgdorferi* s.l., *B. valaisiana, E. canis*
	turanicus		—	*Rickettsia massiliae* Mtu1
Ornithodoros	*coniceps*		WNV	—
	maritimus		Soldado virus	*Borrelia hispanica*
	erraticus		—	
Argas	*reflexus*		WNV	*C. burnetii*
	vespertilionis	—		*B. burgdorferi* s.l.

Note. —: no remark; SINV: Sindbis virus; WNV: West Nile virus; TBEV: tick-borne encephalitis virus; CCHFV: Crimean-Congo haemorrhagic fever virus; OHFV: Omsk haemorrhagic fever virus.

Source: Data presented have been collected by the authors from numerous sources.

Table 10.5. Human pathogenic TBDs that are endemic in North America

Tick genus	Tick species	Common name	Viral pathogens	Bacterial and parasitic pathogens
Ixodes	*scapularis*	Black-legged tick	Deer tick virus (Powassan encephalitis virus)	*B. burgdorferi* s.s., *Borrelia* sp. nov. (relapsing fever group), *A. phagocytophilum*, *B. microti*
	pacificus	Western black-legged tick	—	*B. burgdorferi* s.s.
Dermacentor	*variabilis*	American dog tick	—	*R. rickettsii*
	andersoni	Rocky Mountain wood tick	CTF	*R. rickettsii* (RMSF)
Amblyomma	*americanum*	Lone star tick	—	*E. chaffeensis*, *E. ewingii*, *Borrelia lonestari*
Various hard tick species (or contact with host fluids, aerosols, and the like)			—	*F. tularensis*
Ornithodoros	various species		—	*Borrelia* spp.

Note. —: no remark; CTF: Colorado tick fever; HGA: human granulocytic anaplasmosis; HME: human monocytic ehrlichiosis; RMSF: Rocky Mountain spotted fever.

Source: Data presented have been collected by the authors from numerous sources.

The following sections provide more comprehensive treatments of the most prevalent TBDs in Europe and North America: LB on both continents, TBE in Europe, and RMSF in North America.

10.4. Lyme borreliosis

The clinical features, diagnosis, treatment, pathology, microbiology, ecology, surveillance and management of LB have been extensively reviewed (Ginsberg, 1993; Gray et al., 2002; Piesman & Gern, 2004; Steere, Coburn & Glickstein, 2005). Features relevant to current trends in LB epidemiology in Europe and North America are summarized below.

10.4.1. Public health

LB is the most common TBD in both North America and northern Eurasia. The complex of related pathogenic species, *B. burgdorferi* s.l., the causative agents of LB, are Gram-negative, microaerophilic bacteria that belong to the family Spirochaetaceae. To date *B. burgdorferi* s.l. can be divided into at least 12 species (Fingerle et al., 2005), of which those with human-pathogenic significance are *Borrelia afzelii*, *Borrelia burgdorferi* sensu stricto,

B. garinii, and *Borrelia spielmanii* (Richter et al., 2006). *B. valaisiana* and *Borrelia lusitaniae* may also be pathogenic to people (Ryffel et al., 1999; Collares-Pereira et al., 2004), but firm evidence is currently lacking.

B. burgdorferi s.l. infection can be subclinical or it can have a broad range of clinical presentations (Gern & Falco, 2000). Symptoms apparently depend on the *Borrelia* genospecies involved, the tissues affected, the duration of infection and individual human host factors, including genetic predisposition. There is considerable evidence that infection with different LB genospecies have different clinical outcomes (Gern & Falco, 2000; WHO Regional Office for Europe, 2004). Thus, *B. burgdorferi* s.s. is most often associated with arthritis, particularly in North America, where it is the only known cause of human LB; *B. garinii* is associated with neurological symptoms; and *B. afzelii* is associated with the chronic skin disease acrodermatitis chronica atrophicans (ACA). All four pathogenic *B. burgdorferi* s.l. genospecies, including *B. spielmanii* (formerly named A14S), have been isolated from erythema migrans (EM) lesions (Fingerle et al., 2005). There is evidence in Europe that EM occurs more frequently in *B. afzelii* infections than in those caused by *B. garinii*.

Generally, clinical presentations can be divided into three stages (Gern & Falco, 2000; Steere, Coburn & Glickstein, 2005).

1. The first stage, early localized LB, is characterized by an expanding red rash (EM, often with central clearing) and flu-like symptoms (such as headache and fever) 2–30 days after an infective tick bite, which occurs in about 60% of cases. The rash can be faint and difficult to notice and resolves even without treatment.

2. The second stage, early disseminated LB, varies from patient to patient and can include more severe flu-like illness, secondary skin lesions, facial palsy, aseptic meningitis, mild encephalitis and arthritis with effusion or carditis.

3. The third stage, late LB, is most commonly manifested as Lyme arthritis, typically affecting large joints, especially the knee. Other presentations are ACA, an unusual skin condition, and, rarely, chronic Lyme meningoencephalitis, where sporadic fatalities have been reported. Late stage central nervous system involvement can be severe and difficult to treat. Late LB symptoms can be nonspecific, difficult to diagnose, and can occur in other conditions.

According to treatment guidelines, LB treatment involves different antibiotic regimens in varying concentrations, adapted to specific clinical manifestation (Wormser et al., 2000). Doxycycline is effective in early LB. Amoxicillin and penicillin are also still drugs of choice. Treatment of late-stage disseminated LB requires higher doses, often of ceftriaxone or cefuroxime, and sometimes longer treatment periods. A specific vaccine for people, based on outer surface protein A (OspA), was temporarily available in the United States, but was withdrawn by the manufacturer in 2002. Due to the heterogenicity of *B. burgdorferi* s.l. genospecies in Europe and Asia, an effective vaccine for Europe would most probably require a defined so-called cocktail of immunogenic outer surface proteins.

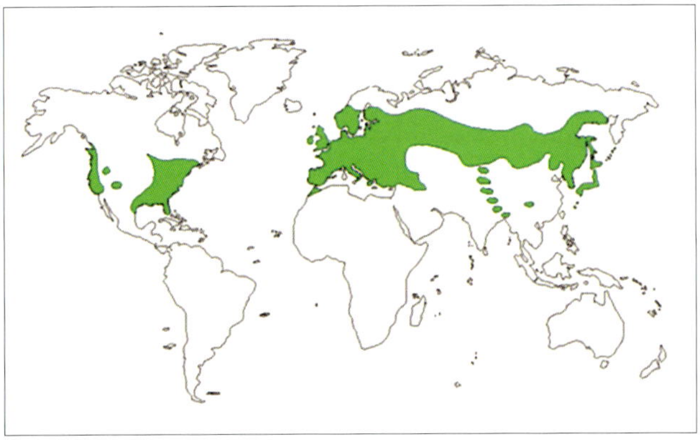

Fig. 10.1. Global distribution of Lyme borreliosis

Source: CDC (2006).

10.4.1.1. LB in Europe and North America

LB is broadly distributed in the northern hemisphere (Fig. 10.1). The prevalence of LB varies considerably among European countries, with estimated average rates between 0.3 case per 100000 population in the United Kingdom and up to 130 cases per 100000 population in parts of Austria. LB tends to be focal, with defined hot spots within countries. In Germany, for example, the average incidence in the Oder-Spree region in the federal state of Brandenburg was estimated to be 89.3 cases per 100000 population in 2003. Within this area, the local incidence of LB varied from 16 cases per 100000 population in Erkner county to 311 cases per 100000 population in Brieskow-Finkenheerd county (Talaska, 2005; Anonymous, 2005a). Therefore, mapping hot spots is an important tool for disease prevention.

Over 23000 cases of LB were reported to the CDC in 2002 (Groseclose et al., 2004), and it has been estimated that this is a small fraction (roughly 10%) of the actual total number of cases in the United States. In one study in Connecticut (Meek et al., 1996), about 16% of diagnosed cases had been reported. Cases follow the geographic distribution of the *Ixodes* vectors (the black-legged tick and the western black-legged tick) (Fig. 10.2), with most cases in the north-eastern, mid-Atlantic and northern Midwest regions (within the range of the black-legged tick), and with some hot spots in California (western black-legged tick) (Dennis et al., 1998; Dennis & Hayes, 2002). As in Europe, the distribution of LB tends to be highly focal, because of nonrandom distributions of tick vectors, reservoir hosts, appropriate habitat types and other ecological conditions. This focal pattern is illustrated by the distribution of cases in 1999, when the national incidence was 6.0 cases per 100 000 population (16273 cases). The number of cases in individual states varied dramatically, with a maximal incidence of 98.0 cases per 100000 population in Connecticut (Dennis & Hayes, 2002).

The costs associated with LB can be significant. Assuming a cost of about €10000 per case of so-called disseminated LB in Europe and, on average, 20–30% disseminated LBs per notified clinical case, with 1800–2000 cases a year in the federal state Brandenburg, an economic impact of €1 million a year can be easily exceeded for that state alone (Talaska, 2003). An economic burden of several €100 million up to 1 billion a year is plausible for Europe. Similarly, in the United States, Meltzer, Dennis & Orloski (1999) estimated costs (including treatment and lost work) of US$ 161 for early LB with no sequelae (previous diseases or injuries), US$ 34 354 for disseminated cases with arthritic symptoms, US$ 61243 for neurological cases and US$ 6845 for cardiac cases. Assuming 83% of cases with effective early treatment, and 17% with disseminated disease (12% with

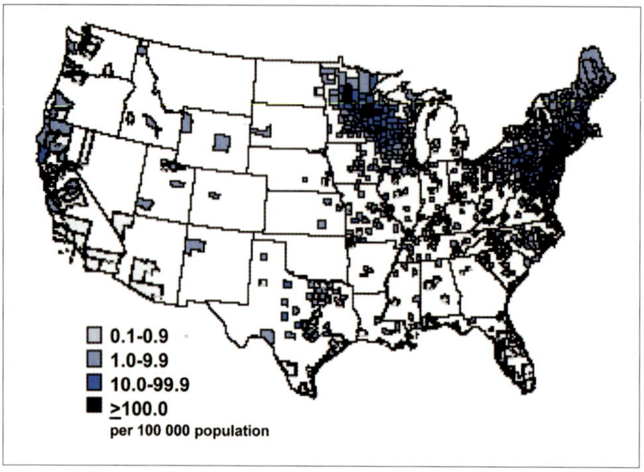

Fig. 10.2. Distribution of Lyme disease in the United States
A. Lyme disease cases

Source: Groseclose et al. (2004).

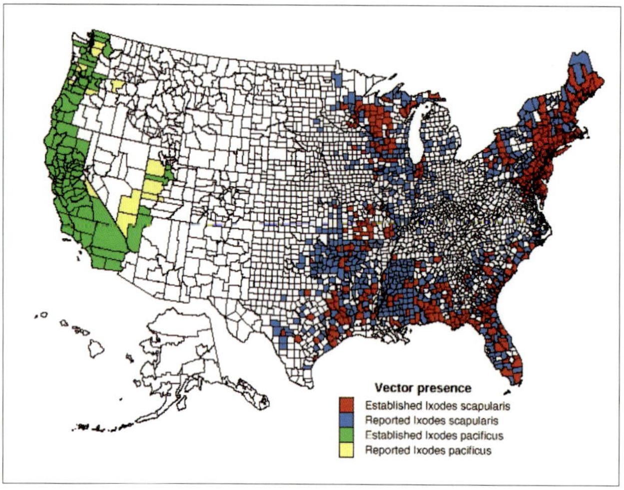

Fig. 10.2. Distribution of Lyme disease in the United States
B. Tick distribution

Source: CDC (2006).

arthritic symptoms, 4% with neurological disease and 1% with cardiac disease), the total of about 23 000 reported cases a year results in about US$150 million in costs. If the number of cases reported is only about 10% of the total number of cases, the actual costs are in US$ billions. These very rough estimates refer primarily to costs of medical treatment. Expenses associated with family accommodations for patients, lost work time and the like would greatly increase these estimates. Zhang and colleagues (2006) used actual cost data to estimate the economic impact of LB (including treatment and loss of productivity) in the Eastern Shore area of Maryland. They estimated a national cost of about US$ 203 million for the 23 763 cases reported in 2002. Again, unreported cases (probably the vast majority of actual cases) would greatly inflate this estimate. Furthermore, the costs of prevention activities associated with LB (such as landscaping and pesticide applications) contribute further to the costs of this disease, including human, economic and environmental costs.

10.4.2. Geographical distribution

The global distribution of human pathogenic *B. burgdorferi* s.l. genospecies includes parts of North America and most of Europe and extends eastward in Asia to Japan (Fig. 10.1 and 10.2). In Europe, LB has been reported throughout the continent (including the European parts of the Russian Federation), except for the northernmost areas of Scandinavia. Taking the limitations of seroprevalence studies into account, LB in Europe shows a gradient of increasing incidence from west to east, with the highest rates of incidence in central-eastern Europe. Simultaneously, LB shows a gradient of decreasing incidence from south to north in Scandinavia and north to south in the European Mediterranean and Balkan countries (Lindgren, Talleklint & Polfeldt, 2000; Faulde et al., 2002; WHO Regional Office for Europe, 2004). The incidence of LB is apparently also increasing eastward in Asia. Infection rates are highest in adult ticks and vary between 10% and 30% in Europe (5–10% in nymphs), reaching up to a 45% positivity rate in adult ticks in hot spots of LB in Germany and Croatia (Hubalek & Halouzka, 1998; Kimmig, Oehme & Backe, 1998; Golubic & Zember, 2001).

In the foreseeable future, the incidence of TBDs, especially LB, seem likely to increase, partly due to man-made environmental changes. For example, some current approaches to urban planning can provide additional ecotopes suitable for castor-bean tick and taiga tick, *Ixodes persulcatus*, infestations (Kriz et al., 2004). In North America, suburbanization has produced extensive suburban and periurban areas that provide an interface between urban and sylvan environments – a so-called border effect. Property sizes in these areas tend to be larger than in urban areas and therefore allow ready access to tick habitats that border infested natural ecosystems. This border effect is more pronounced in North America than in Europe. However, the European landscape is beginning to change. Increasing suburbanization can potentially create conditions similar to those in North America, as recently shown in the federal state of Mecklenburg-Western Pomerania, Germany (Talaska, 2003), potentially leading to greater human exposure to TBDs. Thus, the increase of LB is apparently related to that of urban sprawl, which often results in invasion of residential areas by deer and mice, providing reservoirs, tick hosts, and carriers for the spirochete (Matuschka et al., 1996). Moreover, some studies suggest that climate changes in Europe have resulted in a northern shift in the distributional limit of castor-bean ticks, an increase in their population density in Sweden and a shift into higher altitudes in mountainous areas in the Czech Republic (Lindgren, Talleklint & Polfeldt, 2000; Danielova, 2006). Castor-bean tick nymphs infected with *B. afzelii* were found at altitudes up to 1024 m, and tick populations reached up to 1250–1270m. Thus, the range of LB is apparently increasing in Europe. The prevalence of ticks infected with *B. burgdorferi* s.l. has also increased at some sites (Kampen et al., 2004), possibly due to changes in climate or wildlife management.

In the United States, LB is most common in the north-eastern and mid-Atlantic states and in the northern Midwest, with scattered foci in the south-eastern states and in California (Fig. 10.2). Scattered foci also exist in the Great Lakes region in southern Ontario and possibly other parts of Canada (Barker & Lindsay, 2000). *Borrelia burgdorferi* s.l. has been present in North America at least since the 1800s (Marshall et al., 1994). The increase and expanding range of LB in North America apparently results from a combination of factors: increasing populations of white-tailed deer (*Odocoileus virginianus*), an important host for adult black-legged ticks; habitat modifications that favour dissected second-growth woodlands (following movement of eastern farmers to the Midwest); and suburbanization that has produced excellent tick habitats and brought residents close to ticks (Spielman, Telford & Pollack, 1993). Genetic evidence suggests recent expansion of black-legged tick and *B. burgdorferi* populations in the north-eastern United States (Qiu et al., 2002). *Borrelia burgdorferi* s.s. is a generalist in the north-east, with individual genotypes infecting a variety of mammalian hosts, which may have contributed to its rapid expansion (Hanincová et al., 2006). Its range continues to expand – for example, with the spread of tick populations and LB cases in New Jersey and up the Hudson Valley of New York (White et al., 1991; Schulze, Jordan & Hung, 1998).

10.4.3. Epizootiology and epidemiology

LB is a sylvatic zoonosis. Ticks that are generally associated with temperate deciduous woodlands that include patches of dense vegetation with little air movement and high humidity carry the infective agent. LB is also associated with some coniferous forests, when conditions are suitable for the ixodid tick vectors (Ginsberg et al., 2004). In open habitats in Europe, such as meadows and moorland, the main source of blood-meals is usually livestock, such as sheep and cows. With increasing frequency, ticks also occur in domestic settings when a moist microhabitat is provided by high grass, gardens and rough forest edges. Foliage, decomposing organic matter and litter can give shelter to both ticks and small mammals that act as hosts for immature ticks. Therefore, contemporary trends of suburbanization can potentially increase exposure in the peridomestic environment. Vector ticks are frequently encountered in residential areas (Maupin et al., 1991), and they are also encountered by people recreationally or occupationally exposed to forest habitats (Ginsberg & Ewing, 1989; Rath et al., 1996).

Closed enzootic cycles that involve reservoir-competent hosts and host-specific ticks also have a role in maintaining LB in nature, and the spirochete can be transmitted to people when a bridge vector, such as the castor-bean tick, intrudes into the cycle. An example of this in Europe is the circulation of borreliae between the European hedgehog (*Erinaceus europaeus*) and the hedgehog tick, *Ixodes hexagonus* (Gern & Falco, 2000). Since the castor-bean tick frequently feeds on hedgehogs, the potential is there for the hedgehog tick/hedgehog cycle to have a considerable impact on the eco-epidemiology (the specific association between an ecosystem or habitat and the enzootic transmission chain of reservoir hosts and vectors living therein) of LB in some areas. The widespread recommendation to encourage hedgehogs to live in home gardens, by preparing piles of leaf litter, may therefore contribute to the currently seen so-called urbanization cycle. Also, urban sprawl and invasion of commensal and non-commensal rodents can influence LB epidemiology. Norway rats, *Rattus norwegicus*, and garden dormice, *Eliomys quercinus*, can carry vector ticks and borreliae and can contribute to the urbanization of LB (Matuschka et al., 1996; Richter et al., 2004).

In Europe, the castor-bean tick and the taiga tick serve as vectors to people, while the hedgehog tick transmits spirochetes among medium-sized mammals, and the seabird tick, *Ixodes uriae*, transmits *B. garinii* among seabirds. The prevalence of infection in nymphal sheep ticks averages 10.8% in Europe, with considerable variation among locales (Hubálek & Halouzka, 1998). In North America, the black-legged tick and western black-legged tick act as vectors to people, while *Ixodes dentatus*, *Ixodes spinipalpus* and other species serve as enzootic vectors to small animals, such as rabbits and wood rats (Eisen & Lane, 2002). The prevalence of infection in nymphal black-legged ticks varies from about 15% to 30% in endemic areas of the north-east (Piesman, 2002). A variety of other tick species, as well as some haematophagous insects, have been found to carry borreliae, but are most probably not involved in disease transmission. *B. burgdorferi* s.l. is transmitted transstadially by vector ticks, but transovarial transmission, while it occurs, is relatively rare. Besides these tick-specific transmission modes, a co-feeding effect has been described, in which uninfected ticks can acquire spirochetes while feeding near infected ticks on an uninfected host (Ogden, Nuttall & Randolph, 1997).

Compared with North America, important differences in the ecology of LB in Europe result from the greater diversity of *Borrelia* spp. that cause human disease in Europe. Table 10.6 provides an overview of known genospecies of *B. burgdorferi* s.l., their primary vectors and reservoir hosts, geographical distribution, and virulence in people. In North America, *B. burgdorferi* s.s. is responsible for the vast majority of human cases, while in Europe, *B. afzelii*, *B. garinii* and *B. valaisiana* are most common. The most important reservoir in North America is the white-footed mouse, *Peromyscus leucopus* (Mather et al., 1989), and other rodents can also serve as major reservoirs, including voles (such as the meadow vole, *Microtus pennsylvanicus*), chipmunks (such as the eastern chipmunk, *Tamias striatus*) and rats (such as the Norway rat) (Smith et al., 1993; Markowski et al., 1998). Some North American birds, such as the American robin and the song sparrow, *Melospiza melodia*, can also serve as reservoirs (Richter et al., 2000; Ginsberg et al., 2005). In Europe, on the other hand, different species of *Borrelia* are associated with different wild hosts. The primary reservoirs of *B. afzelii* are rodents, including mice (*Apodemus* spp.) and voles (*Clethrionomys* spp.) (Kurtenbach et al., 2002a; Hanincová et al., 2003a). In contrast, the primary reservoirs of *B. garinii* and *B. valaisiana* are birds, including pheasants and songbirds (Humair et al., 1998; Kurtenbach et al., 1998, 2002b; Hanincová et al., 2003b).

Reservoir competence varies among hosts. Lagomorphs, such as hares (*Lepus* spp.) and rabbits (*Oryctolagus* spp. and *Sylvilagus* spp.), show varying degrees of reservoir capacity. Similarly, carnivorous mammals, such as foxes (the red fox, *Vulpes vulpes*, for example), dogs (the domestic dog, *Canis familiaris*, for example) and cats (the domestic cat, *Felis domesticus*, for example), vary considerably in competence as reservoirs. Borreliae, however, are eliminated in ticks attached to some lizard species (Lane & Quistad, 1998), which apparently limits the importance of LB in areas where ground-dwelling lizards are abundant, such as south-eastern North America. In addition to their roles as reservoirs of some borreliae, many birds can serve as carriers of attached infected ticks when migrating (see Chapter 8). Ungulates (such as deer, sheep, cattle, goats and pigs) feed large numbers of mainly adult ticks in nature and may influence the epidemiology of LB, by increasing tick numbers (and thus the number of ticks per individual reservoir host), even if they themselves are not competent reservoirs.

10.5. TBE

10.5.1. Public health

TBE is caused by the TBE virus (TBEV), a member of the RNA virus family Flaviviridae. Three subtypes can be differentiated. One of them causes central European encephalitis (CEE); this virus subtype was first isolated in 1937, and the castor-bean tick is the main vector. The Siberian and far-eastern subtypes (endemic in eastern Europe and throughout northern Asia) are the causative agents of Russian spring-summer encephalitis (RSSE), which is responsible for a disease similar to CEE, but with a more severe clinical course. The primary vector of RSSE is the taiga tick. Transmission can also occur on an epidemic scale after consumption of raw milk from TBE-infected goats, sheep or cows. Person-to-person transmission has not been reported. However, vertical virus transmission from an infected mother to her fœtus has been described (Hubálek & Halouzka, 1996).

Table 10.6. Overview of known genospecies of *B. burgdorferi* s.l.

***Borrelia* genospecies**	**Literature**	**Geographical distribution**	***Ixodes* tick vector**	**Primary vertebrate host**	**Primary symptoms**
B. burgdorferi s.s.	Baranton et al. (1992)	North America, Europe, N. Africa	*I. scapularis* *I. dammini* *I. pacificus* *I. ricinus* *I. dentatus*	Rodents, insectivores	Arthritis, neuropathy
B. garinii	Baranton et al. (1992)	Worldwide	*I. ricinus* *I. persulcatus* *I. uriae* *I. hexagonus* *I. trianguliceps*	Passerine birds, pheasants	Neuropathy
B. afzelii	Canica et al. (1993)	Eurasia	*I. ricinus* *I. persulcatus* *I. nipponensis*	Rodents	Erythema migrans, skin lesions
B. japonica	Kawabata, Masuzawa & Yanagihara (1993)	Japan	*I. ovatus*	Not determined	—
B. andersonii	Marconi, Liveris & Schwartz (1995)	United States	*I. dentatus*	Cottontail rabbit	—
B. tanukii	Fukunaga et al. (1996)	Japan	*I. tanuki*	Not determined	—
B. turdi	Fukunaga et al. (1996)	Japan	*I. turdus*	Not determined	—
B. lusitaniae	Le Fleche et al. (1997)	Europe, North Africa	*I. ricinus*	Birds	Unclear
B. valaisiana	Wang et al. (1997)	Eurasia	*I. ricinus* *I. columnae* *I. granulatus*	Passerine birds, pheasants	Unclear
B. bissettii	Postic et al. (1998)	United States, Europe	*I. scapularis* *I. pacificus* *I. spinipalpis*	Not determined	Not determined
B. sinica	Masuzawa et al. (2001)	China	*I. ovatus*	Rodents	Unclear
B. spielmanii	Richter et al. (2006)	Central Europe	*I. ricinus*	Garden dormice	Erythema migrans

[a] Because not all *Ixodes* ticks have common names, only the scientific names are given here.

Note. —: no remarks.

Source: CDC (2006).

The incubation period of TBE is usually between 7 and 14 days (sometimes shorter with milk-borne transmission). A characteristic biphasic febrile illness occurs in about 30% of cases, with an initial phase that lasts 2–4 days, which corresponds to the viraemic phase. Symptoms are nonspecific and may include fever, malaise, anorexia, headache, muscle aches and nausea or vomiting (or both). After a remission phase of about 8 days, up to 25% of the patients develop an infection of the central nervous system with symptoms of meningitis (50%), encephalitis or meningoencephalitis (40%) and myelitis (10%). Case fatality rates are generally below 5% in European TBE, but they are up to 50% in some outbreaks of Asian subtypes (Nuttall & Labuda, 2005). In up to 40% of cases, convalescence can be prolonged by sequelae (known as post-encephalitic syndrome), and about 4% of the CEE cases produce a residual paresis (slight or partial motor paralysis).

Treatment depends on the symptoms and often requires hospitalization and intensive care. Anti-inflammatory drugs are sometimes utilized, and intubation and ventilatory support are sometimes necessary. Licensed vaccines (active and passive) that neutralize all three virus subtypes (Rendi-Wagner, 2005) are commercially available, with protection rates exceeding 98%.

10.5.1.1. Public health impact of TBE in Europe

TBE is the most frequent viral TBD in central Europe. Overall, several thousand clinical cases a year occur in Europe: mainly in the Russian Federation (5000–7000 cases a year), the Czech Republic (400–800 cases a year), Latvia (400–800 cases a year), Lithuania (100–400 cases a year), Slovenia (200–300 cases a year), Germany (200–400 cases a year) and Hungary (50–250 cases a year). In 1997, 10208 clinical cases of TBE (with 121 fatalities) were reported from all over Europe. In 2005, a sharp increase of 50% or more in notified clinical cases of TBE was seen in Switzerland (91 cases in 2004 versus 141 cases in 2005; weeks 1–33) (Anonymous, 2005b) and Germany (258 cases in 2004 versus 426 cases in 2005) (Anonymous, 2005b).

Since treatment of this potentially fatal disease depends on the symptoms, vaccination, prevention of infective tick-bite and pasteurization of contaminated milk constitute the first line of defense in preventing TBE. Due to the frequent need for hospitalization (often with intensive care), subsequent prolonged recovery time and neurotropic sequelae, the economic impact of this disease, in addition to its effect on health, is costly. As has been reported in Austria, vaccination programmes can substantially lower the annual incidence of TBE. Vaccination coverage of the Austrian population increased from 6% in 1980 to 86% in 2001, exceeding 90% in some hyperendemic areas (Kunz, 2003). This programme led to a steady decline in cases of TBE, drastically reducing the annual health impact for Austria to less than 10%. For example, in Carinthia, Austria, there were an average of 155 cases a year from 1973 to 1982, while from 1997 to 2001 there were only four cases a year (Kunz, 2003). In Hungary, 3–5% of the population were reported to be vaccinated, and in the southern Bohemia region of the Czech Republic it was 10% (WHO Regional Office for Europe, 2004). For other European countries, the vaccination status is unknown, but is probably low (Kunz, 2003).

10.5.2. Geographical distribution

The currently known geographical distribution of European TBE foci includes much of central and eastern Europe and extends broadly into Asia. Randolph (2001) predicted an eventual future decline in the distribution and incidence of TBE, due to global climate change, but currently both the geographical distribution and incidence of infection are increasing. Therefore, programmes that promote vaccination and prevention of tick bites are essential in highly affected areas. TBE has recently spread in a north-westerly direction from central Europe to western Germany and has moved north to Finland, Norway and Sweden, as well as to higher altitudes in mountainous areas in the Czech Republic (Hillyard, 1996). The north-westward spread of TBE might be explained by:

- the movement of wildlife, migrating birds and domestic animals together with their ticks across the continent;

- landscape changes, resulting from human activities; and

- the result of global warming.

Milder winter temperatures in particular have important effects on tick distribution and can foster shifts into higher latitudes and altitudes (Lindgren, Talleklint & Polfeldt, 2000).

10.5.3. Epizootiology and epidemiology

Ixodid ticks act as both the vector and reservoir for TBEV. This virus can chronically infect ticks and can be transmitted transstadially and transovarially. Small rodents are the main hosts, although viraemia has been reported from insectivores (representing an order of mammals whose members basically feed on insects and other arthropods), goats, sheep, cattle, canids (which include foxes, wolves, dogs, jackals and coyotes) and birds. People are an accidental host, and large mammals are feeding hosts for adult vector ticks, but do not play a significant role in maintaining the natural virus cycle. The infection rates in castor-bean ticks and taiga ticks in endemic foci usually vary from 0.1% to 5%, but can reach up to 10% in hyperendemic foci – for example, in Austria. The rate of infection increases steadily from the larval to the adult stage. Human TBE cases occur mainly during the highest period of vector tick activity, between April and November, peaking from mid-June to early August. Nevertheless, sheep ticks can be active at any temperatures above about 10 °C, even during winter. Thus sporadic clinical cases occur even during wintertime.

TBE is usually contracted in habitats suitable for the vector tick species and primary rodent reservoirs. These include mixed forest, pastoral and mountainous sylvan areas for castor-bean ticks and mixed taiga forest for taiga ticks. During recent years, man-made changes in natural areas have increased the periurban abundance of both tick species. This trend is associated with growing disease transmission, including a tendency towards urban transmission. Urban TBE transmission has been described in Europe and Asia – for example, in Novosibirsk, the Russian Federation (Hubalek & Halouzka, 1996).

Commensal rodents, cats and dogs are known to carry host-seeking ticks into human dwellings in periurban and urban areas. *Ixodes* ticks can survive for several hours and bite humans, but they do not persist in houses or stables.

TBE is most likely to be acquired in forests rich in small mammals, so forest workers, hunters and others highly exposed to this ecotope are at high risk. The seroprevalence of this virus in foresters can reach 12–16% in hyperendemic foci – for example, in Austria and Switzerland. In Germany, seroprevalence rates exceeding 20% have been found in foresters in the Emmendingen and Ludwigsburg counties (Kimmig, Oehme & Backe, 1998). TBE morbidity rates in the Czech Republic and Slovakia averaged 4.2 (1.4–9.9) deaths per 100 000 population between 1955 and 2000. In Switzerland (Thurgau canton) a morbidity rate of 5.4 people per 100 000 population was estimated for 1995. The highest morbidity in Germany was estimated for the federal state of Baden-Württemberg, with 1.1 cases per 100 000 population. In some cases, up to 76% of human TBE infections can result from consumption of raw milk, as was reported in Belarus (Ivanova, 1984).

10.6. RMSF

10.6.1. Public health

RMSF was first recognized in an epidemic in the Bitterroot Valley of Montana, in the United States, in the late 1800s. The etiological agent is *Rickettsia rickettsii*, and the primary vectors are the American dog tick in eastern and central North America and the Rocky Mountain wood tick in the Rocky Mountain region (Sonenshine, Lane & Nicholson, 2002). The number of cases reported to the CDC varies from about 200 to about 1200 a year, with an average incidence from 1985 to 2002 of between 0.24 to 0.32 cases per 100 000 population (Schriefer & Azad, 1994).

RMSF is characterized by the sudden onset of high fever, headache and myalgia, often with nausea and other symptoms (Macaluso & Azad, 2005). A few days after the onset of symptoms, a rash generally appears, beginning as macropapular eruptions on the ankles and wrists that then spread to the entire body, producing a so-called spotted appearance. The rickettsiae are intracellular parasites that affect (in particular) cells of the capillaries and arterioles. Symptoms are often severe, and though early treatment (generally with tetracyclines) is effective, the disease is fatal in around 5% of cases.

10.6.2. Geographical distribution

The distribution of human cases of RMSF, or at least the distribution of recognized cases, has shifted from the Rocky Mountain region in the late 1800s to eastern and central North America today. The incidence of the disease is currently highest in the south-eastern and south-central states (such as the Carolinas and Oklahoma), but cases are scattered throughout the eastern and central regions of North America (Fig. 10.3), with relatively few cases in the Rocky Mountain and western states (Groseclose et al. 2004; Macaluso & Azad, 2005).

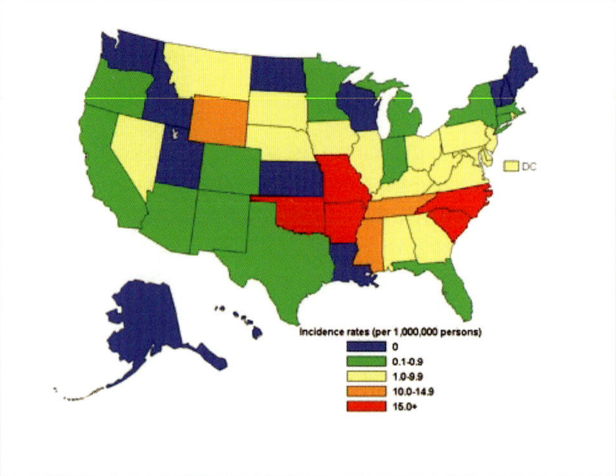

Fig. 10.3. Distribution of human cases of RMSF in the United States, 2002 *Source:* CDC (2006).

10.6.3. Epizootiology and epidemiology

RMSF is generally acquired in rural and suburban areas with woodland and associated open vegetation where the tick vectors are abundant (Sonenshine, Peters & Levy, 1972; Sonenshine, Lane & Nicholson, 2002). However, foci sometimes occur in appropriate habitats within large cities (Salgo et al., 1988). The pathogen is transmitted vertically in the tick (from mother to offspring) and is maintained transstadially, so the tick can act as both vector and reservoir. However, infection with nonpathogenic rickettsiae can interfere with transovarial transmission (Burgdorfer, Hayes & Mavros, 1981). Small mammals also can serve as reservoirs and apparently can contribute to amplification under appropriate circumstances, but occurrence of RMSF does not seem to depend on any particular vertebrate reservoir (Schriefer & Azad, 1994). Larvae and nymphs of American dog ticks and Rocky Mountain wood ticks attach to a variety of small and medium-sized mammals, including mice, voles, rats, ground squirrels, hares and rabbits, many of which can maintain infection with spotted fever group rickettsiae. Adults of these tick species generally attach to larger mammals, including human beings. Infection rates of adults vary considerably from site to site, ranging from less than 1% to about 10%.

10.7. Emerging TBDs

Several TBDs have recently been recognized in Europe and North America. Some of these might represent new introductions of the diseases to these continents, while others were undoubtedly already present, but were recognized recently because of the renewed attention to TBDs that resulted from the recent increase of LB. Also, some diseases that have been rare in the past are apparently expanding in range, along with expanding tick populations. Selected diseases that have recently been recognized in North America and Europe are discussed in this section.

10.7.1. Crimean-Congo haemorrhagic fever

Crimean-Congo haemorrhagic fever (CCHF) was first mentioned by the Tajik physician Abu-Ibrahim Djurdjani in the 12th century and has been extensively studied since the 1944/1945 epidemic in the Crimean Peninsula (Hubalek & Halouzka, 1996). This epidemic resulted in more than 200 human cases, with 10% of them fatal. The disease is caused by the CCHF virus (CCHFV), a *Nairovirus* (family Bunyaviridae) closely related to Dugbe and Nairobi sheep disease viruses and classified as a biosafety level-4 virus (the highest biological security level). The clinical course appears as a haemorrhagic fever with severe typhoid-like symptoms, including fever, chills, headache, myalgia, backache, anorexia, nausea, repeated vomiting, conjunctivitis, pharyngitis, bradycardia, meningitis and encephalitis. Haemorrhagic manifestations can vary from petechiae (pinpoint-sized haemorrhages of small capillaries in the skin) to large haematomas (solid swellings of clotted blood within tissues) on the mucous membranes and skin, and bleeding from the gums, nose and intestines and, less frequently, lungs and kidneys. Case fatality rates are usually between 8% and 30%, but may reach up to 50–60% in cases transmitted from person to person (Hubalek & Halouzka, 1996). Convalescence is slow, but usually without sequelae. Treatment of confirmed human cases requires barrier nursing and special hygienic care to prevent nosocomial infection.

Treatment usually depends on the symptoms, but treatment with ribavirin seems promising during the early stages of the disease (Ozkurt et al., 2006). An inactivated CCHF vaccine was administered to several hundred people in Bulgaria and Ukraine (Rostov oblast), but severe side-effects appeared. Specific immunoglobulins can also be used prophylactically or therapeutically. However, no licensed, safe vaccine is currently available.

CCHF is the most severe TBD in Europe and has the potential to spread quickly from person to person. The disease is probably underreported worldwide, so European and global incidences are unknown. Bulgaria, the southern part of the Russian Federation and Ukraine are among the most highly affected areas within Europe. Cases have also been reported from Bosnia and Herzegovina, Greece, Hungary, Montenegro, the Republic of Moldova, Serbia, and the former Yugoslav Republic of Macedonia. From 1952 to 1970, 865 cases of CCHF were recorded in Bulgaria alone, with a case fatality rate of 17%, and 6% of the cases of nosocomial origin (Vasilenko et al., 1971). In the Rostov region, 312 cases were registered between 1963 and 1969. Human cases sporadically occur in that region, with an outbreak occurring in 1999 (65 cases with 6 fatalities) (Onishchenko et al., 2000). The virus has been detected in almost all south-eastern districts of the Russian Federation, resulting in an additional regional budget of Rub 2.5 million (US$ 872 000) for diagnostic procedures and preventive measures (ProMED Mail, 2005). In 2002, eight cases clustered within families were observed in Albania (Papa et al., 2002). Although the overall incidence for Europe remains unclear, CCHF is a re-emerging disease with an estimated annual incidence far greater than100 cases, especially during outbreaks (Faulde et al., 2002).

The bont-legged tick, *Hyalomma marginatum*, is the principal vector and tick reservoir of CCHFV in Europe. Transstadial, transovarial and venereal transmission occur. This tick species inhabits pastoral steppe ecosystems, and the adult stage frequently feeds on sheep. CCHFV is highly contagious and transmission to people can occur by tick bite, by contact with infected animals (such as during sheep shearing and meat handling) and by person-to-person contact. Laboratory infections have also been reported.

10.7.2. Tick-borne rickettsioses

Several new human-pathogenic tick-borne rickettsioses of the spotted fever group have been reported from Europe during the last decade. Among them, *Rickettsia conorii* and *Rickettsia helvetica* are of greatest concern. *Rickettsia slovaca*, *Rickettsia aeschlimannii* and *Rickettsia mongolotimonae* are also endemic, although with very few human cases reported to date. Novel rickettsioses have recently been described in North America as well.

10.7.2.1. Boutonneuse fever

R. conorii is the causative agent of Boutonneuse fever (BF), also known as tick-borne typhus, Mediterranean spotted fever and South African tick bite fever. Patients usually present with fever, malaise, a generalized maculopapular erythematous rash and a typical black skin lesion, called *tache noir*, at the site of the infected-tick bite. While the disease is usually mild, severe forms, including encephalitis, occur occasionally. Overall, the case fatality rate in Europe is estimated to be less than 2.5%, even if untreated. Fever usually persists for a few days to two weeks, with a specific antibiotic treatment required for no more than two days. The seroprevalence rates in dogs, which are often infested with up to 100 adult brown dog ticks per animal, can be quite high in hyperendemic foci, varying between 35.5% in Italy and 93.3% in Portugal. The annual incidence rate in people has been estimated to be 48 cases per 100 000 population in Corsica, France, whereas 1000 cases have been reported annually from Portugal. Human seroprevalence rates can exceed 70% in hyperendemic foci in Spain (WHO Regional Office for Europe, 2004). However, the overall incidence of BF in Europe is unclear.

R. conorii is widely found in southern Europe and the Mediterranean countries. This disease is spreading northwards, reaching Belgium, Germany and the Netherlands, where antibodies were detected in dogs and people, and *R. conorii* has been isolated from sheep ticks and rodents in Belgium (Jardin, Giroud & LeRay, 1969; Gothe, 1999; WHO Regional Office for Europe, 2004).

The major tick vector of *R. conorii* in Europe is the brown dog tick. Other vectors include the castor-bean tick, the hedgehog tick, the marsh tick (also called the ornate cow tick), *Dermacentor reticulatus*, and the ornate sheep tick, *Dermacentor marginatus*. Besides vector ticks, the primary reservoirs are dogs, rabbits and rodents. Pet dogs can acquire infected ticks during family holidays, and they can carry *R. conorii* with them when they return home further north in Europe. Human infection with BF in urban areas, often in a person's own home, can be caused by skin or eye contamination from rickettsiae-infected dog ticks that are crushed while de-ticking infested dogs (Hillyard, 1996).

10.7.2.2. Rickettsia helvetica

First isolated in Switzerland in 1979, this agent was linked with human disease in 1999, when it was associated with two fatal Swedish cases of chronic perimyocarditis (Nilsson, Lindquist & Pahlson, 1999). *R. helvetica* is now known to have caused chronic interstitial inflammation and pericarditis in people in France, Sweden and Switzerland. A serosurvey of foresters conducted after seroconversion of a 37-year-old man in 1997 in eastern France revealed a seroprevalence rate of 9.2% (Fournier et al., 2000). The disease is transmitted by the castor-bean tick, and initial results show infection rates in ticks between 1.7% in Sweden and 8.2% in northern and central Italy (Nilsson et al., 1999; Beninati et al., 2002). Recent studies indicate that *R. helvetica* is widely distributed throughout Europe and might cause more clinical disease and (even) mortality than is currently recognized (WHO Regional Office for Europe, 2004).

10.7.2.3. HME

HME is caused by the rickettsial pathogen *Ehrlichia chaffeensis*. In North America, this pathogen exists in a tick–deer cycle, with the lone star tick serving as the primary vector (Ewing et al., 1995) and the white-tailed deer serving as the primary reservoir (Lockhart et al., 1997). Human cases are most common in the southern Midwest, with foci along the East Coast (Dawson et al., 2005). In 2001, 142 cases were reported in the United States; in 2002, 216 cases were reported; and in 2003, 321 cases were reported (CDC, 2003; Groseclose et al., 2004; Hopkins et al., 2005). *E. chaffeensis* has also been found to be endemic in Europe – in Belgium, the Czech Republic, Denmark, Greece, Italy and Sweden – but human cases of disease have not been described to date (WHO Regional Office for Europe, 2004; Oteo & Brouqui, 2005).

10.7.2.4. HGA

HGA is caused by the rickettsial pathogen *Anaplasma phagocytophilum* (formerly *Ehrlichia phagocytophila*). Patients present with an acute febrile illness, and most develop leukopenia or thrombocytopenia (or both), and elevated concentrations of C-reactive protein and transaminases, with occasional fatalities occurring. Treatment with tetracycline generally leads to full recovery. The pathogen was first isolated from ticks and people in northern Midwestern United States in the 1990s (Chen et al., 1994; Dumler et al., 2001). The black-legged tick is the vector in the United States, and its mammal hosts, especially the white-footed mouse, serve as reservoirs (Pancholi et al., 1995; Levin & Fish, 2001). The United States distribution includes the Atlantic coastal states, the northern Midwest and California (CDC, 2003; Maurin, Bakken & Dummler, 2003; Brown, Lane & Dennis, 2005). In 2001, 261 cases were reported to the CDC; in 2002, 511 cases were reported (CDC, 2003; Groseclose et al., 2004).

In Europe, HGA in people was first recognized in 1995, when serum antibodies against *A. phagocytophilum* were confirmed. In 1997, the first proven European case of human disease was reported from Slovenia. Through March 2003, about 65 patients with confirmed HGA (and several patients fulfilling criteria for probable HGA) had been reported in Europe (Strle, 2004). Seroprevalence rates in the WHO European Region range from 0% to 28%, and infection rates in adult castor-bean ticks (the recognized tick vector) range from 0% to more than 30%. The relatively high seroprevalence rates in people and the presence of *A. phagocytophilum* in vector ticks in many European countries are discordant with the rather low number of patients with proven HGA. This may be due to an inadequate awareness among European physicians and limited recording and reporting of the disease, or it may be due to the presence of nonpathogenic strains of *A. phagocytophilum* (Strle, 2004).

The castor-bean tick is probably the principal vector in Europe and the taiga tick in north-eastern Europe and Asia, although transmission studies have not been reported to date. HGA is known to cause febrile illness in several domestic animals, including sheep, goats, cattle and horses. A Swiss study stressed the importance of small mammals, with the bank vole, *Clethrionomys glareolus*, wood mouse, *Apodemus sylvaticus*, yellow-necked mouse, *Apodemus flavicollis*, and common shrew, *Sorex araneus*, as likely animal reservoirs in nature (Liz, 2002).

10.7.3. Babesiosis

Human babesiosis, first described in 1957, is a malaria-like illness caused by piroplasms (pear-shaped protozoan organisms that live in red blood cells of mammals), including *B. microti* in North America and *Babesia divergens* in Europe (Homer & Persing, 2005). The primary vectors are the black-legged tick in eastern North America and the castor-bean tick in Europe. Rodents, such as white-footed mice serve as reservoirs (Spielman, 1976; Spielman et al., 1979). Babesiosis is often mild and self-limiting, but can be severe and is undoubtedly underreported. Nevertheless, hundreds of cases have been reported in North America, and 29 in Europe (from England and France). In the United States, cases have been reported primarily in coastal areas of the north-eastern and mid-Atlantic states (Dammin et al., 1981; Spielman et al., 1985).

10.8. Ticks in human dwellings

In Europe, the brown dog tick can persist in long-term infestations of human dwellings with dogs. The European pigeon tick can also occur in dwellings with pigeon infestations or breeding. The fowl tick and *Ornithodoros erraticus* may also occur in houses close to poultry stables (*Argas* spp.) in south-east Europe and pig stables (*Ornithodoros* spp.) in Spain and Portugal.

Ticks found in human dwellings in North America are primarily soft ticks (of the genus *Ornithodoros*) associated with rodents that nest in buildings. The most important human disease transmitted by these ticks is tick-borne relapsing fever, which is caused by various species of the bacterial genus *Borrelia*. The most common pathogens in this group are *B. hermsi* (transmitted by *O. hermsi*) in mountainous areas of the western United States and Canada, and *B. turicatae* (transmitted by *O. turicata*) in desert and scrub habitats in the south-western United States and Mexico (Barbour, 2005). People generally encounter these pathogens recreationally, when occupying rustic cabins that are inhabited by tick-bearing rodents. Recently, specimens of the bat-associated soft tick, *Carios kelleyi* (collected from buildings in Iowa) were found to be infected with spotted fever group *Rickettsia*, relapsing fever group *Borrelia*, and *Bartonella henselae* (the etiological agent of cat scratch disease), but the role of these ticks as vectors of these bacterial pathogens has not been established (Loftis et al., 2005). Also, the brown dog tick can be found in homes, associated with dogs, but generally does not bite people.

10.9. Tick and tick-borne disease surveillance

TBDs that are reportable in the United States include LB, RMSF, HME, HGA, Q fever, and tularaemia (Hopkins et al. 2005). In Europe, where regulations differ among countries, only TBE is widely reportable.

Active surveillance for ticks or TBDs requires purposeful sampling of ticks or samples from wild or domestic hosts, or from people (Nicholson & Mather, 1996; Lindenmayer, Marshall & Onderdonk, 1991). Passive surveillance, on the other hand, utilizes information collected for other purposes, such as data collected from tick laboratories or hospital registries, to assess tick or disease distribution (White, 1993). Active surveillance tends to provide more accurate information, but is expensive and labour intensive. Passive surveillance is less expensive and requires less effort, and it can provide useful information of appropriate types, but the value of the results are sometimes limited by unidentifiable biases in data collection (Johnson et al., 2004). Most current tick surveillance programmes are of the passive type.

10.10. Tick and TBD management

Ticks are controlled for a variety of reasons, including nuisance prevention, commodity protection (to prevent cattle loss, for example) and protection against TBDs. This section briefly reviews tick control methods and then discusses IPM strategies that are appropriate for various purposes of tick control.

10.10.1. Self-protection

10.10.1.1. Avoidance

Ticks can be avoided by refraining from exposure to fields, forests and other hard tick-infested habitats, especially in known disease foci (Ginsberg & Stafford, 2005). Specific habitats to be avoided depend on tick distribution, which can differ for different species and for different stages of the same species. Use of clearly defined paths can help avoid contact with tick-infested vegetation. Bites of soft ticks can be prevented by avoiding old campsites, animal and poultry stables, and infested cabins and mud houses and by taking appropriate precautions when coming in contact with animals that are potentially infested with ticks.

10.10.1.2. Repellents

Effective repellents can prevent ticks from becoming attached to the body and can be applied to clothing or directly on the skin (some products are not labelled for use on skin). Effective skin repellents include N,N-diethyl-3-methylbenzamide (DEET), (N,N-butyl-N-acetyl)-aminopropionic acid-ethyl ester and 1-piperidinecarboxylic acid 2-(2-hydroxyethyl)-1-methylpropylester (picaridin). Depending on the active ingredient and formulation, skin repellents generally do not last longer than a few hours, because of absorption or abrasion.

10.10.1.3. Clothing

Individuals can protect themselves against tick attachment by tucking trousers into boots or socks and tucking shirts into trousers. Light-coloured clothing aids detection of dark-coloured ticks, which can be collected or removed with commercial tape. Most TBDs require a period of attachment (often several hours) before the pathogen is transmitted, so thorough body examination and prompt removal of attached ticks at the end of a day spent in tick-infested areas can minimize exposure to TBD agents.

10.10.1.4. Tick removal

Hard ticks should be removed by grasping the tick where the mouthparts are attached to the skin and then pulling it out slowly, but steadily (Needham, 1985); the use of pointed forceps is preferable, because it avoids contact with fingers and the tick's infective body fluid or excreta. The bite site should be cleansed with antiseptic before and after removal. Soft ticks withdraw their mouthparts when touched with a hot needle tip or when dabbed with chloroform, ether, alcohol or other anaesthetics (Gammons & Salam, 2002).

10.10.1.5. Clothing impregnation

A major advance in the protection of high-risk personnel, such as outdoor workers, hunters, travellers and soldiers, has been the development of residual insecticides that can impregnate clothing, tents and netting (WHO, 2001a, b). Permethrin, a synthetic pyrethroid insecticide, has been widely used for decades as an arthropod contact repellent in fabric impregnation, by spraying or soaking the fabric at final concentrations between 500 mg/m^2 and 1,300 mg/m^2 (Young & Evans, 1998; Faulde, Uedelhoven & Robbins, 2003). Recently, factory-based impregnation methods have been introduced, such as soaking the fabric or using a new polymer coating technique for impregnating clothing and battle dress uniforms. The polymer coating is safe, and the impregnation lasts the life of the fabric (Faulde & Uedelhoven, 2006). Ticks crawling up impregnated fabric quickly fall off. The benefits to people are the bites prevented and the acaricidal activity. This method can also be used to protect against other haematophagous arthropod vectors of public health importance.

10.10.1.6. Vaccination

Of tick-borne diseases endemic in Europe and North America, only TBE can be prevented by the use of a vaccine. TBE vaccination is widely neglected as a public health tool for disease prevention (Austria is an exception). A vaccine for preventing LB was briefly available in North America, but this was specific to *B. burgdorferi* s.s. and would not be efficacious in Europe, where diverse *Borrelia* spp. are associated with LB in people. The manufacturer removed the vaccine from production in 2002, and no vaccine against LB is currently available.

10.10.2. Habitat manipulation and urban design

Ticks have species-specific habitat requirements, often associated with habitats of hosts and the need to avoid desiccation. Therefore, habitats can be manipulated to make them unsuitable for ticks or to minimize encounters between ticks and people (Stafford, 2004). Suburban habitats associated with natural woodlands foster populations of black-legged ticks and castor-bean ticks, because these habitats are excellent for both immature and adult ticks and for vertebrate hosts suitable for all tick stages. Lawns that were cut short and were open to the sun had minimal numbers of deer ticks, while tick densities increased incrementally in gardens, wood edges and forests (Maupin et al., 1991). Therefore, maintaining a short-clipped lawn and establishing barriers to prevent access to the woods can minimize human exposure to ticks in this environment. Mowing and burning vegetation in natural areas lowers tick numbers temporarily, but ticks reinfest treated areas as the vegetation grows back (Wilson, 1986).

Most ticks that are important to human health are rare in highly urbanized environments, but parks with natural patches and appropriate host species, and natural habitats interspersed with human dwellings in suburban areas, foster encounters between ticks and people. These encounters can be minimized with appropriate design features, such as barriers between areas frequently used by people and natural patches, and pathways constructed through natural sites (boardwalks, for example). Medical entomologists and natural resource experts should be consulted, so that urban design appropriate for the local tick species of concern can be incorporated into the planning process. Unfortunately, in the past, TBDs have rarely been considered in urban or suburban design.

10.10.3. Host-centred methods

Domestic animals can be vaccinated to minimize tick attachment (de la Fuente, Rodriguez & Garcia-Garcia, 2000) or to protect them against TBDs (Kocan et al., 2001). House pets, especially dogs, are commonly vaccinated against LB in the United States. Vaccination of wild reservoir species of animals (Tsao et al., 2001) could theoretically interrupt enzootic transmission cycles of tick-borne zoonoses and, in field trials, it has reduced the prevalence of Lyme spirochetes in questing ticks (Tsao et al., 2004), but this approach has not yet been applied to manage the risk of disease.

Manipulation of host populations can also lower tick populations. Excluding deer can lower populations of deer ticks, and deer-proof fencing can contribute to a tick management programme (Daniels, Fish & Schwartz, 1993). Although lowering deer populations by hunting can also lower tick numbers, this approach is not generally practical, because deer populations must be reduced to extremely low levels to have a reliable effect on the transmission of LB (Ginsberg & Stafford, 2005).

10.10.4. Biological control

Ticks have numerous natural enemies, including predators, parasites and pathogens. In the northern hemisphere, predators are generally not specific to ticks. In contrast, wasps of the genus *Ixodiphagus* parasitize ticks, and the most widespread species, *Ixodiphagus hookeri*, has been studied as a possible biocontrol agent. This species was released on an island off the New England coast in the early 1900s, resulting in establishment of the wasp, but no tick control. Inundative releases have shown some promise of efficacy in agricultural settings (Mwangi et al., 1997), and theoretical analyses suggest that with additional research and development widespread releases might eventually be effective in

North America (Knipling & Steelman, 2000). However, considerable problems remain to be overcome before this approach becomes practical.

Numerous pathogens attack ticks, including bacteria, fungi, and nematodes (Samish, Ginsberg & Glazer, 2004). At present, one of the best candidates for tick biocontrol is the entomopathogenic fungus, *Metarhizium anisopliae* (Zhioua et al., 1997; Samish et al., 2001). Preliminary field trials have had modest results; but enhanced tick mortality, from the use of an oil-based carrier solution, compared with a water-based solution (Kaaya & Hassan, 2000), suggests that improved formulations may provide effective control. The pathogens that affect ticks typically also affect other arthropods (Ginsberg et al., 2002), so effects on non-target arthropods must be considered in application strategies of biocontrol materials.

10.10.5. Pesticide applications

Numerous pesticides are effective against ticks, and they are widely used to control ticks and TBDs. Acaricides can be broadcast for area control of ticks or can be targeted at host animals used by the ticks. Broadcast applications have the advantage that they can rapidly lower tick numbers, but timing, chemical distribution and formulation can profoundly influence the effectiveness of treatment. For example, broadcast applications for controlling nymphal deer ticks (the primary vector stage of LB in North America) need to penetrate the leaf litter where the nymphs dwell, while other ticks are better targeted by area sprays. Schulze, Jordan & Hung (2000) found that granular formulations of carbaryl effectively controlled deer tick nymphs (which quest down in the leaf litter where the heavy granules were deposited), but they did not control lone star tick nymphs (which quest up in the shrub layer). Also, most materials used for tick control are broadly toxic to arthropods, so broadcast applications can have substantial effects on non-target species (Ginsberg, 1994). Pesticide applications that are carefully targeted can help minimize these non-target effects. Application concentrations for tick control vary with materials and formulations, but examples of label application concentrations include: carbaryl (43% by weight, 0.17–0.34 g/m^2); cyfluthrin (11.8% by weight, 0.04–0.065 ml/m^2); and permethrin (36.8% by weight, 0.12–0.24 g/m^2).

Pesticides can be targeted at host animals by attracting the hosts (using feed, nesting materials or other attractants) to devices that apply the pesticide to them. Examples include bait boxes, permethrin-treated cotton balls and so-called four-poster devices (Stafford & Kitron, 2002). Four-poster devices, which coat the heads and necks of animals with a pesticide that kills the ticks, have the advantage of well-targeted applications, allowing far lower amounts of pesticide to be applied than in broadcast applications. The effectiveness of the approach taken tends to depend on ecological conditions at the application site. These methods can be important tools in IPM programmes, especially when integrated with other management techniques appropriate for local conditions of tick distribution and transmission dynamics.

Permanent infestations in houses and stables – for example by the brown dog tick or the pigeon tick – require the professional use of acaricides. Governmental European authorities – for example, in Germany – recommend the use of formulations that contain popoxur (1% by weight, 200 ml/m2) and diazinon (2% by weight, 50–100 ml/m2) (Anonymous, 2000). Besides the application of acaricides, dog hosts have to be treated with tick-repelling or -controlling spot-on or dipping formulations, and construction modifications of infested houses and stables are needed to prevent further infestations of pigeons, which are natural hosts of pigeon ticks (Uspensky & Ioffe-Uspensky, 2002).

10.11. IPM

IPM is an approach to the management of arthropod pests that fosters the integration of various pest control methods, so as to minimize reliance on individual environmentally damaging approaches and to provide sustained management of pest populations. IPM was developed for agriculture, where decisions are based on cost–benefit analyses that compare the cost of control with the economic value of crops protected. For vector-borne diseases, decisions are more appropriately based on cost–effectiveness (or cost–efficiency) analyses that integrate management methods, so as to prevent the greatest number of possible human cases of disease at a given cost (Phillips, Mills & Dye, 1993; Ginsberg & Stafford, 2005). Efficient management of TBDs maximizes the number of human cases prevented with available resources and minimizes dependence on broad-spectrum approaches to control that tend to be environmentally damaging. However, these analyses require information from field trials of various management methods and from models of transmission dynamics that use each potential combination of techniques to estimate the costs and the number of cases prevented (Mount, Haile & Daniels, 1997). Given the many tick control techniques currently available and the numerous novel techniques being developed, it is important to develop the theory and practice of efficient integration of methods, so that these techniques can be applied in such a manner as to most effectively prevent human disease.

10.12. Conclusions

The following conclusions can be drawn about public activities, surveillance and management, and research.

10.12.1. Public activities

Conclusions that relate to public activities cover three areas, as follows.

1. Accurate and practical information about ticks and TBDs should be made readily and widely available to health professionals, pest management professionals and the general public. Printed and online information about the effects on health, personal protection and preventive measures would be especially useful, as would information on tick biology and behaviour and on effective control strategies.

2. Specific education and health promotion programmes should be provided for people

with occupational and recreational exposure to ticks and TBDs. These programmes should emphasize the threat of ticks, TBDs of public health importance, personal protection measures against tick bites, tick avoidance, tick removal, available control measures and medical follow-up in case of exposure.

3. Development and design of human residential and recreational areas should routinely consider TBDs as part of the planning effort. Public health experts (including specialists on TBDs) should be consulted early in the planning process.

10.12.2. Surveillance and management

Conclusions that relate to surveillance and management cover two areas, as follows.

1. Reporting programmes should be developed for major endemic TBDs, where these currently do not exist. These programmes can include passive or active disease surveillance, or both.

2. Management programmes should be implemented for TBDs. Such programmes should efficiently target the sites where encounter rates between people and infected ticks are greatest. Surveillance and specific public education should be part of these programmes.

10.12.3. Research

Conclusions that relate to research cover three main areas, as follows.

1. Research is needed on new, emerging, and resurging TBDs, including: epidemiology, vector biology, disease-transmission competence of potential vector and reservoir species, transmission dynamics and geographical distribution; and anthropogenic, environmental, and climatic factors that affect emergence, re-emergence and geographical spread of ticks and TBDs.

2. Research is also needed on principles and strategies of tick and TBD management, including least toxic approaches, strategies for well-targeted integrated tick management and optimal approaches in urban, periurban and rural areas, especially in hyperendemic disease foci.

3. Research should be encouraged and carried out on new vaccination strategies, chemoprophylaxis and treatment regimens for TBDs of public health importance.

References[1]

Ammon A (2001). Risikofaktoren für Lyme-Borreliose: Ergebnisse einer Studie in einem Brandenburger Landkreis. *Robert-Koch-Institut Epidemiologisches Bulletin*, 21:147–149.

Anonymous (2000). Bekanntmachung der geprüften und anerkannten Mittel und Verfahren zur Bekämpfung von tierischen Schädlingen nach § 10c Bundes-Seuchengesetz. *Bundesgesundheitsblatt, Gesundheitsforschung, Gesundheitsschutz*, 43 (Suppl. 2):S61–S74.

Anonymous (2005a). Neuerkrankungen an Lyme-Borreliose im Jahr 2004. *Robert-Koch-Institut Epidemiologisches Bulletin*, 32:285–288

Anonymous (2005b). Aktuelle Statistik meldepflichtiger Infektionskrankheiten. *Epidemiologisches Bulletin*, 48:453–456.

Anonymous (2005c). Zeckenenzephalitis (FSME): deutliche Zunahme der gemeldeten Fälle. *Schweizerisches Bundesamt für Gesundheit Bulletin*, 38:671–673 (http://www.bag.admin.ch/themen/medizin/00682/00684/01114/index.html?lang=de, accessed 5 January 2007).

Baranton G et al. (1992). Delineation of *Borrelia burgdorferi* sensu stricto, *Borrelia garinii* sp. nov., and group VS461 associated with Lyme borreliosis. *International Journal of Systematic Bacteriology*, 42:378–383.

Barbour AG (2005). Relapsing fever. In: Goodman JL, Dennis DT, Sonenshine DE, eds. *Tick-borne diseases of humans*. Washington, DC, ASM Press: 268–291.

Barker IK, Lindsay LR (2000). Lyme borreliosis in Ontario: determining the risks. *Canadian Medical Association Journal*, 162:1573–1574.

Barker SC, Murrell A (2004). Systematics and evolution of ticks with a list of valid genus and species names. *Parasitology*, 129:S15–S36.

Beninati T et al. (2002). First detection of spotted fever group rickettsiae in *Ixodes ricinus* from Italy. *Emerging Infectious Diseases*, 8:983–986.

Brown RN, Lane RS, Dennis DT (2005). Geographic distributions of tick-borne diseases and their vectors. In: Goodman JL, Dennis DT, Sonenshine, DE, eds. *Tick-borne diseases of humans*. Washington, DC, ASM Press: 363–391.

[1] Sources cited in this review are nearly all from peer-reviewed scientific literature. Some CDC and WHO reports, several review articles and book chapters, and some recent information (on epidemiological trends, for example) from presentations at scientific conferences and from web sites of broadly recognized organizations (such as ProMED) are also cited.

Burgdorfer W, Hayes SF, Mavros AJ (1981). Nonpathogenic rickettsiae in *Dermacentor andersoni*: a limiting factor for the distribution of *Rickettsia rickettsii*. In: Burgdorfer W, Anacker RL, eds. *Rickettsiae and rickettsial diseases.* New York, Academic Press: 585–594.

Canica MM et al. (1993). Monoclonal antibodies for identification of *Borrelia afzelii* sp. nov. associated with cutaneous manifestations of Lyme borreliosis. *Scandinavian Journal of Infectious Diseases*, 25:441–448.

CDC (2003). Summary of notifiable diseases – United States, 2001. *Morbidity and Mortality Weekly Report*, 50:1–108.

CDC (2006). Lyme disease: introduction and global distribution [web site]. Atlanta, Georgia, Centers for Disease Control and Prevention (http://www.cdc.gov/ncidod/dvbid/lyme/who_cc/index.htm#over, accessed 24 October 2006).

Chen SM et al. (1994). Identification of a granulocytotropic *Ehrlichia* species as the etiologic agent of human disease. *Journal of Clinical Microbiology*, 32:589–595.

Collares-Pereira M et al. (2004). First isolation of *Borrelia lusitaniae* from a human patient. *Journal of Clinical Microbiology*, 42:1316–1318.

Dammin GJ et al. (1981). The rising incidence of clinical *Babesia microti* infection. *Human Pathology*, 12:398–400.

Danielova V et al. (2006). Extension of *Ixodes ricinus* ticks and agents of tick-borne diseases to mountain areas in the Czech Republic. *International Journal of Medical Microbiology*, 296 (Suppl. 40):48–53.

Daniels TJ, Fish D, Schwartz I (1993). Reduced abundance of *Ixodes scapularis* (Acari: Ixodidae) and Lyme disease risk by deer exclusion. *Journal of Medical Entomology*, 30:1043–1049.

Dautel H, Scheurer S, Kahl O (1999). The pigeon tick (*Argas reflexus*): its biology, ecology, and epidemiological aspects. *Zentralblatt für Bakteriologie: International Journal of Medical Microbiology*, 289:745–753.

Dawson JE et al. (2005). Human monocytotropic ehrlichiosis. In: Goodman JL, Dennis DT, Sonenshine DE, eds. *Tick-borne diseases of humans*. Washington, DC, ASM Press: 239–257.

de la Fuente J, Rodriguez M, Garcia-Garcia JC (2000). Immunological control of ticks through vaccination with *Boophilus microplus* gut antigens. *Annals of the New York Academy of Sciences*, 916:617–621.

Dennis DT, Hayes EB (2002). Epidemiology of Lyme borreliosis. In: Gray JS et al., eds. *Lyme borreliosis: biology, epidemiology and control*. Oxon, CABI Publishing: 251–280.

Dennis DT et al. (1998). Reported distribution of *Ixodes scapularis* and *Ixodes pacificus* (Acari: Ixodidae) in the United States. *Journal of Medical Entomology*, 35:629–638.

Dumler JS et al. (2001). Reorganization of genera in the families Rickettsiaceae and Anaplasmataceae in the order Rickettsiales: unification of some species of *Ehrlichia* with *Anaplasma*, *Cowdria* with *Ehrlichia*, and *Ehrlichia* with *Neorickettsia*, descriptions of six new species combinations and designation of *Ehrlichia equi* and 'HGE agent' subjective synonyms of *Ehrlichia phagocytophila*. *International Journal of Systematic and Evolutionary Microbiology*, 51:2145–2165.

Ebel GD, Spielman A, Telford SR 3rd (2001). Phylogeny of North American Powassan virus. *The Journal of General Virology,* 82:1657–1665.

Eisen L, Lane RS (2002). Vectors of *Borrelia burgdorferi* sensu lato. In: Gray JS et al., eds. *Lyme borreliosis: biology, epidemiology and control*. Oxon, CABI Publishing: 91–115.

Ewing SA et al. (1995). Experimental transmission of *Ehrlichia chaffeensis* (Rickettsiales: Ehrlichieae) among white-tailed deer by *Amblyomma americanum* (Acari: Ixodidae). *Journal of Medical Entomology*, 32:368–374.

Faulde MK, Uedelhoven WM, Robbins RG (2003). Contact toxicity and residual activity of different permethrin-based fabric impregnation methods for *Aedes aegypti* (Diptera: Culicidae), *Ixodes ricinus* (Acari: Ixodidae), and *Lepisma saccharina* (Thysanura: Lepismatidae). *Journal of Medical Entomology*, 40:935–941.

Faulde M, Uedelhoven W (2006). A new clothing impregnation method for personal protection against ticks and biting insects. *International Journal of Medical Microbiology*, 296 (Suppl. 40):225–229.

Faulde M et al. (2002). Tiere als Vektoren und Reservoire von Erregern importierter lebensbedrohender Infektionskrankheiten des Menschen. *Bundesgesundheitsblatt, Gesundheitsforschung, Gesundheitsschutz*, 45:139–151.

Fingerle V et al. (2005). Detection of the new *Borrelia burgdorferi* s.l. genospecies A14S from patient material and ticks. In: *Programme and Compendium of Abstracts, VIII International Potsdam Symposium on Tick-borne Diseases*, Jena, Germany, 10–12 March 2005.

Fournier PE et al. (2000). Evidence of *Rickettsia helvetica* infection in humans, eastern France. *Emerging Infectious Diseases*, 6:389–392.

Fukunaga M et al. (1996). *Borrelia tanukii* sp. nov. and *Borrelia turdae* sp. nov. found from ixodid ticks in Japan: rapid species identification by 16S rRNA gene-targeted PCR analysis. *Microbiology and Immunology*, 40:877–881.

Gammons M, Salam G (2002). Tick removal. *American Family Physician*, 66:643–645.

Gern L, Falco RC (2000). Lyme disease. *Revues Scientifique et Technique de l'Office Internationale des Epizooties,* 19:121–135.

Ginsberg HS, ed. (1993). *Ecology and environmental management of Lyme disease*. New Brunswick, NJ, Rutgers University Press.

Ginsberg HS (1994). Lyme disease and conservation. *Conservation Biology*, 8:343–353.

Ginsberg HS, Ewing CP (1989). Habitat distribution of *Ixodes dammini* (Acari: Ixodidae) and Lyme disease spirochetes on Fire Island, New York. *Journal of Medical Entomology*, 26:183–189.

Ginsberg HS, Stafford KC 3rd (2005). Management of ticks and tick-borne diseases. In: Goodman JL, Dennis DT, Sonenshine DE, eds. *Tick-borne diseases of humans*. Washington, DC, ASM Press: 65–86.

Ginsberg HS et al. (2002). Potential nontarget effects of *Metarhizium anisopliae* (Deuteromycetes) used for biological control of ticks (Acari: Ixodidae). *Environmental Entomology*, 31:1191–1196.

Ginsberg HS et al. (2004). Woodland type and spatial distribution of nymphal *Ixodes scapularis* (Acari: Ixodidae). *Environmental Entomology*, 33:1266–1273.

Ginsberg HS et al. (2005). Reservoir competence of native North American birds for the Lyme disease spirochete, *Borrelia burgdorferi*. *Journal of Medical Entomology*, 42:445–449.

Golubic D, Zember S (2001). Dual infection: tularemia and Lyme borreliosis acquired by single tick bite in northwest Croatia. *Acta Medica Croatica*, 55:207–209.

Gothe R (1999). *Rhipicephalus sanguineus* (Ixodidae): Häufigkeit der Infestation und der vektoriell an diese Zeckenart gebundene Ehrlichien-Infektionen bei Hunden in Deutschland; eine epidemologische Studie und Betrachtung [*Rhipicephalus sanguineus* (Ixodidae): frequency of infestations and ehrlichial infections transmitted by this tick species in dogs in Germany: an epidemiological study and consideration]. *Wiener Tierärztliche Monatsschrift*, 86:49–56 (in German).

Gray JS et al., eds (2002). *Lyme borreliosis: biology, epidemiology, and control*. Oxon, CABI Publishing.

Groseclose SL et al. (2004). Summary of notifiable diseases – United States, 2002. *Morbidity and Mortality Weekly Report*, 51:1–84.

Hanincová K et al. (2003a). Association of *Borrelia afzelii* with rodents in Europe. *Parasitology*, 126:11–20.

Hanincová K et al. (2003b). Association of *Borrelia garinii* and *B. valaisiana* with songbirds in Slovakia. *Applied and Environmental Microbiology*, 69:2825–2830.

Hanincová K et al. (2006). Epidemic spread of Lyme borreliosis, northeastern United States. *Emerging Infectious Diseases*, 12:604–611.

Hillyard PD (1996). *Ticks of north-west Europe*. Shrewsbury, Field Studies Council.

Homer MJ, Persing DH (2005). Human babesiosis. In: Goodman JL, Dennis DT, Sonenshine DE, eds. *Tick-borne diseases of humans*. Washington, DC, ASM Press: 343–360.

Hopkins RS et al. (2005). Summary of notifiable diseases – United States, 2003. *Morbidity and Mortality Weekly Report,* 52:1–85.

Hubalek Z, Halouzka J (1996). Arthropod-borne viruses of vertebrates in Europe. *Acta Scientiarum Naturalium Academiae Scientiarum Bohemicae Brno*, 30:10–22.

Hubalek Z, Halouzka J (1998). Prevalence rates of *Borrelia burgdorferi* sensu lato in host-seeking *Ixodes ricinus* ticks in Europe. *Parasitology Research*, 84:167–172.

Humair PF et al. (1998). An avian reservoir (*Turdus merula*) of the Lyme borreliosis spirochete. *Zentralblatt für Bakteriologie*, 287:521–538.

Ivanova LM (1984). [Current epidemiology of natural focus infections in the RSFSR]. *Medicinskaia Parazitologiia i Parazitarnye Bolezni*, 62:17–21 (in Russian).

Jardin J, Giroud P, LeRay D (1969). Presence de rickettsies chez *Ixodes ricinus* en Belgique. In: Evans GO, ed. *Proceedings of the 2nd International Congress of Acarology*, Sutton Bonington, England, 19–25 July1967. Budapest, Akademiai Kiado: 615–617.

Johnson JL et al. (2004). Passive tick surveillance, dog seropositivity, and incidence of human Lyme disease. *Vector Borne and Zoonotic Diseases*, 4:137–142.

Jongejan F, Uilenberg G (2004). The global importance of ticks. *Parasitology*, 129:S3–S14.

Kaaya GP, Hassan S (2000). Entomogenous fungi as promising biopesticides for tick control. *Experimental and Applied Acarology*, 24:913–926.

Kampen H et al. (2004). Substantial rise in the prevalence of Lyme borreliosis spirochetes in a region of western Germany over a 10-year period. *Applied and Environmental Microbiology*, 70:1576–1582.

Kawabata H, Masuzawa T, Yanagihara Y (1993). Genomic analysis of *Borrelia japonica* sp. nov. isolated from *Ixodes ovatus* in Japan. *Microbiology and Immunology*, 37:843–848.

Kimmig P, Oehme R, Backe H (1998). Epidemiologie der Frühsommer-Meningoenzephalitis (FSME) und Lyme-Borreliose in Südwestdeutschland. *Ellipse*, 14:95–105.

Knipling EF, Steelman CD (2000). Feasibility of controlling *Ixodes scapularis* ticks (Acari: Ixodidae), the vectors of Lyme disease, by parasitoid augmentation. *Journal of Medical Entomology*, 37:645–652.

Kocan KM et al. (2001). Immunization of cattle with *Anaplasma marginale* derived from tick cell culture. *Veterinary Parasitology*, 102:151–161.

Kriz B et al. (2004). Socio-economic conditions and other anthropogenic factors influencing tick-borne encephalitis incidence in the Czech Republic. *International Journal of Medical Microbiology*, 293 (Suppl. 37):63–68.

Kunz C (2003). TBE vaccination and the Austrian experience. *Vaccine*, 21 (Suppl. 1):S50–S55.

Kurtenbach K et al. (1998). Differential transmission of the genospecies of *Borrelia burgdorferi* sensu lato by game birds and small rodents in England. *Applied and Environmental Microbiology*, 64:1169–1174.

Kurtenbach K et al. (2002a). *Borrelia burgdorferi* sensu lato in the vertebrate host. In: Gray JS et al., eds. *Lyme borreliosis: biology, epidemiology and control.* Oxon, CABI Publishing: 117–148.

Kurtenbach K et al. (2002b). Differential survival of Lyme borreliosis spirochetes in ticks that feed on birds. *Infection and Immunity*, 70:5893–5895.

Lane RS, Quistad GB (1998). Borreliacidal factor in the blood of the western fence lizard (*Sceloporus occidentalis*). *The Journal of Parasitology*, 84:29–34.

Le Fleche A et al. (1997). Characterization of *Borrelia lusitaniae* sp. nov. by 16S ribosomal DNA sequence analysis. *International Journal of Systematic Bacteriology*, 47: 921–925.

Levin ML, Fish D (2001). Interference between the agents of Lyme disease and human granulocytic ehrlichiosis in a natural reservoir host. *Vector Borne and Zoonotic Diseases*, 1:139–148.

Lindenmayer JM, Marshall D, Onderdonk AB (1991). Dogs as sentinels for Lyme disease in Massachusetts. *American Journal of Public Health*, 81:1448–1455.

Lindgren E, Talleklint L, Polfeldt T (2000). Impact of climatic change on the northern latitude limit and population density of the disease-transmitting European tick *Ixodes ricinus*. *Environmental Health Perspectives*, 108:119–123.

Liz JS (2002). Ehrlichiosis in *Ixodes ricinus* and wild mammals. *International Journal of Medical Microbiology*, 291 (Suppl. 33):104–105.

Lockhart JM et al. (1997). Isolation of *Ehrlichia chaffeensis* from wild white-tailed deer (*Odocoileus virginianus*) confirms their role as natural reservoir hosts. *Journal of Clinical Microbiology*, 35:1681–1686.

Loftis AD et al. (2005). Detection of *Rickettsia*, *Borrelia*, and *Bartonella* in *Carios kelleyi* (Acari: Argasidae). *Journal of Medical Entomology*, 42:473–480.

Macaluso KR, Azad AF (2005). Rocky Mountain spotted fever and other spotted fever group rickettsioses. In: Goodman JL, Dennis DT, Sonenshine DE, eds. *Tick-borne diseases of humans*. Washington, DC, ASM Press: 292–301.

Marconi RT, Liveris D, Schwartz I (1995). Identification of novel insertion elements, restriction fragment length polymorphism patterns, and discontinuous 23S rRNA in Lyme disease spirochetes: phylogenetic analyses of rRNA genes and their intergenic spacers in *Borrelia japonica* sp. nov. and genomic group 21038 (*Borrelia andersonii* sp. nov.) isolates. *Journal of Clinical Microbiology*, 33:2427–2434.

Markowski D et al. (1998). Reservoir competence of the meadow vole (Rodentia: Cricetidae) for the Lyme disease spirochete, *Borrelia burgdorferi*. *Journal of Medical Entomology*, 35:804–808.

Marshall WF 3rd et al. (1994). Detection of *Borrelia burgdorferi* DNA in museum specimens of *Peromyscus leucopus*. *The Journal of Infectious Diseases*, 170:1027–1032.

Masuzawa T et al. (2001). *Borrelia sinica* sp. nov., a Lyme disease-related *Borrelia* species isolated in China. *International Journal of Systematic and Evolutionary Microbiology*, 51:1817–1824.

Mather TN et al. (1989). Comparing the relative potential of rodents as reservoirs of the Lyme disease spirochete (*Borrelia burgdorferi*). *American Journal of Epidemiology*, 130:143–150.

Matuschka FR et al. (1996). Risk of urban Lyme disease enhanced by the presence of rats. *The Journal of Infectious Diseases*, 174:1108–1111.

Maupin GO et al. (1991). Landscape ecology of Lyme disease in a residential area of Westchester County, New York. *American Journal of Epidemiology*, 133:1105–1113.

Maurin M, Bakken JS, Dummler JS (2003). Antibiotic susceptibilities of *Anaplasma (Ehrlichia) phagocytophilum* strains from various geographic areas in the United States. *Antimicrobial Agents and Chemotherapy*, 47:413–415.

McLean RG et al. (1981). The ecology of Colorado tick fever in Rocky Mountain National Park in 1974. I. Objectives, study design, and summary of principal findings. *The American Journal of Tropical Medicine and Hygiene*, 30:483–489.

McQuiston JH, Childs JE (2002). Q fever in humans and animals in the United States. *Vector Borne and Zoonotic Diseases* 2:179–191.

Meek JI et al. (1996). Underreporting of Lyme disease by Connecticut physicians, 1992. *Journal of Public Health Management and Practice*, 2:61–65.

Mehnert WA (2004). Erkrankungen an Lyme-Borreliose in den sechs östlichen Bundesländern in den Jahren 2002 und 2003. *Robert Koch-Institut Epidemiologisches Bulletin*, 28:219–222.

Meltzer MI, Dennis DT, Orloski KA (1999). The cost effectiveness of vaccinating against Lyme disease. *Emerging Infectious Diseases*, 5:321–328.

Mount GA, Haile DG, Daniels E (1997). Simulation of management strategies for the blacklegged tick (Acari: Ixodidae) and the Lyme disease spirochete, *Borrelia burgdorferi*. *Journal of Medical Entomology*, 34:672–683.

Mwangi EN et al. (1997). The impact of *Ixodiphagus hookeri*, a tick parasitoid, on *Amblyomma variegatum* (Acari: Ixodidae) in a field trial in Kenya. *Experimental & Applied Acarology*, 21:117–126.

Needham GR (1985). Evaluation of five popular methods for tick removal. *Pediatrics*, 75:997–1002.

Nicholson MC, Mather TN (1996). Methods for evaluating Lyme disease risks using Geographic Information Systems and geospatial analysis. *Journal of Medical Entomology*, 33:711–720.

Nilsson K, Lindquist O, Pahlson C (1999). Association of *Rickettsia helvetica* with chronic perimyocarditis in sudden cardiac death. *Lancet*, 354:1169–1173.

Nilsson K et al. (1999). *Rickettsia helvetica* in *Ixodes ricinus* ticks in Sweden. *Journal of Clinical Microbiology*, 37:400–403.

NPMA (2006). PestWorld for Kids: Rocky Mountain spotted fever: 2002 cases [web site]. Fairfax, Virginia, National Pest Management Association, Inc. (http://www.pestworld-forkids.org/images/RMSFever_map.gif, accessed 24 October 2006).

Nuttall PA, Labuda M (2005). Tick-borne encephalitis. In: Goodman JL, Dennis DT, Sonenshine DE, eds. *Tick-borne diseases of humans.* Washington, DC, ASM Press: 130–163.

O'Connell S et al. (1998). Epidemiology of European Lyme borreliosis. *Zentralblatt für Bakteriologie*, 287:229–240.

Ogden NH, Nuttall PA, Randolph SE (1997). Natural Lyme disease cycles maintained via sheep by co-feeding ticks. *Parasitology*, 115:591–599.

Onishchenko GG et al. (2000). [Crimean-Congo hemorrhagic fever in Rostov Province: the epidemiological characteristics of an outbreak]. *Zhurnal Mikrobiologii, Epidemiologii i Immunobiologii*, March–April:36–42 (in Russian).

Oteo JA, Brouqui P (2005). Ehrlichiosis y anaplasmosis humana [Ehrlichiosis and human anaplasmosis]. *Enfermedades Infecciosas y Microbiologia Clinica*, 23:375–380 (in Spanish).

Ozkurt Z et al. (2006). Crimean-Congo hemorrhagic fever in Eastern Turkey: clinical features, risk factors and efficacy of ribavirin therapy. *The Journal of Infection*, 52:207–215.

Pancholi P et al. (1995). *Ixodes dammini* as a potential vector of human granulocytic ehrlichiosis. *Journal of Infectious Diseases*, 172:1007–1012.

Papa A et al. (2002). Crimean-Congo hemorrhagic fever in Albania, 2001. *European Journal of Clinical Microbiology & Infectious Diseases*, 21:603–606.

Phillips M, Mills A, Dye C (1993). *PEEM guidelines 3 – guidelines for cost-effectiveness analysis of vector control.* Geneva, World Health Organization. (document number: WHO/CWS/93.4;
http://www.who.int/docstore/water_sanitation_health/Documents/PEEM3/english/peem3toc.htm, accessed 15 October 2006).

Piesman J (2002). Ecology of *Borrelia burgdorferi* sensu lato in North America. In: Gray JS et al., eds. *Lyme borreliosis: biology, epidemiology and control.* Oxon, CABI Publishing: 223–249.

Piesman J, Gern L (2004). Lyme borreliosis in Europe and North America. *Parasitology*, 129:S191–S220.

Postic D et al. (1998). Expanded diversity among Californian *Borrelia* isolates and description of *Borrelia bissettii* sp. nov. (formerly *Borrelia* group DN127). *Journal of Clinical Microbiology*, 36:3497–3504.

ProMED Mail [web site] (2005). Crimean-Congo hemorrhagic fever – Russia (Southern Federal District). Brookline, MA, International Society for Infectious Diseases (archive number: 20051003.2891;
http://www.promedmail.org/pls/promed/f?p=2400:1202:5299434327567973779::NO::F2400_P1202_CHECK_DISPLAY,F2400_P1202_PUB_MAIL_ID:X,30566, accessed 20 October 2006).

Qiu WG et al. (2002). Geographic uniformity of the Lyme disease spirochete (*Borrelia burgdorferi*) and its shared history with tick vector (*Ixodes scapularis*) in the northeastern United States. *Genetics*, 160:833–849.

Randolph SE (2001). The shifting landscape of tick-borne zoonoses: tick-borne encephalitis and Lyme borreliosis in Europe. *Philosophical Transactions of the Royal Society of London. Series B, Biological Sciences*, 356:1045–1056.

Rath PM et al. (1996). Seroprevalence of Lyme borreliosis in forestry workers from Brandenburg, Germany. *European Journal of Clinical Microbiology & Infectious Diseases*, 15:372–377.

Rendi-Wagner P (2005). Risk and prevention of tick-borne encephalitis in travellers. In: *Programme and Compendium of Abstracts, VIII International Potsdam Symposium on Tick-borne Diseases*, Jena, Germany, 10–12 March 2005.

Richter D et al. (2000). Competence of American robins as reservoir hosts for Lyme disease spirochetes. *Emerging Infectious Diseases*, 6:133–138.

Richter D et al. (2004). Relationships of a novel Lyme disease spirochete, *Borrelia spielmani* sp. nov., with its hosts in central Europe. *Applied and Environmental Microbiology*, 70: 6414–6419.

Richter D et al. (2006). Delineation of *Borrelia burgdorferi* sensu lato species by multilocus sequence analysis and confirmation of the delineation of *Borrelia spielmanii* sp. nov. *International Journal of Systematic and Evolutionary Microbiology*, 56:873–881.

Ryffel K et al. (1999). Scored antibody reactivity determined by immunoblotting shows an association between clinical manifestations and presence of *Borrelia burgdorferi* sensu stricto, *B. garinii*, *B. afzelii*, and *B. valaisiana* in humans. *Journal of Clinical Microbiology*, 37:4086–4092.

Salgo MP et al. (1988). A focus of Rocky Mountain spotted fever within New York City. *The New England Journal of Medicine*, 318:1345–1348.

Samish M, Ginsberg H, Glazer I (2004). Biological control of ticks. *Parasitology*, 129:S389–S403.

Samish M et al. (2001). Pathogenicity of entomopathogenic fungi to different developmental stages of *Rhipicephalus sanguineus* (Acari: Ixodidae). *The Journal of Parasitology*, 87:1355–1359.

Schriefer ME, Azad AF (1994). Changing ecology of Rocky Mountain spotted fever. In: Sonenshine DE, Mather TN, eds. *Ecological dynamics of tick-borne zoonoses*. New York, Oxford University Press: 314–326.

Schulze TL, Jordan RA, Hung RW (1998). Comparison of *Ixodes scapularis* (Acari: Ixodidae) populations and their habitats in established and emerging Lyme disease areas in New Jersey. *Journal of Medical Entomology*, 35:64–70.

Schulze TL, Jordan RA, Hung RW (2000). Effects of granular carbaryl applications on sympatric populations of *Ixodes scapularis* and *Amblyomma americanum* (Acari: Ixodidae) nymphs. *Journal of Medical Entomology*, 37:121–125.

Smith RP Jr et al. (1993). Norway rats as reservoir hosts for Lyme disease spirochetes on Monhegan Island, Maine. *The Journal of Infectious Diseases*, 168:687–691.

Sonenshine DE, Lane RS, Nicholson WL (2002). Ticks (Ixodida). In: Mullen G, Durden L, eds. *Medical and veterinary entomology*. Amsterdam, Academic Press: 517–558.

Sonenshine DE, Peters AH, Levy GF (1972). Rocky Mountain spotted fever in relation to vegetation in the eastern United States, 1951–1971. *American Journal of Epidemiology*, 96:59–69.

Spielman A (1976). Human babesiosis on Nantucket Island: transmission by nymphal *Ixodes* ticks. *The American Journal of Tropical Medicine and Hygiene*, 25:784–787.

Spielman A, Telford SR 3rd, Pollack RJ (1993). The origins and course of the present outbreak of Lyme disease. In: Ginsberg, HS, ed. *Ecology and environmental management of Lyme disease*. New Brunswick, NJ, Rutgers University Press: 83–96.

Spielman A et al. (1979). Human babesiosis on Nantucket Island, USA: description of the vector, *Ixodes* (*Ixodes*) *dammini* n. sp. (Acarina: Ixodidae). *Journal of Medical Entomology*, 15:218–234.

Spielman A et al. (1985). Ecology of *Ixodes dammini*-borne babesiosis and Lyme disease. *Annual Review of Entomology*, 30:439–460.

Stafford KC 3rd (1994). Survival of immature *Ixodes scapularis* (Acari: Ixodidae) at different relative humidities. *Journal of Medical Entomology*, 31:310–314.

Stafford KC 3rd (2004). *Tick management handbook: a integrated guide for homeowners, pest control operators, and public health officials for the prevention of tick–associated disease*. New Haven, The Connecticut Agricultural Experiment Station (http://www.caes.state.ct.us/SpecialFeatures/TickHandbook.pdf, accessed 20 October 2006).

Stafford KC 3rd, Kitron U (2002). Environmental management for Lyme borreliosis control. In: Gray JS et al., eds. *Lyme borreliosis: biology, epidemiology and control*. Oxon, CABI Publishing: 301–334.

Steere AC, Coburn J, Glickstein L (2005). Lyme borreliosis. In: Goodman JL, Dennis DT, Sonenshine DE, eds. *Tick-borne diseases of humans*. Washington, DC, ASM Press: 176–206.

Strle F (2004). Human granulocytic ehrlichiosis in Europe. *International Journal of Medical Microbiology*, 293 (Suppl. 37):27–35.

Talaska T (2003). Borreliose-Epidemiologie unter besonderer Berücksichtigung des Bundeslandes Brandenburg. In: Janata O, Reisinger E, eds. *Infektiologie – Aktuelle Aspekte, Jahrbuch 2003/2004*, Vienna, Austria, pm-Verlag: 119–125.

Talaska T (2005). Zur Lyme-Borreliose im Land Brandenburg. *Epidemiologisches Bulletin*, 20:173–178.

Tsao J et al. (2001). OspA immunization decreases transmission of *Borrelia burgdorferi* spirochetes from infected *Peromyscus leucopus* mice to larval *Ixodes scapularis* ticks. *Vector Borne and Zoonotic Diseases*, 1:65–74.

Tsao J et al. (2004). An ecological approach to preventing human infection: vaccinating wild mouse reservoirs intervenes in the Lyme disease cycle. *Proceedings of the National Academy of Science of the United States of America*, 101:18159–18164.

Uspensky I, Ioffe-Uspensky I (2002). The dog factor in brown dog tick *Rhipicephalus sanguineus* (Acari: Ixodidae) infestations in and near human dwellings. *International Journal of Medical Microbiology*, 291 (Suppl. 33):156–163.

Varma MGR (1989). Tick-borne diseases. In: WHO. *Geographical distribution of arthropod-borne diseases and their principal vectors*. Geneva, World Health Organization (document number: WHO/VBC/89.967;
http://whqlibdoc.who.int/hq/1989/WHO_VBC_89.967.pdf_, accessed 20 October 2006): 55–70.

Vasilenko SM et al. (1971). Investigation of Crimean haemorrhagic fever in Bulgaria. In: Chumakov MP, ed. *Viral haemorrhagic fevers*. Moscow, IPVE AMN SSSR: 100–111.

Wagner B (1999). Borreliose und FSME: Gefahr durch Zeckenstiche! *Der Hausarzt*, 8:34.

Wang G et al. (1997). Genetic and phenotypic analysis of *Borrelia valaisiana* sp. nov. (*Borrelia* genomic groups VS116 and M19). *International Journal of Systematic Bacteriology*, 47:926–932.

White DJ (1993). Lyme disease surveillance and personal protection against ticks. In: Ginsberg HS, ed. *Ecology and environmental management of Lyme disease*. New Brunswick, NJ, Rutgers University Press: 99–125.

White DJ et al. (1991). The geographic spread and temporal increase of the Lyme disease epidemic. *The Journal of the American Medical Association*, 266:1230–1236.

WHO (2001a). Vectors of diseases: hazards and risks for travellers: part I. *Weekly Epidemiological Record*, 76:189–194 (http://www.who.int/docstore/wer/pdf/2001/wer7625.pdf, accessed 22 October 2006).

WHO (2001b). Vectors of diseases: hazards and risks for travellers: part II. *Weekly Epidemiological Records*, 26:201–203 (http://www.who.int/docstore/wer/pdf/2001/wer7626.pdf, accessed 22 October 2006).

WHO Regional Office for Europe (2004). *The vector-borne human infections of Europe: their distribution and burden on public health*. Copenhagen, WHO Regional Office for Europe. (document number: EUR/04/5046114; www.euro.who.int/document/E82481.pdf, accessed 11 October 2006).

Wilson ML (1986). Reduced abundance of adult *Ixodes dammini* (Acari: Ixodidae) following destruction of vegetation. *Journal of Economic Entomology*, 79:693–696.

Wormser GP et al. (2000). Practice guidelines for the treatment of Lyme disease. The Infectious Diseases Society of America. *Clinical Infectious Diseases*, 31 (Suppl. 1):S1–S14.

Young D, Evans S (1998). Safety and efficacy of DEET and permethrin in the prevention of arthropod attack. *Military Medicine*, 163:324–330.

Zhang X et al. (2006). Economic impact of Lyme disease. *Emerging Infectious Diseases*, 12:653–660.

Zhioua E et al. (1997). Pathogenicity of the entomopathogenic fungus *Metarhizium anisopliae* (Deuteromycetes) to *Ixodes scapularis* (Acari: Ixodidae). *The Journal of Parasitology*, 83:815–818.

11. Mosquitoes

Helge Kampen and Francis Schaffner

Summary

Mosquitoes are relevant to public health when their density is great enough to make them a nuisance and when they transmit disease agents. Depending on climate and the availability of breeding sites, mosquitoes can become tremendously annoying nuisances regularly after seasonal mass reproduction. Except for a few examples, such as WNV and SLEV, mosquito-borne infections are still exceptional in Europe and North America. However, concern is rising as international travel and trade increasingly introduce both vectors and pathogens. In terms of people infected and geographical range, West Nile fever is the most widespread mosquito-borne disease in Europe and North America.

Of mosquito species, the Asian tiger mosquito (*Aedes albopictus*) is the most famous, because of its recent geographical spread. It is a vector of at least 22 arboviruses, including those that cause dengue fever and chikungunya fever, which lately became epidemic on some Indian Ocean islands. Risk assessments of such diseases being introduced to (and established in) countries where the Asian tiger mosquito occurs are presently being performed. In addition to viruses, mosquitoes may transmit malaria parasites and dirofilarial worms in Europe and North America. Though still uncommon, incidences of locally acquired (autochthonous) viral, parasitic and filarial infections appear to be increasing, but reliable data are scarce.

To minimize their numbers, so that mosquitoes are much less of a nuisance and a risk to health, pest management is important at personal and public levels. In the urban environment, proper sanitation and water management are key factors. This includes preventing mosquitoes from breeding in the immediate surroundings of dwellings and housing estates, by ensuring that there are no potential breeding places. Larviciding, adulticiding or both can accomplish mosquito control. While various approaches to larviciding are possible, including biological and biochemical treatment, adulticiding is confined to pesticide treatment (indoor application of residual pesticides and outdoor pesticide fogging). Therefore, mosquito management should consider the health hazards of pesticides and aim to promote environmental changes detrimental to the development of mosquitoes, rather than treating mosquitoes with pesticides – that is, prevention rather than control.

11.1. Introduction

Unlike many tropical and subtropical regions of the world, mosquitoes in Europe and North America are more important as pest nuisances than as vectors of disease. In Europe and North America, the bites of large numbers of mosquitoes can cause problems to both people and livestock. For some decades, contact between people and mosquitoes has become more intense, due to the expansion of suburbs into previously undisturbed natural areas that provide more numerous and varied breeding places than do inner-city areas. Also, because of the presence of mosquitoes, the use of gardens and public recreation areas near wetland breeding sites may be reduced considerably for certain parts of the year. In urban areas, however, mosquito populations are relatively small, due to sanitation, water management and control programmes.

Control practices notwithstanding, there have always been locally acquired (autochthonous) mosquito-borne diseases, both endemically and epidemically, in Europe and North America. In the presence of local sources of infection and a suitable climate, the likelihood of mosquito-borne infections increases with the abundance of mosquitoes; transmission also may be possible with low or moderate densities of mosquitoes. As our climate and environment change, mosquitoes can spread unnoticed to new areas, and with them come the *emerging and resurging vector-borne diseases* they transmit. Such dissemination may not only be caused by active mosquito migration, but may also be caused by passive transportation, such as by motor vehicles, trains and aircraft. The explosive increases in international (and even intercontinental) travel and in various trade imports, including livestock and pets, allow the accidental importation of mosquito vectors and vertebrate reservoir hosts. Thus, control programmes will shift from a pest control strategy to a vector control strategy, including the adaptation of control methods and improved efficiency – for example, supplementing or replacing larvicide with adulticide for better and faster control.

Finally, improper use of insecticides during mosquito-control operations has an indirect effect on public health, as it increases the risk of damage to people and the environment.

11.2. Mosquito biology

With some exceptions (Antarctica and some islands), mosquitoes (family Culicidae) are distributed worldwide. They are present in nearly all climates and occur both at high altitudes (up to 5500m) and below sea level (in mines and pits, down to 1250m). The family Culicidae contains about 3500 species, of which many are vectors of disease. Its members transmit viruses (by many species and genera of mosquitoes), protozoa (malaria parasites by *Anopheles* spp.) and filarial worms (such as *Brugia* spp., *Wuchereria bancrofti* and *Dirofilaria* spp., by *Aedes* spp., *Culex* spp., *Anopheles* spp. and *Mansonia* spp.). The genus *Aedes* (*Ochlerotatus* is treated here as a subgenus of *Aedes* and not as a genus (Reinert, 2000; Reinert, Harbach & Kitching, 2004; JME Editors, 2005) contains important pest species found in central Europe and North America. However in the tropics and subtropics some species are vectors of yellow fever, dengue fever, various encephalitis viruses and other arboviruses. Also, some important viral vector species are presently becoming established in Europe and North America (see section 11.5).

Most mosquitoes are haematophagous and feed on the blood of vertebrates. However, only females need blood for egg production, while males, which cannot bite, live on nectar and sugary plant juices. Depending on the species, gravid females produce some 30–300 eggs within several days to two weeks after a blood-meal. The eggs are laid on water surfaces (such as those of *Culex* spp., *Anopheles* spp. and *Coquillettidia* spp.), attached to the underside of leaves of aquatic plants (such as those of *Mansonia* spp.) or are deposited on damp substrates adjacent to water bodies, such as leaf litter or mud (such as those of *Aedes* spp., *Psorophora* spp. and *Haemagogus* spp.). Eggs of aedine mosquitoes usually enter a state of diapause (a suspension of development) and hatch, often in instalments, when the eggs are flooded with water several weeks to even years after being oviposited.

Larval and pupal development is aquatic. Larvae are filter feeders, often feeding on plankton, or are browsers, scraping food from submerged vegetative matter, such as rotting leaves. When resting, anopheline larvae lay parallel to the water surface – they do not have a siphon – whereas culicine larvae have a respiratory siphon and generally hang down from the water surface. Both anopheline and culicine mosquito larvae take in atmospheric oxygen. Anopheline larvae take it in through posterior abdominal spiracles, while culicine larvae breath through their siphon. Some culicine species, belonging to the genera *Mansonia* and *Coquillettidia*, have a modified siphon, which is short and conical; with it, they attach themselves to submerged roots or stems of aquatic plants, from which they obtain oxygen. Anopheline and some culicine larvae remain at the water surface during feeding, but some culicine larvae also dive to the bottom of their habitats to browse.

Mosquitoes have four larval stages, and larval development often lasts from about five to seven days or up to several weeks, depending on nutrition and temperature. Some species, however, overwinter as larvae. Pupae do not feed, but they are mobile. They also rest and breathe while clinging to the water surface. Pupal development takes two to three days, sometimes more in cooler water.

Some mosquito species prefer to feed on people (anthropophilic species), and others prefer to feed on animals (zoophilic species). However, a third group has no feeding preference (indiscriminative biters) and is thus predestined to transmit zoonotic diseases, as bridge vectors.

Temporal biting activity is species specific. For example, *Anopheles* spp. almost exclusively bite at night, whereas *Aedes* spp. bite during the daytime as well, with peak biting activity at dawn and dusk. Some species enter human dwellings to take a blood-meal (endophagic behaviour), while others bite outdoors (exophagic behaviour). Species that stay indoors for blood digestion and egg maturation are called endophilic, while those that leave the dwelling shortly after feeding on blood are called exophilic. These behavioural characteristics may be, and in practice actually are, exploited for controlling mosquitoes.

11.3. Mosquitoes as a pest nuisance

Depending on the availability of breeding sites, on climatic factors and on the species, mosquitoes may occur and bite throughout the year. In Europe and the United States, they can present a permanent annoyance from early spring to late autumn, but their density in urbanized areas is usually moderate (Schaffner, 2003). When they are encountered in large numbers, they generally have nearby breeding sites, as mosquitoes only exceptionally cover long distances. In city outskirts and rural residential areas, meadows, wetlands or swampy forests may provide breeding sites. In urban peridomestic areas, *Culex* spp. may predominate, and these often adapt to smaller (sometimes man-made) container habitats that hold organically polluted water.

Seasonal mass reproduction of mosquitoes may occur when large areas of land are flooded. This may occur, for example, after snow melts in the mountains during springtime, after heavy and enduring summer rainfalls and after river inundations or man-made artificial flooding as in rice fields and marshes – to attract ducks for hunting, for example. The floods provide a vast number of breeding sites for mosquitoes, such as the floodwater mosquito species *Aedes vexans* and *Aedes sticticus*, especially when isolated ponds are formed. Floods and the subsequent mass emergences of mosquitoes are certainly not modern phenomena, but what is relatively new are the large human populations that live near flooded areas.

During the last decade, several dramatic river floods have affected much of Europe – as a consequence of snow melting in the Alps, in combination with prolonged heavy rainfalls – particularly the floods of 1997 and 2002 in the Czech Republic, Germany and Poland or those of 2005 in Austria and Romania. The riverbeds could not drain the large volumes of water produced in a relatively short time, and huge areas of land, including settlements, were flooded. Similar flooding occurred a few years ago in the United States, in the Mississippi River Basin (Illinois, Missouri) and also in Florida. In the autumn of 2005, as a consequence of breached levees, Hurricane Katrina caused (directly and indirectly) terrible flooding in New Orleans, Louisiana. The expected large mosquito populations and subsequent epidemics of mosquito-borne diseases in this stronghold area of SLEV transmission failed to appear, due both to post-flooding mosquito control and to the devastation-related lack of animal and human mosquito hosts (S. Straif-Bourgeois, Louisiana Office of Public Health, personal communication, 2006).

Such catastrophes are generally accepted as being principally caused by the disappearance of natural river meadow landscapes, accompanied by natural riverbanks being replaced by artificial fortifications; by rivers being straightened; and by the lack of drainage on settlements and industrial estates. These environmental changes not only create great physical damage to buildings and harm to people and livestock, but they also sometimes create very large populations of mosquitoes that require prompt and extensive control efforts.

Furthermore, mosquitoes may change their behaviour and adapt to new breeding habitats that allow for mass reproduction. This appears to be the case with *Anopheles plumbeus* in Europe. Typically, *An. plumbeus* is a tree-hole breeder and usually does not occur at high densities. For some years, however, it has adopted several significant nuisances, using slurry pits, liquid manure pits and the like as breeding sites (Schaffner et al., 2001a). Recently, *An. plumbeus* was found at high larval densities in rainwater pools in used tyres (Schaffner, van Bortel & Coosemans, 2004).

11.4. Mosquitoes as vectors of disease

Mosquitoes are important vectors of viruses, parasitic protozoa (plasmodia) and filarial worms (Table 11.1). Species receptive to one or more pathogens have been indigenous to Europe and the United States, but because the pathogens have either been absent or have only rarely become apparent, they mostly have been long neglected. Because of changes in travel and commerce, however, vector species and pathogens may be spread actively (or be introduced passively) to areas and become established.

11.4.1. West Nile fever

West Nile fever is a viral, sometimes neuroinvasive mosquito-borne disease. Due to its ongoing spread in the western hemisphere, it is becoming increasingly important. While about 80% of human infections with WNV are apparently asymptomatic, about 20% of them result in symptomatic illness, with 1% leading to encephalitis, meningitis or acute flaccid paralysis (Mostashari et al., 2001). A substantial portion of people who develop severe neuroinvasive West Nile fever will suffer from long-term disability or else will die as a result of the infection (Pepperell et al., 2003; Klee et al., 2004).

Since the 1960s, Europe has faced West Nile fever epidemics that involve people and horses (Panthier, 1968). However, morbidity and mortality rates were significantly higher during the outbreaks of the last decade – in Romania (1996/1997), the Russian Federation (1999), France (2000) and Italy (2002) (Tsai et al., 1998; Platonov et al., 2001; Murgue et al., 2001; Autorino et al., 2002) – than in earlier times, probably due to the circulation of new virus strains (Gratz, 2004). The 1996/1997 Romanian and 1999 Russian epidemics, for example, which mainly affected human beings, had case fatality rates of nearly 10% (Tsai et al., 1998; Platonov et al., 2001).

Birds are important reservoir hosts for the virus, and migrating birds play an important role in the epidemiology of West Nile fever, as they regularly transport the virus from endemic areas in sub-Saharan Africa to Europe (Lundström, 1999). Certain mechanisms, however, obviously prevent the virus from becoming endemic in Europe. As there is no observed association between bird mortality and WNV infection, European birds, for example, seem to have adapted immunologically to most of the circulating virus variants and may contribute to virus elimination (Buckley et al., 2003). WNV activity has been found directly (virus isolation) or indirectly (antibody detection) in southern and eastern Europe mainly (Hubálek & Halouzka, 1999); however little is known about the situation elsewhere, probably because it has been explored inadequately.

Table 11.1. Pathogen transmission by mosquitoes in Europe and North America

Infection	Pathogen	Clinical manifestations	Countries	Important vectors
Inkoo virus	*Bunyavirus* (Bunyaviridae) Inkoo virus	No visible signs	Northern Europe	*Aedes communis* *Aedes punctor*
Lednice virus	*Bunyavirus* (Bunyaviridae) Lednice virus	No visible signs	Central Europe	*Culex modestus*
Batai (Calovo) virus	*Bunyavirus* (Bunyaviridae) Batai (Calovo) virus	No visible signs	Northern and central Europe	*Anopheles maculipennis* sensu lato
Tahyna virus	*Bunyavirus* (Bunyaviridae) Tahyna virus	Sometimes mild febrile illness; rarely meningitis	Germany and eastern Europe	*Ae. vexans* *Aedes caspius* *Aedes dorsalis*
Sindbis virus	*Alphavirus* (Togaviridae) Sindbis virus	Severe headache, muscle ache, dengue-like symptoms	Mediterranean Basin	*Culex* spp. *Aedes* spp.
Ockelbo virus	*Alphavirus* (Togaviridae) Ockelbo type	Febrile disease with rash and polyarthralgia	Finland, Norway, Russian Federation and Sweden	*Cx. pipiens* *Culex torrentium* *Aedes cinereus*
WNV	*Flavivirus* (Flaviviridae) WNV	Mild, flu-like illness; sometimes meningitis	Europe and North America	*Cx. pipiens* *Cx. modestus* *Culex quinquefasciatus* *Culex restuans* *Culex tarsalis*
St Louis encephalitis	*Flavivirus* (Flaviviridae) SLEV	Usually mild, febrile illness; rarely meningitis	Throughout the United States	*Cx. pipiens* *Culex nigripalpus* *Cx. tarsalis*
Eastern equine encephalitis	*Alphavirus* (Togaviridae) EEEV	Usually mild, flu-like illness; sometimes encephalitis, coma, death	Eastern United States	*Ae. vexans* *Aedes canadensis* *Aedes sollicitans* *Culex salinarius* *Culiseta melanura* *Coquillettidia perturbans*
Western equine encephalitis	*Alphavirus* (Togaviridae) WEEV	Usually mild, flu-like illness; sometimes encephalitis, coma, death	Western and central United States	*Cx. tarsalis* *Aedes melanconion*
La Crosse encephalitis	*Bunyavirus* (Bunyaviridae) La Crosse encephalitis virus	Usually mild, febrile illness; rarely seizures, coma	Upper mid-western and mid-Atlantic United States	*Aedes triseriatus*
Malaria	*Plasmodium ovale* *Plasmodium malariae* *Plasmodium falciparum*	Fever attacks, chills; in *P. falciparum* infection, often renal failure, coma, death	Southern Europe, southern North America	*An. sacharovi* *An. atroparvus* *An. labranchiae* *An. freeborni* *An. pseudopunctipennis* *An. quadrimaculatus*
Filariasis	*D. immitis* *D. repens* *D. tenuis*	Dogs: cardiovascular filariasis; human beings: pulmonary or subcutaneous lesions	European Mediterranean countries, throughout the United States	*Ae. caspius* *Ae. detritus* *Ae. vexans* *Cx. pipiens* *An. maculipennis* s.l. *Mansonia* spp.

Sources: Schaffner (2003); CDC (2004b).

Unlike Europe, the United States was free of WNV before 1999. Then, between 1999 and 2003, the virus spread rapidly in the United States, from the east coast to the west coast (Gould & Fikrig, 2004). There have been thousands of equine and human infections with hundreds of deaths and, unlike in Europe, birds, predominantly crows (*Corvus* spp.), also have been affected badly. The CDC human case counts total 19,655 infections for the years 1999–2005, including 782 fatalities (CDC, 2006). At present, WNV has been detected in most of Canada and Central America and in the Caribbean. Most likely, the virus was brought to North America by illegally imported birds. Molecular analysis has shown that the North American virus strain is related most closely to a strain isolated in Israel, which is much more virulent than the European strains (Lanciotti et al., 1999; Brault et al., 2004).

According to Higgs, Snow & Gould (2004), 73 mosquito species have been found naturally infected with WNV, could be experimentally infected or have even been shown to transmit the virus after experimental infection. At present, more than 15 mosquito species in Europe (Higgs, Snow & Gould, 2004; Medlock, Snow & Leach, 2005) and at least 60 species in numerous genera in the United States are suspected to be vectors of WNV (CDC, 2005). Most important among them are species that feed on different types of hosts, such as those of the *Culex pipiens* complex, which were involved in several outbreaks, particularly in urban areas where they breed peridomestically in small man-made water containers. As with other indiscriminate biters, the northern house mosquito (*Cx. pipiens*) is both ornithophilic and mammophilic, thus serving as a bridge vector between birds and mammals, including human beings (Fonseca et al., 2004; Kilpatrick et al., 2005).

11.4.2. Other mosquito-borne viral infections that cause encephalitides

For Europe, six mosquito-borne viruses have been described as active during the last 50-year period: Inkoo, Lednice, Batai (Calovo), Tahyna, Sindbis/Ockelbo and West Nile viruses. The Tahyna, Sindbis/Ockelbo and West Nile viruses are pathogenic in people (Lundström, 1994). A seventh virus (Usutu virus), whose pathogenic potential for people is unknown, was detected in 2002 in Austria, where it was responsible for mass mortality in birds, mainly common blackbirds (*T. merula*) (Weissenböck et al., 2003). Antibodies to Usutu virus were also recently found in birds in the United Kingdom (Buckley et al., 2003).

The course of most European mosquito-borne virus infections in people is considered to proceed asymptomatically or to present as mild summer flu-like symptoms. Only occasionally do these infections result in meningitis or encephalitis, the etiologies of which usually are not pursued. A study conducted in English hospitals revealed that about 60% of all registered viral encephalitides are of unknown etiological origin (Davison et al., 2003). Diagnosis usually is based on the symptoms and some clinical parameters; virus isolation or the detection of specific viral RNA or DNA is seldom attempted. To date, comprehensive scientific studies on the activities of mosquito-borne viruses in Europe are almost completely lacking.

In the United States, in addition to WNV, four other mosquito-borne encephalitis viruses are of epidemiological interest: St. Louis, eastern equine, western equine and La Crosse encephalitis viruses. As with WNV, SLEV has been involved in large urban epidemics, resulting in thousands of human deaths since 1933 (CDC, 2001; Day, 2001).

11.4.3. Dengue (haemorrhagic) fever and yellow fever

Globally, dengue (haemorrhagic) fever is the most important human viral disease transmitted by mosquitoes (*Aedes* spp.). It is endemic in tropical and subtropical regions of Africa, Asia, Australia and South America, but at present only imported cases are found in Europe and the United States. This was not always the case, and the potential for the return of active transmission still exists, as global dengue virus transmission increases rapidly, especially in Central America and South America (Malavige et al., 2004). In the United States, dengue epidemics occurred in the 1800s and the first half of the 1900s. While there apparently was no indigenous transmission between the late 1940s and the 1970s, autochthonous cases were again registered between 1980 and 1999 along the border between Texas and Mexico (Reiter et al., 2003). During the 1927/1928 dengue epidemic in Europe, Greece was badly affected, with about a million people infected and more than 1500 deaths (Cardamatis, 1929; Rosen, 1986). Ninety percent of the population of Athens was infected with the virus. However, the appearance of some level of immunity among infected people, combined with vector control programmes that targeted the yellow fever mosquito (*Aedes aegypti*), successfully eradicated the disease from Europe.

A somewhat similar situation occurred with yellow fever, which is endemic in sub-Saharan Africa and South America. Although Europe and the United States had experienced yellow fever epidemics for a while (Eager, 1902; Tomlinson & Hodgson, 2005), they are only affected by imported cases at present. However, unlike the situation with dengue, an efficient vaccination keeps the number of imported yellow fever cases quite low. Also, few imported cases have been reported. Travellers imported only two cases to Europe during the last four years (Bae et al., 2005), and only three imported cases have been reported in the United States since 1996 (Hall et al., 2002).

Nowadays, the advent of the Asian tiger mosquito (an efficient dengue virus vector) in yellow fever mosquito-free areas of Europe and North America and the proximity of the United States to the southern dengue endemic countries have evoked a climate of uncertainty about the potential return of dengue fever. Preventing the further spread of this mosquito species, by using effective control measures, is now imperative. The classical yellow fever vector and also the most efficient vector of the dengue virus, the yellow fever mosquito, could never be completely eradicated from the United States and adds to the list of possible viral vectors. Despite occasional reports from Italy (Callot & Delécolle, 1972) and Turkey (B. Alten, Hacettepe University, Ankara, personal communication, 2001), the yellow fever mosquito is supposed to have been eradicated from Europe for decades.

11.4.4. Chikungunya fever

Since first being isolated in the early 1950s, from a human patient in Tanzania (Ross, 1956), chikungunya virus has caused numerous outbreaks in both Africa and South-East Asia, involving hundreds of thousands of people, mainly in the 1960s (Rao, 1966; Halstead et al., 1969a,b; Moore et al., 1974). It is a febrile, but usually not fatal disease transmitted by *Aedes* spp., especially the subgenera *Aedimorphus*, *Diceromyia* and *Stegomyia* (Jupp & McIntosh, 1988; Diallo et al., 1999). Recently, epidemics of chikungunya fever have occurred on several islands in the Indian Ocean, such as the Union of the Comoros, Madagascar, Mayotte, Mauritius, Seychelles and Réunion, as well as in southern Asia (India and Malaysia) (Institut de Veille Sanitaire, 2006; WHO, 2006a). As of 29 March 2006, the Réunion outbreak has affected more than a fourth of the total population and has brought about extraordinarily severe cases, including meningoencephalitis and multi-organ failure, and even 155 deaths associated directly or indirectly with the infection (ProMED Mail, 2006). As the virus vector in the Indian Ocean is the Asian tiger mosquito, which had been introduced at least a century ago (Fontenille & Rodhain, 1989; Salvan & Mouchet, 1994), there is great concern that the virus will be imported by travellers and be transmitted and established in Europe by the Asian tiger mosquito. In fact, 160 cases of chikungunya disease were imported to France between April 2005 and 31 January 2006 (Institut de Veille Sanitaire, 2006), and several others were imported to Germany, Italy, Norway and Switzerland (WHO, 2006b). As a consequence, the ECDC, together with a group of experts, is at present conducting a risk assessment for Europe.

11.4.5. Malaria

With a high level of morbidity, but moderate mortality, malaria pestered Europe (even northern Europe) until the middle of the 20th century. The Second World War provided an opportunity to install wide-ranging control programmes and to rehabilitate previously neglected areas. Thus, countries in southern Europe saw a large reduction of the endemic disease, which finally disappeared, due to such factors as improvement of sanitary conditions, drainage of swampy areas, effective medical treatment, separation of animal and human dwellings, and the instability of the *Plasmodium* cycle in temperate climate zones.

Malaria was declared eradicated from Europe in the early 1970s (Bruce-Chwatt & de Zulueta, 1980). In contrast, it has never disappeared from the south-eastern part of Turkey where, although there were considerably fewer cases in the 1970s, malaria increased to the former endemic level after control programmes were largely discontinued (Çağlar & Alten, 2000). In some eastern countries of the WHO European Region, such as Armenia, Azerbaijan and Tajikistan, the number of cases of malaria increased after the dissolution of the Union of Soviet Socialist Republics, owing to the collapse of the infrastructure and public health services (Majori, Sabatinelli & Kondrachine, 1999).

Unlike the endemic situation in the eastern European countries, there are sporadic cases of autochthonous malaria in central and southern Europe and in the United States. These locally acquired infections result mainly from bites of *Anopheles* mosquitoes that have become infected by feeding on malarial tourists, immigrants, seasonal workers and other sources. In particular, travellers often return home with a case of malaria when they have underestimated the risk of becoming infected in tropical holiday destinations, have disregarded chemical malaria prophylaxis regimens or have been misinformed about the malaria situation at their holiday destination by travel agencies. Moreover, during the last decade, malaria infections have occurred in some tropical holiday resorts that had been previously free of malaria and where no prophylaxis had been necessary, such as parts of the Dominican Republic (Kay et al., 2005). On the other hand, visitors from malaria-endemic regions often ignore the possibility that they might be infected with malaria, as they are not sick due to the semi-immunity acquired from repeated infection since childhood. Due to indigenous vector-competent *Anopheles* spp. in Europe (for example, *Anopheles sacharovi*, *Anopheles atroparvus* and *Anopheles labranchiae* that take their blood-meals on gametocyte-carrying people) single locally acquired malaria cases have been occurring in Bulgaria, Greece, Germany, Italy and Spain during the last 17 years (Sartori et al., 1989; Nikolaeva, 1996; Baldari et al., 1998; Krüger et al., 2001; Cuadros et al., 2002; Kampen et al., 2002).

In the United States, *Anopheles freeborni, Anopheles quadrimaculatus* and *Anopheles pseudopunctipennis* are considered to be the primary malaria vectors. During the last two decades, autochthonous malaria cases have occurred in California (Maldonado et al. 1990), New Jersey (Brook et al., 1994; Shah et al., 2004), Houston (CDC, 1995), New York City (Layton et al., 1995), Virginia (Shah et al., 2004), Florida (Selover et al., 2004) and Maryland (Eliades et al., 2005).

Sometimes, imported infected mosquitoes may provoke cases of so-called airport malaria or baggage malaria. Between 1969 and 1997, 63 cases of malaria associated with airports were recorded in western Europe (Isaacson, 1989; Castelli et al., 1993; Mouchet, 2000). Such infected *Anopheles* mosquitoes originate from airports located in zones of endemic transmission, mostly sub-Saharan Africa. Local transmissions in the proximity of destination airports in non-endemic regions may occur during the summer and, particularly, during so-called hot years. The infections are extremely difficult to diagnose when people who fall ill have no recent history of travelling to malarial areas. Disinfection of aircrafts can minimize the risk of such cases (WHO, 1995; Guillet et al., 1998; Gratz, Steffen & Cocksedge, 2000).

11.4.6. Dirofilariasis

Human dirofilariasis seems to be a more significant problem in Europe than in North America. It is caused by an opportunistic infection with filarial worms (*Dirofilaria tenuis* and *Dirofilaria immitis* in North America; *Dirofilaria repens* and *D. immitis* in Europe), leading to heartworm disease in their natural canine hosts and sometimes in felines. Possible vectors are various species of the mosquito genera *Culex*, *Anopheles and Aedes* (Pampiglione, Canestri Trotte & Rivasi, 1995), including the Asian tiger mosquito (Cancrini et al., 2003a,b).

While there are few human cases in the United States, canine dirofilariasis is endemic in southern Europe. Also, human infections with the major parasite *D. repens* have been increasing for several decades in France, Greece and Spain, and especially in Italy (Pampiglione, Canestri Trotte & Rivasi, 1995; Muro et al., 1999; Pampiglione & Rivasi, 2000). Until the middle of the last century, human dirofilariasis was considered exceptional, and from its first demonstration in 1864 until 1995, a total of only 181 human cases of *D. repens* infection had been reported for Italy. However, between 1995 and 2000, 117 more cases were registered in Italy (Pampiglione, Canestri Trotte & Rivasi, 1995; Pampiglione & Rivasi, 2000). Of these recent cases, no human deaths have been reported and some human infections were asymptomatic; for the infections that were symptomatic, the symptoms were usually subcutaneous and pulmonary nodules or parenchymal lesions. Evidence from serological data, however, suggests that people living in endemic areas are infected to a similar extent (up to 30% and more) as the canine populations (Simón et al., 2005). Because imported dogs are seldom examined for parasitic infections, the increasing commercialized (and sometimes illegal) import of dogs from southern to central Europe – especially to Germany, where there are vector-competent mosquito species (Rossi et al., 1999) – gives reason to fear cases of autochthonous canine and human dirofilariasis in central Europe in the near future. In fact, the first case of autochthonous infection in a dog was recently demonstrated for Germany (Hermosilla et al., 2006).

11.5. The spread of mosquitoes and mosquito-borne pathogens

The significant increase in international travel and trade in consumer goods and animals offers multiple chances for insects and infected vertebrate hosts to be transported across natural barriers, such as mountains, oceans and climate zones. Examples already mentioned are: the assumed illegal import of WNV-infected birds, which led to the emergence of West Nile fever in the Americas; the intercontinental travel of *Plasmodium*-infected *Anopheles*-mosquitoes by aircraft, which possibly caused cases of airport or baggage malaria; and the import of dogs infected with filaria, which caused heartworm disease.

People who travel can also harbour vector-borne pathogens obtained from endemic areas, such as malaria parasites, dengue virus or chikungunya virus. In 2003, the CDC received reports of 1278 cases of malaria among people in the United States, 1268 of them imported (Eliades et al., 2005). For the same year, 11 573 imported cases of malaria in Europe were reported to the WHO Regional Office for Europe (2006), with major contributions by France (6392 estimated cases, 3511 of them confirmed), the United Kingdom (1722 confirmed cases) and Germany (819 confirmed cases). For the same year, 166 travel-related cases of dengue fever were documented in the WHO Regional Office for Europe centralized information system for infectious diseases (CISID) data bank for Europe. As with dengue fever, however, several countries do not report their cases regularly, and France, which made no report in 2003, counted 3157 cases of dengue fever in 2004. In the United States, the CDC identified 77 cases of dengue fever in 37 states and the District of Columbia, from 2001 to 2004 (Beaty et al., 2005). However, as dengue fever is not a nationally notifiable disease in the United States, this figure is likely subject to underreporting. Only recently, more than 160 cases of chikungunya disease were imported in less than a year to several European countries from Indian Ocean islands (see subsection 11.4.4) – particularly to France, due to traditionally close ties.

Although the number of imported cases of mosquito-related infections is not dramatic from an epidemiological perspective, it is sufficient to expect sporadic local transmission by indigenous vector-competent mosquitoes that become infected by biting local people who have acquired infections abroad. Also, although endemic malaria has been eradicated from Europe and the United States, the mosquitoes capable of transmitting malaria are still widely distributed (see subsection 11.4.5). The primary dengue virus vector, the yellow fever mosquito, has been successfully eradicated from Europe (see subsection 11.4.3), but not from the United States. Recently, some non-indigenous mosquito species (such as the Asian tiger mosquito) that are vector-competent for dengue virus, chikungunya virus and several other arboviruses have been introduced to Europe and the Unites States through trade in used tyres and lucky bamboo (*Dracaena* spp.). After introduction, these non-indigenous mosquito species have often survived and spread into other areas of their new country.

Anthropogenic transportation and active spread are the causes of the emergence of the Asian tiger mosquito in the Western world. It was introduced from its South-East Asian home by the used tyre trade and, after a period of adaptation, began to spread (Knudsen, Romi & Majori, 1996). In the United States, it was first detected in Texas, in 1985 (Sprenger & Wuithiranyagool, 1986), and in 2000 it was established in 866 counties in 26 states (CDC, 2000). In 2003, however, it was recorded for the first time in Colorado (Bennett et al., 2005). In Europe, the Asian tiger mosquito first appeared in Albania, in 1979 (Adhami & Reiter, 1998), and about 10 years later, in 1990, in Italy (Sabatini et al., 1990). The Italian strain of Asian tiger mosquito was imported from the United States, where it adapted to more moderate climate conditions by diapause overwintering after its introduction from South-East Asia (Rightor, Farmer & Clarke, 1987; Knudsen, Romi & Majori, 1996). It soon became widely distributed in southern Europe, and in 1999 it was found in Normandy, in northern France (Schaffner et al., 2001b). A year later, in 2000, it was found in Belgium (Schaffner, van Bortel & Coosemans, 2004), which had been predicted previously as an area of minor distribution risk (Knudsen, Romi & Majori, 1996). More recently, it was recorded in Montenegro, in 2001 (Petrić et al., 2001); in Switzerland, in 2003 (Flacio et al., 2004); in Croatia and Spain, in 2004 (Aranda, Eritja & Roiz, 2006; Klobučar et al., 2006); and in Bosnia and Herzegovina, Greece, the Netherlands and Slovenia, in 2005 (Samanidou-Voyadjoglou et al., 2005; Petrić et al., 2006; Scholte et al., 2006) (Fig. 11.1).

The Asian tiger mosquito is an efficient vector of numerous arboviruses, including dengue, chikungunya and (probably) West Nile viruses (Mitchell, 1995). Dengue fever used to be endemic in Europe some time ago and was associated with high mortality rates (Cardamatis, 1929; Rosen, 1986).

The rock pool mosquito (*Aedes atropalpus*), which is a vector of EEEV in the United States, has been introduced to southern Europe by used tyres from North America. However, it was then eradicated from two introductory locales in central Italy (Romi, di Luca & Majori, 1999; Snow & Ramsdale, 2002) and western France (F. Schaffner, personal communication, 2003) during Asian tiger mosquito control campaigns.

The Asian bush mosquito (*Aedes japonicus japonicus*) is a vector of Japanese encephalitis virus and has also been found infected with WNV. It has invaded Europe and the United

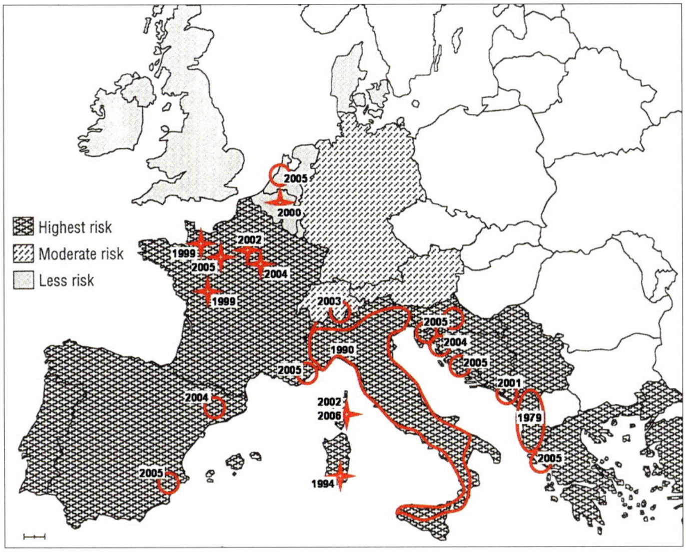

Fig. 11.1. Climate-based distribution risk areas for the Asian tiger mosquito in Europe, 1996, and subsequent detection of the Asian tiger mosquito

Note. Mosquito specimen detection (red markings) demonstrated by year. Eradicated foci are shown as stars, proliferation areas are encircled.

Source: Climate-based distribution risk areas: Knudsen, Romi & Majori (1996); current detection spots and areas of Asian tiger mosquito: synthesis by the authors.

States from Asia. In the United States, it was recorded for the first time in the states of New York and New Jersey in 1998 (Peyton et al., 1999). In Europe, it was recorded for the first time in north-western France in 2000 (Schaffner, Chouin & Guilloteau, 2003) and in Belgium in 2002 (F. Schaffner, personal communication, 2002).

11.6. National reporting and notification

In the United States, the CDC is an efficient governmental institution that collects data on both vectors and vector-borne diseases. Data are supplied by state and local health departments and by numerous specialized mosquito control associations. Together with the EPA, the CDC and subordinate departments are responsible for: monitoring disease cases, mosquito distributions and abundances; providing instruction and informative material; and implementing and coordinating nationwide control activities. Also, the American Mosquito Control Association, a scientific and educational institution, supports control measures and provides basic biological data and services to public agencies and their principal staff members engaged in mosquito control, mosquito research and related activities.

The situation in Europe is different. In Europe, an institution equivalent to the American CDC, the ECDC, is presently being set up in Sweden. Its primary aim is to collect epidemiological data on infectious diseases and coordinate preventive measures on a Europe-wide scale. Another organization, the European Mosquito Control Association (EMCA), focuses on mosquitoes and mosquito-borne diseases; it is comprised of institutions from 22 European countries. These institutions have a governmental, scientific or commercial background. In some European countries, public mosquito control agencies also contribute. These agencies, under the auspices of local governments, have entomological expertise, a mosquito control mission and can be contacted. Most of them are gathered within the European association of public authorities and operators involved in mosquito control and the management of natural areas where mosquitoes have been controlled, which is also known as the EDEN association.

Generally, local, district or (even) state health authorities are to be informed in the event of mosquito-related problems. These authorities, however, are usually unprepared for such problems and are forced to contact specialists affiliated with research institutes, universities and pest control companies. Also, in many countries, few research institutions that deal with mosquito biology and ecology and mosquito-borne diseases are left, and because of the lack of data these institutions appear to be nonexistent. Usually, mosquito control is governmentally organized (at the state level) only when there is a proven transmission of mosquito-borne pathogens.

In the United States, while malaria, West Nile fever, yellow fever, and eastern equine, western equine, St. Louis and La Crosse encephalitis are notifiable diseases on a national level, dengue fever is not. In the various countries of Europe, there is no consistent notification duty for mosquito-borne infections. In most European countries, however, the national health authorities are to be notified about cases of malaria, West Nile fever, yellow fever, haemorrhagic fevers (such as dengue haemorrhagic fever) and all viral meningoencephalitides, which are sometimes caused by Tahyna and Sindbis virus infections. The increasing risk of further mosquito-borne diseases being introduced from related overseas territories could result in new notification guidelines, as was implemented for chikungunya fever and dengue fever in France, in April 2006.

11.7. Mosquito monitoring

Surveillance and detection are the basis for any efficient mosquito control programme. The extent and origin of the infestation, as well as the mosquito species involved, must be determined (Schaffner et al., 2001a; Becker et al., 2003). To be efficient, any intervention has to be preceded by on-site inspection and sample collection, to identify the culprit. Identifying the species helps to indicate the location of breeding sites, which are then quantified and recorded on a site map (numbers of mosquitoes and surface area), which helps to organize control procedures and estimate costs. Mapping a site before an infestation occurs gives an indication of the location of potential breeding sites for further monitoring and permits more efficient sampling, population assessment and follow-up surveys. Standardized methods for surveillance exist (Service, 1993) and consist primarily of larval sampling in breeding sites, as well as adult trapping methods, including dry-ice traps, light traps, baited traps, gravid traps and oviposition traps. In many instances, combinations of these techniques are used, but some programmes may use only one method, due to budgetary constraints or lack of personnel capable of interpreting certain types of collected data.

Due to species-specific biology, behaviour and ecology, which have a great impact on the mosquito control measures to be undertaken, experts must identify the mosquitoes collected. In some cases, when sibling species – that is, species extremely similar in appearance, but incapable of interbreeding – are involved, differentiation cannot remain confined to morphological features, but needs more sophisticated methods, such as cytotaxonomy (banding pattern analysis of giant chromosomes), zymotaxonomy (analysis of electrophoretic characteristics of corresponding enzymes) or molecular biological techniques (White, 1984; Kampen, 2004).

Attempts have been made to list, on a literature research base, all mosquito species for all countries belonging to the WHO European Region, but much of the data are outdated and several countries do not have data available. According to Schaffner and colleagues (2001a), Snow & Ramsdale (2003), and Becker and colleagues (2003), there are about 100 species of mosquitoes in Europe. For Canada and the United States, more than 160 mosquito species are recorded (Darsie & Ward, 1981). Owing to continuous mosquito monitoring, mainly done by mosquito control associations, states in the United States are entomologically far more up to date than most European countries and can react more quickly and more efficiently to emergencies.

11.8. Mosquito control and management

Current challenges posed by the emergence of mosquito-borne pathogens in the western hemisphere illustrate the importance of cooperation and partnership at all levels of government to protect public health. In the United States, the EPA and the CDC work closely with each other and with other federal, state and local agencies to manage mosquito-borne diseases. Although this functional unity cannot prevent pathogens, especially not those borne by arthropods, from being introduced into the country, it is able to take quick and appropriate measures to minimize the impact on public health. In Europe, recent developments within the European Union (EU) have led to the creation of an agency similar to the CDC – the ECDC. However, there is no EU-wide policy on mosquito control. Instead, most European countries address this problem at the state or local level, or both.

In the United States, the CDC, working closely with state and local health departments, monitors the potential sources and outbreaks of mosquito-borne diseases and provides advice and consultation on the prevention and control of these diseases. The CDC works with a network of experts in human and veterinary medicine, entomology, epidemiology, zoology and ecology to obtain quick and accurate information on emerging trends, which they develop into national strategies that reduce the risk of transmitting diseases.

In the United States, the EPA ensures that state and local mosquito control departments have access to effective mosquito control tools they can use without posing an unacceptable risk to human health and the environment. The Agency also educates the public through outreach efforts that encourage the proper use of insect repellents and pesticides. Moreover, the EPA's rigorous pesticide review process is designed to ensure that registered chemicals, when used according to label directions and precautions, can reduce disease-carrying mosquito populations.

In Europe and North America, state and local government agencies play a critical role in protecting the public from mosquito-borne diseases. They serve at the front, providing information through their outreach programmes to the medical and environmental surveillance networks that first identify possible outbreaks. They also manage the mosquito control programmes that carry out prevention, public education and vector population management.

Integrated control aims to synergize a range of measures intended to reduce and control the nuisance or the vectorial transmission. It aims to reduce the density and the longevity of vector insects by adapting measures to the environmental and epidemiological local conditions. The monitoring and reappraisal of field conditions allow permanent readjustments of these measures, with the objective of better effectiveness and less damage to the environment and human health. Integrated control proposes:

- sanitation, environmental modification measures and biological control, to reduce the development of vectors and to maintain their density under a minimum threshold;

- larviciding with chemical or biochemical insecticides, for a rapid reduction of the likelihood of mosquito proliferation;

- residual adulticiding, to reduce the longevity of vectors and to stop the cycles of transmission; and

- passive protection measures, to reduce host–vector contacts.

Guidelines for surveillance, prevention and control plans are already developed in several countries for managing the risk of WNV (CDC, 2003; CDHS/M&VCAC/UC, 2004; SSSQ, 2006; MSS/MAP/MMEDD, 2005).

11.8.1. Sanitation and water management

Sanitation and water management, such as source reduction, are key to any solution of a mosquito problem. Depending on the species, culicid mosquitoes develop in many different types of natural and artificial waters. Mosquito larvae can be found in clean and polluted, fresh and brackish, and stagnant or slow-flowing waters, such as marshes. They can also be found in river deltas, rice fields, swamps, tidal floodwaters, lakes, puddles, pools, ponds, tree holes, rock holes and creeks, as well as in gutters, flowerpots, tin cans, buckets, dishes, tyres, pits and cellars.

Source reduction is therefore the only long-term solution to mosquito infestation, especially in urban areas (Schaffner, 2003). This method requires modification of aquatic habitats – for example, some natural habitats could be modified to reduce the production of mosquitoes by a relative stabilization of the water level. Preventing, or at least identifying, stagnant waters is crucial in artificial areas that may provide breeding sites for the northern house mosquito, the Asian tiger mosquito, or the yellow fever mosquito) (Fig. 11.2). This can be accomplished by avoiding stagnation in sewers, draining stagnant water from subfloor crawl spaces (or closing air gaps with mosquito nets), covering tyre stocks (or storing tyres in warehouses), removing containers that hold water or at least regularly replacing the water (Fig. 11.3a–c).

As Fig. 11.2 illustrates, there are five sanitary measures that can reduce mosquito problems considerably.

1. Flooding of subfloor crawl spaces, cellars and basements in houses or apartment buildings plays an important role in the establishment of breeding sites for the northern house mosquito. To prevent such breeding sites, it is important that effective drainage and water-proofing measures be implemented, that residual water be pumped out and that potentially floodable space be filled with gravel, up to the highest level of the water table.

Fig. 11.2. Some sanitary measures that can considerably reduce mosquito problems

Note. The numbers encircled in red correspond to the following areas where sanitary measures can be used:

1. flooding of subfloor crawl spaces, cellars and basements in houses or apartments;
2. cesspits and cisterns, wells, watering and leisure pools, rainwater tanks, swimming pools not in use, construction site excavations, potted plants with saucers, cemetery flower vase and diverse containers;
3. rainwater channels;
4. water treatment plants; and
5. wastewater treatment ponds.

Source: Schaffner (2003).

2. For cesspits and cisterns, mosquito nets should be installed in aeration gaps, to ensure that covers do not leak. For septic tanks, open discharge should not be allowed. Instead an underground purification bed should be installed; once it is linked to a sewer system, the old pit should be filled or destroyed. Also, wells, watering and leisure pools, rainwater tanks, swimming pools not in use, construction site excavations, potted plants with saucers, cemetery flower vases and diverse containers must not be prone to mosquito development. To avoid it, any water reservoirs should be emptied every 10 days or be covered with mosquito nets. Moreover, all receptacles not in use should be eliminated. Furthermore, goldfish can be introduced in leisure ponds. Finally, containers exposed to the weather should be emptied, turned over or discarded.

3. For rainwater channels, all obstacles to the flow of water should be removed.

4. Water treatment plants working on a part time basis can create problems, if they are too big for the amount of water to be treated. Also, abandoned plants that still retain rainwater can create problems.

5. Wastewater treatment ponds should be clear of rooted vegetation, and the water height should be maintained at more than 80 cm. Ponds that use vegetation as a purification measure should be sufficient in number, so as to allow for more than one month of total dryness a year (in winter), to prevent *Coquillettidia* spp. from becoming established. Also, embankments must be covered with cement or a geotextile fabric, and ponds must be designed in a manner that avoids zones with stagnant waters.

Fig. 11.3. Water-filled uncontrolled containers are potential mosquito breeding sites

Note. The photos illustrate potential breeding sites in: (a) farms and gardens, (b) terraces and balconies and (c) next to warehouses and on waste sites.

Source: Photos by F. Schaffner/EID Méditerranée.

These efforts are labour intensive, but they lend themselves to community-wide involvement, through public education campaigns.

11.8.2. Larviciding

Treatment efficiency and economic and ecological costs influence the final choice of intervention methods. The golden rule is to get to the root of the problem: to control larval development (Fig. 11.4). This strategy has usually proved effective. To maximize efficiency, the actual aquatic habitats should be identified accurately. This often involves the aid of a map, to identify breeding sites that support mosquito production. In Europe, the active ingredients used most frequently in larviciding at present (Table 11.2) are chemi-

Table 11.2. Active insecticide ingredients available for larvicide treatments

Active ingredient (AI)	Dosage (AI/ha)	Residual time	Formulations
Organophosphates			
Temephos	56–112 g	2–4 weeks	EC, G, S
Bioinsecticides			
Bacillus thuringiensis var. *israelensis* (strain H14)	250 g	5–10 days	B, G, IG, S, WDP
Bacillus sphaericus	250 g	1–3 weeks	S, WDP
Growth regulators			
Diflubenzuron	25–100 g	1–4 weeks	WDP
Methoprene	100–1000 g	4–8 weeks	SRS
Pyriproxyfen	100 g	4–8 weeks	G
Surface oils	3–5 litres	2–10 days	S

Note. B: briquette; EC: emulsifiable concentrate; G: granulate; IG: ice granules; S: solution; SRS: slow release suspension; WDP: water dispersible powder.

Source: WHO (1996).

cal (temephos), biochemical (such as *Bacillus thuringiensis* var. *israeliensis* and *Bacillus sphaericus*) or IGR (such as diflubenzuron and methoprene), depending on compliance with local regulations. Biochemicals and growth regulators have the advantage of being more specific, whereas chemicals are less expensive and easier to use, especially for very large habitats (WHO, 1996; Chavasse & Yap, 1997).

Fig. 11.4. Larviciding against northern house mosquitoes in manholes

Source: Photo by EID Méditerranée.

11.8.3. Adulticiding

Treating adult mosquitoes can reinforce control, if efficacy in treating breeding sites is poor or if larviciding is not possible. To increase control efficiency, in the framework of controlling a vector or urban pest (or both), it can be advantageous to do a few treatments of this type, limiting both the number of applications and the area covered. Furthermore, such a strategy may avoid large-scale repetitive treatments in a sensitive natural environment that would, otherwise, increase pest control costs. However, adult control, which is usually done by fogging, must be carefully applied, due to its low specificity and to risks (such as allergies or damage to vehicle paintwork) that may result from the product being used. Such applications are often restricted to critical situations, such as the mass production of adult mosquitoes that have not been or could not be controlled at the aquatic immature stages, or situations that result in a risk to health.

Adult mosquitoes can be killed outdoors and indoors, depending on where they rest. When done indoors, it is usually through spraying residual insecticides (generally a

Fig. 11.5. Adulticiding against Asian tiger mosquitoes in an urban environment

Source: Photo by F. Schaffner/EID Méditerranée

pyrethroid; WHO, 2002) on the walls of residences, so that either unfed or recently blood-fed mosquitoes will die when they rest on the sprayed walls. This type of control has a long-lasting effect (up to two months or more), but will be effective only against endophilic species (such as some biotypes of the northern house mosquito and some *Anopheles* spp.). It is important that specially trained and equipped professionals do the spraying.

In some cases, the treatment of adults can be done outdoors – that is, against exophilic mosquitoes – by using ULV applications produced by cold (emulsion) or thermal (diesel suspension) foggers, mounted on a vehicle (Fig. 11.5) or aircraft. This has no long-lasting effect and must be repeated daily during either periods of high risk of disease outbreak or periods of severe nuisance biting, originating from areas inaccessible to larviciding.

The active ingredients available (Tables 11.3 and 11.4) are organophosphates (such as fenitrothion and malathion) or pyrethroids (such as deltamethrin and permethrin), depending on compliance with local regulations. Spraying operations are usually carried out early in the morning – before mosquitoes become active (around 05:00 in summer) and people leave their dwellings – and target the resting places of adult mosquitoes (such as hedges and groves close to human habitations) (WHO, 1996; Chavasse & Yap, 1997; WHO, 2003). Aerial spraying can also be undertaken in the evening and at night, especially if it targets *Culex* spp.; the disadvantage of this is that it is likely that more people will be out of doors and thus exposed to insecticides, and it may also be more difficult or expensive (or both) to get spraying personnel.

Table 11.3. Active insecticide ingredients available for outdoor adulticide treatments

AI	Dosage (AI/ha) cold fog	Dosage (AI/ha) thermal fog
Organophosphates		
Fenitrothion	250–300 g	270–300 g
Fenthion	112 g	—
Malathion	112–693 g	500–600 g
Pyrethroids		
Bioresmethrin	5–10 g	20–30 g
Deltamethrin	0.5–1.0 g	—
Permethrin	5–10 g	—

Source: WHO (1996).

Table 11.4. Active insecticide ingredients for residual (indoor) adulticide treatments

AI	Dosage (AI/m²)	Residual effect	Mode of action
Organophosphates			
Fenitrothion	1–2 g	3 months or more	Contact and ingestion
Malathion	1–2 g	2–3 months	Contact
Pyrethroids			
Cypermethrin	0.5 g	4 months or more	Contact
Deltamethrin	0.05 g	2–3 months	Contact
Permethrin	0.5 g	2–3 months	Contact

Source: WHO (1996).

11.8.4. Use of predators (biological control)

Worldwide, research on about 40 biocontrol agents has received the support of a WHO research programme; these agents include bacteria, fungi, protozoa, nematodes, viruses, fish, insects, snails and plants (Schrieber & Jones, 2000). In some cases, the introduction of fish was successfully used as a biological control agent. The top minnow (*Gambusia affinis*) has been introduced into ditches, rice fields and marshes of southern Europe; its persistence depends on the water surface not freezing. An average of six fish per $10 m^2$ seems to be sufficient for control. While the effectiveness of this predator remains limited in natural environments, because these fish seem to feed on mosquito larvae only in the absence of other food sources, it remains very effective in artificial sites, such as ornamental ponds. For such ponds, some people might prefer more decorative fish, such as cyprinids (such as goldfish), the guppy (*Poecilia reticulata*) or tilapia (*Oreochromis mossambicus*), which are also efficient predators of mosquito larvae.

Other predators have been tested in the United States (Rose, 2001). Some have yet to become available, some have production and storage problems and some are still just candidates. The EPA has registered the entomopathogenic fungus *Lagenidium giganteum* for mosquito control, but products have not become readily available. For technical reasons, the pathogenic protozoon *Brachiola* (*Nosema*) *algerae* has also been unavailable. Entomoparasitic nematodes, such as *Romanomermis culicivorax* and *Romanomermis iyengari*, are effective and do not require EPA registration; however, they are not easily produced and have storage viability limitations. Also, the predacious copepod *Mesocyclops longisetus* preys on mosquito larvae and is a candidate for larval control in water containers under semitropical conditions.

11.8.5. Passive protection

Passive methods of protection may not necessarily eliminate the pest problem, but they can limit its impact. Protection can be achieved by avoiding vector-infested areas, by using physical barriers, such as clothing, screens and nets, by using space repellents, or by applying repellents to skin or clothing, or both (Barnard, 2000).

11.8.5.1. Space repellents

The smoke from burning basil-type herbs, seeds of the neem tree (*Azadirachta indica*), or tree wood and resin of aromatic trees has been used with various levels of success to repel insects. Burning mosquito coils that contain pyrethrum significantly reduce mosquito biting rates outdoors, such as on terraces (they should not be used indoors); other types of pyrethrum dispensers include electric vaporizing mats, electric liquid vaporizers and aerosol spray cans.

11.8.5.2. Personal repellents

Personal repellents are useful when visiting areas where severe biting is likely to be encountered. Among the most efficient repellents applied in most parts of the EU and United States are DEET (*N,N*-diethyl-meta-toluamide), lemon eucalyptus oil (*p*-menthane-3,8-diol), ethyl butylacetylaminopropionate (3-[*N*-acetyl-*N*-butyl]aminopropionic acid ethyl ester) and icaridin (1-piperidinecarboxylic acid 2-(2-hydroxyethyl)-1-methylpropylester); of course, for safe use, some precautions have to be followed. Under special and extreme circumstances, a hooded jacket and fur clothing treated with permethrin are efficient repellents, as was demonstrated during studies undertaken in the Florida Everglades (Schreck, Haile & Kline, 1984).

11.8.5.3. Bednets

Bednets are simple to install, lightweight and durable. To be effective, they should cover sleepers completely yet be large enough so that sleepers do not come in contact with the fabric. Holes in bednets should be mended immediately, and nets should be let down before darkness and tucked beneath the sleeping mat or mattress.

WHO has initiated programmes to reduce the transmission of tropical diseases, using insecticide-impregnated bednets (Rozendaal, 1997; WHO, 2001). Insecticide-treated bednets have been successful in reducing the number of malaria infections in villages where the level of transmission of malaria is low or moderate. In some villages where the level of transmission is high, the use of insecticide-impregnated bednets has reduced the parasitic load of infected people and, as a consequence, reduced morbidity by 50% and mortality by 20%.

11.8.6. Genetic control and transgenic mosquitoes

Diverse approaches have been considered for genetic control of nuisance or vector mosquito populations. Genetic control consists of the release of genetically modified individuals in the field, to reduce or modify the composition of natural populations of target insects. It is based on a number of approaches (Asman, McDonald & Prout, 1981; Grover, 1985; Alphey et al., 2002):

- the introduction of sterility in males by irradiation or chemicals (the so-called sterile insect technique);

- sterility induced by hybridization (sterility inherited by crossing sibling species);

- cytoplasmic incompatibility (natural incompatibility of some allopatric populations of the same species); and
- chromosomal translocation (sterility inherited by crossing normal individuals with heterozygotes having a translocation).

Except for some trials that used irradiated males, the theoretical prediction of the long-term impact of these methods on suppressing nuisance mosquitoes was not conclusive in field applications. Among the various possible reasons for this failure are:

- an incomplete mixing of the populations released with the natural populations or behavioural differences in reproduction between these two populations;
- the density threshold that allows regulation of the population after the introduction of the genetic modification was not reached;
- the absence of compensation, by an increase in the number of released genetically modified individuals when the natural population grows or the lack of competitiveness of the released sterile males; and
- immigration of fertile females towards the trial area from nearby areas.

At present, research is concentrating on the production of transgenic mosquitoes. These are genetically modified to:

- prevent pathogen development in (and transmission by) the insects (Blair, Adelman & Olson, 2000; Christophides, 2005);
- foster insecticide susceptibility or prevent insecticide resistance (Carlson et al., 1995; Collins & James, 1996; Hemingway, 1999); or
- reduce mosquito populations (Benedict & Robinson, 2003; Pates & Curtis 2005).

Apart from ethical concerns about the release of uncontrollable transgenic organisms, it is doubtful whether the modified mosquitoes are sufficiently fit to replace the natural populations (Spielman, 1994; Catteruccia, Godfray & Crisanti, 2003).

11.9. Economic burden of mosquitoes

The economic burden of controlling mosquitoes and mosquito-borne diseases is very difficult to assess, and few data from North America and Europe are available. Doubtless, expenditures in the United States have increased considerably since the advent of WNV, as control and preventive measures have been intensified in many states.

As with other pests, various items contribute to their economic burden – such as mosquito control, medical treatment and health care (in the case of associated diseases), and loss of productivity – and additional costs may be incurred for preventive medicinal measures. For example, health authorities throughout the United States recently introduced mandatory screening for WNV in blood from blood donors (Bren, 2003). Despite the lack of systematic studies, representative data are available to illustrate the huge economic impact of mosquitoes.

Mosquitoes can affect people's livelihoods and property values, as illustrated in the following example from Germany. The German Mosquito Control Association is a semi-private and semi-public institution responsible for mosquito control in the Upper Rhine Valley of Germany. The area it covers stretches 300 km along the Rhine and covers about 6000 km^2. With regard to mosquito breeding sites, it is a particularly sensitive region, because of the varying river levels and its vast floodplains. Several times a year, some 2.5 million residents are at risk for mosquito nuisances. The Association has an annual budget of about €2 million, derived from public taxes. According to a cost–effectiveness analysis, the economic loss to be expected in the absence of mosquito control is a factor of 3.7 greater than this budget or about €7.5 million (N. Becker, German Mosquito Control Association, personal communication, 2006). Without mosquito control, this economic burden would be due mainly to losses in the gastronomic trade and recreational sectors and to expenditures for private mosquito control. Without mosquito control, business volume in gastronomy would be reduced by 25% during the peak summer season and by about 10% for the whole year. Moreover, due to mosquito control, property values have increased in this region, although figures are not available.

At present, the most important disease transmitted by mosquitoes in Europe and North America is West Nile fever. In 2002, 329 of a total of 4156 West Nile fever cases reported in the United States were from Louisiana. Of the people infected in Louisiana, 204 had an illness that involved the central nervous system and 24 of them died. According to a conversion of hospital charges to economic costs, using Medicare cost-to-charge ratios, the estimated cost of the Louisiana WNV epidemic was US$ 20.1 million from June 2002 to February 2003 (Zohrabian et al., 2004). This figure can be divided into two components: (1) a US$ 10.9 million cost of illness, comprising US$ 4.4 million in medical costs and US$ 6.5 million in nonmedical costs (such as loss of productivity, illness-related costs for transport and child care costs.); and (2) a US$ 9.2 million cost for the public health response, such as control of the epidemic.

The spread of WNV and evidence of transmission by transfusion (CDC, 2004a) prompted the United States Food and Drug Administration to institute mandatory screening of the nation's blood supplies and to regulate the routine screening of blood donations for the virus. A prospective cost analysis, on the basis of 2 million transfusions, was calculated to be between US$ 7 million and US$ 19 million, depending on the specific test applied and on the period of the year the blood samples were to be screened – that is, year-round screening or only screening during months of high incidences of WNV. However, screening by a blood donor questionnaire alone in low-transmission areas with short WNV seasons was the most cost-effective strategy, whereas in areas with high levels of WNV transmission seasonal screening of individual samples and restricting screening to blood donations designated for immunocompromised recipients was most cost effective (Korves, Goldie & Murray, 2006).

Efforts are in progress to develop a vaccine against WNV (Chang et al., 2004; Hall & Khromykh, 2004). However, a cost–effectiveness analysis calculated for the United States indicated that universal vaccination would be unlikely to result in societal monetary savings, compared with the cost per case of illness under the present disease incidence rates (Zohrabian, Hayes & Petersen, 2006). According to this analysis, at a cost of US$ 8.7 billion in a hypothetical population of 100 million people, vaccination would prevent 256 000 cases of illness (including neuroinvasive diseases, lifetime disabilities and deaths) from WNV during a 10-year period, given an average cost per case of illness of about US$ 34 000. Under these assumptions, a universal vaccination programme would be cost effective only when the incidence of disease increased substantially or the costs of vaccination were below US$ 12 per person (in the analysis, baseline vaccination costs were assumed to be US$ 100).

The economic impact of mosquitoes, in particular of mosquito-borne diseases, must also be expanded to the animal industry. Horses are highly susceptible to infection with WNV, and many infections end in their death (Murgue et al., 2001). Thus, for example, WNV cost the equine industries in Colorado and Nebraska more than US$ 1.25 million in 2002, with an additional US$ 2.75 million estimated for preventive measures (USDA, 2003). Older data, from 1965, tell that American losses from mosquito attacks on livestock reached US$ 25 million that year, including a US$ 10 million decline in milk production (Rodhain & Perez, 1985).

11.10. Benchmarks

Mosquitoes are virtually ubiquitous – that is, they are adapted to a wide range of environmental conditions and develop in numerous different types of water bodies, including organically polluted ones and small and often concealed aquatic habitats. Also, the ability of adult females, including infected ones, to disperse some distance from their larval habitats in search of a blood-meal threatens many localities with their bites, disease transmission or both. According to the species, biting can take place indoors or outdoors, during the day, and in the evening or at night-time. Control measures can usually only reduce mosquito abundances and mosquito–human contacts. This can be achieved through a variety of methods, including personal protection through the use of bednets, window screens, appropriate clothing and insect repellents.

Public efforts to control mosquitoes, however, should be implemented when:

- autochthonous mosquito-borne infections in people have been identified and endemic transmission poses a threat; or
- an intolerable level of mosquito biting is reached.

The borderline between acceptable and unacceptable infestation levels may vary from region to region, from time to time and from culture to culture and must be determined (and occasionally reassessed) according to actual requirements.

With some exceptions, the mosquito nuisance in Europe and the United States is usually rather limited, because of reasonable water management systems within urban areas. In more rural residential areas that have insufficient water management or have experienced floods, the mass emergence of adult mosquitoes may follow, which may cause increased biting or disease transmission, both in rural areas and urban or suburban areas. As a consequence, mosquito control programmes should be implemented. However, the appropriate facilities for controlling such outbreaks of mosquitoes are not commonly available. The choice of management strategy for controlling mosquitoes should be made according to the specific situation and according to preset national or federal guidelines. It always should be based on an IPM philosophy – that is, various approaches (physical, biological, biochemical and chemical) should be combined, and applications of pesticides should be minimized as much as possible. Experts are therefore essential for both planning control programmes and implementing them. These programmes must include:

- monitoring and identifying mosquito species;
- locating their breeding sites;
- continuously assessing adult or larval (or both) densities; and
- having a clear strategy of control goals.

Information (such as the risks associated with the application of various pesticides) and educating the public (such as on how to prevent unintentional provision of breeding sites in the urban residential environment) are fundamental for effective control.

Also, ongoing climate and environmental changes should be observed carefully, as they may ameliorate conditions for mosquito development – for example, mosquito seasons may be extended, and mosquito habitats and densities may increase, which may lead to a greater impact on health. Furthermore, increased international travel and traffic in consumer items and animals increase the risk of introducing pathogens. This is especially true of viruses, when competent mosquito vectors are already present or when, in addition to the introduction of new viruses, mosquito vectors are also introduced. Permanent surveillance for autochthonous vectors and also for introduced potential vectors and pathogens will allow better risk assessment and management. The complex interactions between mosquitoes, people and the environment, and the impact mosquitoes may have on public health (Fig. 11.6), calls for more international cooperation in legislative, executive and research matters.

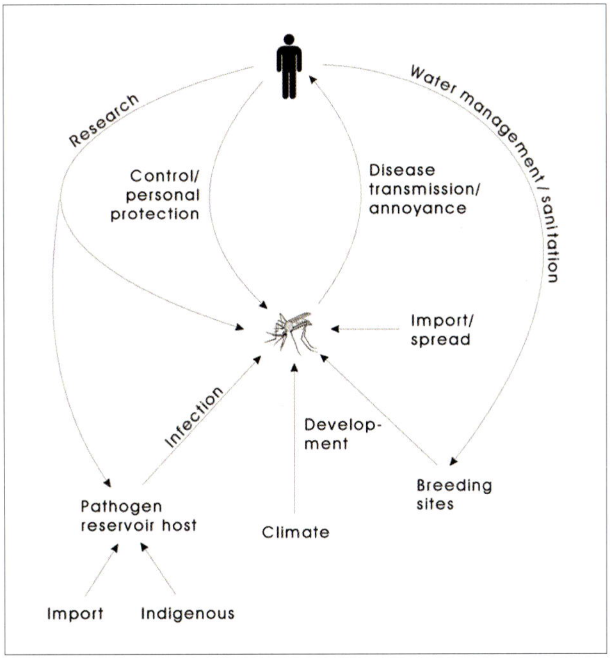

Fig. 11.6. Public health related interactions between mosquitoes, environment and people

11.11. Conclusions

For decades, in the high- and middle-income parts of the world, the field of medical entomology has been neglected in many countries in temperate climate zones, because no urgent demands for specialists were apparent. However, for various and partly unknown reasons and for some time, we have been encountering increasingly more problems with arthropod nuisances and disease transmission. It is therefore essential to re-intensify research in medical entomology and to train medical entomologists to deal with vector-borne diseases and their control. These arguments apply particularly to mosquitoes, which are (next to ticks) the most ubiquitous and medically important arthropods. Therefore, not only cases of mosquito-borne diseases, but also the distribution and abundance of mosquitoes have to be monitored regularly by specialized, government-authorized institutions.

With regard to the distribution of possible vectors and the occurrence of vector-borne diseases, it is necessary to intensify international and European collaboration, both on the legislative and executive levels. In Europe, notification systems should be standardized and notifications reported to a central agency, such as the ECDC. Like the CDC in the United States, this institution should be further promoted, and its duties should be extended, to collect and analyse vector-related epidemiological data and to disseminate them to the public. Its duties may also include coordinating public health research programmes. For example, more epidemiological data about mosquito-borne virus activity in Europe, such as that about WNV, are urgently needed. Moreover, there is hardly any information on the temporal and spatial circulation of encephalitis viruses; thus, the etiology of cases of summer encephalitides should be followed up.

To minimize the risk of establishing and spreading newly introduced exotic vectors in the WHO Europe Region, such as the Asian tiger mosquito, it is necessary to support collaboration between European surveillance teams, which is getting under way within the EMCA. Furthermore, it is very important to provide legal requirements for adequate tyre storage (indoors or at places not exposed to rainfall) and tyre traceability. It is also necessary to implement more efficient control of international animal transport and to strengthen guidelines and directives for such control. Moreover, aircraft and ship disinfection should be reassessed, and compulsory international rules should be established.

Mosquito control management, as a consequence of mosquito monitoring, should be efficient and at the same time ecologically sound. Governments would benefit from establishing a network of centres in each country or state to gather information and form monitoring and operational (control) tasks. To avoid the formation of mosquito breeding sites by city management or landscaping (such as through the restoration of large tracts of land along rivers to a natural state), it is essential for building authorities to collaborate with biologists who are knowledgeable about the occurrence, biology and ecology of the indigenous culicid species. It is strongly recommended to harmonize mosquito control practices, to reduce unwelcome environmental impacts. Ecological risks posed by insecticides can be checked in the laboratory, but risk monitoring in the field will provide further knowledge for making decisions about the implementation of mosquito control in the framework of sustainable management of wetlands.

Within the EU, guidelines for insecticide review and application should be harmonized between the Member States. Formation of an insecticide panel, to facilitate effective control of a biting nuisance or vector transmission, must be guaranteed for the future. However, possible health hazards caused by control activities must not exceed those of the pests they are intended to control.

References[1]

Adhami J, Reiter P (1998). Introduction and establishment of *Aedes (Stegomyia) albopictus* Skuse (Diptera: Culicidae) in Albania. *Journal of the American Mosquito Control Association*, 14:340–343.

Alphey L et al. (2002). Malaria control with genetically manipulated insect vectors. *Science*, 298:119–121.

Aranda C, Eritja R, Roiz D (2006). First record and establishment of *Aedes (Stegomyia) albopictus* in Spain. *Medical and Veterinary Entomology*, 20:150–152.

Asman SM, McDonald PT, Prout T (1981). Field studies of genetic control systems for mosquitoes. *Annual Review of Entomology*, 26:289–318.

Autorino GL et al. (2002). West Nile virus epidemic in horses, Tuscany region, Italy. *Emerging Infectious Diseases*, 8:1372–1378.

Bae HG et al. (2005). Analysis of two imported cases of yellow fever infection from Ivory Coast and The Gambia to Germany and Belgium. *Journal of Clinical Virology*, 33:274–280.

Baldari M et al. (1998). Malaria in Maremma, Italy. *Lancet*, 351:1246–1247.

Barnard DR (2000). *Global collaboration for development of pesticides for public health: repellents and toxicants for personal protection: position paper*. Geneva, World Health Organization (document WHO/CDC/WHOPES/GCDPP/2000.5; http://whqlibdoc.who.int/hq/2000/WHO_CDS_WHOPES_GCDPP_2000.5.pdf, accessed 5 October 2006).

Beaty ME et al. (2005). Travel-associated dengue infections – United States, 2001–2004. *Morbidity and Mortality Weekly Report*, 54:556–558.

Becker N et al. (2003). *Mosquitoes and their control*. New York, Kluwer Academic/Plenum Publishers.

Benedict MQ, Robinson AS (2003). The first releases of transgenic mosquitoes: an argument for the sterile insect technique. *Trends in Parasitology*, 19:349–355.

Bennett JK et al. (2005). New state record for the Asian tiger mosquito, *Aedes albopictus* (Skuse). *Journal of the American Mosquito Control Association*, 21:341–343.

[1] The authors selected the reference works that appear here according to their best knowledge. Whenever possible, peer-reviewed journals were cited. In very few cases, information or data were available only through other documents, but renowned institutions or organizations or authors published these.

Blair CD, Adelman ZN, Olson KE (2000). Molecular strategies for interrupting arthropod-borne virus transmission by mosquitoes. *Clinical Microbiology Reviews*, 13:651–661.

Brault AC et al. (2004). Differential virulence of West Nile strains for American crows. *Emerging Infectious Diseases*, 10:2161–2168.

Bren L (2003). West Nile virus: reducing the risk. *FDA Consumer*, 37:20–27.

Brook JH et al. (1994). Brief report: malaria probably locally acquired in New Jersey. *The New England Journal of Medicine*, 331:22–23.

Bruce-Chwatt LJ, de Zulueta J (1980). *The rise and fall of malaria in Europe.* Oxford, Oxford University Press.

Buckley A et al. (2003). Serological evidence of West Nile virus, Usutu virus and Sindbis virus infection of birds in the UK. *The Journal of General Virology*, 84:2807–2817.

Çağlar SS, Alten B (2000). Malaria situation and its vectors in Turkey. In: Çağlar SS, Alten B, Özer N, eds. *Proceedings of the 13th European Society for Vector Ecology Meeting, Belek-Antalya, Turkey, 24–29 September 2000.* Ankara, DTP:234–245.

Callot J, Delécolle JC (1972). Notes d'entomologie – VI. Localisation septentrionale d'*Aedes aegypti*. *Annales de Parasitologie*, 47:665.

Cancrini G et al. (2003a). *Aedes albopictus* is a natural vector of *Dirofilaria immitis* in Italy. *Veterinary Parasitology*, 118:195–202.

Cancrini G et al. (2003b). First finding of *Dirofilaria repens* in a natural population of *Aedes albopictus*. *Medical and Veterinary Entomology*, 17:448–451.

Cardamatis JP (1929). La dengue en Grèce. *Bulletin de la Société Pathologie Exotique*, 22:272–292.

Carlson DA et al. (1995). Molecular genetic manipulation of mosquito vectors. *Annual Review of Entomology*, 40:359–388.

Castelli F et al. (1993). 'Baggage malaria' in Italy: cryptic malaria explained. *Transactions of the Royal Society of Tropical Medicine and Hygiene*, 87:394.

Catteruccia F, Godfray HCJ, Crisanti A (2003). Impact of genetic manipulation on the fitness of *Anopheles stephensi* mosquitoes. *Science*, 299:1225–1227.

CDC (1995). Local transmission of *Plasmodium vivax* malaria – Houston, Texas, 1994. *Morbidity and Mortality Weekly Report*, 44:295, 301–303.

CDC (2000). Map: distribution of *Aedes albopictus* in the United States, by county, 2000 [web site]. Fort Collins, CO, National Center for Infectious Diseases, Centers for Disease Control and Prevention (http://www.cdc.gov/ncidod/dvbid/arbor/albopic_97_sm.htm, accessed 4 October 2006).

CDC (2001). Arboviral encephalitides [web site]. Fort Collins, CO, National Center for Infectious Diseases, Centers for Disease Control and Prevention (http://www.cdc.gov/ncidod/dvbid/arbor/cases-sle-1964-2000.htm, accessed 3 October 2006).

CDC (2003). Epidemic/enzootic West Nile virus in the United States: guidelines for surveillance, prevention and control, 3rd revision [online monograph]. Fort Collins, CO, National Center for Infectious Diseases, Centers for Disease Control and Prevention (http://www.cdc.gov/ncidod/dvbid/westnile/resources/wnv-guidelines-aug-2003.pdf, accessed 6 December 2006).

CDC (2004a). Update: West Nile virus screening of blood donations and transfusion-associated transmission – United States, 2003. *Morbidity and Mortality Weekly Report*, 53:281–284.

CDC (2004b). Mosquito-borne diseases [web site]. Fort Collins, CO, National Center for Infectious Diseases, Centers for Disease Control and Prevention (http://www.cdc.gov/ncidod/diseases/list_mosquitoborne.htm, accessed 8 October 2006).

CDC (2005). West Nile virus: entomology [web site]. Fort Collins, CO, National Center for Infectious Diseases, Centers for Disease Control and Prevention (http://www.cdc.gov/ncidod/dvbid/westnile/mosquitoSpecies.htm, accessed 3 October 2006).

CDC (2006). West Nile virus: statistics, surveillance, and control [web site]. Fort Collins, CO, National Center for Infectious Diseases, Centers for Disease Control and Prevention (http://www.cdc.gov/ncidod/dvbid/westnile/surv&control.htm, accessed 3 October 2006).

CDHS/M&VCAC/UC (2004). *California mosquito-borne virus surveillance & response plan*. Sacramento, CA, California Department of Health Services, Mosquito & Vector Control Association of California, and University of California. (http://westnile.ca.gov/website/publications/2005_ca_mosq_response_plan.pdf, accessed 12 January 2007).

Chang GJ et al. (2004). Recent advancement in flavivirus vaccine development. *Expert Review of Vaccines*, 3:199–220.

Chavasse DC, Yap HH (1997). *Chemical methods for the control of vectors and pests of public health importance*. Geneva, World Health Organization (document WHO/CTD/WHOPES/97.2; http://whqlibdoc.who.int/hq/1997/WHO_CTD_WHOPES_97.2.pdf, accessed 5 October 2006).

Christophides GK (2005). Transgenic mosquitoes and malaria transmission. *Cellular Microbiology*, 7:325–333.

Collins FH, James AA (1996). Genetic modification of mosquitoes. *Science & Medicine*, 3:52–61.

Cuadros J et al. (2002). *Plasmodium ovale* malaria acquired in central Spain. *Emerging Infectious Diseases*, 8:1506–1508.

Darsie RF Jr, Ward RA (1981). Identification and geographical distribution of the mosquitoes of North America, north of Mexico. *Mosquito Systematics Supplement*, 1:1–313.

Davison KL et al. (2003). Viral encephalitis in England, 1989–1998: what did we miss? *Emerging Infectious Diseases*, 9:234–240.

Day JF (2001). Predicting St. Louis encephalitis virus epidemics: lessons from recent, and not so recent, outbreaks. *Annual Review of Entomology*, 46:111–138.

Diallo M et al. (1999). Vectors of Chikungunya virus in Senegal: current data and transmission cycles. *The American Journal of Tropical Medicine and Hygiene*, 60:281–286.

Eager JM (1902). Yellow fever in France, Italy, Great Britain and Austria, and bibliography of yellow fever in Europe. *Yellow Fever Institute Bulletin*, 8:25–35.

Eliades MJ et al. (2005). Malaria surveillance – United States, 2003. *Morbidity and Mortality Weekly Report: Surveillance Summaries*, 54(2):25–40.

Flacio E et al. (2004). Primo ritrovamento di *Aedes albopictus* in Svizzera [First detection of *Aedes albopictus* in Switzerland]. *Bollettino della Società Ticinese di Scienze Naturali*, 92:141–142 (in Italian).

Fonseca DM et al. (2004). Emerging vectors in the *Culex pipiens* complex. *Science*, 303:1535–1538.

Fontenille D, Rodhain F (1989). Biology and distribution of *Aedes albopictus* and *Aedes aegypti* in Madagascar. *Journal of the American Mosquito Control Association*, 5:219–225.

Gould LH, Fikrig E (2004). West Nile virus: a growing concern? *Journal of Clinical Investigation*, 113:1102–1107.

Gratz NG (2004). *The vector-borne human infections of Europe: their distribution and burden on public health*. Copenhagen, WHO Regional Office for Europe (document EUR/04/5046114; www.euro.who.int/document/E82481.pdf, accessed 7 October 2006).

Gratz NG, Steffen R, Cocksedge W (2000). Why aircraft disinsection? *Bulletin of the World Health Organization*, 78:995–1003 (http://www.who.int/docstore/bulletin/pdf/2000/issue8/99-0285.pdf, accessed 7 October 2006).

Grover KK (1985). Chemosterilization trials against *Aedes aegypti*. In: Laird M, Milnes JJ, eds. *Integrated mosquito control methodologies. Vol. 2.* London, Academic Press: 79–116.

Guillet P et al. (1998). Origin and prevention of airport malaria in France. *Tropical Medicine & International Health*, 3:700–705.

Hall P et al. (2002). Fatal yellow fever in a traveller returning from Amazonas, Brazil, 2002. *Morbidity and Mortality Weekly Report*, 51:324–325.

Hall RA, Khromykh AA (2004). West Nile virus vaccines. *Expert Opinion on Biological Therapy*, 4:1295–1305.

Halstead SB et al. (1969a). Dengue and chikungunya virus infection in man in Thailand, 1962–1964. IV. Epidemiologic studies in the Bangkok metropolitan area. *The American Journal of Tropical Medicine and Hygiene*, 18:997–1021.

Halstead SB et al. (1969b). Dengue and chikungunya virus infection in man in Thailand, 1962–1964. V. Epidemiologic observations outside Bangkok. *The American Journal of Tropical Medicine and Hygiene*, 18:1022–1033.

Hemingway J (1999). Insecticide resistance in malaria vectors: a new approach to an old subject. *Parassitologia*, 41:315–318.

Hermosilla C et al. (2006). First autochthonous case of canine ocular *Dirofilaria repens* infection in Germany. *The Veterinary Record*, 158:134–135.

Higgs S, Snow K, Gould EA (2004). The potential of West Nile virus to establish outside of its natural range: a consideration of potential mosquito vectors in the United Kingdom. *Transactions of the Royal Society of Tropical Medicine and Hygiene*, 98:82–87.

Hubálek Z, Halouzka J (1999). West Nile fever – a re-emerging mosquito-borne viral disease in Europe. *Emerging Infectious Diseases*, 5:643–649.

Institut de Veille Sanitaire (2006). Actualités [web site]. Saint-Maurice, Institut de Veille Sanitaire (http://www.invs.sante.fr/actualite/index.htm, accessed 3 October 2006).

Isaacson M (1989). Airport malaria: a review. *Bulletin of the World Health Organization*, 67:737–743.

JME Editors (2005). Journal policy on names of aedine mosquito genera and subgenera. *Journal of Medical Entomology*, 42:511.

Jupp PG, McIntosh BM (1988). Chikungunya virus disease. In: Monath TP, ed. *The arboviruses: epidemiology and ecology. Vol. II*. Boca Raton, CRC Press: 137–157.

Kampen H (2004). Die Differenzierung von *Anopheles*-Zwillingsarten (Diptera: Culicidae). *Denisia*, 13:497–513.

Kampen H et al. (2002). Individual cases of autochthonous malaria in Evros Province, northern Greece: serological aspects. *Parasitology Research*, 88:261–266.

Kay C et al. (2005). Transmission of malaria in resort areas – Dominican Republic, 2004. *Morbidity and Mortality Weekly Report*, 53:1195–1198.

Kilpatrick AM et al. (2005). West Nile virus risk assessment and the bridge vector paradigm. *Emerging Infectious Diseases*, 11:425–429.

Klee AL et al. (2004). Long-term prognosis for clinical West Nile virus infection. *Emerging Infectious Diseases*, 10:1405–1411.

Klobučar A et al. (2006). First record of *Aedes albopictus* in Croatia. *Journal of the American Mosquito Control Association*, 22: 147–148.

Knudsen AB, Romi R, Majori G (1996). Occurrence and spread in Italy of *Aedes albopictus*, with implications for its introduction into other parts of Europe. *Journal of the American Mosquito Control Association*, 12:177–183.

Korves CT, Goldie SJ, Murray MB (2006). Cost-effectiveness of alternative blood-screening strategies for West Nile virus in the United States. *PLoS Medicine*, 3:e21.

Krüger A et al. (2001). Two cases of autochthonous *Plasmodium falciparum* malaria in Germany with evidence for local transmission by indigenous *Anopheles plumbeus*. *Tropical Medicine & International Health*, 6:983–985.

Lanciotti RS et al. (1999). Origin of the West Nile virus responsible for an outbreak of encephalitis in the northeastern United States. *Science*, 286:2333–2337.

Layton M et al. (1995). Malaria transmission in New York City, 1993. *Lancet*, 346:729–731.

Lundström JO (1994). Vector-competence of western European mosquitoes for arboviruses: a review of field and experimental studies. *Bulletin of the Society for Vector Ecology*, 19:23–36.

Lundström JO (1999). Mosquito-borne viruses in western Europe: a review. *Journal of Vector Ecology*, 24:1–39.

Majori G, Sabatinelli G, Kondrachine AV (1999). Re-emerging malaria in the WHO European Region: control priorities and constraints. *Parassitologia*, 41:327–328.

Malavige GN et al. (2004). Dengue viral infections. *Postgraduate Medical Journal*, 80:588–601.

Maldonado YA et al. (1990). Transmission of *Plasmodium vivax* malaria in San Diego County, California, 1986. *The American Journal of Tropical Medicine and Hygiene*, 42:3–9.

Medlock JM, Snow KR, Leach S (2005). Potential transmission of West Nile virus in the British Isles: an ecological review of candidate mosquito bridge vectors. *Medical and Veterinary Entomology*, 19:2–21.

Mitchell CJ (1995). The role of *Aedes albopictus* as an arbovirus vector. *Parassitologia*, 37:109–113.

Moore DL et al. (1974). An epidemic of chikungunya fever at Ibadan, Nigeria, 1969. *Annals of Tropical Medicine and Parasitology*, 68:59–68.

Mostashari F et al. (2001). Epidemic West Nile encephalitis, New York, 1999: results of a household-based seroepidemiological survey. *Lancet*, 358:261–264.

Mouchet J (2000). Airport malaria: a rare disease still poorly understood. *Euro Surveillance*, 5:75–76.

MSS/MAP/MMEDD (2005). *Guide de procédures de lutte contre la circulation du virus West Nile en France métropolitaine*. Paris, Ministère de la Santé et des Solidarités, Ministère de l'Agriculture et de la Pêche, Ministère de l'Ecologie et du Développement Durable (http://www.sante.gouv.fr/htm/pointsur/zoonose/guide_WestNil_0507.pdf, accessed 7 December 2006).

Murgue B et al. (2001). West Nile outbreak in southern France, 2000: the return after 35 years. *Emerging Infectious Diseases*, 7:692–696.

Muro A et al. (1999). Human dirofilariasis in the European Union. *Parasitology Today*, 15:386–389.

Nikolaeva N (1996). Resurgence of malaria in the former Soviet Union (FSU). *Society of Vector Ecology Newsletters*, 27:10–11.

Pampiglione S, Canestri Trotte G, Rivasi F (1995). Human dirofilariasis due to *Dirofilaria (Nochtiella) repens*: a review of world literature. *Parassitologia*, 37:149–193.

Pampiglione S, Rivasi F (2000). Human dirofilariasis due to *Dirofilaria* (*Nochtiella*) *repens*: an update of world literature from 1995 to 2000. *Parassitologia*, 42:231–254.

Panthier R (1968). Épidemiologie du virus West Nile. Étude d'un foyer en Camargue: I. Introduction. *Annales de l'Institut Pasteur*, 114:518–520.

Pates H, Curtis C (2005). Mosquito behaviour and vector control. *Annual Review of Entomology*, 50:53–70.

Pepperell C et al. (2003). West Nile virus infection in 2002: morbidity and mortality among patients admitted to hospital in southcentral Ontario. *Canadian Medical Association Journal*, 168:1399–1405.

Petrić D et al. (2001). *Aedes albopictus* (Skuse, 1894), nova vrsta komarac (Diptera: Culicidae) u entomofauni Jugoslavije [*Aedes albopictus* (Skuse, 1894), a new mosquito species (Diptera: Culicidae) in the insect fauna of Yugoslavia]. *Abstract Volume of Symposia of Serbian Entomologists, 2001. Entomological Society of Serbia, Goc, 26–29 September 2001*: 29 (in Serbian).

Peyton EL et al. (1999). *Aedes (Finlaya) japonicus japonicus* (Theobald), a new introduction into the United States. *Journal of the American Mosquito Control Association*, 15:238–241.

Platonov AE et al. (2001). Outbreak of West Nile virus infection, Volgograd region, Russia, 1999. *Emerging Infectious Diseases*, 7:128–132.

ProMED Mail [web site] (2006). Chikungunya – Indian Ocean update (11): islands, India. Brookline, MA, International Society for Infectious Diseases (archive number: 20060330.0961;
http://www.promedmail.org/pls/askus/f?p=2400:1001:424240::::F2400_P1001_BACK_PAGE,F2400_P1001_ARCHIVE_NUMBER,F2400_P1001_USE_ARCHIVE:1001,20060330.0961,Y, accessed 3 October 2006).

Rao TR (1966). Recent epidemics caused by Chikungunya virus in India, 1963–1965. *Scientific Culture*, 32:215.

Reinert JF (2000). New classification for the composite genus *Aedes* (Diptera: Culicidae: Aedini), elevation of subgenus *Ochlerotatus* to generic rank, reclassification of the other subgenera, and notes on certain subgenera and species. *Journal of the American Mosquito Control Association*, 16:175–188.

Reinert JF, Harbach RE, Kitching IJ (2004). Phylogeny and classification of Aedini (Diptera: Culicidae), based on morphological characters of all life stages. *Zoological Journal of the Linnean Society*, 142:289–368.

Reiter P et al. (2003). Texas lifestyle limits transmission of dengue virus. *Emerging Infectious Diseases*, 9:86–89.

Rightor JA, Farmer BR, Clarke JL Jr. (1987). *Aedes albopictus* in Chicago, Illinois. *Journal of the American Mosquito Control Association*, 3:657.

Rodhain F, Perez C (1985). *Précis d'entomologie médicale et vétérinaire; notions d'épidémi-*

ologie des maladies à vecteurs. Paris, Maloine.

Romi R, di Luca M, Majori D (1999). Current status of *Aedes albopictus* and *Ae. atropalpus* in Italy. *Journal of the American Mosquito Control Association*, 15:425–427.

Rose RI (2001). Pesticides and public health: integrated methods of mosquito management. *Emerging Infectious Diseases*, 7:17–23.

Rosen L (1986). Dengue in Greece in 1927 and 1928 and the pathogenesis of dengue hemorrhagic fever: new data and a different conclusion. *The American Journal of Tropical Medicine and Hygiene*, 35:642–653.

Ross RW (1956). Original isolation and characteristics of Chikungunya virus. *Journal of Hygiene*, 54:192–200.

Rossi L et al. (1999). Quattro specie di culicidi come possibili vettori di *Dirofilaria immitis* nella risaia piemontese [Four species of mosquito as possible vectors for *Dirofilaria immitis* in Piedmont rice fields]. *Parassitologia*, 41:537–542 (in Italian).

Rozendaal JA (1997). Vector *control: methods for use by individuals and communities*. Geneva, World Health Organization (www.who.int/whopes/resources/vector_rozendaal/en/, accessed 5 October 2006).

Sabatini A et al. (1990). *Aedes albopictus* in Italia: possibile diffusione della specie nell'area mediterranea [*Aedes albopictus* in Italy: possible diffusion of the species into the Mediterranean area]. *Parassitologia*, 32:301–304 (in Italian).

Salvan M, Mouchet J (1994). *Aedes albopictus* et *Aedes aegypti* à l'île de La Réunion. *Annales de la Société Belge de Médecine Tropicale*, 74:323–326.

Samanidou-Voyadjoglou A et al. (2005). Confirmation of *Aedes albopictus* (Skuse) (Diptera: Culicidae) in Greece. *European Mosquito Bulletin*, 19:10–12.

Sartori M et al. (1989). A case of autochthonous malaria in Italy. *Scandinavian Journal of Infectious Diseases*, 21:357–358.

Schaffner F (2003). *The mosquitoes and the community*. Copenhagen, WHO Regional Office for Europe (Local Authorities, Health and Environment Briefing Pamphlets Series, No. 39).

Schaffner F, Chouin S, Guilloteau J (2003). First record of *Ochlerotatus* (*Finlaya*) *japonicus japonicus* (Theobald, 1901) in metropolitan France. *Journal of the American Mosquito Control Association*, 19:1–5.

Schaffner F, van Bortel W, Coosemans M (2004). First record of *Aedes* (*Stegomyia*) *albopictus* in Belgium. *Journal of the American Mosquito Control Association*, 20:201–203.

Schaffner F et al. (2001a). *The mosquitoes of Europe. An identification and training programme* [CD-ROM]. Montpellier, France, IRD Éditions & EID Méditerranée.

Schaffner F et al. (2001b). *Aedes albopictus* (Skuse, 1894) established in metropolitan France. *European Mosquito Bulletin*, 9:1–3.

Scholte EJ et al. (2007). The Asian tiger mosquito (*Aedes albopictus*) in the Netherlands: should we worry? *Proceedings of the Netherlands Entomological Society Meeting*, 18:131–136.

Schreck CE, Haile DG, Kline DL (1984). The effectiveness of permethrin and DEET, alone or in combination, for protection against *Aedes taeniorhynchus*. *The American Journal of Tropical Medicine and Hygiene*, 33:725–730.

Schrieber ET, Jones C (2000). *Mosquito control handbook: an overview of biological control*. Gainesville, Institute of Food and Agricultural Sciences, University of Florida (http://edis.ifas.ufl.edu/pdffiles/IN/IN06100.pdf#search=%22Schrieber%20ET%2C%20Jones%20C%20(2000).%20Mosquito%20control%20handbook%3A%20an%20overview%20of%20biological%20control.%20Gainesville%2C%20Institute%20of%20Food%20and%20Agricultural%20Sciences%2C%20University%20of%20Florida%22, accessed 5 October 2006).

Selover C et al. (2004). Multifocal autochthonous transmission of malaria – Florida, 2003. *Morbidity and Mortality Weekly Report*, 53:412–413.

Service MW (1993). *Mosquito ecology – field sampling methods*, 2nd ed. London, Elsevier Applied Science.

Shah S et al. (2004). Malaria surveillance – United States, 2002. *Morbidity and Mortality Weekly Report: Surveillance Summaries*, 53:21–34.

Simón F et al. (2005). What is happening outside North America regarding human dirofilariasis? *Veterinary Parasitology*, 133:181–189.

Snow K, Ramsdale C (2002). Mosquitoes and tyres. *Biologist (London, England)*, 49:49–52.

Snow K, Ramsdale C (2003). A revised checklist of European mosquitoes. *European Mosquito Bulletin*, 15:1–5.

Spielman A (1994). Why entomological antimalaria research should not focus on transgenic mosquitoes. *Parasitology Today*, 10:374–376.

Sprenger D, Wuithiranyagool T (1986). The discovery and distribution of *Aedes albopictus* in Harris County, Texas. *Journal of the American Mosquito Control Association*, 2:217–219.

SSSQ (2006). Plan d'intervention de protection de la santé publique contre le virus du Nil occidental. Montreal, Direction des communications du Ministère de la Santé et des Services Sociaux du Québec (http://publications.msss.gouv.qc.ca/acrobat/f/documentation/2006/06-211-03W.pdf, accessed 11 December 2006).

Tomlinson W, Hodgson RS (2005). Centennial year of yellow fever eradication in New Orleans and the United States, 1905–2005. *The Journal of the Louisiana State Medical Society*, 157:216–217.

Tsai TF et al. (1998). West Nile encephalitis epidemic in southeastern Romania. *Lancet*, 352:767–771.

USDA (2003). *Economic impact of West Nile virus on the Colorado and Nebraska equine industries: 2002. Animal and Plant Health Protection Service (APHIS) Info Sheet, April 2003.* Fort Collins, CO, United States Department of Agriculture, Animal and Plant Health Protection Service, Veterinary Services Centers for Epidemiology and Animal Health (www.aphis.usda.gov/vs/ceah/ncahs/nahms/equine/wnv2002_CO_NB.pdf, accessed 8 October 2006).

Weissenböck H et al. (2003). Usutu virus activity in Austria, 2001–2002. *Microbes and Infection*, 5:1132–1136.

White GB (1984). Needs and progress in the application of new techniques to mosquito identification. In: Newton BN, Michal F, eds. *New approaches to the identification of parasites and their vectors: proceedings of a symposium on application of biochemical and molecular biology techniques to problems of parasite and vector identification held in Geneva, Switzerland, 8–10 November 1982.* Basel, Schwabe & Co. AG (Tropical Diseases Research Series, No. 5):293–332.

WHO (1995). *Report of the informal consultation on aircraft disinsection.* Geneva, World Health Organization (document WHO/PCS/95.51; http://whqlibdoc.who.int/hq/1995/WHO_PCS_95.51_Rev.pdf, accessed 5 October 2006).

WHO (1996). *Operational manual on the application of insecticides for control of the mosquito vectors of malaria and other diseases.* Geneva, World Health Organization, Division of Control of Tropical Diseases (document WHO/CTD/VBC/96.1000; http://whqlibdoc.who.int/hq/1996/WHO_CTD_VBC_96.1000_(part1).pdf, accessed 5 October 2006).

WHO (2001). *Specifications for netting materials – report on an informal consultation, 8–9 June 2000.* Geneva, World Health Organization (document WHO/CDS/RBM/2001.28; http://whqlibdoc.who.int/hq/2001/WHO_CDS_RBM_2001.28.pdf, accessed 5 December 2006).

WHO (2002). *Manual for indoor residual spraying: application of residual sprays for vector control.* Geneva, World Health Organization, Department of Communicable Disease Prevention, Control and Eradication (document WHO/CDS/WHOPES/GCDPP/2000.3

Rev.1; (http://whqlibdoc.who.int/hq/2000/WHO_CDS_WHOPES_GCDPP_2000.3.Rev.1.pdf, accessed 5 October 2006).

WHO (2003). *Space spray application of insecticides for vector and public health pest control. A practitioner's guide.* Geneva, (document WHO/CDS/WHOPES/GCDPP/2003.5; (http://whqlibdoc.who.int/hq/2003/WHO_CDS_WHOPES_GCDPP_2003.5.pdf, accessed 5 October 2006).

WHO (2006a). *Chikungunya in Mauritius, Seychelles, Mayotte (France) and La Reunion island (France) – update.* Geneva, World Health Organization (http://www.who.int/csr/don/2006_03_01/en/index.html, accessed 3 October 2006).

WHO (2006b). *Chikungunya and Dengue in the south west Indian Ocean.* Geneva, World Health Organization (http://www.who.int/csr/don/2006_03_17/en/index.html, accessed 3 October 2006).

WHO Regional Office for Europe (2006). Malaria [online database]. Copenhagen, WHO Regional Office for Europe (http://data.euro.who.int/cisid/?TabID=102202, accessed 4 October 2006).

Zohrabian A, Hayes EB, Petersen LR (2006). Cost-effectiveness of West Nile virus vaccination. *Emerging Infectious Diseases*, 12:375–380.

Zohrabian A et al. (2004). West Nile virus economic impact, Louisiana, 2002. *Emerging Infectious Diseases*, 10:1736–1744.

12. Commensal rodents

Stephen Battersby, Randall B. Hirschhorn and Brian R. Amman[1]

Summary

Commensal rodents, such as the brown rat (*Rattus norvegicus*), the roof rat (*Rattus rattus*) and the house mouse (*Mus musculus*), present a great risk to human health, especially to people whose health is already compromised. Studies show that rats and mice can be infected with a large variety of parasites and zoonotic agents, which elevates their status from mere nuisances to public health pests. In addition to being reservoirs for zoonotic diseases, commensal rodents are also linked to medical problems associated with asthma and indoor allergic reactions, which further supports the need for effective pest management practices for controlling them.

Various aspects of the behaviour and biology of commensal rodents, such as enormous reproductive potential, trap avoidance and feeding behaviour, contribute to the failure of many rodent control programmes. To achieve acceptable results, all aspects of the biology and behaviour of commensal rodents should be understood and taken into account by those involved in pest management and control.

Also, a legal framework is necessary to support measures that secure effective practices for the control and prevention of commensal rodents that are urban pests and to simultaneously safeguard the health and safety of practitioners. Legislation that supports efficient rodent control programmes, while providing regulatory powers for overseeing these programmes (such as the use of rodenticides), is of paramount importance.

[1] Disclaimer: The findings and conclusions in this article are those of the author and do not necessarily reflect the views of the United States Department of Health and Human Services.

12.1. Introduction

There are 2227 species of rodents (Carleton & Musser, 2005) that inhabit every land mass of our planet, except Antarctica and a few oceanic islands. This chapter is primarily concerned with three rodent species: the brown rat (*Rattus norvegicus*), the roof rat (*Rattus rattus*) and the house mouse (*Mus musculus*). These three species are considered to be the most serious pests in urban environments. The taxonomy used in this chapter for both common and scientific names follows Wilson & Reeder (2005).

Rats and mice are thought of as commensal rodents because of their close association with human activity. In an ecological sense, the term commensalism refers to a symbiotic condition in which one participant benefits while the other is neither benefited nor harmed. Etymologically, commensalism refers to a sharing of one's table. These rodents benefit from their association with people in that they share dwellings with human occupants and, metaphorically speaking, eat from the same table. People, however, not only do not benefit from an association with these rodents, but they also may in fact suffer harm. The primary concern for environmental health practitioners and pest control professionals are synanthropic rodents – that is, rodents associated with people or human dwellings – because both rats and mice are known to carry a variety of disease-causing pathogens, either as a reservoir or as hosts to ectoparasite vectors. In addition, commensal rodents are responsible for considerable property damage and financial losses.

The fear and loathing of these commensal rodents is embedded in many cultures. For rats, this may be due to their ancient association with plague. After its arrival in Europe in 1348, this disease led to the death of nearly a third of the human population, and today it is still causing illness and death in many parts of the world (Keeling & Gilligan, 2000). Although not as closely associated with plague, the house mouse is still an unwelcome pest in any household, and it carries with it the social stigma of uncleanliness and squalor.

12.2. Biology of commensal rodents

12.2.1. Description and natural history

12.2.1.1. Brown rat

The brown rat (also known as the Norway rat and sewer rat) is typically a large bodied rat with a long, sparsely haired, scaly tail. Its dorsal pelage (the hair, fur or wool that covers an animal) is coarse and varies from brown to brownish-black, while the venter (the abdomen or belly) is usually a paler grey or tan. Its ears are hairless, and the tail is shorter than the length of its body. The measurements of the brown rat range from 315 mm to 460 mm for the total length; 195 mm to 245 mm for the head and body; 120 mm to 215 mm for the tail; and 140 g to 500 g for the weight. Males are typically larger than females.

Following a series of introductions, the brown rat, originally from Asia, now has a worldwide distribution that covers every continent, except Antarctica (Lund, 1994; Reid, 1997; Nowak, 1999; Myers & Armitage, 2004). It is typically more terrestrial than the roof rat

(Eisenberg & Redford, 1999); however, the brown rat is also an accomplished climber (Reid, 1997) and swimmer (Keeling & Gilligan, 2000; Myers & Armitage, 2004). The brown rat constructs extensive underground burrows that contain long branching tunnels with one or more exits and numerous rooms for nesting and food storage (Nowak, 1999). Once a native to forests and brushy areas, the brown rat now prefers a variety of habitats, including open fields and hedgerows, woodlands, garbage dumps, sewers, human dwellings and almost any place adequate resources can be obtained (Myers & Armitage, 2004). Essentially, any area occupied by humans is a potential habitat for the brown rat.

12.2.1.2. Roof rat

The roof rat (also known as the house rat, black rat and ship rat) is a medium-sized rat with large hairless ears and a long naked, scaly tail, which is always longer than the head and body. Its coarse pelage ranges from black to dark brown or brownish-grey on the dorsum and is a paler grey on the venter. The measurements of the roof rat range from 315 mm to 460 mm for the total length; 155 mm to 205 mm for the head and body; 160mm to 255 mm for the tail; and 115 g to 350 g for the weight. Males are larger than females, and this species is typically smaller than the brown rat.

Because this species is most commonly spread through human seafaring, the roof rat is most abundant in coastal areas (Gillespie & Myers, 2004). It is also more common than the brown rat in tropical areas, but in temperate zones the more aggressive brown rat has essentially excluded the roof rat from favourable habitat on the ground and in lower levels of buildings (Nowak, 1999). This has occurred to such an extent that the roof rat has been declared endangered in many areas (Handley, 1980; Lund, 1994).

The roof rat prefers dry areas above ground and is known to build nests constructed from grass and twigs in trees (Lund, 1994) or other elevated locations, such as roofs and attics of buildings and houses (Nowak, 1999), a habit that earned the species its common name in the United States. It is more agile than the brown rat and can run between buildings by using telephone wires, and it can climb building walls easily (Lund, 1994). Ewer (1971) reports the roof rat is capable of running along a wire only 1.6 mm in diameter. They are skilled climbers and can also swim well. These rats are more likely to live within dwellings than the burrowing brown rat.

Both of the aforementioned species of rats are nocturnal; however, they may be seen during the day where populations are large (Reid, 1997; Battersby, 2002; Sullivan, 2006). The brown rat may also be active at dusk (Gillespie & Myers, 2004).

12.2.1.3. House mouse

The commensal house mouse is comprised of four subspecies, according to its geographic location: the western European house mouse (*M. musculus domesticus*), the south-western Asian house mouse (*M. musculus bactrianus*), the south-eastern Asian house mouse (*M. musculus castaneus*) and the eastern European house mouse (*M. musculus musculus*) (Nowak, 1999; Lundrigan, Janser & Tucker, 2002; Carleton & Musser, 2005). In this chapter, however, all of these subspecies will be treated as the house mouse, *M. musculus*.

The house mouse is considerably smaller than the brown rat and roof rat. Its tail is long, with circular rows of scales, and it has very little fur. Its pelage varies from a nearly uniform greyish-brown to a grey-brown dorsum and a pale grey to buff venter. The measurements of the house mouse range from 125 mm to 200 mm for the total length; 65 mm to 95 mm for the head and body; 60 mm to 105 mm for the tail; and 12 g to 30 g for the weight.

The house mouse is distributed worldwide and tends to have a close association with people. Commensal house mice occupy a number of man-made structures, including houses, barns and granaries (Ballenger, 1999), and it prefers to nest behind rafters, in woodpiles, in storage areas or anywhere close to a source of food (Nowak, 1999). Because of their association with people, house mice are able to occupy such areas as tundra and desert regions, which they would not be able to inhabit independently (Ballenger, 1999).

Wild house mice occupy cultivated fields, fencerows and wooded areas, but they seldom stray far from buildings, with some individuals moving from one habitat to the other with the changing seasons (Berry, 1970; Ballenger, 1999). They live in cracks in rocks and walls or construct extensive burrows with many rooms and exits (Berry, 1970). Wild house mice have even been reported in coal mines at depths of up to 550 m (Bronson, 1979). Although house mice are good swimmers (Nowak, 1999), they avoid water and damp conditions, as they have difficulty maintaining their body temperature when damp or wet (WHO Regional Office for Europe, 1998).

House mice are generally nocturnal; however, some may be active during the day in human dwellings (Ballenger, 1999). They can live in almost any available space and can squeeze through extremely small openings, living and breeding under floors, in wall cavities and ceiling voids, and behind skirting boards (baseboards).

12.2.2. Reproduction and life cycle of commensal rodents

The potential breeding rate of brown rats is enormous. For example, a story in a weekly agribusiness newsletter in the United States suggested that the young from just one pair of rats could be responsible for bringing forth 3.5 million more rats in three years (Anonymous, 1984). The brown rat is capable of reproducing year-round in some wild populations, but there are usually spring and autumn peaks (Berry, 1970). Capable of mating within 18 hours of giving birth, the polyestrous females may bear anywhere from 1 to 12 litters a year of up to 22 pups per litter, with an average of 8–9 pups. Gestation is typically 21–26 days, and the young are born naked and blind. After two weeks, they are fully furred and open their eyes. The pups are weaned at around three weeks and leave the nest at that time. They reach sexual maturity at 2–3 months of age (Whitaker, 1980; Lund, 1994; Nowak, 1999). While huge numbers of offspring are mathematically possible, predation, limited resources and unfavourable climate, as well as behavioural characteristics, limit the realization of this rat's reproductive potential (Whitaker, 1980).

Calhoun (1962) observed that on or near the time when female brown rats are capable of conception there is a precopulatory phase where the behaviour of both sexes changes. In this phase the female wanders more than normal beyond the limits of her home range as she actively seeks males. This increased activity may lead to more sightings and complaints from the public.

The lifespan of wild brown rats is difficult to assess. The maximum known longevity in captivity is 3 years, but in the wild their lifespan is probably less than 18 months (Meehan, 1984). In a farm population of brown rats, only about 5% of the population at the start of the year was alive at the end (Meehan, 1984). Calhoun (1962) concluded that, for weaned rats, learning to cope with the environment reduces the proportion that survives. There is then a levelling off of the rate of attrition, but eventually individuals reach an age when learned behaviours fail to compensate for the debilities of ageing, which then increases the probability of death.

Reproduction in the roof rat can occur throughout the year, with varying seasonal peaks. Gestation ranges from 21 to 29 days and litter size ranges from 1 to 11 pups, with an average of 8 pups. Females can produce up to five litters in a year. The pups are born naked and blind and are weaned and independent at 21–28 days. Females are capable of giving birth at 3–5 months of age (Whitaker, 1980; Nowak, 1999; Myers & Armitage, 2004).

Annual mortality for roof rats that live in the wild is 91–97%, with most living only a year. Captive roof rats have been known to live for 4 years (Nowak, 1999).

House mice are prolific breeders. They are capable of reproducing throughout the year, although seasonal breeding may occur in wild populations. Females may experience postpartum oestrus 12–18 hours after giving birth. The gestation period is 19–21 days, but may be longer if the female is lactating. Litter size ranges from 3 to 12 pups, with an average of 5–6 pups. Pups are born naked and blind and are fully furred at 10 days; eyes are open at 14 days, and they are weaned and independent at 21 days. There is usually 60–70% mortality before independence is reached. Sexual maturity is reached at 5–7 weeks. On average, captive mice live 2 years, while the normal lifespan of wild mice is 12–18 months (Ballenger, 1999; Nowak, 1999).

12.2.3. Feeding

Throughout their distribution, brown rats typically are associated with human settlements, colonizing areas with adequate resources. They are opportunistic omnivores, utilizing any food that becomes available. Kingdon (1974) states that brown rats will eat everything people eat and much more, including soap, hides, paper and beeswax. In urban settings, the brown rat relies heavily on discarded human food (Myers & Armitage, 2004). As an accomplished swimmer and diver, this rodent has been known to prey on ducklings (Lund, 1994) and is quite adept at catching fish (Grzimek, 1975; Lund, 1994; Nowak, 1999; Myers & Armitage, 2004). The brown rat is more carnivorous than the roof rat and has been known to prey on mice, poultry, young lambs and piglets, insects, birds, and small reptiles (Reid, 1997; Nowak, 1999; Myers & Armitage, 2004).

The roof rat, although omnivorous, relies less on animal matter and prefers a diet that consists of fruits, grains and seeds, when available (Lund, 1994). It will also feed on insects, carrion, refuse and faeces.

Like rats, commensal house mice will eat any accessible human food, as well as paste, glue, soap and other household materials (Ballenger, 1999). Wild populations will consume a variety of available food, including leaves, seeds, roots, stems, and some insects and meat, when available (Ballenger, 1999; Nowak, 1999).

12.2.4. Behaviour relevant to control

Much of the behaviour exhibited by commensal rodents allows them to exploit their environment to its fullest, which can adversely affect the lifestyles of people. Knowledge of this behaviour, as it relates to interactions with people, is a prerequisite for effective control, either by using rodenticides or by securing environmental changes to make the habitat unsuitable.

Young rats can squeeze through openings of less than 25 mm (Lund, 1994), and house mice can get through openings of half that size (WHO Regional Office for Europe, 1998) – a factor that should be taken into account when rodent proofing buildings. These rodents can also create their own points of entry in walls and containers, gaining access to resources and contaminating food stores in the process.

Rodents gnaw continually to keep their ever-growing incisors sharp. They gnaw through the insulation of electrical wires, causing fires, and occasionally puncture lead pipes and concrete dams (Nowak, 1999). This continual gnawing causes physical damage to timber and other building materials.

Brown rats are skilled burrowers. The complexity of their system of burrows may vary with the density of the population in the colony (Nieder, Cagnin & Parisi, 1982). In loose soil, these burrows can reach depths of 3 m, although they do not usually exceed 0.5 m. The burrow system may be short in length (less than a metre), but in established colonies burrows may interconnect, creating an extensive network of interconnecting tunnels.

Four factors influence the location and entrance of a brown rat burrow. The first has been termed the thigmotropic response, whereby rats, on direct contact with a solid surface or object, prefer to move against a vertical or under a horizontal surface (Calhoun, 1962). This movement, characterized by close contact with walls and other surfaces, leaves a revealing sign in the form of a smudge. This smudge, in addition to other signs, is indicative of an infestation and may prove useful in identifying well-used runs for control efforts.

The second factor is overhead cover, such as shrubs, low trees and floors of buildings. This appears to favour the location of a burrow, because it reduces both direct light and visibility of entrances to predators.

The third factor is the slope of the terrain, as sloping terrain appears to be preferred to flat terrain. Brown rats prefer to move downward with the force of gravity, displaying a tendency for *positive geotaxis* (Meehan, 1984).

The fourth factor that influences the location of the entrance to a burrow is the proximity of major resources, such as food and water. Calhoun (1962) reported that the shortest routes are not always those travelled to such goals. The orientation of the run is from one vertical object to another or along a continuous vertical object. Once rats have learned the location of targets and objects in the immediate environment, runs become well established (Meehan, 1984).

Unlike house mice, both species of commensal rats exhibit a behaviour called *neophobia*. This avoidance of the novel is especially prevalent with new food items and may have more to do with the rat's naturally timid demeanour than with food selection, but it may facilitate the association of adverse effects with the eating of new food (MacDonald, Mathews & Berdoy, 1999). Neophobia can confound control efforts and must be compensated for by pre-baiting (MacDonald, Mathews & Berdoy, 1999). Pre-baiting allows the rat population to become accustomed to the novel food at the bait station before it is replaced with poisoned food.

Barnett (1975) suggested that in the wild the choice of food is not influenced socially and that there is no evidence of true imitation or observational learning by young rats. The choice of food is not learned from parents, but is a matter of habit. Rats tend to consume a small quantity of something with which they are unfamiliar and see what happens, and then they eat a little more if all is well, eventually feeding more freely (Quy, 2001). However laboratory studies have shown that rats can be influenced by the scent of another rat that has eaten a particular food (Galef & Wigmore, 1983; Posadas-Andrews & Roper, 1983) and that information about the food, such as palatability or toxicity, can be transferred from mother to offspring, even through mother's milk (Galef & Clark, 1972; Bond, 1984; Hepper, 1990).

Neophobia also extends to novel objects that suddenly appear in the rat's environment. Bait boxes and traps are often avoided because they are new to the immediate surroundings. Even after rats have become accustomed to the presence of the traps, it is often only juveniles or socially low ranking adults that are caught (Calhoun, 1962), as these low ranking members of the population are forced to seek food in alternative places. Thus, it seems that without a well-organized strategy, it will be largely the socially inferior rats that are caught and killed. The more dominant and virile rats will survive to breed and eventually restore the population to its original size.

In contrast to neophobic behaviour, rats may exhibit *neophilic* tendencies (curiosity) in selecting unfamiliar objects to gnaw (Quy, 2001). This is evidenced by their gnawing such structures as pipework for no apparent reason. Barnett (2001) suggested that in an unstable environment, such as a waste landfill site where everything is new, neophobic behaviour is replaced by neophilic behaviour, as rats explore their ever-changing environment.

12.2.5. Population growth and socialization of commensal rodents

It has been suggested that two major factors govern the ultimate size of a rodent population: the amount of harbourage (or cover) and food available. In an urban environment, the situation is probably somewhat more complex than that. For example, social status within a colony and exclusion of inferior males will be a factor (Calhoun, 1962). Nevertheless, it has also been suggested that when food or harbourage are reduced, there will be migration (Twigg, 1975). Usually, rodent populations attain equilibrium, when deaths are balanced by births, until (and unless) there is a change in habitat or food supply. Furthermore, it has been suggested that rat populations may fluctuate over a 10-year cycle (Swift, 2001). In the assessment of the effectiveness of any control strategy, the affect of any natural fluctuations in population would need to be taken into account.

With respect to commensal rat control, Barnett (2001) concluded that there is no one density-related factor that can be identified as the key to keeping population size down. People, however, may exercise varying degrees of control over a number of interrelated factors, including the availability (or scarcity) of food and water, and lack of shelter, predation, pathogens and social interaction. Changes of the environment caused by people will be addressed in section 12.5, "Control of commensal rodents".

As noted, the social ranking of individual rats influences behaviour. Rats exhibit varying degrees of aggressive behaviour, and to avoid the possibility of aggression rats of lower social status avoid contact with higher-status rats by visiting food sources at different times and for very short periods. In a colony, individuals that become social outcasts between weaning and sexual maturity show a slow growth rate and lower adult weight, and they are more likely to enter traps (Calhoun, 1962). Social outcasts, as a means of avoiding conflict, exhibit a tendency to shelter in less favourable areas with greater exposure to the weather. They also form non-reproducing male cohorts, which occasionally include females incapable of breeding or rearing young.

Incomplete extermination of a rat population can lead to increased reproduction within the population. When most members of a rat population have been killed, the remainder may breed more quickly, thus increasing the population to its original level (Greaves, Hammond & Bathard, 1968; Barnett, 1975).

A similar increased growth rate is observed when rats colonize a new and favourable habitat. At first, the reproductive rate is high, but it gradually declines as the population reaches its optimal size. Control strategies should recognize this pattern, particularly during urban redevelopment, so that redundant lengths of sewers and drains are removed or sealed and potential harbourage sites and food sources are minimized. The aim should be to make the environment less favourable and to monitor the rodent population regularly.

Although predation has an effect on the behaviour of rats (MacDonald, Mathews & Berdoy, 1999), it does not appear to have a significant effect on population density. Calhoun (1962) reported from observations in urban apartment blocks in the state of Maryland that predation by dogs and cats and sporadic attacks by people did not have any appreciable effect on the density of the rat population. Indeed, general observations indicated that city blocks with dogs and feral cats also had high-density populations of rats. The presence of free-ranging cats and rats in urban areas appears to be positively related, perhaps due to a common benefit derived from access to waste food (Langton, Cowan & Meyer, 2001). Also, the presence of pets may lead to the provision of food and shelter for rodents (Langton, Cowan & Meyer, 2001). For example, in England, infestations of commensal rodents inside and outside homes are higher in those properties where pets or livestock are kept in the garden (Langton, Cowan & Meyer, 2001; DEFRA, 2005). Because predation by dogs and cats has no appreciable effect on rodent densities in urban areas, rodent control is likely to be the primary mechanism by which population densities can be kept low.

12.2.6. Movement

Rats, particularly the brown rat, do not normally move great distances, especially in urban areas where streets act as barriers (Twigg, 1975). This may not be the case in rural areas, where rats have been reported to move as many as 3.3 km at speeds of 0.5–1.1 km an hour in one night (Taylor & Quy, 1978). Involuntary dispersal may also result when rats are transported with goods in vehicles. Habitat destruction will also cause movement when rats are forced to seek alternate shelter.

The diameter of the normal home range of the brown rat varies from 25 m to 150 m (Grzimek, 1975). MacDonald, Mathews & Berdoy (1999) reported that farming activity caused home ranges of brown rats around farms to fluctuate, with males having larger home ranges (679 linear metres) when crops were in the field and smaller home ranges (90 linear metres) after the harvest. Farming activity, however, appeared to have no effect on females who, in general, had smaller home ranges.

Resource availability also had an effect on home range. When resources were plentiful, the home range of females was smaller (85 linear metres) than when resources were not as plentiful (428 linear metres); similar effects were noted for males (MacDonald, Mathews & Berdoy, 1999).

The home range of the roof rat is never more than about 100 m^2. This species often has smaller territories that surround the food sources it defends (Gillespie & Myers, 2004).

Commensal house mice have been recorded as travelling over 2 km, but this is exceptional. Typically, they will not move more than 3–10 m in buildings (WHO Regional Office for Europe, 1998). Murphy, Williams & Hide (2005) used DNA analysis to assess relatedness among house mice that colonized terraced housing in England and found that mice colonizing adjoining buildings were related, but the mice in infested properties on other streets and terraces were genetically different. Their work indicated that each block represented individual breeding units and that migration rates between blocks were very low, because mice moved easily between adjoining properties, but moved little between non-adjacent housing blocks (Murphy, Williams & Hide, 2005). The focus of control should therefore be the blocks of properties, not individual houses.

12.3. Association with urban infrastructure

12.3.1. Rodents in housing

Dwellings that were more susceptible to rodent infestations were identified in the English House Condition Survey (EHCS) of 1996 (Langton, Cowan & Meyer, 2001). A high correlation was found between rat infestations and areas where problems of litter, vandalism, dishevelled gardens, neglect and vacant buildings were widespread (Langton, Cowan & Meyer, 2001; Murphy & Oldbury, 2002). Domestic mouse infestations were most likely to occur where there was poor structural maintenance, poor hygiene and ample internal harbourage (Murphy & Oldbury, 2002). The 2001 EHCS (DEFRA, 2005) revealed that 2.9% of 16 676 occupied homes in England had rat infestations outside, representing a 70% increase over results reported in 1996. The EHCS also revealed that infestations inside the home in 2001 were 1.4% and 0.3% for mice and rats, respectively. This is a slight decrease from the 1.8% and 0.4% for mice and rats reported for 1996. It should be noted, however, that a previous rodent survey within an inner-city area, using a different methodology, found considerably higher levels of infestations (Meyer et al., 1995), indicating that within urban areas, there will be hot spots and that national surveys will not identify the true level of infestations in specific areas. Such a hot spot was described in Manchester, England, where 50% of the terraced properties had mouse infestations (Murphy, Lindley & Marshall, 2003). This was substantially higher than infestation rates reported in the EHCS and supports the notion that infestations of commensal rodents may be clumped.

Housing density is one important factor that influences urban rodent infestations. Because the home range of rats can encompass more than one dwelling, the higher the density of homes in an area, the more likely it is that rodents infesting one home will disperse and colonize the surrounding dwellings. Dispersal by both rats and mice is also more likely to be successful over short distances.

The age of housing is another factor that influences rodent infestations in urban areas, with rat infestations significantly more common in older properties (Langton, Cowan & Meyer, 2001) and with infestation rates higher in dilapidated structures. In the United States, the presence of rats in urban areas is taken to be a common indicator of a degraded environment (Colvin in Martindale, 2001). There is also some evidence that defective drains, an artefact of an ageing community infrastructure, are linked to outdoor rat infestations (Langton, Cowan & Meyer, 2001; Battersby, 2002).

Other factors associated with the urban environment have a significant effect on the size of the populations of rats and mice. Pest control practitioners have raised concerns that excess litter, carelessly discarded food waste and inadequate sewer baiting contribute to the existence of above-ground rat infestations (Battersby, 2002).

12.3.2. Sewers and drains

Urban sewers are the perfect man-made rat habitat. They minimize temperature fluctuations, with cooler conditions in the summer and warmer conditions in the winter; they provide a steady influx of food, as wastes of all types are flushed through the system; and they greatly reduce or eliminate predation. Due to the more stable climate, breeding continues year-round without seasonal fluctuations. Heavily infested sewers and drains can act as reservoirs of rats, which restock surface areas where control efforts have been undertaken. Thus, sewer systems are very important harbourages for urban rat colonies and should be considered in conjunction with surface control efforts (Twigg, 1975).

In urban areas of the United Kingdom, at least a quarter of the surface infestations in over a half of the local authority districts are due to defects in the sewer system. In some local authorities, over half the surface infestations are attributable to defects below ground (Battersby, 2002).

Rats do not normally live in active drains and sewers, but instead they live in disused pipes, in excavations adjacent to cracks or bad joints in pipelines, or in the dry parts of the network, such as benching (raised ledges) at manholes and inspection chambers (Hall & Griggs, 1990). Colonies normally are located in one area and sorties of limited extent are made in search of food (Bentley, 1960).

The home range of rats in sewers may be extremely limited. This limited range may be unique to sewers that are particularly favourable to rats or to situations where there is regular traffic between sewer and surface. These favourable conditions and restricted ranges can inhibit control efforts. Bentley (1960) reported that rat colonies that experienced the aforementioned conditions may subsist between (but not include) two manholes. These rats do not depend on the sewage flowing from upstream for a food supply; rather, the waste is discharged from a source between the manholes. Bait placed at these manholes will therefore not be effective. Nevertheless, the distance moved may depend on the availability of food. When circumstances require them to do so, rats in sewers may move great distances (140–200 m) (Bentley, 1960).

12.4. Commensal rodents and human health concerns

12.4.1. Zoonoses of rats

Commensal rodents have been associated with a variety of zoonoses. This is of particular concern because of their close association with people. Over the last 10 centuries, rat-borne diseases may have taken more lives than all of the wars ever fought (Nowak, 1999).

Gratz (1984) included schistosomiasis as one of about 40 diseases rats carry, and Nowak (1999) reported that as many as 200 million people worldwide are infected with this disease. Rats can also spread murine typhus, plague, salmonellosis, leptospirosis, trichinellosis and rat-bite fever (Nowak, 1999). Webster & MacDonald (1995) found that brown rats in the United Kingdom were infected with 13 different endoparasitic organisms and zoonotic agents, with some having infections of up to 9 of these simultaneously (see Table 12.1 for a list of zoonoses associated with commensal rodents). MacDonald, Mathews &

Berdoy (1999) reported that rats exhibited behavioural changes when infected with *T. gondii*, making them more susceptible to predation by cats and making transmission of *T. gondii* to cats possible, which further increases the risk of transmission to people.

In addition to the 13 species of endoparasites, Webster & MacDonald (1995) also reported finding three types of ectoparasites – mainly arthropods that live on the rat's body. Of the 510 brown rats sampled, all (100%) carried fleas, 67% carried mites and 38% carried lice. None, however, carried ticks. Such ectoparasites act as vectors for serious diseases that affect people in many countries. Bubonic plague is the most widely known example, where the primary vector of the pathogen *Y. pestis* is the Asiatic rat flea, *X. cheopis*. Other diseases for which rodent ectoparasites are vectors include murine typhus, rickettsial pox, spotted fevers, LBRF and tick-borne relapsing fever (Dennis, 1998; Nowak, 1999; Padovan, 2006).

With regard to rat-borne parasites and the risk they present to public health in an urban environment, the significance of the presence of rats in and around homes depends on the numbers and prevalence of parasitic species among urban rats. Battersby, Parsons & Webster (2002) reported that the number and prevalence of parasitic species detected in urban rats tended to be lower than that previously obtained from rural rats: *Capillaria* spp., *Toxocara cati*, *Hymenolepis nana*, *Hymenolepis diminuta*, *Taenia taeniaeformis* and *T. gondii* all had a lower prevalence in urban rats; *Listeria* spp. and *Y. enterocolitica* just failed to reach significant levels of infection; and *Pasteurella* spp. and *Pseudomonas* spp. were not detected at all among urban rats. The only species that showed significantly higher prevalence among urban rats were *Trichuris* spp. (Battersby, Parsons & Webster, 2002).

12.4.2. Zoonoses of mice

Typically, mice have been seen merely as a nuisance, because of the spoilage of foodstuffs and the damage they cause in homes. Mice do, however, carry several zoonotic agents and should be treated as a potential threat to public health. As warm-blooded mammals, they have the potential to carry ectoparasites into a home and unwittingly assist in the dissemination of murine typhus and rickettsial pox; however, house mice are known to transmit lymphocytic choriomeningitis (Lehmann-Grube, 1975; Buchmeier et al., 1980). Although lymphocytic choriomeningitis is not usually a serious threat to healthy individuals, this viral disease causes severe illness in immunocompromised people and can cause severe birth defects when contracted during pregnancy (Fischer et al., 2006; Amman et al., 2007). Also, Williams and colleagues (2005) reported finding *T. gondii* in house mice at a prevalence of 58.5%; results from this study, in Manchester, England, indicated that of 200 mice, 4 tested positive for *Cryptosporidium* spp., and 2 tested positive for *Chlamydia* spp. Recent studies have found that house mice also carry the mouse mammary tumour virus. This virus may be linked to breast cancer in people (Stewart et al., 2000; Indik et al., 2005).

12.4.3. Other health concerns

In addition to zoonoses, commensal rodent infestations in homes present other human

Table 12.1. Zoonoses associated with commensal rodents

Human disease	Vector, pathogen or both
	Ectoparasites
Bubonic plague	Asiatic rat flea – *Y. pestis*
LBRF	Body louse – *B. recurrentis*
Tick-borne relapsing fever	Ticks *(Ornithodoros hermsi) – Borrelia spp.*
Lyme disease	Ticks *(Ixodes spp.) – B. burgdorferi*
Rickettsial pox [a]	Rodent mite *(Liponyssoides sanguineus) – Rickettsia akari*
Murine typhus [a]	Asiatic rat flea – *R. typhi* Body louse – *R. typhi*
	Endoparasites
Capillariasis	*Capillaria* spp.
Toxocariasis	*Toxocara* spp.
Rat tapeworm infection	*Hymenolepis nana*
Diarrhoeal disease	*Trichuris* spp.
Diarrhoeal disease	*Hymenolepis* spp.
Diarrhoeal disease	*Taenia* spp.
Schistosomiasis ***	*Schistosoma* spp.
Trichinellosis*	*Trichinella* spp.
Cryptosporidiosis [a]	*C. parvum*
Toxoplasmosis [a]	*T. gondii*
Babesiosis	*Babesia* spp.
Sarcosporidiosis	*Sarcocystis* spp.
Coccidiosis	*Coccidia (Eimeria* spp.*)*
Amoebic dysentery	*Entamoeba* spp.
	Bacteria
Leptospirosis [a]	*Leptospira* spp.
Listeriosis	*Listeria* spp.
Yersiniosis	*Y. enterocolitica*
Pasteurellosis	*Pasteurella* spp.
Rat-bite fever [a]	*Streptobacillus moniliformis* and *Spirillum minus*
Melioidosis	*Pseudomonas* spp.
Q fever	*C. burnetii*
Salmonellosis [a] **	*Salmonella* spp.
Diarrhoeal disease	*Vibrio* spp.
Tularemia*	*F. tularensis*
	Viruses
Hantaan fever	Hantavirus
Lymphocytic choriomeningitis [b] ****	Lymphocytic choriomeningitis virus

[a] Indicates zoonoses of house mice and *Rattus* spp.; b. indicates zoonosis of house mice.

Source: All (Webster & Macdonald (1995), Battersby (2002), and Battersby, Parsons & Webster (2002)), except *Nowak (1999), **Seguin et al. (1986), Hilton, Willis & Hickie (2002), ***Gratz (1984) and ****Lehmann-Grube (1975).

health-related issues. Carrer, Maroni & Cavallo (2001) reported that the presence of rodents in the home may contribute to increased levels of indoor allergens, causing allergic asthma and rhinoconjunctivitis. Other studies confirmed asthma attacks as being associated with the presence of rat and mouse allergens in the home (Perry, Matsui & Merriman, 2003; Cohn et al., 2004; also, see Chapter 1). It should also be recognized that the awareness of rats and mice in and around a dwelling can be a source of anxiety for its occupants (WHO Regional Office for Europe, 1998; Battersby, Parsons & Webster, 2002; Williams et al., 2005). Thus, the presence of rats and mice also affect mental health. In particular, in decaying urban areas, this can further stress people whose health status is already compromised by poverty and social exclusion (Battersby, Parsons & Webster, 2002). Rat bites in urban settings are also an important health concern and will be discussed in subsection 12.4.4.1.

12.4.4. Public health risks in urban areas

With the exception of sewers, rat colonies in urban areas tend to be much smaller than those on farms. In these urban areas, public health risks from zoonoses may not be as great as in rural areas where rat population numbers are high (Battersby, 2002).

The discrepancy between the high parasite loads previously identified among rural brown rats found in England and the low rates in their urban counterparts may have at least two explanations (Battersby, 2002). The first relates to the impact of wildlife and domestic livestock on driving cycles of zoonotic infection, and the second relates to the different population densities between the two habitat types (Battersby, Parsons & Webster, 2002).

In a rural environment, soil and water contaminated by infected excreta from domestic livestock may spread infection to the neighbouring rat population, thereby maintaining or even initiating rodent reservoirs of infection. Indeed, high rates of infection with *C. burnetii* among rodent populations have been found on livestock farms, but low or zero rates of infection have been reported for rats on arable farms where commercial livestock were absent (Webster, Lloyd & MacDonald, 1995). In urban settings where veterinary care limits the impact of zoonoses on domestic animals, the opportunity for contributions from domestic animals to the infection of commensal rodents is limited, so domestic animals, such as cats, generally play a smaller role than domestic livestock in driving zoonotic cycles.

The second explanation for the discrepancy, which is directly applicable to inadequate control, relates to differences between rodent population densities in urban and rural habitats. Evidence shows that rat population densities are often very high in rural environments, while they are generally very low (and with restricted inter-group social interaction) in the modern, developed urban environment. The latter has been achieved primarily through pest control programmes and enhanced sanitation (Twigg, 1975). The high rural population densities favour the transmission of zoonoses and parasites, particularly those spread by direct contact or aerosolization over a short distance (Anderson, 1993), whereas the lower population densities in urban areas do not. However, this is not to say that in urban environments where rodent control measures are inadequate and low levels of predation exist rodent population densities will not increase. Because the population growth of commensal rodents can be explosive, the potential exists for an increase in zoonotic diseases within those populations. Thus, to be effective and to prevent commensal rodent population densities from increasing, urban rodent control strategies should not rely solely on public complaints.

Another public health issue of great concern is bites inflicted by rats in urban environments. This threat to public health has medical, social and emotional dimensions (Anderson, 1993). The continuing presence of commensal rats in and around many urban and rural residences in the United States results in reports of hundreds of rat bites each year. This number is said to be underreported by factor of at least ten (Hirschhorn & Hodge, 1999; Hirschhorn, 2005).

Rats are often found in substandard dwellings where the density of buildings is very high and the construction design denies rats outdoor burrowing opportunities. In these situations, rats colonize basements and kitchens, and attempt to coexist with their human occupants. Each rat bite has the potential to spread infection, and the ectoparasites associated with the rodent can spread additional infectious organisms. The following case study illustrates the identification of risk factors.

12.4.4.1. Case study 1 – rat bites in Philadelphia: identifying the factors contributing to risk
In this study (Hirschhorn & Hodge, 1999), investigators examined rat-bite reports of 622 urban victims. To determine risk factors, the study assessed demographic characteristics and environmental factors. Cases were divided into two groups by date: 1974–1984 and 1985–1996 (Fig. 12.1). The study used United States Census Bureau data for Philadelphia, from 1980 and 1990. Each rat bite was investigated, and the following data were examined from each case reported: characteristics of the victim, place and time, number of rat bites, circumstances of the bite, environmental conditions, and location of the bite on the victim. Rat-bite incidence was 2.12 bites per 100 000 people in the first period and decreased by 54%, to 1.39 bites per 100 000 people in the second period.

Most bites were received in the home. Of all the rat bites reported, 67% occurred in single-family dwellings, 24% in multiple-family dwellings, and only 8% in laboratories and schools. About half the rat-bite victims lived in housing that was in poor repair. Rat bites also showed a seasonal pattern and were most likely to occur in the summer months between the hours of 0:00 and 6:00. Nearly 50% of the bites reported were received on the hands, 20% on the head and the remainder on other extremities. Nearly all victims were asleep when bitten, and the majority were in bed. Socioeconomic factors were important contributors to the risk of being bitten. Most victims were living in poverty, as defined by the United States Census Bureau.

The typical rat-bite victim in Philadelphia was an impoverished child less than 5 years old (Fig. 12.2), with many less than a year old, living in substandard housing in close proximity to brown rat infestations. The rat bite was usually to the hands or head, and it occurred during the night, while the child was sleeping. These results indicate that sur-

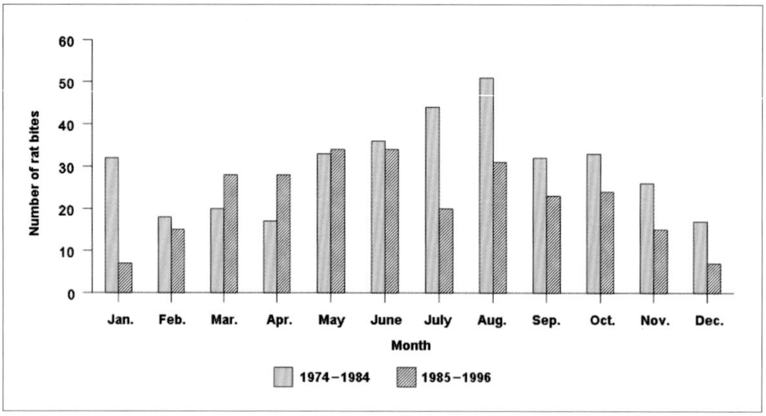

Fig. 12.1. Frequency of rat bites throughout the year: case study 1

Source: Hirschhorn & Hodge (1999).

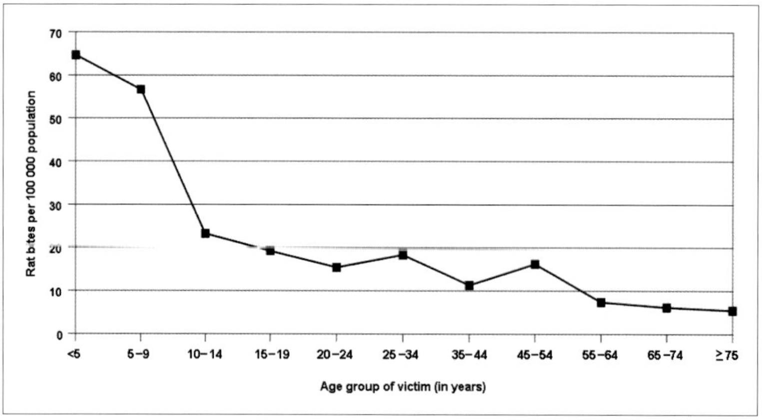

Fig. 12.2. Frequency of rat bites by age of victims: case study 1, 1974–1994

Source: Hirschhorn & Hodge (1999).

veying the rat population alone does not provide the data necessary to determine the risk of a rat attack. Instead, age, race, location, income level, season and temporal factors were of equal or greater importance in determining the risk. Table 12.2 indicates the percentage of rat bites, by gender of the victim. If prevention or intervention programmes are to be effective, all risk factors must first be identified and then addressed. The results of the Philadelphia study agree closely with studies done in New York City and Baltimore (Hirschhorn & Hodge, 1999).

Table 12.2. Percentage of rat-bite victims by gender: case study 1

Gender	Percentage (by period)	
	1974–1984	1985–1996
Male	48%	42.6%
Female	52%	56.5%

Note. P > 0.05, no significant difference between time periods.

Source: Hirschhorn & Hodge (1999).

12.5. Control of commensal rodents

To develop an effective control strategy, the true rate and location of infestations need to be assessed. Other risk factors and any assessment of the possible size of the rat population are generally secondary considerations. Also, an increase in the number of premises infested does not necessarily mean the overall rat population has increased. Because rat populations may fluctuate (Swift, 2001), an increase in complaints over time does not confirm a consistent increase in the size of the rat population.

The choice of control strategy at a site where an infestation has been identified will depend on several factors, including type of rodent, age and extent of the infestation, type and design of premises, and presence of non-target species. The initial intention of a control strategy must be to control existing pest populations, which can then be followed by implementing strategies to prevent reinfestation. Too often, strategies are limited to responding to complaints, with the assumption that a lack of complaints indicates an absence of rodents. Where there are defects in the sewage system, there may be regular movement of rats between the sewerage infrastructure and the surface, without the public being aware of their presence (Bradshaw, 1999).

When a population of sewer-dwelling rats has been reduced via rodenticide treatment, its numbers will recover. An estimate of the inherent growth rate of rat populations is about 3% a week. However, rates of up to 11–12% a week have been observed in some sewage systems, with immigration possibly contributing to this higher rate (Bentley, 1960). A mean rate of increase of a sewer population of about 20% has also been recorded after attempts to rid a sewer section of rats (Greaves, Hammond & Bathard, 1968).

Perfunctory, unplanned poisoning merely kills some of the population, leaving rat numbers to recover quickly. It was found that a more thorough and planned operation with two treatments reduced a population in a sewer to a small fraction of its original size (Barnett, 2001). When surface infestations were dealt with at the same time, recovery of the rat population was slow (Barnett, 2001).

One poisoning technique, the pulse baiting technique (Dubock, 1982), employs a series of baiting rounds, where bait laced with a single-dose anticoagulant is set out for a period of 3–13 days and then removed for 7 days. Death is usually delayed by three or more days. After the seven-day period, the bait is set out again. This technique accommodates the hierarchical feeding common to rats, by allowing the older dominant rats to die before replacing the bait, providing the less dominant rats with an opportunity to eat the bait. Bait conservation is also achieved, in that rats that have already ingested enough poisoned bait to insure death do not continue to consume bait at the stations. Three baiting pulses can remove almost an entire population (CIEH, 2003).

Current practices of rodent control in Europe and North America, based on complaints by members of the public, are not the best and most sustainable pest management practices available to protect public health. However, it has been suggested that, where the rate of infestation is 1% or less, it is not worthwhile replacing a system that relies on complaints with a more structured approach of systematic surveys, because the costs of securing any significant improvement in the rate at which infestations are discovered and treated would be prohibitive (Drummond, 1970). Thus, priority areas should be those where infestation rates are above 1%. This implies, however, that the municipal authority has adequate information on which to make that assessment.

As part of a coherent strategy, rodenticides are an essential means of effectively controlling rat populations in sewers (Colvin, Swift & Fothergill, 1998), when coupled with maintenance and repair of the underground sewerage infrastructure. The successful use of rodenticides over the past 50 years has been managerially convenient, so that rodent control can be reduced to routine, fixed-price procedures with an apparently predictable outcome. In many countries, rodenticides are seen as an immediate, economical and relatively easy way of addressing infestations in urban areas. This practice has unfortunately led to an over-reliance on this approach. This over-reliance is understandable, however, in complex situations where there are many different agencies, in addition to individual homeowners, involved in securing the necessary changes to the environment to reduce infestations in the longer term. Reliance on this technical solution has thus led to a minimization of strategic thinking.

Apart from surveillance of specific individual premises, such as high-risk food shops and restaurants, rodent management is essentially driven by a complaint procedure within municipal authorities (Richards, 1989). The reported level of infestations in the urban ecosystem depends on public perception. In effect, this means that if the residents of some urban areas are more tolerant of rats than others, a less stringent (or perhaps no control) strategy will be implemented. The failure to have an effective control strategy means the overall level of control will not be optimized (Meyer & Drummond, 1980). Furthermore, little attention has been paid to deciding what level of infestation, if any, represents satisfactory control.

The problems caused by rodents should not be viewed as only belonging to the individuals affected, but they should be seen as community problems that need to be addressed by the community as a whole. Attempts to solve rodent infestation problems across wider areas need to be addressed on a community-wide basis if the effort invested is to be cost effective. Control of rodent infestations should not be treated simply as a question of killing the rodents. The problem must be seen more broadly, as an infestation within a vulnerable (and more often than not), degraded urban environment. Once the risk factors that contribute to infestations and subsequent public health threats are identified, as in case study 1 (subsection 12.4.4.1), a plan to provide long-term solutions can be developed and implemented. The following case study provides an example of threat assessment and the actions required for an effective intervention.

12.5.1. Case study 2: Fairhill case study

The Fairhill neighbourhood of Philadelphia was selected for intervention because of the existence of a number of risk factors for rat bites. The area included 50 square blocks with 1520 premises, consisting of residential, commercial and vacant buildings. Single-family attached row houses dominated the neighbourhood, and about 10 000 people resided in it. The population was young, with less than 10% of the residents over 55 years of age (Hirschhorn, 2005).

Three objectives were the basis of this intervention:

1. to deliver a comprehensive environmental improvement, rodent control and safety programme to residential properties in the target area;

2. to coordinate with other city agencies and community partners, to improve the quality of life in the target area; and

3. to solicit community cooperation, to improve substandard conditions and maintain a healthy and safe environment.

A survey was conducted to provide an accurate assessment of conditions and problems. All exterior areas and 60% of the interior areas were inspected. Twenty environmental and safety factors were examined as part of the survey. Surveys of exterior areas examined rodent infestations, structural damage, unapproved refuse storage, abandoned automobiles and large accumulations of refuse. Surveys of interior areas focused on evidence of rat, mouse and insect infestations, points of rodent entry and harbourage, actual or potential food sources, building condition, and points of disrepair. City health department staff performed repairs inside the residential units, to seal off points of entry and harbourage for vermin, and used non-chemical methods to eliminate active rodent and cockroach infestations. At the same time, public education was provided to help the residents understand how they could combat the rodent problems and reduce the environmental and safety hazards that contributed to the risk of infestation. Also, local building and sanitation codes were explained and enforced, and sewers were inspected and repaired. The contributions to rat bites of various housing and environmental conditions are shown in Fig. 12.3.

This abatement and intervention effort required the cooperation of many city agencies. As a result the programme reduced rat complaints by more than 50%, reduced unintentional injuries in residential areas and led to safety improvements, including protection from fires.

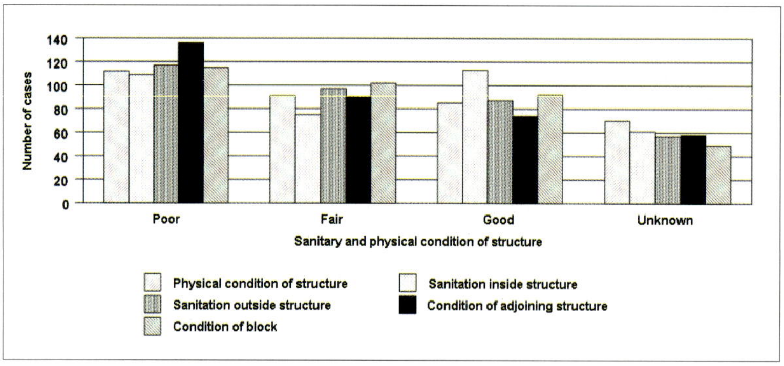

Fig. 12.3. Rat-bite cases and housing and environmental conditions: case study 2

Source: Hirschhorn & Hodge (1999).

The study demonstrated that a prerequisite for an effective public health rodent control programme is cooperation between agencies and the cooperation of the community with those agencies. Public education and information are important. This in turn demonstrates the need for a high level of management and commitment within public authorities.

12.6. Legal framework

Any legal framework reflects the value judgements of a society and is a means of giving expression to those judgements; it also sets the norms of behaviour within society. By itself, law is not generally sufficient to shape the behaviour of society towards accepted goals, such as achieving more effective urban pest management. Law also provides the framework within which public authorities operate, and it emanates from the procedures established in the relevant constitution.

With respect to pest management, legal responsibilities for effective control lie at three levels: national or state governments; local authorities or municipalities and other public agencies that exercise some local control or influence; and individuals (including business enterprises) who own or occupy land and buildings. Many different agencies and organizations will therefore have a role to play in effective urban pest management, and these players will normally require a legislative mandate that they contribute to any commensal rodent control or integrated rodent management strategy. Public authorities will also need the appropriate legal powers and sanctions to ensure that these mandates are met.

The legislation that covers urban commensal rodents normally is directed at two levels:

1. legislation that establishes the powers to control and prevent infestations

2. provisions for the regulation of the pest control operations themselves.

These provisions will include approved methods and materials, which take into account operator safety and protection of non-target species. The legislation should provide for controls over the use of rodenticides, including approval of formulations, and prescribe the circumstances and safe manner in which they can be used. Within the EU, the European Biocidal Products Directive (98/8/EC) (European Commission, 1998) has introduced an authorization scheme for placing biocidal products (which include rodenticides) on the market and for their subsequent use. The key aims of the Directive are to establish a single European market in biocidal products and, at the same time, to ensure that people and the environment are highly protected.

As an example, in the United Kingdom, before any person (other than a householder in their own premises) can carry out any rodent control work, they must conduct an assessment, as required by the Control of Substances Hazardous to Health Regulations 2002 (the COSHH assessment) (HSE, 2003). The purpose of the COSHH assessment is to make sure that any product selected will control rodents effectively when used in accordance with the specified method, while at the same time minimizing risks to both the operator and any other person or animal (non-target species) that might come in contact with the rodenticide.

The legal framework that relates to the presence of commensal rodents on land (and beneath the surface) will depend on the legal history and the legal system within the state. For example, in England, the law that relates to the control of rats and mice (Prevention of Damage by Pests Act of 1949) was drafted at a time of relative food shortage and was aimed at protecting foodstuffs from damage. It gave power to local authorities to require occupiers to deal with infestations, but did not impose a statutory duty on the authority to provide a pest control service. In practice, most local authorities in the United Kingdom do provide such a service, but there is no duty to do so. On behalf of the authority, a commercial pest control company may deliver this service under contract. In some cases, local authorities in England and Wales have also introduced charges for what had previously been a free service. The tariff and charging structure varies from authority to authority, with so-called public health pests treated free of charge and a service charge made for the treatment of nuisance pests (Battersby, 2002; Murphy, 2002; CIEH, 2003). Across authorities, there is no general agreement as to what is a *public health pest*, so that in some municipalities domestic mice are seen as public health pests and in other areas they are not. This confused situation makes effective urban rodent control more difficult (Murphy & Battersby, 2005).

Where private companies perform public services under contract, the terms of the contracts are critical. For example, in England and Wales, much of the sewerage network is the responsibility of privatized companies, which have no clear legal obligation as part of this service to control rats in their sewers. Where they invest in such controls, these private companies incur the direct costs, but do not gain any direct benefits. Also, these companies do not incur the costs that result from increased rat infestations above ground, even though these may be the result of their failure to control infestations and maintain the sewerage infrastructure.

States governments normally have in place some control, via building codes, over the design and construction of new buildings, particularly new dwellings. Although building codes may not be viewed as pest control laws, such codes, when properly enforced, can contribute to effective pest management by containing provisions for designing out potential deficiencies that can lead to future rodent problems. Municipal authorities will also have provisions available to address issues of repair and maintenance of existing houses and other buildings, and they may apply somewhat different standards or criteria to justify interventions. These provisions should also take into account the need to exclude rodents. Provisions for hygiene, including refuse storage, should apply to existing buildings as well as new buildings.

While public education is a necessary part of the control programme, there will be times when municipal authorities need to resort to the law. It is futile to deal with some infested buildings and not others in an area. It is also futile to deal with part of a building and leave other parts infested. When there are clear legal requirements and obligations, it is also a form of education when municipal authorities advise building occupants of these obligations and responsibilities.

Any legal framework must be appropriate for society's needs, but it should recognize the need for an integrated approach to urban pest management. For this to be effective, however, requires properly trained enforcement personnel within the regulatory agencies. Such a legal framework should also address the need for adjoining urban municipalities to operate similarly and cooperatively. Rodents do not recognize administrative boundaries, so there is little gained from one authority implementing a comprehensive strategy while the adjoining authority does little to manage urban commensal rodents.

The foundation for assessing threats and emerging diseases from vectors is disease and pest surveillance. To strengthen this foundation, efforts should be made at the international level to establish and improve networks able to quickly gather and share information on the emergence or spread of communicable and novel diseases. Also, at national and regional levels, systems must be in place to ensure that diseases and conditions that can threaten public health and that occur within their jurisdiction are reportable to public health authorities by physicians, hospitals and laboratories.

12.7. Economic issues and the economic justification for effective control

12.7.1. Rodent damage

Unlike most spoilage of foodstuffs and crops, much of the rodent damage to urban infrastructure that causes economic loss is hidden from the public and may not always be attributed to rodent activity. Because of this, it is difficult to quantify the costs of rodent infestations. Thus, while a cost–benefit analysis can be applied to determine the best response to a problem, in this instance it lacks important data. Also, making an investment can be difficult when present costs are compared with future benefits (Sandmo, 2000).

Rats are known to cause damage to buildings and installations, with a significant risk of fire and electrocution as the result of damage to cables (Colvin in Martindale, 2001; Hall & Griggs, 1990). Burrowing rats can cause landslides on embankments; they can also cause the collapse of banks of canals and ditches, leading to flooding (Meehan, 1984). The direct and indirect costs of structural damage caused by rats can be substantial (DEFRA, 2006). The annual bill for rodent control in the United States in the early 1970s was estimated at US$ 100 million (Brooks, 1973); at that time, commensal rodents in the United States caused between US$ 500 million and US$ 1 billion in damage annually (Pratt, Bjornson & Littig, 1977). The cost associated with damage and loss caused just by rats in the United States is now estimated at close to US$ 19 billion (Pimentel et al., 2000).

On farms in the United Kingdom, all sources of damage amounted to an estimated £10–20 million a year (at 1989 prices) (Battersby, 2004). Richards (1989) reported fire as the most significant form of economic damage that occurred on farms, where roughly 50% of fires reported resulted from rats gnawing electrical cables.

A model was constructed that estimated in the United Kingdom, the costs to the economy of damage to the infrastructure by rats could be between £61.9 million and £209.0 million. It was concluded that, based on the size of the rodent control industry, the higher figure was more likely (Battersby, 2004). Moreover, the damage caused by rats in Budapest, Hungary, was estimated at between US$ 6.4 million and US$ 8.5 million annually between the years1978 and 1985 (WHO Regional Office for Europe, 1998).

In the agricultural sector, the International Rice Research Institute (IRRI) (2006) estimates that in rice-growing regions, rodents cause annual pre-harvest losses of between 5% and 17%. A 6% loss in rice production amounts to approximately 36 tons, which would be enough rice to feed 215 million people (roughly the population of Indonesia) for one year (IRRI, 2006). Brown & Singleton (2002) reported that in a 1994–1995 house mouse plague (more than 1000 mice per hectare), house mice caused an estimated US$ 60 million in damage to crops, livestock industries and rural communities in Australia.

12.7.2. Economics of poor health

Despite evidence that rats are infected with a range of zoonotic agents, little published data are available for assessing the costs to society of ill health due to commensal rodents. The paucity of data may be attributable to a lack of surveillance, diagnosis or awareness on the part of medical practitioners.

One method of calculating the benefits of reducing or eliminating a disease is to estimate the current costs that will be averted by doing so. These costs include medical care, losses of current production, and the pain and discomfort caused by disease (Mishan, 1994). In the context of current rat control activity, such a calculation appears to be impossible. The potential remains, however, for enumerating direct adverse effects on health that have a negative economic impact. These costs could include lost time at work, lost production or increasing demands on medical services. Regardless of such estimates, it cannot be denied that areas with substantial rodent infestations will be associated with ill health and stress that will have a negative impact on both individual and national economies.

12.7.3. Other factors

Who pays for pest treatments (and how) may be reflected in the relative success of that treatment, especially in poorer urban environments. In England and Wales, municipal authorities have begun to charge for pest treatments on domestic properties, and in many cases complaints declined when charges were introduced (CIEH, 2003). This reduction in requests for service may reflect complainants not valuing a rat-free environment as highly as may be supposed. Alternatively, they may be attempting to treat the problem themselves, with undoubtedly less effectiveness, which leads to increases in both current expenditures and future control costs. Any charging regime should be assessed for its potential impact on the rodent population and the control strategy, and it should be supported with free information about what householders can do to prevent rodent infestations.

Any economic assessment of rat infestations should take into account that rats are, or can be, a reflection of lower income housing and poor environmental quality. It has been suggested that the public should be concerned about a rising rat population, mainly because the presence of rats is an economic issue that reflects urban degeneration and an environment that will be unattractive to investment, with businesses and developers being less willing to operate or stay in rat infested areas (Colvin in Martindale, 2001). Rat infestations both reflect and add to the spiral of decline.

A poor quality urban environment, such as one with substantial rat infestations, also means additional stress for its inhabitants. This could contribute to social unrest. This factor should be taken into account when considering the quality of the environment. Thus, the cost to society should include the social cost (including health and well-being) of rat infestations.

The economics of commensal rodent control must also take into account the notion of commensal rodent infestations being a reflection of poor environmental quality – endured by those who are already economically disadvantaged. In such areas, dealing only with the rodent problem may merely be treating a symptom, rather than taking a long-term, more sustainable approach to improve the urban environment and reduce economic disadvantage. In turn, this would lead to better responses from residents, so that over time infestations and treatment costs would be reduced.

As noted, the costs to society attributable to commensal rodents arise from a number of sources: direct damage and adverse effects on health. Fig. 12.4 illustrates the composition of the costs to society of commensal rodent infestations. The relative impact or importance of these costs may depend on the structure and economic strength of the society affected.

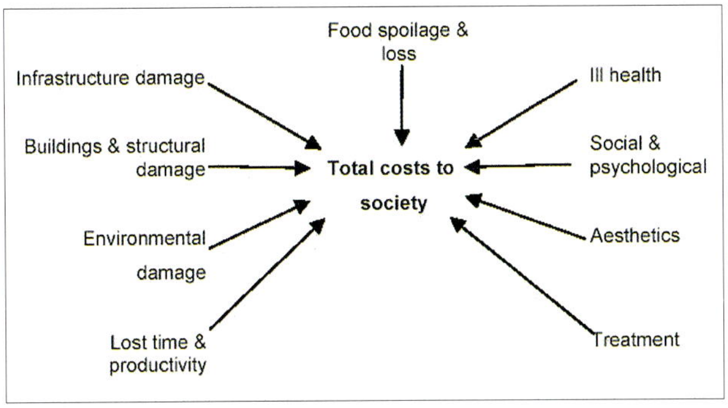

Fig. 12.4. Composition of costs to society of commensal rodent infestations

Source: Battersby (2002).

12.8. Conclusions

A number of steps can be taken to ameliorate the problem of commensal rodents in the urban environment.

- States should consider establishing more effective surveillance mechanisms for identifying the contribution of commensal rodents to the spread of disease.

- Addressing the need for environmental change as a measure to control commensal rodents will have both positive economic and health benefits for residents. The emphasis must be centred on environmental changes designed to remove sources of food and shelter. This will include effective collection, storage and management of waste, improved inspection, repair of buildings and other aspects of the urban infrastructure (such as sewers and drains), and reduced harbourage through landscape removal or management.

- The legal framework should reflect the need for an integrated approach to the control of commensal rodents, as well as the need to regulate the use of rodenticides. Public authorities should be equipped with necessary powers to intervene where voluntary action on the part of other organizations and individuals is inadequate. The legal framework should also support the need for environmental change.

- The level of management within public authorities, as well as a properly trained and equipped workforce, should be adequate to ensure effective control and management of urban commensal rodent populations.

- Adequate time must be allowed for operational staff to carry out thorough surveys and address infestations properly. Also, an integrated programme of control for all urban areas must be rigorously and consistently monitored to assess its effectiveness.

References[2]

Amman BR et al. (2007). Pet rodents and fatal lymphocytic choriomeningitis in transplant patients. *Emerging Infectious Diseases*, 13:719–725 (http://www.cdc.gov/EID/content/13/5/719.htm, accessed 18 May 2007).

Anderson RM (1993). Epidemiology. In: Cox FEG, ed. *Modern parasitology*. Oxford, Blackwell Science:75–117.

Anonymous (1984). No rats? Look again. *Feedstuffs* (Minnetonka, Minnesota), 28 May 1984:14.

Ballenger L (1999). *Mus musculus*. Animal Diversity Web [web site]. Ann Arbor, MI, University of Michigan Museum of Zoology (http://animaldiversity.ummz.umich.edu/site/accounts/information/Mus_musculus.html, accessed 3 September 2006).

Barnett SA (1975). *The rat: a study in behavior*. Chicago, University of Chicago Press.

Barnett SA (2001). *The Story of rats – their impact on us and our impact on them*. Crows Nest, New South Wales, Allen & Unwin.

Battersby SA (2002). *Urban rat infestations: society's response and the public health implications* [PhD thesis]. Guildford, United Kingdom, University of Surrey.

Battersby SA (2004). Public health policy – can there be an economic imperative? An examination of one such issue. *Journal of Environmental Health Research*, 3:19–28 (http://www.cieh.org/library/Knowledge/Public_health/JEHR/JEHRVol3Iss1-PublicHealthPolicy.pdf, accessed 8 March 2007).

Battersby SA, Parsons R, Webster JP (2002). Urban rat infestations and the risk to public health. *Journal of Environmental Health Research*, 1:57–65 (http://www.cieh.org/library/Knowledge/Public_health/JEHR/JEHRv1i2-1-urbanrats.pdf, accessed 27 March 2006).

Bentley EW (1960). *Control of rats in sewers*. London, Her Majesty's Stationery Office (MAFF Technical Bulletin No 10).

[2] Wherever possible peer-reviewed papers and texts from published books have been cited. Where not otherwise available, use has been made of information from learned conference proceedings. Moreover, where possible, relevant use has been made of publications from government departments, agencies and international agencies, so that all sources are authoritative. Use has also been made of interventions at WHO meetings of temporary advisers or experts. Last, where there is no alternative, use has been made of non-peer-reviewed articles in journals of professional bodies that have long-standing respectability. Only one reference that is from a less authoritative source has been used for illustrative purposes (Anonymous, 1984).

Berry RJ (1970). The natural history of the house mouse. *Field Studies*, 3:219–262.

Bond NW (1984). The poisoned partner effect in rats: some parametric considerations. *Animal Learning and Behaviour*, 12:89–96.

Bradshaw J (1999). Know your enemy. *Environmental Health*, 107:126–128.

Bronson FH (1979). The reproductive ecology of the house mouse. *The Quarterly Review of Biology*, 54:265–299.

Brooks JE (1973). A review of commensal rodents and their control. *CRC Critical Reviews in Environmental Control*, 3:405–453.

Brown PR, Singleton GR (2002). Impacts of house mice on crops in Australia: costs and damage. In: Clark L, ed. *Human conflicts with wildlife: economic considerations*. Fort Collins, CO, United States Department of Agriculture, Animal and Plant Health Inspection Service, Wildlife Services, National Wildlife Research Center:23–33.

Buchmeier MJ et al. (1980). The virology and immunobiology of lymphocytic choriomeningitis virus infection. *Advances in Immunology*, 30:275–331.

Calhoun JB (1962). *The ecology and sociology of the Norway rat*. Bethesda, MD, United States Department of Health, Education and Welfare, Public Health Service (document no.1008).

Carleton MD, Musser GG (2005). Order Rodentia. In: Wilson DE, Reeder DM, eds. *Mammal species of the world,* 3rd ed. *Vol. 2.* Baltimore, MD, Johns Hopkins University Press:745–752.

Carrer P, Maroni M, Cavallo D (2001). Allergens in indoor air: environmental assessment and health effects. *The Science of the Total Environment*, 270:33–42.

CIEH (2003). *The role of pest management in environmental health: a guidance document for local authorities*. London, Chartered Institute of Environmental Health (http://www.ciehnpap.org.uk/documents/The_Role_of_Pest_Management_in_Environmental_Health__A_Guidance_Document_for_Local_Authorities.pdf#search=%22Colvin%20BA%2C%20Swift%20TB%2C%20Fothergill%20FE%20(1998).%20Control%20of%20Norway%20Rats%20in%20sewer%20and%20utility%20systems%20using%20pulsed%20baiting%20methods.%20%20%22, accessed 6 September 2006).

Cohn RD et al. (2004). National prevalence and exposure risk for mouse allergen in US households. *The Journal of Allergy and Clinical Immunology*, 113:1167–1171.

Colvin BA, Swift TB, Fothergill FE (1998). Control of Norway rats in sewer and utility systems using pulsed baiting methods. In: Baker RO, Crabb AC, eds. *Proceedings of the 18th Vertebrate Pest Conference*. Davis, University of California at Davis:247–253.

Dennis DT (1998). Borreliosis (relapsing fever). In: Palmer SR, Lord Soulsby EJL, Simpson DIH, eds. *Zoonoses*. Oxford, Oxford University Press:17–21.

DEFRA (2005). *Rodent infestations in domestic properties in England, 2001. A report arising from the English House Condition Survey 2001*. London, Department for the Environment, Food and Rural Affairs (http://www.defra.gov.uk/wildlife-countryside/vertebrates/reports/English-house-survey-rodent-report.pdf#search=%22Rodent%20infestations%20in%20domestic%20properties%20in%20England%2C%202001%20%E2%80%93%20a%20report%20arising%20from%20the%20English%20House%20Condition%20Survey%202001.%20%22, accessed 6 September 2006).

DEFRA (2006). *Rats: options for controlling infestations*, 3rd ed. London, Department for the Environment, Food and Rural Affairs (Rural Development Service, Technical Advice Note 34; http://www.defra.gov.uk/rds/publications/technical/TAN_34.pdf, accessed 15 February 2007).

Drummond DC (1970). Rat free towns. The strategy of area control. *Royal Society of Health Journal*, 90:131–133, 169.

Dubock AC (1982). Pulsed baiting. A new technique for high potency, slow acting rodenticides. In: Marsh RE, ed. *Proceedings of the Tenth Vertebrate Pest Conference, 23–25* February 1982, Monterey, California:123–136 (http://digitalcommons.unl.edu/cgi/viewcontent.cgi?article=1010&context=vpc10, accessed 15 February 2007).

Eisenberg JF, Redford KH (1999). *Mammals of the Neotropics: the Central Neotropics. Vol. 3. Ecuador, Peru, Bolivia, Brazil*. Chicago, University of Chicago Press.

Ewer RF (1971). The biology and behaviour of a free-living population of black rats (*Rattus rattus*). *Animal Behavior Monographs*, 4:127–174.

European Commission (1998). Directive 98/8/EC of the European Parliament and of the Council of 16 February 1998 concerning the placing of biocidal products on the market. *Official Journal of the European Communities*, L123/1 (http://ec.europa.eu/environment/biocides/pdf/dir_98_8_biocides.pdf#search=%22European%20Biocidal%20Products%20Directive%20(98%2F8%2FEC)%20%22, accessed 5 September 2006).

Fischer SA et al. (2006). Transmission of lymphocytic choriomeningitis virus by organ transplantation. *The New England Journal of Medicine*, 354:2235–2249.

Galef BG Jr, Clark MM (1972). Mother's milk and adult presence: two factors determining initial dietary selection by weanling rats. *Journal of Comparative and Physiological Psychology*, 78:220–225.

Galef BG Jr, Wigmore SW (1983). Transfer of information concerning distant foods: a laboratory investigation of the "information centre" hypothesis. *Animal Behaviour*, 31:748–758.

Gillespie H, Myers P (2004). *Rattus rattus*, Animal Diversity Web [web site]. Ann Arbor, MI, University of Michigan Museum of Zoology (http://animaldiversity.ummz.umich.edu/site/accounts/information/Rattus_rattus.html, accessed 3 September 2006).

Gratz NG (1984). The global public health importance of rodents. In: Dubock AC, ed. *Proceedings of a Conference on the Organisation and Practice of Vertebrate Pest Control*, Geneva, World Health Organization:413–435.

Greaves JH, Hammond LE, Bathard AH (1968). The control of re-invasion by rats of part of a sewer network. *The Annals of Applied Biology*, 62:341–351.

Grzimek B, ed. (1975). *Grzimek's animal life encyclopaedia: mammals, I–IV.* Vols 10–13. New York, Van Nostrand–Reinhold.

Hall J, Griggs J (1990). *Rats in drains*. Watford, Building Research Establishment (BRE Information Paper 6/90).

Handley CO Jr (1980). Mammals. In: Lindzey DW, ed. *Endangered and threatened plants and animals of Virginia*. Blacksburg, VA, Center for Environmental Studies, Virginia Polytechnic Institute:483–621.

Hepper PG (1990). Fetal olfaction. In: MacDonald DW, Muller-Schwarze D, Natynczuk SE, eds. *Chemical signals in vertebrates*. Oxford, Oxford University Press:282–287.

Hilton AC, Willis RJ, Hickie SJ (2002). Isolation of *Salmonella* from urban wild brown rats (*Rattus norvegicus*) in the West Midlands, UK. *International Journal of Environmental Health Research*, 12:163–168.

Hirschhorn R (2005). Fairhill case study [6]: urban rodent control and environmental improvement and safety project [web site] (http://www.healthyhomestraining.org/On-Line/docs/Fairhill.htm, accessed 14 May 2007).

Hirschhorn R, Hodge R (1999). Identification of risk factors in rat-bite incidents involving humans. *Pediatrics*, 104:e35.

HSE (2003). *Urban rodent control and the safe use of rodenticides by professional users*. London, Health and Safety Executive (HSE Information Sheet Misc. 515; http://www.hsebooks.com/Books/product/product.asp?catalog_name=HSEBooks&category_name=Home%3A%3AYour+Industry%3A%3A&product_id=4386&cookie%5Ftest=1, accessed 6 September 2006).

Indik S et al. (2005). Mouse mammary tumor virus infects human cells. *Cancer Research*, 65:6651–6659.

International Rice Research Institute (2006). *Ecologically based rodent management.* Manila, Philippines, Irrigated Rice Research Consortium, International Rice Research Institute (http://www.irri.org/irrc/rodents/index.asp, accessed 3 September 2006).

Keeling MJ, Gilligan CA (2000). Metapopulation dynamics of bubonic plague. *Nature*, 407:903–906.

Kingdon J (1974). *East African mammals: an atlas of evolution in Africa. Vol. II, Part B: hares and rodents*. London, Academic Press.

Langton SD, Cowan DP, Meyer AN (2001). The occurrence of commensal rodents in dwellings as revealed by the 1996 English House Condition Survey. *Journal of Applied Ecology*, 38:699–709.

Lehmann-Grube F (1975). Lymphocytic choriomeningitis virus. In: Gard S, Hallauer C, Meyer KF, eds. *Virology Monographs. Vol 10.* New York, Springer-Verlag:1–173.

Lund M (1994). Commensal rodents. In: Buckle AP, Smith RH, eds. *Rodent pests and their control.* Wallingford, CABI Publishing:23–43.

Lundrigan BL, Jansa SA, Tucker PK (2002). Phylogenetic relationships in the genus *Mus*, based on paternally, maternally, and biparentally inherited characters. *Systematic Biology*, 51:410–431.

MacDonald DW, Mathews F, Berdoy M (1999). The behavior and ecology of *Rattus norvegicus*: from opportunism to kamikaze tendencies. In: Singleton GR et al., eds. *Ecology-based rodent management*. Monograph No. 59. Canberra, Australian Centre for International Agricultural Research:49–80.

Martindale, D (2001). The rat catcher. *New Scientist*, 169:40 (http://www.newscientist.com/article/mg16922754.500-the-rat-catcher.html, accessed 2 May 2007).

Meehan AP (1984). *Rats and mice: their biology and control*. East Grinstead, The Rentokil Library, Rentokil Ltd., Brown Knight and Truscott Ltd.

Meyer AN, Drummond DC (1980). Improving rodent control strategies in Lambeth, *Environmental Health*, 88:77–81.

Meyer AN et al. (1995). National Commensal Rodent Survey 1993. *Environmental Health*, 103:127–135.

Mishan EJ (1994). *Cost-benefit analysis: an informal introduction*, 4th ed. London, Routledge.

Murphy RG (2002). Rats and mice: is there a public health threat? In Bonnefoy X,

Rusticali F, eds. *Proceedings of the International Housing and Health Symposium*, Forli, Italy, 21–23 November 2002. Copenhagen, WHO Regional Office for Europe:122–123.

Murphy RG, Battersby SA (2005). Local Authority Pest Management Services in the UK. In: Lee CY, Robinson WH, eds. *Proceedings of the Fifth International Conference on Urban Pests,* Singapore, 11–13 July 2005. Singapore, International Conference on Urban Pests:53–57 (http://www.icup.org.uk/reports%5CICUP009.pdf, accessed 12 February 2007).

Murphy RG, Lindley B, Marshall P (2003). Controlling mouse infestations in domestic properties. *Structural Survey*, 5:190–195.

Murphy RG, Oldbury DJ (2002). Rat control by local authorities within the UK. In: Jones SC, Zhai J, Robinson W, eds. *Proceedings of the Fourth International Conference on Urban Pests*, Charleston, South Carolina, USA, 7–10 July 2002. Charleston, SC, International Conference on Urban Pests:413–420 (http://www.icup.org.uk/reports%5CICUP246.pdf, accessed 12 February 2007).

Murphy RG, Williams RH, Hide G (2005). Population biology of the urban mouse (*Mus domesticus*) in the UK. In: Lee CY, Robinson WH, eds. *Proceedings of the Fifth International Conference on Urban Pests,* Singapore, 11–13 July 2005. Singapore, International Conference on Urban Pests:351–355 (http://www.icup.org.uk/reports%5CICUP054.pdf, accessed 12 February 2007).

Myers P, Armitage D (2004). *Rattus norvegicus*. Animal Diversity Web [web site]. Ann Arbor, MI, University of Michigan Museum of Zoology (http://animaldiversity.ummz.umich.edu/site/accounts/information/Rattus_norvegicus.html, accessed 3 September 2006).

Nieder L, Cagnin M, Parisi V (1982). Burrowing and feeding behaviour in the rat. *Animal Behaviour*, 30:837–844.

Nowak RM (1999). *Walker's mammals of the world,* 6th ed. *Vol. II*. Baltimore, MD, Johns Hopkins University Press.

Padovan D (2006). *Infectious diseases of wild rodents*. Anacortes, WA, Corvus Publishing Company.

Perry T et al. (2003). The prevalence of rat allergen in inner-city homes and its relationship to sensitisation and asthma morbidity. *The Journal of Allergy and Clinical Immunology*, 112:346–352.

Pimentel D et al. (2000). Environmental and economic costs of nonindigenous species in the United States. *BioScience*, 50:53–65.

Posadas-Andrews A, Roper TJ (1983). Social transmission of food preferences in adult rats. *Animal Behaviour*, 31:265–271.

Pratt HD, Bjornson BF, Littig KS (1977). *Control of domestic rats and mice*. Atlanta, GA, United States Public Health Service (Publication No. (CDC) 77-841).

Quy RJ (2001). *Rats and bait boxes: a question of avoidance.* Pest Ventures Seminar on *Rodent Control: A Modern Perspective*, Kegworth Notts, United Kingdom, 27 March 2001. Conference Abstracts. Melton Mowbray, Leicestershire, Pest Ventures:8–9.

Reid FA (1997). *A field guide to the mammals of Central America and Southeast Mexico*. New York, Oxford University Press.

Richards CGJ (1989). The pest status of rodents in the United Kingdom. In: Putman RJ, ed. *Mammals as pests*. London, Chapman & Hall Ltd:21–33.

Sandmo A (2000). *The public economics of the environment*. Oxford, Oxford University Press.

Seguin B et al. (1986). Bilan épidémiologique d'un échantillon de 91 rats (*Rattus norvegicus*) capturés dans les égouts de Lyon. *Zentralblatt für Bakteriologie, Mikrobiologie und Hygiene, Serie A*, 261:539–546.

Stewart T et al. (2000). Breast cancer incidence highest in the range of one species of house mouse, *Mus domesticus*. *British Journal of Cancer*, 82:446–451.

Sullivan R (2006). *Rats: a year with New York's most unwanted inhabitants*. London, Granta Books.

Swift JD (2001). Rat population dynamics – is there a 10 year cycle? *International Pest Control*, 43:156–159.

Taylor KD, Quy RJ (1978). Long distance movements of a common rat (*Rattus norvegicus*) revealed by radio tracking. *Mammalia*, 42:63–71.

Twigg G (1975). *The brown rat*. Devon, David & Charles (Holdings) Ltd.

Webster JP, Lloyd G, Macdonald DW (1995). Q fever (*Coxiella burnetii)* reservoir in wild brown rat (*Rattus norvegicus*) populations in the UK. *Parasitology*, 110:31–35.

Webster JP, Macdonald DW (1995). Parasites of wild brown rats (*Rattus norvegicus*) on UK farms. *Parasitology*, 111:247–255.

Williams RH et al. (2005). The urban mouse (*Mus domesticus*) and its role in the transmission of *Toxoplasma gondii* infection. In: Lee CY, Robinson WH, eds. *Proceedings of the Fifth International Conference on Urban Pests,* Singapore, 11–13 July 2005. Singapore, International Conference on Urban Pests:357–361 (http://www.icup.org.uk/reports%5CICUP055.pdf, accessed 12 February 2007).

Wilson DE, Reeder DM (2005). *Mammal species of the world,* 3rd ed. *Vol. 2.* Baltimore, MD, Johns Hopkins University Press.

Whitaker JO (1980). *The Audubon Society field guide to North American mammals.* New York, Alfred A. Knopf Inc.

WHO Regional Office for Europe (1998). *Rodents.* Copenhagen, WHO Regional Office for Europe (Local Authorities, Health and Environment Briefing Pamphlet Series, No.14).

13. Non-commensal rodents and lagomorphs

Kenneth L. Gage and Michael Y. Kosoy[1]

Summary

Although most problems caused by small mammals in urban environments involve commensal rats and mice, non-commensal rodents and ecologically similar rabbits and hares can threaten the health of people living, working or participating in recreational activities in areas undergoing development and conversion from rural landscapes to ones that are more urbanized. As rural landscapes are initially developed, human contact with non-commensal rodents is likely to increase, which can lead to an increased risk of exposure to the various zoonotic disease agents carried by these animals. People also can come in contact with non-commensal rodents in areas that are only moderately urbanized and still contain significant patches of natural rodent habitat.

To reduce the threat of rodent-related disease, steps must be taken to raise awareness among local residents, health care providers, veterinary staff and public health workers. Before beginning development activities, appropriate planning should aim to minimize the risk of exposing people to the disease agents carried by non-commensal rodents. These steps can include relocating some developments or managing the environmental aspects of the remaining natural rodent habitats, home sites, work sites and recreational areas, to reduce the attractiveness of these areas to rodents. Effective management of non-commensal rodents and prevention of the diseases they carry also will require implementation of regionally appropriate rodent and disease surveillance programmes. In some instances, pesticide use might be advised, to reduce the risk of disease.

[1] Disclaimer: The findings and conclusions in this article are those of the authors and do not necessarily reflect the views of the United States Department of Health and Human Services.

13.1. Introduction

Many species of non-commensal or so-called wild rodents, as well as various rabbits and hares, occur as interesting and usually harmless residents of wilderness and rural regions of Europe and North America. In such settings, these animals usually have negligible impacts on human health and economic well-being, although a few species occasionally do cause damage to crops, pastures and forest plantations. Unfortunately, when these same animals occur near human habitations, work sites or recreational areas, they can serve as sources of disease and cause significant damage to homes or other sites. Based on current economic development and housing trends, it appears inevitable that an increasing number of people will come in contact with non-commensal rodents, rabbits and hares found in rural or wilderness landscapes undergoing conversion to more urbanized environments.

This chapter discusses the major types of rodents, rabbits and hares likely to be encountered in areas undergoing urbanization. It also discusses the health and economic risks these animals pose, how aspects of their biology affect these risks, and what steps can be taken to reduce human exposure to non-commensal rodents, rabbits and hares, and the diseases they carry. Additional information on the health-related effects of these animals in Europe and North America can be found in a recently published book by Gratz (2006)

13.2. Types, distribution and abundance

The following subsections discuss briefly the distribution and biology of major rodent groups (order: Rodentia; families: Muridae, Sciuridae, Myoxidae and Castoridae) most likely to be encountered in areas undergoing urban sprawl in Europe or North America. Emphasis will be placed on the factors that affect the abundance of these animals and the likelihood they will come in contact with people. Unless stated otherwise, basic information on the habitats, distributions, abundance, reproduction and foods of these animals, as well as their interactions with people, was obtained from the following standard mammalian reference works: Cahalane (1961); Grzimek (1968); van den Brink (1968); Hall (1981); MacDonald (1984); Nowak (1991); MacDonald & Barrett (1993); Alderton (1996); Nowak (1999); Nechay (2000); MacDonald (2001). Additional citations specific to each type of rodent are also provided in some instances. Similar discussions that use information obtained from the same literature sources listed above for rodents are provided for ecologically similar rabbits and hares (order: Lagomorpha; family: Leporidae) – hereafter referred to as lagomorphs.

13.2.1. Tree squirrels

These attractive and highly visible animals (primarily *Sciurus* spp.) provide a classic example of the successful use of urbanized environments by rodents. The European red squirrel (*Sciurus vulgaris*) is the most common European tree squirrel and is widespread across much of the continent. However, it has been displaced in some areas, such as in the United Kingdom, by the gray squirrel (*Sciurus carolinensis*), a species introduced from

North America (Lloyd, 1983; Reynolds, 1985). Following introductions, from 1876 to1929, gray squirrels largely replaced red squirrels throughout much of England and Wales. Recent research suggests that this replacement occurred not as a result of direct competition, but rather as a result of gray squirrels carrying a novel poxvirus that is lethal to red squirrels but causes little or no serious illness in gray squirrels (Thomas et al., 2003). Populations of gray squirrels also occur in Scotland, Ireland and northern Italy (Po Valley), where they were introduced during the late 1940s. Gray squirrels are most common in deciduous and mixed woodlands, parks and gardens. They are considered a pest species in some countries, though others welcome them as interesting garden visitors. Common tree squirrels in North America include the gray squirrel and fox squirrel (*Sciurus niger*) in the eastern United States. Both species are common in urban environments, and the latter has been introduced into various cities in the western United States. Other kinds of tree squirrels, including American red squirrels (*Tamiasciurus hudsonicus*), Douglas squirrels (*Tamiasciurus douglasii*) and western gray squirrels (*Sciurus griseus*), are less likely to be found in urban areas, but *Tamiasciurus* spp. occasionally invade attics and wall spaces in mountainous areas.

Although tree squirrels forage on the ground, they rarely stray far from a tree, where they can flee to safety. Common foods for these squirrels include nuts, berries, buds, fungi, insects, and (sometimes) bird eggs or small animals. Their nests are built in trees from twigs, shredded bark, leaves and sometimes moss. When available, hollows in tree trunks are often used as nesting sites. Densities of red squirrels can range from 0.2 squirrel/ha to 1.6 squirrels/ha in coniferous or broadleaved forests. These squirrels also can occur at high densities in parks and gardens, raising the likelihood of contact with people. European red squirrels produce 1–2 litters a year, depending on latitude, with 3–4 young per litter; reproductive rates for the invasive gray squirrels are similar, as are those of fox squirrels.

13.2.2. Flying squirrels

Flying squirrels, such as the southern flying squirrel (*Glaucomys volans*) and the northern flying squirrel (*Glaucomys sabrinus*), are common nocturnal inhabitants in many forests and some semi-rural and suburban landscapes in North America. These animals do not actually fly but are able to glide with the help of a large skin membrane (patagium) that runs alongside the body from the forelimbs to the hindlimbs (Eisenberg, 1981). As the flying squirrel stretches out its limbs, the membrane becomes taut and forms an air-resistant surface that allows the animal to remain aloft for many seconds. The relatively common southern flying squirrel frequently lives in attics or wall spaces (Banfield, 1974), but it often goes unnoticed because of its secretiveness and nocturnal habits. Its food consists primarily of nuts, seeds, fruits, bark, fungi, lichens and insects. Flying squirrel females can mate twice a year and typically give birth to 1–6 young after a gestation period of about 40 days. Births usually occur from late winter to early spring and from early summer to midsummer. Another flying squirrel species, the Siberian flying squirrel (*Pteromys volans*), exists in some regions of Europe, but has little impact on people, because it rarely enters human environs and is experiencing population declines, as tracts of birch, spruce and pine forest are destroyed to make way for human development (Grzimek, 1975).

13.2.3. Ground squirrels, antelope ground squirrels and prairie dogs

The squirrel family contains a number of burrowing species, many of which are medically and economically important. Because members of the genus *Spermophilus* (ground squirrels) are large and active during daylight hours, people frequently notice them, and some enjoy having them near their homes, while others consider them destructive vermin that should be eliminated. In general, ground squirrels usually give birth to a litter of as many as 12 young a year, but certain species that live in southern regions can have two litters a year. Ground squirrels eat primarily plant foods, including nuts, seeds, roots, leaves and fungi, but insects and even an occasional bird or mouse-sized mammal also can be consumed. Many of these species exhibit varying degrees of social behaviour and live in colonies or loose aggregations (Murie & Michener, 1984).

Many more species of ground squirrels occur in North America than in Europe, but Europe is home to two well-known ground squirrels: the European ground squirrel (*Spermophilus citellus*) and the spotted souslik (*Spermophilus suslicus*). Neither of the European species poses a serious health risk to people, in part because they are relatively uncommon in urbanized areas, although they can cause agricultural damage.

North American species of ground squirrels are found primarily in the western grasslands, mountains and deserts, although a couple of species have ranges that extend into the eastern half of the continent. In many instances, these ground squirrels live in wilderness or highly rural areas, but a few species occur regularly near human dwellings and city parks, where they can damage structures, gardens, orchards, crops and other items. Foremost among the ground squirrels encountered in peridomestic environments are two closely related species, the rock squirrel (*Spermophilus variegatus*) and the Beechey (California) ground squirrel (*Spermophilus beecheyi*). Rock squirrels occur throughout much of the south-western United States and north-eastern Mexico. Beechey ground squirrels are found in many areas of California, western Nevada and southern Oregon. Both species behave quite similarly and often dig burrows under concrete slabs, woodpiles or other sites near people's homes (Oaks et al., 1987; Jameson & Peeters, 2004). Some have suggested that rock squirrel numbers in the south-western United States have increased as a result of home building and other human activities that provide these animals with novel sources of food (such as pet foods, seeds from bird feeders and water from dripping faucets) and shelter (such as rock piles and walls) (Barnes, 1982). A third species of ground squirrel, the golden-mantled ground squirrel (*Spermophilus lateralis*), can occur near human dwellings in mountainous areas of western North America. It is frequently encountered in recreational sites, including heavily used campgrounds in California and adjoining areas, as well as in many regions of the Rocky Mountains. Other species of North American ground squirrels cause agricultural damage and occur occasionally near human habitations, but they generally have limited impacts on people's lives in suburban environments.

Although less often associated with people than the above-mentioned squirrels, antelope ground squirrels (*Ammospermophilus* spp.) can live in close proximity to urbanized areas, particularly those in the south-western United States where urban sprawl has resulted in newly constructed human dwellings being scattered among the large patches of native vegetation preferred by these animals. The species most likely to be found near human habitations is the white-tailed antelope ground squirrel (*Ammospermophilus leucurus*), which rarely causes significant damage to residential vegetation or property (Belk & Smith, 1991). These animals prefer sites with gravelly soils covered by grasses and sagebrush, greasewood, shadscale or creosote bush. Their burrows can be recognized by the radiating pathways that merge at the burrow entrance, which lacks a mound and can be somewhat inconspicuous (Zeveloff, 1988). They feed primarily on seeds and the green portions of forbs and grasses. Typically, a litter of 5–14 young is born in the spring and, occasionally, a second litter will be produced in a given breeding season.

The so-called prairie dogs are not actually dogs, but rather are large burrowing squirrels of the genus *Cynomys*. Their unusual name comes from the bark-like calls they make to warn other colony members of potential dangers. Each of the five North American species exhibits highly complex social behaviour and lives in large colonies that can include thousands of individuals living in well-defined family groups, called coteries (Clark, Hoffman & Nadler, 1971; Pizzimenti & Hoffmann, 1973; Pizzimenti & Collier, 1975; Ceballos & Wilson, 1985; Hoogland, 1995). Their food consists primarily of grasses and other plants, but insects and even smaller rodents are eaten occasionally. Each year females give birth to a single litter of about four pups. Mexican and Utah prairie dogs (*Cynomys mexicanus* and *Cynomys parvidens*, respectively) have the most limited distributions and live in fairly remote areas (Long, 2002). Both are considered threatened and are rarely encountered by people. White-tailed prairie dogs (*Cynomys leucurus*) are quite abundant and widespread, but also typically live in environments far removed from major urbanized areas (Clark, Hoffmann & Nadler, 1971). The closely related Gunnison's prairie dog (*Cynomys gunnisoni*) is found on the Colorado Plateau and surrounding regions of the south-western United States. Unlike the above three species, Gunnison's prairie dogs often establish colonies near human dwellings. The final species is the black-tailed prairie dog (*Cynomys ludovicianus*) of the plains grasslands located east of the Rocky Mountains (Hoogland, 1995).

Among the five prairie dog species, the black-tailed prairie dog is most likely to occur in close proximity to people, being fairly common in many suburban and even some urban areas, particularly those along the Front Range of the Colorado Rocky Mountains, a region that includes the Denver Metropolitan Area and numerous smaller cities. In some instances, small colonies of this species occur in isolated patches of habitat that are almost completely surrounded by urban development. When living in urbanized environments, prairie dogs can damage shrubs or other plants that are eaten for food or instinctively cropped, to reduce the risk of being ambushed by predators, such as coyotes (*Canis latrans*) or American badgers (*Taxidea taxus*).

13.2.4. Chipmunks

Chipmunks are members of the genus *Tamias*. The Siberian chipmunk (*Tamias sibiricus*), which can be found frequently in parks and towns, is widely distributed in parts of the Russian Federation (the north-eastern part of Europe, Siberia and the far-eastern part). Occasionally, escaped pet Siberian chipmunks have succeeded in establishing more or less permanent populations in Austria, Finland, France, Germany, Italy and the Netherlands (Amori & Gipoliti, 1995; Chapuis, 2005). The 22 species of North American chipmunks occur primarily in the mountain forests and nearby sagebrush habitats of the western third of the continent. A single species, the eastern chipmunk (*Tamias striatus*), occurs abundantly in the deciduous forest regions of the eastern United States and south-eastern Canada, routinely entering yards and gardens in many suburban areas (Mahan & O'Connell, 2005). Although chipmunks spend the bulk of their time on the ground, the eastern chipmunk sometimes nests in hollows in trees, as do a few western species. Species that nest on the ground make a network of shallow burrows under stones or logs. Some western species occasionally invade homes, where they often build nests in attics or wall spaces, sometimes damaging these structures in the process. Invaded spaces are often partially filled with large stockpiles of nuts, pine cones and other edible items. Although chipmunk nesting and hoarding activities can cause some damage, they are of little economic importance. Their primary foods are fruits, nuts, berries, seeds and occasional invertebrates. Depending on conditions, females can produce two litters a year, with the first appearing in February to April and the second typically appearing in June to August; each litter usually consists of 2–8 pups.

13.2.5. Beavers

Beavers are large aquatic rodents that exist in both North America and the Eurasian continent. The European beaver (*Castor fiber*) is confined to isolated pockets in Scandinavia, eastern Germany and certain other sites where it has been recently introduced or re-established (Veron, 1992). Its preferred habitats are broad river valleys with extensive flood-plains. The Canadian (North American) beaver (*Castor canadensis*) was introduced from North America to Finland, Poland and the Russian Federation, and now is distributed widely in other parts of Europe. It readily occupies artificial ponds and ditches that provide suitable food and shelter.

Beavers typically build a conical lodge of branches with an underwater entrance and above-water chambers within the lodge. The outside of the lodge is covered with a layer of mud. Beavers that live in small streams construct extensive stick dams that result in the formation of a small pond. When heavily hunted, beavers may live in burrows in banks or construct less conspicuous lodges. Beavers eat shrubs, bulrushes, tree buds and roots of aquatic and semi-aquatic plants. In winter, they consume considerable quantities of bark. Beavers do not become sexually mature until they are 3–4 years of age, and mating occurs from January to March, with 2–4 pups being born sometime between April and July. Beaver lodges can contain up to three generations from the same family, although the previous season's offspring are often driven away by the parents to make room for the current season's litter.

Originally decimated by the fur trade (Banfield, 1974), the Canadian beaver is making a strong comeback in much of North America, as a result of attempts to re-establish former populations, natural population growth and reduced demand for beaver fur. Due either to their timber-cutting or dam-building activities, recovering beaver populations have come into increasing conflicts with people. Beaver ponds do, however, provide considerable benefits to wildlife and fish populations. They also help retain water in stream drainages and recharge groundwater reservoirs.

13.2.6. Hamsters

The common hamster (*Cricetus cricetus*) is found in the grassy steppes and field edges of eastern Europe, including Austria, Belarus, Bulgaria, the Czech Republic, Hungary, Poland, the former Yugoslavia and the Russian Federation. Pockets of these animals also exist in Belgium, north-eastern France, western Germany, the Netherlands and Switzerland. During periods of peak population densities, hamsters sometimes emigrate, often appearing in gardens and around houses, as has been recorded in the Czech Republic, Germany, Hungary and the Netherlands (Nechay, 2000). In the eastern part of their range, common hamsters often live in close proximity to people (Poljakov, 1968). Gray hamsters (*Cricetulus migratorius*), which occur on cultivated grasslands and are common in gardens, range eastward from Greece, Bulgaria and Romania towards Asia. Other European hamster species include Eversmann's hamster (*Allocricetulus eversmanni*), the Dobrudja hamster (*Mesocricetus newtoni*), the Turkish hamster (*Mesocricetus brandti*) and the Daghestan hamster (*Mesocricetus raddei*). The level of contact between these species and people depends on hamster population densities, land use patterns, and ecological and people-related factors. Their primary food sources are plant materials, including cereal grains, fruits, roots and leaves, although insects are eaten occasionally. Hamsters hoard sizeable quantities of seeds and other foodstuffs and can be significant agricultural pests. Females produce 2–3 litters a year, with each litter consisting of 2–3 young.

13.2.7. Voles

Voles are among the northern temperate region's most abundant rodents (Elton, 1942). Bank voles (*Clethrionomys glareolus*) are common in the deciduous woodland regions of Europe, occurring as far north as the Arctic Circle and as far south as the Pyrenees and Alps, with a few relict populations present in Italy. Bank voles require ground cover and are found most frequently in hedgerows, banks, field edges, woodlands and other well-vegetated areas, although in the northern reaches of Europe they can be found on relatively open ground, and in Norway they have been reported to enter homes. Bank voles rarely travel more than about 50m and largely confine their activities to runways built just under the ground or in ground-hugging vegetation. Bank voles feed primarily on vegetation, including plant stems, leaves, roots, bulbs, fruits and seeds, although insects make up about a third of their diet. They are often considered pests, because of their habit of stripping bark from small trees, especially larch, elder and young conifers. Bank vole populations often fluctuate dramatically over cycles of 3–4-year duration, sometimes becoming serious pests at high densities. Although most voles die within a year after being born, their high reproductive rates easily compensate for their low survivability. Voles can breed from April to October, producing 4–5 litters a year, each of which contains 3–6 pups. Another species of *Clethrionomys*, the southern red-backed vole (*Clethrionomys gapperi*), can be found in mesic coniferous, deciduous and mixed forests in the mountains of the western United States and thoughout most of Canada.

More than 40 species of *Microtus* voles are known to exist. Important European species include the common vole (*Microtus arvalis*), root vole (*Microtus oeconomus*) and the field or short-tailed vole (*Microtus agrestis*). The common vole is predominant in the eastern half of the continent, but the most frequently encountered vole in many western European countries, except Ireland, is the field vole, a species that has declined in some areas but remains abundant in open grasslands and damp pastures. These three European species of *Microtus* voles also occur often in gardens and open woodlands, as they seek out their primary food sources, which consist of grasses, as well as the stems, roots and bark of other plants. The foraging activities of field voles also can occasionally result in damage to newly sown cornfields. As these animals travel through the grass and low-lying vegetation of their home ranges, which are about 100–150m in diameter, they form a characteristic network of runways that are shared by other voles and provide some protection from predators.

Field vole populations also fluctuate greatly from year to year, though not as dramatically as those of the related lemmings. Even among rodents, *Microtus* voles are notable for their ability to quickly increase their reproductive rates in response to the availability of new resources or improved habitats (Elton, 1942; Begon, Harper & Townsend, 1996). In some instances, their numbers increase so greatly that they cause extensive damage to pastures, cornfields or forest plantations. Densities of voles during such so-called vole plagues have reached as high as 4,900 voles/ha. The field vole breeding season is from February to September, after which females produce as many as 10 litters a year, with 3–7 offspring per litter. Females can become sexually mature as soon as 3 weeks after their birth. These animals eat primarily grasses, stems, new plant shoots, roots, bulbs and bark, as well as some insects.

At least 17 species of *Microtus* voles occur in North America. Among the most important of these are the meadow vole (*Microtus pennsylvanicus*), California vole (*Microtus californicus*), prairie vole (*Microtus ochrogaster*), montane vole (*Microtus montanus*) and long-tailed vole (*Microtus longicaudus*). The meadow vole occurs in moist grassy fields and meadows over much of the northern two thirds of North America. Some have claimed that it is the most prolific mammal on earth (Kays & Wilson, 2002) and, like most voles, its populations can experience dramatic fluctuations in density (Krebs & Myers, 1974; Reich, 1981). At high densities, meadow voles can cause serious damage to woody vegetation, especially in fruit orchards (Byers, 1979). The California vole occurs throughout much of California and southern Oregon in low-elevation grasslands, wet meadows, coastal wetlands and open oak savannahs with adequate ground cover (Kays & Wilson, 2002). Prairie voles occur in grasslands in the centre of the continent, and montane and long-tailed voles are found in the mountains of western North America.

The water vole (*Arvicola terrestris*) is widespread throughout much of Europe, but is absent from Ireland and most of the Iberian Peninsula. In some areas, such as the United Kingdom, it usually is closely associated with fresh water (such as flooded ditches, slow rivers and lakes) and favours steep riverbanks with abundant grass and layered vegetation. Non-aquatic, fossorial (digging or burrowing) populations of water voles occur in some regions (Meylan, 1975), including central Europe, but in the southern parts of their range water voles seem to prefer pastures over more typical riparian habitats. Due to predation by introduced American mink (*Mustela vison*), loss of suitable riverside habitat and water pollution, water vole populations in the United Kingdom have declined by as much as 90% over the past 25 years. These animals eat various grasses, sedges and other plants found growing near the edge of their aquatic homes. Water voles also consume cereal grains and fruits, including apples. Breeding takes place from April to October, with females producing 3–4 litters of 2–7 young per litter. Few water voles live more than a year. If these animals are numerous, their burrowing activities can damage dikes and canals.

Muskrats (*Ondatra zibethicus*) are basically large aquatic voles that are native to North America, but they have been transplanted to many regions in Europe. They are quite abundant throughout much of North America and occur frequently in ponds and waterways in suburban and largely urban areas. Muskrats require aquatic habitats with suitable aquatic vegetation for food and shelter. They are valued for their fur, and their tunnelling activities can cause damage to dikes and stream banks (Danell, 1977; Corbet, 1978).

13.2.8. Old World mice

Mice of the genus *Apodemus* are quite widespread and abundant. They eat primarily grasses, seeds, nuts, fruits and other plant materials, as well as some insects and snails. Their nests are made from finely shredded grasses. *Apodemus* mice can produce up to five litters a year with each litter containing about six young. Striped field mice (*Apodemus agrarius*) and wood mice (*Apodemus sylvaticus*) are the most ubiquitous and persistent small mammal species in many urban areas of Europe (Montgomery, 1976; Babinska-Werka, Gliwicz & Goszczynski, 1981). The striped field mouse is widespread in eastern Europe, with western populations reaching Germany and Northern Italy. These mice can be very abundant in fields, scrub and woodlands associated with damp habitats and river valleys. They also may enter houses, barns and stables and have been reported to colonize such highly urbanized areas as Warsaw, Poland (Andrzejewski et al., 1978). The related wood mouse is widespread in woodlands, scrubland and dune areas throughout continental Europe, the British Isles and southern Scandinavia, having a range that extends as far as the southern part of western Siberia, northern Kazakhstan and the mountains of central Asia. Being highly adaptable, they often occupy gardens and city parks and will enter houses in winter, particularly when house mice are absent. Another species of *Apodemus*, the yellow-necked mouse (*Apodemus flavicollis*) is common in woodlands, hedgerows, field margins, orchards and wooded gardens. Yellow-necked mice are more common than wood mice in alpine coniferous forests, but are less common than the latter in open scrub and fields. Yellow-necked mice also frequently enter homes as winter approaches, but typically depart by spring. They are known to store caches of nuts in small spaces and under floorboards.

13.2.9. Dormice

These squirrel-like rodents occur over much of Europe, although deforestation has severely impacted some species (Pucek, 1989; Amori, Cantini & Rota, 1995). Although other species also occur on the continent, the ones most likely to be encountered by people are the fat or edible dormouse (*Myoxus glis*, formerly *Glis glis*), the common or hazel dormouse (*Muscardinus avellanarius*), the garden dormouse (*Eliomys quercinus*), and the forest dormouse (*Dryomys nitedula*). Dormice typically live in wooded areas, hedgerows and rocky places, but will enter gardens or other areas near human dwellings. They find shelter in hollow trees, rock crevices and abandoned burrows, where they build a nest of plant materials. Given the opportunity, these rodents also will build nests in building attics or barns. Dormice are normally nocturnal, but can be active at dawn or dusk. Their diet consists primarily of nuts, fruits, insects, eggs and small vertebrates. In the northern portions of their range, dormice put on considerable fat in the fall and hibernate during the winter, waking only occasionally to feed on stored food items. Females give birth to one or two litters a year, with each litter containing 2–10 young. Dormice are believed to live for 5–6 years in the wild. Although they occasionally can become nuisances and damage wine grapes and fruits, dormouse populations are typically so small that they have little impact on people.

13.2.10. New World rats and mice

The New World murine subfamily Sigmodontinae contains a wide variety of species, including some that occur near human habitations. The most important genera are *Peromyscus* (deer mice and their allies), *Neotoma* (wood rats) and *Sigmodon* (cotton rats).

Carlton (1989) recognized 53 species of *Peromyscus*, and all of these are likely to invade human dwellings under certain circumstances (Cahalane, 1961). However, the two species of *Peromyscus* most likely to be encountered by people are the widespread deer mouse (*Peromyscus maniculatus*) and white-footed mouse (*Peromyscus leucopus*) (King, 1968; Kays & Wilson, 2002). Both species are abundant over large areas of North America, with the deer mouse occupying all but the south-eastern portion of temperate North America and the white-footed mouse occurring over the eastern half of the continent and in portions of the south-western United States. In many respects, including behaviour, appearance and general ecology, these mice resemble European species of *Apodemus* mice and can be considered ecological equivalents. *Peromyscus* mice consume a variety of seeds, other vegetable matter and often insects. Deer mice are particularly common in grasslands or mixed grass and brush habitats; white-footed mice are more likely to occur in woodland or mixed woodland and brush habitats. Both species will enter homes and other buildings, particularly as winter approaches. Although they can be quite abundant, these mice are nocturnal and, therefore, rarely observed. Their nests usually consist of a mass of grass, leaves and other soft debris, although deer mice, like certain other *Peromyscus* spp., occasionally construct short, shallow burrows. Females give birth to up to four litters a year, with 1–9 young per litter. Lifespans of *Peromyscus* mice rarely exceed 2 years. Their gnawing near nest entry points on homes or other structures can cause limited damage to wood siding and their excreta can create an unsanitary situation.

Although wood rats (*Neotoma* spp.) occur in both the eastern and western portions of temperate North America, the diversity of species is greatest in the western half of the continent. These rats feed on a variety of seeds, berries and other vegetable matter, as well as on occasional invertebrates. Their name comes from the distinctive nests they build, which consist of large piles of sticks that are often placed at the base of a tree or other sizeable plant, although the nests of Mexican wood rats (*Neotoma mexicana*) are constructed in large cracks in cliff faces, along the walls of caves or under large rocks. Many species in the western United States and Mexico cover their stick nests with pieces of spine-bearing cactus to provide additional protection against predators. If left undisturbed, wood rats will continue to add items to their nests until they become quite large. Bonaccorso & Brown (1972) reported that a single desert wood rat (*Neotoma lepida*) could build a complete nest 40 cm high and 100 cm wide over a period of 7–10 days. Wood rats are highly territorial, and the valuable nest sites are defended vigorously, rarely remaining unoccupied for long periods. In some instances, nests can exist for many decades and perhaps even longer, being enlarged by each new resident. Several species of *Neotoma* are known to build their nests in the walls or crawl spaces of homes, garages or other buildings. The gnawing activities of these rats, as well as the extensive piles of excreta associated with their nests, can result in damage to homes or other property and cause an unsightly mess. Female wood rats in the northern part of the continent typically produce a litter a year, while those living in more southerly areas sometimes have two litters a year. The average litter size is 2–6 pups. Wood rats do not become sexually mature for 7–8 months after birth, and they live longer-than deer mice; one wood rat in captivity lived for nearly 8 years, although lifespans of 3–5 years are probably more typical in the wild.

Cotton rats, of the genus *Sigmodon*, are common in grassy and weedy fields in many areas of southern North America. The most important species in the temperate regions of the continent is the hispid cotton rat (*Sigmodon hispidus*), which is often extremely abundant in thick grassy habitats in the south-eastern and south-central United States (Cameron & Spencer, 1981). In many ways, the behaviour and ecology of cotton rats resemble those of voles, which are more common in the northern parts of the continent. These similarities include not only types of habitats selected, but also include the construction of grass nests and runways, extremely high reproduction rates and populations that often fluctuate dramatically from year to year. In the warmer portions of its range, the hispid cotton rat can breed year-round. Females produce several litters a year, with up to 15 young in each litter. Most cotton rats that survive to breed live less than 2 years. Their activity typically peaks around dawn and dusk, during which time they forage on various types of green vegetation, as well as on the occasional insect or bird egg. Cotton rats can be quite destructive to some crops, including sugar cane and sweet potatoes.

13.2.11. Rabbits and hares

Rabbits and hares are not rodents, but rather belong to the family Leporidae within the mammalian order Lagomorpha, which also includes pikas. The most notable difference between lagomorphs and rodents is the possession of an extra pair of incisors in the former group. Rabbits and hares are extremely widespread, occupying habitats from the tropics to the Arctic, including forests, deserts, grasslands, wetlands and mountainous areas. All species are long-eared, grazing animals with elongated limbs for running quickly over open ground. Their large eyes, placed on the side of the head, improve their chances of detecting approaching predators, even in dim light. These animals also have a keen sense of smell, which provides additional protection from predators. Some species, such as the European rabbit (*Oryctolagus cuniculus*) construct a system of burrows, referred to as a warren; other species rest and sometimes hold their young in shallow depressions, termed forms, that are often hidden in brush and can be lined with fur and soft plant materials. Although rabbits and hares can be differentiated based on anatomical, behavioural and developmental characteristics, both are largely similar in appearance and their differences have little significance for the purposes of this chapter.

Like rodents, rabbits and hares have high reproductive potential, becoming mature in about 3 months and having multiple litters per year. Gestation periods are about 40 days in the genus *Lepus*, but closer to 30 days in other genera. Litter sizes range from 1–9 offspring, typically being larger in *Sylvilagus* spp. and *Oryctolagus* spp. than in *Lepus* spp. Although many species and genera exist, the species of greatest importance to people belong to the above three genera.

The European rabbit was originally found in central Europe and the Carpathians, but is now widespread in western and central Europe, although its numbers have been reduced by myxoma virus infections. These rabbits occur on sandy and light clay soils in woods, particularly those dominated by conifers. Although they occur in hilly country, these rabbits are not found in high mountains. When present in high numbers, they can be significant agricultural pests.

Another important species is the European (or brown) hare (*Lepus europaeus*), which is found not only in Europe, but is also found as introduced populations in some areas of North America. In Europe, it occurs in flat country (often near cultivated fields), deciduous woods, moors and dune areas, and it will occupy sites higher in mountains than will the European rabbit. Like the European rabbit, the European hare commonly enters areas inhabited by people and can cause agricultural damage. This hare is primarily nocturnal but also can be active during daylight hours. It rests in forms constructed in well-sheltered woods, in scrub areas, along ditches and in open fields.

The North American rabbit and hare species of greatest importance to people are the various cottontail rabbits (*Sylvilagus* spp.) and jackrabbits – primarily, the black-tailed jackrabbit (*Lepus californicus*) and the white-tailed jackrabbit (*Lepus townsendii*). Different species of cottontail rabbits can be found from southern Canada into South America, occupying forests, grasslands, marshes, swamps, deserts and beaches. In general, these rabbits prefer brushy sites or shrub-filled forest clearings, but they also can be common in cultivated areas, parks and well-vegetated home sites. Jackrabbits are residents of the more wide open landscapes of western North America and are less likely to occur in close proximity to people, although some claim they cause agricultural damage.

The most important cottontail rabbits are the eastern cottontail (*Sylvilagus floridanus*), desert cottontail (*Sylvilagus audubonii*), and mountain cottontail (*Sylvilagus nuttalli*). Eastern cottontails occur throughout the eastern half of the continent and northern South America (Chapman, Hockman & Ojeda, 1980). Desert cottontails and mountain cottontails are found in the western half of the continent, with the former species occurring at lower elevation sites than the latter, which is more likely to be found in less arid and more mountainous habitats (Chapman, 1975; Chapman & Willner, 1978).

13.3. Biological factors of relevance to human health

The establishment of rodents and lagomorphs in urban environments depends on a variety of factors, especially those that affect the availability of key food resources and shelter. Obviously, heavily developed areas that are almost completely devoid of vegetation, water sources or other essential environmental factors will have few non-commensal rodents. On the other hand, at least a few non-commensal rodents are likely to be found in those urban areas that lie adjacent to rural areas, are crossed by stream corridors, or contain sizeable parks with native vegetation. Indeed, the abundance of non-commensal species can be surprisingly high in these areas. Many rodents and lagomorphs will invade sites that offer food or shelter, including those near urban areas (MacDonald, 1984). Usually these movements simply allow a particular species to take advantage of valuable but transient resources. If favourable conditions persist, however, these animals are likely to remain in the area, perhaps breeding successfully and establishing more or less permanent populations. The ability of rodents to invade and quickly exploit newly available habitats is favoured not only by their behavioural adaptability, but also is favoured by their high reproductive rates, which allow them to rapidly populate and utilize important resources found in these areas.

The temporary invasion of normally marginal habitats by either rodents or lagomorphs, or the favourable alteration by humans of habitats otherwise unsuitable for these animals, can have significant implications for human health. As contact increases between these animals and people, the risk that people will be exposed to rodent-related disease agents also is likely to increase. For example, in 1993, a previously unrecognized hantavirus (Sin Nombre virus, SNV) caused an outbreak of nearly 50 cases of a severe and often fatal disease in the American South-west (Childs et al., 1994). This outbreak occurred in the wake of an El Niño event that resulted in enhanced food sources and tremendous population increases of deer mice, which were found to be the primary reservoir of SNV. As deer mouse populations increased to extraordinary levels, the mice frequently invaded homes and other buildings, posing high levels of risk to people living or working in these structures (Childs et al., 1994, 1995; Zeitz et al., 1995).

Also, as mouse populations decreased dramatically in the drier years that followed this El Niño event, so did the risk of people being exposed to hantavirus. In this instance, people were exposed when individual mice from the expanding mouse populations entered homes, storage buildings or other enclosed spaces in search of food, nest sites or winter refuge.

13.4. Human diseases and non-commensal rodents, rabbits or hares

The numerous viruses, bacteria, rickettsiae, protozoa and helminths carried by non-commensal rodents and lagomorphs (Table 13.1) pose a serious threat to human health. Many of these agents can cause serious and occasionally life-threatening illnesses in people, including haemorrhagic fever with renal syndrome (HFRS), hantavirus pulmonary syndrome (HPS), LB, TBE, plague, tularaemia, HGA and various rickettsioses transmitted by ticks or fleas (Childs et al., 1994; Gage, Ostfeld & Olson, 1995; Weber, 2001; Blanco & Oteo, 2002; Piesman, 2002; Vapalahti et al., 2003; Gage & Kosoy, 2005; Parola, Davoust & Raoult, 2005; Petersen & Schriefer, 2005). Although many of the agents given in Table 13.1 have been known since the first half of the 20th century to cause human disease, others, such as those responsible for various hantaviral illnesses, LB and HGA have been identified as human pathogens only within the past three decades. The relatively recent discovery of these agents suggests that other as yet unrecognized rodent-related disease agents are likely to be found in the future.

The transmission of disease agents from non-commensal rodents or lagomorphs to people can occur through many routes. People are believed to become infected with various hantaviral agents through inhalation or direct contact with rodent excreta. Human exposures to Puumala virus, a hantaviral agent found in voles, probably result from people handling or inhaling virus particles aerosolized from hay, woodpiles or other materials contaminated with infectious rodent urine (Olsson et al., 2003; Vapalahti et al., 2003). Other epidemiological evidence suggests that human exposures to SNV, the causative agent of HPS, occur when SNV particles are aerosolized by people sweeping rodent-infested spaces (or otherwise disturbing infectious rodent excreta), especially when such disturbances take place in spaces that have remained enclosed and darkened for long periods (Childs et al., 1995; Zeitz et al., 1995; Mills et al., 2002). Another relatively common source of exposure for some agents, such as plague or tularaemia, is the handling of infected rodents or lagomorphs by hunters or other people (Gage, Ostfeld & Olson, 1995; CDC, 1996, 2002; Mills et al., 2002; Petersen & Schriefer, 2005). About 19% of exposures in one series of human plague cases in the United States were thought to be due to the handling of infected animals, including rabbits, prairie dogs and ground squirrels, as well as domestic cats and wild carnivores (CDC, 1996). In other instances, people become infected with rodent- or lagomorph-related pathogens as a result of drinking or eating contaminated water or food, respectively. This last means of exposure has been reported for human infections that involve the causative agents of tularaemia (*F. tularensis*) (Petersen & Schriefer, 2005) and yersiniosis (*Y. pseudotuberculosis*) (Naktin & Beavis, 1999).

Non-commensal rodents and lagomorphs also play indirect, but nevertheless important roles in the natural cycles and epidemiology of certain vector-borne disease agents. Particularly important is the ability of many rodent and lagomorph species to serve as sources for infecting blood-feeding ticks, fleas and other ectoparasites with various pathogens. Ticks of the genus *Ixodes*, particularly those species belonging to the *Ixodes ricinus* complex, are especially significant disease vectors in Europe and North America (Humair & Gern, 2000; Piesman, 2002; Hayes & Piesman, 2003; Charrel et al., 2004;

Parola, Davoust & Raoult, 2005). The castor-bean tick, in particular, is the primary vector of the agents of LB (*B. burgdorferi* s.l.), HGA (*A. phagocytophilum*), babesiosis (*B. microti*) and TBE (TBEV) in Europe. Although TBE is not found in North America, two closely related tick species on this continent (the deer tick and the western black-legged tick) transmit the first three agents. The deer tick, as well as the woodchuck tick (*Ixodes cookei*) and the squirrel tick (*Ixodes marxi*), also carry a virus (Powassan virus) that causes a rare encephalitis in people in the north-eastern United States and in south-eastern Canada. The major reservoirs of this virus are thought to be groundhogs (*Marmota monax*, also known as woodchucks), American red squirrels and white-footed mice. In the case of LB, *Ixodes* ticks can transfer the LB agent from spirochete-infected mice or other non-commensal rodent hosts to people. Still other tick species that feed on rodents or lagomorphs are vectors of spotted fever group rickettsiae in Europe or North America. Foremost among these are species of *Dermacentor*, which transmit the causative agent of RMSF (*R. rickettsii*) in North America.

Non-commensal rodents and lagomorphs also can affect the risk of human disease by providing the blood-meals that vectors need to successfully develop and reproduce. In most instances, a positive correlation exists between the density of rodent or lagomorph hosts and the abundance of the arthropod vectors that feed on these animals. The risk to people of exposure to rodent- or lagomorph-related disease agents transmitted by vectors can be expected to increase as vector populations also increase, particularly when the vector populations undergoing expansion occur in areas heavily utilized by people.

In other instances, non-commensal rodents or lagomorphs contribute to the increased risk of disease in people by acting as intermediate hosts for parasites. The small fox tapeworm (*Echinococcus multilocularis*), which causes the emerging disease alveolar echinococcosis, uses non-commensal rodents as intermediate hosts for its immature stages. Dogs, cats and foxes, which act as definitive hosts for adult fox tapeworms, become infected as a result of eating rodents bearing the cysticercus (larval) stage of this tapeworm (Eckert & Deplazes, 2004). People become infected not through contact with the rodent intermediate hosts, but rather through ingesting eggs shed by worms living in the definitive carnivore hosts. The relatively recent rise in fox densities in some urbanized areas of Europe, probably as a result of rabies control, has led to and increased risk to people of alveolar echinococcosis. Rodents can contribute indirectly to this risk, as demonstrated by a report that indicates that the occurrence of alveolar echinococcosis in people was strongly correlated with the density of water voles, which are major intermediate hosts for this parasite (Viel et al., 1999).

13.4.1. Major rodent- and lagomorph-related viral agents

The following subsections discuss the major rodent- and lagomorph-related viral pathogens that are believed to cause illness in people in Europe or North America.

13.4.1.1. Hantaviruses

Hantaviral infections have been reported for many species of rodents, as well as for other animals and people (Zeier et al., 2005). Clement and colleagues (1997) provided estimates

Table 13.1. Major disease agents reported in European or North American rodents or transmitted by their ectoparasites

Disease	Agent	Continent	Major rodent reservoir(s), vector host(s) or intermediate host(s)	Modes of transmission
Viral illnesses				
HPS	SNV	North America	Deer mouse	Primarily airborne rodent urine, faeces, saliva
HFRS	Puumala virus	Europe	Bank vole	Primarily airborne rodent urine, faeces, saliva
HFRS	Dobrava (Belgrade) virus	Europe	Yellow-necked mouse	Primarily airborne rodent urine, faeces, saliva
HFRS	Saaremaa virus	Europe	Striped field mouse	Primarily airborne rodent urine, faeces, saliva
TBE	TBE virus	Europe	Wood mouse, yellow-necked mouse, bank vole, red-backed vole, grey-sided vole *(Clethrionomys rufocanus)* and other small rodents	Castor-bean tick bite
CTF	CTF virus	North America	Ground squirrels, chipmunks, deer mice, wood rats, North American porcupine *(Erithozon dorsatum)*	Rocky Mountain wood tick bite
California serogroup virus	La Crosse virus	North America	Eastern chipmunk	Bite of eastern treehole mosquito
California serogroup virus	Tahyna virus	Europe	European rabbit and brown hare	Mosquito bite (*Ae. vexans* and perhaps other *Aedes* spp.) The virus also might overwinter in *Culex* spp. or *Culiseta* spp.
Cowpox virus infection	Cowpox virus	Europe	Wood mouse, bank vole, field vole and probably other rodents	Direct contact
Rabies	Rabies virus	Europe and North America	Rodents are not major reservoirs and act only as incidental hosts	Bites or scratches; very rarely airborne in enclosed spaces (caves with infected bats)
Bacterial illnesses				
LB	*B. burgdorferi* s.l.	Europe and North America	White-footed mouse, eastern chipmunk and dusky-footed wood rat (North America) Field mice and wood mice (Europe) Other non-commensal rodents	Ixodid (hard) tick (*Ixodes* spp.) bite
Tick-borne relapsing fever	*Borrelia* spp.	North America	Primarily chipmunks and American red squirrel	Argasid (soft) tick bite
Plague	*Y. pestis*	North America and extreme southeastern Europe	Non-commensal mice, rats, voles, chipmunks, ground squirrels and prairie dogs	Fleabite (many flea species) Direct contact; inhalation (rare)
Yersiniosis (non-plague)	*Y. pseudotuberculosis*	Europe and North America	Many rodent species	Faecal–oral transmission through ingestion of contaminated food and water Direct contact with infected animals
Tularaemia	*F. tularensis*	North America: types A and B Europe: type B only	Voles, muskrats, beavers, ground squirrels and other small mammals	Ixodid (hard) tick, deer fly or mosquito bites Contaminated food or water
Bartonellosis	*Bartonella* spp.	Europe and North America	Deer mice, ground squirrels and other rodents (North America) Bank voles and wood mice (Europe)	Uncertain (fleabite? contact?)
Leptospirosis	*Leptospira* spp.	Europe and North America	Most commonly identified in commensal rats Also found in non-commensal species	Skin or mucous membrane contact with rodent urine-contaminated water, soil or vegetation Ingestion of foods or water contaminated with rodent urine

Disease	Agent	Continent	Major rodent reservoir(s), vector host(s) or intermediate host(s)	Modes of transmission
Rat-bite fever	*Spirillum minus, Streptobacillus moniliformis*	Europe and North America	Primarily in commensal rats Also found in non-commensal rodents	Rodent bite
Pasteurellosis	*P. multocida*	Europe and North America	Many rodent species, rabbits and other mammals	Bite of infected animals
Salmonellosis	*Salmonella* spp.	Europe and North America	Various non-commensal rodents	Ingestion of foods contaminated by faeces of rodents or other hosts
Rickettsial illnesses				
Spotted fever group rickettsioses (tick-borne typhus)	*Rickettsia* spp.	Europe and North America	Various non-commensal rodents	Ixodid tick bite: *Dermacentor* spp. (North America) *Rhipicephalus* spp. and *Dermacentor* spp. (Europe)
Murine typhus	*R. typhi*	Europe and North America	Primary cycle involves *Rattus* spp. Secondary cycles can involve other rodents	Fleabite
HGA	*A. phagocytophilum*	Europe and North America	*Peromyscus* spp. (North America) *Apodemus* spp. (Europe) Other non-commensal rodents	Ixodid tick (*Ixodes* spp.) bite
Q fever	*C. burnetii*	Europe and North America	Many non-commensal rodents reported to be infected	Airborne dissemination of resistant stage of *C. burnetii*; Handling infectious tissues of host animals Rarely by vectors
Protozoal illnesses				
Babesiosis	*Babesia* spp.	Europe and North America	*Peromyscus* spp., *Microtus* spp. and *Apodemus* spp.	Ixodid tick bite: Castor-bean tick (Europe) Deer tick and western black-legged tick (North America)
Leishmaniasis	*Leishmania* spp.	Europe and North America	Non-commensal rodent reservoirs	Sandfly (*Phlebotomus* spp.) bites
Cryptosporidiosis	*Cryptosporidia parva*	Europe and North America	Many rodents species Other wild or domestic animals	Ingestion of infective sporulated oocysts in contaminated food or water
Giardiasis	*Giardia* spp.	Europe and North America	Many rodent species Other wild or domestic mammals	Drinking water contaminated with cysts Faecal–oral contamination.
Toxoplasmosis	*T. gondii*	Europe and North America	Many rodents species serve as intermediate hosts Felines are definite hosts	Ingestion of materials contaminated with cat faeces containing sporulated oocysts Ingestion of undercooked meat containing tissue cysts
Helminth infestations				
Echinococcosis	*E. multilocularis*	Europe and North America	Common vole Other small rodents	Ingestion of intermediate host (rodent) by definitive host (foxes and other carnivores) Humans ingest eggs
Capillariasis	*Capillaria hepatica*	Europe and North America	Primarily commensal rats Also, some non-commensal rodents, including *Peromyscus* spp., *Microtus* spp. Rodents are definitive hosts for these parasites	Ingestion of *Capillaria* eggs
Trichinellosis (trichinosis)	*Trichinella spiralis*	Europe and North America	Many mammals, including rodents	Consumption of undercooked, infested meat

Note. ?: denotes high degree of uncertainty; CTF: Colorado tick fever; HGA: human granulocytic anaplasmosis; HPS: hantaviral pulmonary syndrome; HFRS: haemorrhagic fever with renal syndrome; TBE: tick-borne encephalitis.

Source: The information for this table was derived from the references cited in the text of this chapter.

of the numbers of cases of hantaviral illness in humans in Europe, noting that at least 1000 serologically confirmed cases of Puumala virus infection occur each year in Finland. The countries of the former Yugoslavia also experience about 1000 cases a year, and hundreds of cases occur annually in Sweden. Other countries reporting significant numbers of cases include Belgium, France, Germany, Greece and the Netherlands.

The various hantaviruses exhibit many epidemiological and ecological similarities. Adult rodents develop persistent infections and secrete virus for long periods with no obvious illness. Viral particles can be found in the lungs, spleen, kidneys and urine of these rodent hosts. Saliva also can be contaminated with active virus, and saliva contamination of bite wounds from fighting appears to be important in the horizontal transmission of the virus between rodents. The most commonly encountered hantaviral illness in humans in Europe is a relatively mild form of HFRS called nephropathia epidemica (Settergren, 2000; Sauvage et al., 2002; Vapalahti et al., 2003). The bank vole carries the causative agent, Puumala virus (Sauvage et al., 2002; Olsson et al., 2003, 2005). Following the distribution of its host, Puumala virus can be found throughout much of Europe, excluding the Mediterranean coastal regions and most of the Iberian Peninsula and Greece. In some of these regions, human Puumala virus infections appear to be on the rise (Rose et al., 2003; Ulrich et al., 2004; Mailles et al., 2005; Schneider & Mossong, 2005).

The dynamics of populations of the bank vole hosts of Puumala virus differ geographically, a factor that has important epidemiological consequences. In northern Europe, densities of these voles typically cycle over a 3–4-year period. During the late increase and peak phases of the population cycle, bank vole densities will remain high from late summer of the year of increase until late winter a year and a half later, a pattern that allows Puumala virus to spread efficiently among these voles; this pattern can result in HFRS epidemics in humans in the fall and early winter of years with the highest vole densities. In the more temperate parts of Europe, populations of bank voles are more stable, showing a clear seasonal variation, with a brief peak in autumn. Also, forests in the temperate parts of Europe are much more fragmented, which can lead to more local and patchy occurrences of Puumala virus. Consequently, transmission of Puumala to people under these circumstances is infrequent compared with the more northerly reaches of Europe. However, occasional heavy mast years (heavy seed crops of oak and beech) further south can lead to increased abundances of seed-eating rodents, including bank voles. These heavy mast years, induced by higher than average summer temperatures, can be synchronous over large areas, giving rise to hantavirus epidemics in humans, such as were observed in 1990, 1993, 1996, 1999 and 2001 in the Ardennes, and 1986, 1989, 1995 and 2002 in the Balkans. Unlike in northern Europe, human HFRS cases in the Ardennes and the Balkans, caused by Puumala virus, occurred in the spring or summer of peak rodent years, although a minor peak also was observed in early winter. When winter survival of rodents is good, rodents start to breed earlier than in normal years, resulting in high densities early in summer.

The prototype hantavirus, Hantaan virus, which causes classical HFRS, does not occur in western and central Europe, but rather occurs in China, the Democratic People's Republic of Korea, the Republic of Korea and the far eastern region of the Russian

Federation, where it is carried by the striped field mouse. HFRS caused by Hantaan virus is a much more severe ailment than nephropathia epidemica caused by Puumala virus. However, another hantavirus (Dobrava virus, also referred to as Belgrade virus), carried by the yellow-necked mouse, causes a very severe form of HFRS in southern Europe (Vapalahti et al., 2003).

A third hantavirus (Saaremaa virus), carried by the striped field mouse, is found in Estonia and nearby, in the Russian Federation (Vapalahti et al., 2003), where it causes a relatively mild form of HFRS. It should be noted that the distribution of the yellow-necked mouse, which carries Dobrava virus, and the striped field mouse, which is a reservoir for Saaremaa virus, overlap over most of Europe. The latter mouse, which also is the primary carrier of Hantaan virus, is absent in far western Europe. It is not yet clear if other *Apodemus* spp., such as wood mice, play significant roles as carriers of Dobrava- and Saaremaa-like viruses. In those regions where the predominant local type of hantaviral agent is Saaremaa virus carried by striped field mice, such as in Estonia and the Russian Federation, the clinical course of illness seems to be milder than in the Balkans, where human cases are associated with Dobrava infections acquired from yellow-necked mice.

Another hantavirus, Tula virus, is widespread in central and eastern European populations of the common vole (Vapalahti et al., 2003). At least two cases of Tula virus infection have been reported in people, but the association of this agent with human disease has not been proven unequivocally.

Numerous hantaviruses have been identified in North American rodents during the past 40 years (Schmaljohn & Hjelle, 1997). The most medically significant of these is SNV, which is found in deer mice in Canada, Mexico and the United States (Childs et al., 1994). In people, SNV causes severe HPS. Additional cases of HPS have been associated with infections caused by: Black Creek Canal virus, which was first identified in hispid cotton rats in Florida (Rollin et al., 1995); Bayou virus, which is found in rice rats (*Oryzomys palustris*) in Louisiana (Ksiazek et al., 1997); and New York virus, which was identified from white-footed mice in New York (Song et al., 1994). Additional genotypes have been isolated from other North American rodent species, including Prospect Hill virus, which is found in meadow voles, but their role as agents of human disease is uncertain.

13.4.1.2. TBEV

TBE, a serious febrile illness with neurological complications, is caused by TBEV, which belongs to the viral family Flaviviridae. Castor-bean ticks and taiga ticks act both as vectors and reservoirs for TBEV, which is an important pathogen in many parts of Europe and Asia, having a distribution that corresponds to the geographic range of its ixodid tick reservoirs (Charrel et al., 2004). The main hosts for immature vector ticks are various small rodents, and these same rodents serve as important infection sources for larvae when co-feeding alongside with infected nymphs (Randolph et al., 1999). Recent increases in human cases in central Europe and the Baltic states have been attributed to changes in human behaviour, but it also should be noted that human impacts on landscapes have allowed vector castor-bean tick populations to increase (Randolph, 2001). TBE is discussed in greater detail in Chapter 10.

13.4.1.3. California group viruses (primarily La Crosse virus)

La Crosse virus causes encephalitis and aseptic meningitis, particularly in children, in the Midwestern and mid-Atlantic regions of the United States (Beaty, 2001). The eastern treehole mosquito (*Aedes triseriatus*) transmits this virus. Unlike many arthropod-borne viruses (arboviruses), La Crosse and other California serogroup viruses can be transmitted transovarially and transstadially in their mosquito vector. Eastern chipmunks and gray squirrels are the principal vertebrate hosts and serve as sources for infecting mosquitoes. Another California serogroup virus, Tahyna virus, occurs primarily in central and eastern Europe, where it occasionally causes an acute influenza-like disease that occurs primarily in children, although most infections are believed to be unapparent (Labuda, 2001). The primary vectors are the mosquito *Ae. vexans*, although other *Aedes* spp. might play a role in transmission, and the virus purportedly can overwinter in *Culex* and *Culiseta* mosquitoes. The European rabbit and European (or brown) hare are susceptible to infection, produce high viraemias and can serve as sources for infecting mosquitoes with Tahyna virus.

13.4.1.4. CTF virus

This virus causes fever, chills, headache and myalgia, but is rarely fatal. It is fairly common in some mountainous regions of western North America, where it is transmitted by the Rocky Mountain wood tick. Rodents, some larger mammals and certain birds are important hosts for this virus and can serve as sources of infection for feeding ticks, although ticks also can transmit CTF virus transovarially, as well as transstadially (Callisher, 2001). People walking in areas where adult Rocky Mountain wood ticks quest are at risk of becoming infected with this virus. Immature stages of this tick do not feed on people, but rather infest rodents and rabbits. Although most cases are acquired while camping or hiking, Rocky Mountain wood ticks and important rodent hosts can occur near towns in some mountainous areas.

13.4.1.5. Cowpox virus

It is generally accepted that the reservoir hosts of cowpox virus are wild rodents, although direct evidence for this is lacking throughout most of the geographic range of the virus. Chantrey and colleagues (1999) demonstrated that the main hosts in Great Britain are bank voles, wood mice and short-tailed field voles. They also suggested that wood mice may not be able to maintain infection alone, thus explaining the absence of cowpox from Ireland, where voles are generally absent. Infection in wild rodents varies seasonally, and this variation probably underlies the marked variable seasonal incidence of infection in accidental hosts, such as people and domestic cats. Although rarely reported, a recent case of cowpox in a 16-year-old boy was attributed to handling a rat (Honlinger et al., 2005). An increase in human cases from various sources can be expected because people are no longer routinely immunized with *Vaccinia* virus to protect against smallpox, a procedure that also provides cross-protection against cowpox infection.

13.4.1.6. Rabies virus

Rabies virus causes an acute viral encephalopathy that is virtually always fatal, unless people exposed to it are promptly given appropriate treatment (Beran, 1994). In nature, the most common hosts of the various strains of rabies virus are dogs, wild carnivores and bats, and bites from these animals are the most common sources of human infection. Rodents and lagomorphs also occasionally contract rabies from these sources and potentially could pass these infections to people handling these animals, although the risk is thought to be very low (Cappucci, Emmons & Sampson, 1972; Fleming & Caslick, 1978; Roher et al., 1981; Dowda & DiSalvo, 1984; Fishbein et al., 1986; Moro et al., 1991; Childs et al., 1997; Zaikovskaia et al., 2005).

13.4.2. Major rodent- and lagomorph-related bacterial and rickettsial agents

The following subsections discuss the major rodent-related bacterial and rickettsial agents that are believed to cause illness in people in Europe or North America.

13.4.2.1. *Francisella tularensis*

Tularaemia occurs endemically in much of Europe and North America (Petersen & Schriefer, 2005). The causative organism, *F. tularensis*, is a Gram-negative bacterium that occurs in a large variety of wild rodent species, as well as in rabbits and hares (Hopla, 1974; Morner, 1992; Hopla & Hopla, 1994; Ellis et al., 2002). People are likely to become infected through handling infected animals (especially rabbits and hares), ingesting contaminated water or food, or being bitten by an infectious arthropod vector, most often a tick, although biting flies and mosquitoes also have been implicated as vectors in the western United States and Europe, respectively. In the United States, lagomorphs (particularly *Sylvilagus* spp.) are commonly infected and, along with their ticks, are particularly important in the ecology of type A tularaemia strains (*F. tularensis tularensis*), which are often vector-borne. This form of the tularaemia bacterium, certain strains of which cause the most severe cases of tularaemia, occurs only in North America and is most likely to be vector-borne. In addition to their association with lagomorphs, tularaemia cases in people in North America have also been frequently associated with outbreaks in muskrats, Canadian beavers and voles (*Microtus* spp.). Tularaemia strains found in these animals are typically type B (*F. tularensis holarctica*), which are often waterborne, cause less severe disease than many type-A strains and are found throughout much of North America, northern Asia and Europe. *Francisella tularensis* has also been recognized in ground squirrels and prairie dogs in North America (Jellison, 1974; Peterson et al., 2004). The species most commonly found infected in urban settings and surrounding areas in Europe are rabbits and voles, including water voles and common voles (*Microtus* spp.). In North America, tularaemia infections also have occurred in primates that live in zoos, and the most likely source of infection in one of these incidents was thought to be Richardson's ground squirrels (*Spermophilus richardsoni*) that lived within the zoo (Nayar, Crawshaw & Neufeld, 1979).

Outbreaks of tularaemia in people have been reported frequently from certain European nations and regions, including Spain in 1997 and 1998 (585 and 19 cases, respectively) and Sweden in 2000 (270 cases) (Anda et al., 2001; Bossi et al., 2004). The 1997 and 1998 Spanish outbreaks involved the handling of infected hares and crayfish, respectively (Anda et al., 2001). The Swedish outbreaks have been linked to contact with infected hares or voles, or the bites of infectious mosquitoes. In the United States, a total of 1368 cases were reported between 1990 and 2000, although most public health officials believe the disease is underreported (CDC, 2002).

13.4.2.2. *Borrelia burgdorferi* s.l.

LB is the most commonly reported vector-borne illness in the United States, with about 20 000 cases reported annually during recent years (Bacon et al., 2004; Hopkins et al., 2005; Jajosky et al., 2006), and some have stressed the threat this disease poses to people who live in urban areas (Steere, 1994; Junttilla et al., 1999). Forms of LB also are common in Europe, and Randolph (2001) has noted that populations of castor bean ticks, the vector of LB, have increased in Europe as a result of human impacts on landscapes, which suggests that the risk is likely to remain high on this continent for the foreseeable future. Although LB is covered more thoroughly in Chapter 10, on tick-borne diseases, a brief mention should be made of the essential role rodents play in the maintenance of this disease, by acting as tick hosts and as sources for infecting vector ticks with different genotypes of *B. burgdorferi* s.l. in European and American foci of infection (Kurtenbach et al., 2002). Indeed, the presence of non-commensal rodents has been considered a prerequisite for LB in an area of endemicity (Junttilla et al., 1999). In Europe, voles and non-commensal mice, including yellow-necked mice, wood mice and bank voles, act as important hosts of different genotypes of LB spirochetes, serving as sources of infection for feeding castor bean ticks and as hosts for immature ticks. White-footed mice and eastern chipmunks play similar roles in various regions of the major LB foci found in the northeastern and upper Midwestern United States, as do dusky-footed wood rats (*Neotoma fuscipes*) in California foci (Piesman, 2002).

13.4.2.3. Tick-borne relapsing fever borreliae

Certain spirochetes, including *Borrelia hermsii, B. turicatae* and *B. parkeri* cause relapsing fevers in western North America. All are transmitted by species of *Ornithodoros* ticks that feed primarily on rodents. Some of these rodent hosts develop high levels of spirochetemias and can serve as sources for infecting vector ticks (Burgdorfer & Mavros, 1970; Burgdorfer, 1976). The vectors of *B. turicatae* and *B. parkeri* (the ticks *O. turicata* and *O. parkeri*, respectively) feed primarily on burrow-dwelling rodents and other rodents that occur most often in highly rural areas and rarely come in close contact with people (Burgdorfer, 1976; Gage et al., 2001). However, the tick *O. hermsi*, the vector of *B. hermsii*, frequently feeds on American red squirrels and chipmunks, both of which can invade homes and build tick-containing nests in attics and wall spaces (Thompson et al., 1969; Trevejo et al., 1998; Schwan et al., 2003). If the hosts of these ticks die, they are likely to leave these nests to find other hosts, which can result in people being bitten by *B. hermsii*-infected ticks.

13.4.2.4. Rickettsiae

Non-commensal rodents and rabbits act as hosts for the vectors of various rickettsial agents in Europe and North America (Burgdorfer, Friedhoff & Lancaster, 1966; McDade et al., 1980; McDade & Newhouse, 1986; Gage, Burgdorfer & Hopla, 1990; Gage, Hopla & Schwan, 1992; Gage, Ostfeld & Olson, 1995; Parola & Raoult, 2001; Parola, 2004; Parola, Davoust & Raoult, 2005; Parola, Paddock & Raoult, 2005). In some instances these same animals also act as amplifying hosts for infecting ticks or other vectors with these same rickettsiae.

The most important and severe rickettsial disease in the United States is RMSF, which is caused by *R. rickettsii* (Dumler, 1994; Sexton, 2001). This species of *Rickettsia* is transmitted primarily by the American dog tick (*D. variabilis*) in the eastern United States and by the Rocky Mountain wood tick in the central and northern Rocky Mountains of the United States and south-western Canada. Non-commensal rodents act as both tick hosts and sources of infection for the immature stages of the above two tick species.

Serological evidence of human infection with *A. phagocytophilum*, the agent that causes HGA, has been identified in several European countries (Strle, 2004). Non-commensal rodents act as reservoir hosts and sources for infecting ticks with *A. phagocytophilum* (Parola & Raoult, 2001; Strle, 2004; Parola, Davoust & Raoult, 2005). The primary vector of this agent is the castor bean tick, which feeds heavily on non-commensal rodents, including wood mice, yellow-necked mice and bank voles (Panchola et al., 1995). The primary vector of HGA in the eastern United States is the deer tick, which feeds on rodents, birds and reptiles (Piesman 2002). In the Pacific states, the primary vector is the western black-legged tick, which feeds during its immature stages on non-commensal rodents, birds and lizards (Kramer et al., 1999; Piesman 2002) Additional information on tick-borne rickettsiae can be found in Chapter 10, on ticks and TBDs.

13.4.2.5. Yersiniae

Yersinia pseudotuberculosis causes intestinal illness and occasionally reactogenic arthritis in people. This bacterium is commonly identified in non-commensal rodents and has been implicated in numerous outbreaks that involve the consumption of raw vegetables contaminated with rodent excreta (Naktin & Beavis, 1999; Chesnokova et al., 2003; Jalava et al., 2004). Recent outbreaks of *Y. pseudotuberculosis* infection have been reported from Finland and the Russian Federation. People who live in suburban areas and plant small vegetable gardens accessible to non-commensal rodents could be at risk for yersiniosis.

Plague is a vector-borne rodent-related zoonosis caused by *Y. pestis*, a bacterium closely related to *Y. pseudotuberculosis*, but is able to cause much more severe illness in people than the latter agent (Gage, Ostfeld & Olson, 1995; Gage, 1998; Levy & Gage, 1999; Gage et al., 2000; Gage & Kosoy, 2005). The disease is maintained in nature through cycles that involve transmission between bacteraemic rodent hosts and their fleas. Most human cases are acquired through the bites of infectious rodent fleas, including not only those found on commensal rats, but also those found on non-commensal rodents. Additional cases have been acquired through handling infected animals, including non-commensal rodents and lagomorphs, as well as domestic cats and certain wild carnivores. *Yersinia pestis* probably first arose in central Asia, but has been dispersed along human trade routes to many other parts of the world, including North America, which has extensive wild rodent foci in the western third of the United States, limited portions of south-western Canada and probably areas of northern Mexico (Gage & Kosoy, 2005). Although this organism was the cause of the Black Death in the Middle Ages, *Y. pestis* currently exists only in the extreme south-eastern portion of Europe near the Caspian Sea, where it occurs in enzootic foci that are of little epidemiological significance and rarely, if ever, serve as sources of human cases (Tikhomirov, 1999).

13.4.2.6. Leptospirae

Leptospirosis is an important zoonosis worldwide and is well known in rural and urban settings in Europe (Rosicky & Sebek, 1974; Daiter et al., 1993; Handysides, 1999; Perra et al., 2002). The infection causes a systemic illness that often leads to renal and hepatic dysfunction (Levett, 2001). Urban dwellers are at increased risk because of sporadic exposures to rat urine as inner cities deteriorate. Reportedly, the incidence of leptospirosis is increasing in urban children, but most cases occur in adults and are acquired as a result of occupational exposures. The causative agent, *Leptospira interrogans*, is subdivided into numerous serovars that can differ by geographic location and primary host. Most leptospiral serovars have their primary reservoir in mammals, which can result in continual re-infection of commensal and non-commensal rodent populations in urban settings (Daiter et al., 1993). At least 12 species of small mammals, predominantly those rodents that live in highly moist environments, are known to carry *Leptospira*. Reported seroprevalences in small mammals were 3–12% in water voles, 11% in muskrats and 5% in *Microtus* voles (Tokarevich et al., 2002). The most common serogroups of *Leptospira* found in wild rodents were grippotyphosa (65%), javanica (21%), and pomona (12%). Recent serological surveys of leptospirosis in animals in Croatia indicated that 13% of small rodents were seropositive (Cvetnic et al., 2003). In this study, antibodies to different serovars of leptospires were identified in 15% of field voles, 11% of yellow-necked mice, 9% of bank voles and 7% of wood mice.

13.4.2.7. Bartonellae

With the recognition of several novel or re-emerging diseases caused by *Bartonella* spp., interest in bacteria of this genus has been increasing (Boulouis et al., 2005). The distribution of certain *Bartonella* spp. appears to be closely linked to the distribution patterns of their hosts or arthropod vectors. *Bartonella elizabethae* was isolated originally from a patient with endocarditis in Massachusetts. Subsequently, identical or similar organisms were found in rats in America, Asia and Europe. It is possible that *B. elizabethae* first spread from Asia, because most *Rattus* spp. probably originated there, and only later were carried to other continents with the dispersal of two domestic *Rattus* spp. Two other putative human pathogens, *Bartonella washoensis* and *Bartonella vinsonii* subsp. *arupensis*, are thought to be associated with American rodents (ground squirrels (*Spermophilus* spp.) and deer mice and their allies (*Peromyscus* spp.), respectively) (Welch et al., 1999; Kosoy et al., 2003). *Bartonella vinsonii* subsp. *arupensis* also has been reported for a human case of endocarditis in a French hospital (Fenollar, Sire & Raoult, 2005). Although this bacterium has been associated with non-commensal rodents, the patient was an urban dweller and had no obvious rodent exposures. Three other species, *Bartonella grahamii, Bartonella taylorii,* and *Bartonella doshiae*, circulate in woodland rodent communities in Europe. *Bartonella grahamii* infection has been associated with neuroretinitis in Europe (Kerkhoff et al., 1999; Rothova et al., 1999).

13.4.2.8. Rat-bite fever agents

Rat-bite fever is an illness caused by either of two bacteria, *Streptobacillus moniliformis* and *Spirillum minus* (also called *Spirillum minor*) (Heymann, 2004). The disease is thought to be relatively rare in Europe and North America. However, since reporting this disease is not mandatory, reliable estimates of case numbers are not available. Most cases of rat-bite fever in the United States are due to infection with *S. moniliformis*. The name of the disease is derived from the fact that infection usually develops after a person is bitten or scratched by an infected rat. Other rodents, such as mice, squirrels and gerbils, may also spread the infection to people. Sometimes, infection can result from handling infected rats, with no reported bite or scratch. Infections also result from ingesting food or drink (such as milk or water) contaminated with rodent excrement.

13.4.3. Major rodent- and lagomorph-related parasitic agents

The following subsections discuss the major rodent- and lagomorph-related parasitic agents that are believed to cause illness in people in Europe or North America.

13.4.3.1. Toxoplasma

Toxoplasmosis is caused by the protozoan parasite *T. gondii*. The definitive host of *T. gondii* is the domestic cat, but non-commensal rodents, other mammals and some birds play important roles as intermediate hosts for this parasite (Marquardt, Demaree & Grieve, 2000). Cats can become infected with *Toxoplasma* through consumption of infected intermediate hosts, especially rodents. People typically become infected through contact with infectious oocysts shed in cat faeces, an event that can occur while cleaning cat litter boxes. Infections also occasionally arise from eating raw or undercooked meat, particularly mutton or pork, that contains infectious cysts. Pregnant women who become infected can pass *T. gondii* to their developing fetus, an event that can lead to the death of the fetus or other severe consequences, including brain damage with intracerebral calcification, microcephaly, hydrocephaly, jaundice and convulsions at birth (Marquardt, Demaree & Grieve, 2000; Heymann, 2004). Infections of children and adults are often asymptomatic, but can appear acute with lymphadenopathy. In some instances *T. gondii* infections resemble mononucleosis, with lymphadenopathy, fever and lymphocytosis.

13.4.3.2. Toxocara

Infestation with larval forms of *Toxocara* spp. can cause a chronic (but usually mild) disease in people. Infestations are most common in children, but adults also can become infested. People acquire infestations by ingesting worm eggs contained in faecal-contaminated materials, particularly those shed by dogs or cats, which are definitive hosts of *Toxocara canis* and *T. cati*, respectively. Rodents become infested by ingesting eggs containing infective larvae. Once infected, these rodents can act as transport hosts, maintaining the larval worms in their tissues for long periods. If an appropriate definitive host eats an infested transport host, the worms can mature and complete their life-cycle. Dubinsky and colleagues (1995) found anti-*Toxocara* antibodies in 30.4% and 25.0% of striped field mice and harvest mice (*Micromys minutus*), respectively, in Slovakia, demonstrating that small mammals can play a role in maintaining toxocariasis foci in urban biotopes.

13.4.3.3. Babesiae

Rodent-related *B. microti* and other highly similar agents cause a rare, but potentially severe and even fatal illness in people who live in Europe and North America (Goethert & Telford, 2003b; Meer-Scherrer et al., 2004; Telford & Goethert, 2004). The disease is likely to be severe in immunocompromised individuals, such as those who have undergone splenectomies. The taxonomical and etiological status of these *B. microti*-like agents is currently a topic of much debate, but there seems to be general agreement that this complex of parasitic protozoa poses a risk to people in certain tick-infested areas, including those undergoing urbanization. The primary vectors of these agents are various ticks of the genus *Ixodes*, including the castor bean tick in Europe and the deer tick and the western black-legged tick in North America. Because these same vectors also transmit LB and HGA, areas at risk for these last two diseases also are likely to be at risk for babesiosis caused by *B. microti* or closely related agents. The major rodent hosts for *B. microti*-like agents in Europe are various voles (primarily *Microtus* spp.) and *Apodemus* mice; North American hosts include voles (primarily *Microtus* spp.) and *Peromyscus* mice.

13.4.3.4. Echinococci

Alveolar echinococcosis is a rare, but severe zoonosis due to the hepatic development of the fox tapeworm. The definitive hosts are various carnivores, including wild foxes. The intermediate hosts are various rodents, including voles, lemmings and mice. Viel and colleagues (1999) demonstrated that densities of water voles are a risk factor for human alveolar echinococcosis in France. Domestic dogs and cats can be sources of human infection, if these animals are allowed to hunt and consume non-commensal rodents infected with cysts bearing the larval stages of *E. multilocularis*.

13.4.3.5. Capillaria

Capillaria hepatica is a nematode that affects a wide range of mammal hosts, including humans. In people, it causes a rare, but potentially fatal infestation of the liver that typically manifests as acute hepatitis (hepatic capillariasis) with marked eosinophilia. *Capillaria hepatica* has been found nearly worldwide, primarily as a parasite of the hepatic parenchyma of a number of rodent species. In the Canton of Geneva, Switzerland, Reperant & Deplazes (2005) trapped 664 rodents that belonged to five non-commensal species and found a significant cluster of *C. hepatica* infections in three rodent species in rural and urbanized areas in the northern part of the canton. These rodents included the yellow-necked mouse, in which the overall prevalence of infestations was 7%, while in some sites prevalences reached 20% in the bank vole and the northern water vole.

13.4.4. Disease associations with particular types of rodents

The following subsections list certain noteworthy risks of disease posed by the rodent groups listed in section 13.2 of this chapter.

13.4.4.1. Tree squirrels

Several cases of leptospirosis, rickettsial infections and TBE have been reported in various tree squirrels, but these animals usually do not play important roles as reservoirs for these infections. Squirrels can also carry many species of ticks, mites and fleas, but their tree-dwelling habits prevent them from exchanging ectoparasites with most ground-dwelling rodents. With the few exceptions noted below, tree squirrels rarely have been reported to be involved in epizootics or to serve as sources of disease agents that infected people. One report noted that tularaemia infections in Siberia had spread from hares to red squirrels, causing deaths among the latter, a situation that could be epidemiologically important for people residing in the areas affected. A number of cities in central Colorado and nearby Cheyenne, Wyoming, also have experienced occasional plague epizootics among fox squirrel populations that live in the urban deciduous forests found in these cities, and a single human plague case was associated with these animals in Denver, Colorado (Hudson et al., 1971). It is believed that fox squirrels, which are not native to the region, acquire plague bacteria from rock squirrels or other native species that are common hosts of plague in this region. In another report, a gray squirrel kept as a pet was implicated as the source of an infectious bite that resulted in a human tularaemia case in Arkansas (Magee et al., 1989). Acute *T. gondii* infections have caused fatalities among gray squirrels, but the epidemiological significance of this observation is unknown (Roher et al., 1981). Gray squirrels also were found naturally infected with LB spirochetes in the United Kingdom (Craine et al., 1997). Another report suggested that five patients in Kentucky had contracted spongiform encephalopathy as a result of eating the brains of tree squirrels (species not given) that perhaps contained a prion-like agent (Berger, Weisman & Weisman, 1997). American red squirrels in western North America have been implicated as important hosts for a soft tick (*O. hermsi*) that transmits relapsing fever spirochetes (*B. hermsii*) to people. These same squirrels also are susceptible to infection with *B. hermsii* and are thought to be sources of infection for feeding *O. hermsi* (Burgdorfer & Mavros 1970; Burgdorfer 1976). Although rarely reported in rodents, a case of rabies was identified in a fox squirrel (Cappucci, Emmons & Sampson, 1972).

13.4.4.2. Flying squirrels

In North America, the southern flying squirrel was implicated as a source of typhus caused by a *R. prowazekii*-like agent. Transmission of this rickettsial agent to people, presumably occurred through the bites of flying squirrel ectoparasites (fleas or lice), which can be found on these squirrels and in the nests they build in attics or wall spaces (McDade et al., 1980).

13.4.4.3. Ground squirrels and antelope ground squirrels

Numerous ground squirrel species (*Spermophilus* spp.) and white-tailed antelope ground squirrels have been implicated as hosts of *Y. pestis*, the causative agent of plague in North America (Gage, Ostfeld & Olson, 1995). Among the most important of these species are the rock squirrel and the closely related Beechey (California) ground squirrel. Both species occur frequently in peridomestic environments, and Beechey ground squirrels are common in many recreational sites in California and nearby areas. Rock squirrels and Beechey ground squirrels are considered the most significant sources of *Y. pestis* infection in people in the United States, due both to their presence in peridomestic environments and to one of their common fleas, *O. montana* (formerly *Diamanus montanus*), which is the primary vector of plague in people in this country. Other western ground squirrels (*Spermophilus* spp.) have been reported to be infected with *Y. pestis* or *F. tularensis* (Hopla, 1974; Jellison, 1974; Barnes, 1982). Probably the most important among these is the golden-mantled ground squirrel, which is often mistaken for a large chipmunk and frequently digs its burrows near home sites or recreational sites in mountainous areas of western North America. *Yersinia pestis*-positive samples have been reported frequently from these ground squirrels, and in certain California mountain ranges they have been implicated as likely sources of infectious fleabites for human cases of plague. The white-tailed antelope ground squirrel, which can be abundant near the homes of people in the south-western United States, is commonly infected with *Y. pestis* and carries fleas, particularly *Thrassis bacchi*, that can transmit plague bacteria to people (Montman, Barnes & Maupin, 1986). Recently, *B. washoensis* was isolated from Beechey ground squirrels in the Sierra Nevada mountains of western Nevada (Kosoy et al., 2003). Prior to this report, *B. washoensis* had been reported only in a man with myocarditis, and it was suggested that this bacterium might have been the cause of his illness and might represent the etiological agent of a previously unrecognized zoonosis that is maintained in ground squirrels. Richardson's ground squirrels and prairie dogs also have been reported to harbour *Bartonella* spp. (Stevenson et al., 2003; Jardine et al., 2005).

13.4.4.4. Chipmunks

The Siberian chipmunk is considered to be an important host for larval and nymphal stages of ixodid ticks. Siberian chipmunks can carry many infectious and parasitic disease agents in Europe or Asia, including those that cause TBE, tick-borne rickettsioses, Q fever, tularaemia, pseudotuberculosis, pasteurellosis, listeriosis, erysipelas and toxoplasmosis (Popov & Fedorov, 1958; Olsuf'ev & Dunaeva, 1960; Pestryakova et al., 1966; Astorga et al., 1996). In North America, various species of chipmunks are important hosts for a variety of human disease agents, including Lyme disease spirochetes (*B. burgdorferi*), plague (*Y. pestis*), relapsing fever spirochetes (*B. hermsii*) and tularaemia bacteria (*F. tularensis*) (Burgdorfer & Mavros, 1970; Hopla, 1974; Jellison, 1974). These rodents also serve as important hosts for the blood-feeding stages of major disease vectors. In some areas of the eastern United States, the eastern chipmunk is an important host of not only *B. burgdorferi* spirochetes, but also of the immature stages of their primary vectors (deer ticks) (Piesman, 2002).

A variety of mountain-dwelling chipmunk species invade homes in mountainous areas of western North America. These same species also serve as hosts for *O. hermsi*, a relapsing fever vector in this region (Burgdorfer, 1976; Trevejo et al., 1998). Many of these same chipmunk species also act as hosts for the flea *Eumolpianus eumolpi*, which is an important vector of plague in this region and has been implicated as a likely source of *Y. pestis* infection in people (Nelson, 1980; Barnes, 1982). Yellow-pine chipmunks (*Tamias amoenus*) in the Rocky Mountains of the northern United States and southern Canada also play an important role in the ecology of RMSF rickettsiae (*R. rickettsii*), by acting as tick hosts and sources of rickettsial infection for feeding vector ticks (Burgdorfer, Friedhoff & Lancaster, 1966). Chipmunks in Canada also have been found infected with *P. multocida*. Pet chipmunks imported to Europe from Asia were infected with *Cryptosporidium muris*, a species reported to infect people (Hurkova, Hadjusek & Modry, 2003; Gatei et al., 2006). Eastern chipmunks in New York State also were found infected with *Cryptosporidium parvum* (Perz & Le Blancq, 2001). Also, an eastern chipmunk was found to be infected with rabies (Dowda & DiSalvo, 1984).

13.4.4.5. Beavers

Although encountered only occasionally near urban areas, beavers can play an important role in tularaemia cycles (Hopla, 1974; Jellison, 1974). These animals also have been reported to be infected with various species of *Giardia*, *Toxoplasma, Sarcocystis*, *Salmonella* and *Pasteurella*, as well as helminth species, that cause alveolar echinococcosis, clonorchiasis and other ailments (Monzingo & Hibler, 1987; Marquardt, Demaree & Grieve, 2000; Dunlap & Thies, 2002; Jordan et al., 2005; Lawson et al., 2005). An outbreak of blastomycosis in Wisconsin was associated with children having contact with a beaver lodge and picking up soil associated with the lodge (Klein et al., 1986; Gaus, Baumgardner & Paretsky, 1996).

13.4.4.6. Voles and other microtine rodents

Voles and other microtine rodents are important hosts of many zoonotic and vector-borne diseases. Species of *Microtus* voles have been found infected with *F. tularensis*, *Y. pestis*, *Y. pseudotuberculosis*, *R. rickettsii*, *L. monocytogenes*, *Bordetella bronchiseptica, Pneumocystis carinii*, *Pasteurella* spp., *Brucella* spp., *Salmonella* spp. and *Streptococcus* spp. (Hopla, 1974; Gage, Ostfeld & Olson, 1995; Soveri et al., 2000). Voles are particularly important as hosts of tularaemia, which has been detected in *Microtus* spp. in Austria, Belgium, the Czech Republic, France, Norway, Poland, Romania, the Russian Federation, Slovakia and Turkey (Olsuf'ev & Dunaeva, 1960). Common voles are very sensitive to tularaemia and can die 4–10 days after being inoculated with only a few *F. tularensis* organisms. The high level of bacteraemia (10^7–10^8 *F. tularensis*/ml of blood) commonly experienced by infected voles also enables these animals to serve as sources for infecting feeding tick vectors. During winter epizootics, *F. tularensis* can be transmitted between voles through cannibalism. *Microtus* voles infected with *Y. pestis* have been identified near human habitations in California. Other voles, especially the meadow vole, are considered important hosts of the etiological agents of tularaemia (*F. tularensis*) and RMSF (*R. rickettsii*). Meadow voles also are common hosts of larval and nymphal American dog ticks and wood ticks, which are primary vectors of both tularaemia and RMSF in different regions of North America. Northern water voles also are important hosts of tularaemia in Europe, including the former Union of Soviet Socialist Republics (Hopla, 1974). Finally, muskrats frequently have been found infected with *F. tularensis holarctica* (type B tularaemia strains) in Europe and North America (Hopla, 1974).

Other pathogens have been discovered in microtine rodents in Europe and North America. Among common voles in Europe, the prevalences of *Listeria*, *Y. pseudotuberculosis* and erysipeloid bacteria were 0.6–1.6%, 0.1–0.5%, and 0.3–2.2%, respectively. The common vole was infected predominantly with the grippotyphosa serotype of *Leptospira interrogans* (>90%), and the root vole carried a variety of *L. interrogans* serotypes, including javanica, pomona, hebdomadis and grippotyphosa (Karaseva, 1963, 1971; Rosicky & Sebek, 1974). Viel and colleagues (1999) also demonstrated that population densities of northern water voles are a risk factor for alveolar echinococcosis in people. Other microtine rodents, including *Microtus* voles and lemmings (*Lemmus* spp.), can also serve as intermediate hosts for *E. multilocularis*, the agent that causes alveolar echinococcosis. In Alberta, Canada, southern red-backed voles, meadow voles and long-tailed voles were found to be infected with *Giardia* spp. (Wallis et al., 1984). Microtine rodents are also commonly infected with the causative agent of yersiniosis (*Y. pseudotuberculosis*). Notably, the prevalence of *Y. pseudotuberculosis* in voles in urban and suburban areas was higher than that observed in commensal rodents (Iushchenko, 1970). Recently, a number of outbreaks of yersiniosis, primarily in the Russian Federation, have been linked to eating raw vegetables contaminated with *Y. pseudotuberculosis*-infected rodent urine (Daiter,

Polotskii & Tsareva, 1987). *Microtus* voles also are reported as hosts of *L. interrogans* serotypes in Bulgaria, the Czech Republic, Denmark, Germany, Hungary, the Netherlands, Poland, the Russian Federation, Slovakia and Switzerland.

13.4.4.7. Old World mice

Species of *Apodemus* are particularly important hosts for many tick-borne pathogens, and at least 23 species of ticks are reported on the wood mouse. The most epidemiologically important of these ticks is the castor bean tick, which is the primary vector of LB and TBE in Europe (Humair & Gern, 2000; Huegli et al., 2002; Charrel et al., 2004). Among the pathogens associated with wood mice are those that cause LB, TBE, Omsk haemorrhagic fever, Q fever, tularaemia, leptospirosis, bartonellosis, toxoplasmosis, salmonellosis and infections with hantaviruses. Yellow-necked mice also serve as hosts for the pathogens that cause leptospirosis, TBE, lymphocytic choriomeningitis, toxoplasmosis and a severe form of HFRS caused by Dobrava virus. The yellow-necked mouse also is a host of *B. garinii* (Huegli et al., 2002), a genospecies that has been cultivated frequently from the cerebrospinal fluid of LB patients from Denmark, Germany, the Netherlands and Slovenia. These mice also are important hosts of larval and nymphal castor-bean ticks.

13.4.4.8. Dormice

Although less important as hosts of rodent-related disease agents than the voles or Old World mice, the edible or fat dormouse has been implicated as the source of *F. tularensis* infection for a human tularaemia case in which the person was bitten on the finger by an infected dormouse (Friedl, Heinzer & Fankhauser, 2005). Garden dormice also are hosts of a recently described spirochete (*B. spielmani*) that reportedly causes LB in people (Richter et al., 2004). Others have proposed that dormice might act as hosts for the murine typhus agent (*R. typhi*), after this rickettsia was identified in fleas (*Monopsyllus sciurorum sciurorum*) taken from the nests of these animals (Trilar, Radulovic & Walker, 1994).

13.4.4.9. New World rats and mice

Although many species of rats and mice exist in the New World murine subfamily Sigmodontinae, only a few are likely to be found near human habitations or to pose a significant risk of disease to people. Notable among these are various species of native mice (*Peromyscus* spp.), wood rats (*Neotoma* spp.) and cotton rats (*Sigmodon* spp.). Among the species of *Peromyscus*, the deer mouse and the white-footed mouse are widespread and particularly important. The deer mouse is the major source of SNV infection in people (Childs et al., 1994) and often is considered to be a significant enzootic host of plague. The white-footed mouse is the major host of the LB spirochete (*B. burgdorferi*) and its primary tick vector (the deer tick) throughout the north-eastern and much of the upper Midwestern United States, as well as parts of south-eastern Canada (Piesman, 2002). It is also an important host for the HGA agent (*A. phagocytophilum*), which is also transmitted by deer ticks (Parola & Raoult, 2001; Strle, 2004; Parola, Davoust & Raoult, 2005). Moreover, *Peromyscus* mice are important hosts of the immature stages of the American dog tick, the primary tick vector of the RMSF agent (*R. rickettsii*) in the eastern United States.

Another important group of sigmodontine rodents are the wood rats (*Neotoma* spp.), which occur throughout much of North America. A number of species in the western United States are commonly found to be infected with *Y. pestis* and act as significant hosts for this bacterium and certain species of fleas that transmit it (Gage, Ostfeld & Olson, 1995). Wood rats are also hosts of a recently recognized arenavirus (Whitewater Arroyo virus), which is of unknown significance to human health (Fulhorst et al., 1996; Kosoy et al., 1996). Moreover, these rats are important hosts in the far western United States for the immature stages of western black-legged ticks, which transmit LB spirochetes (*B. burgdorferi*), as well as the HGA agent (*A. phagocytophilum*) (Piesman, 2002). *Borrelia bissetti,* another spirochete, which was initially confused with *B. burgdorferi* s.s., occurs in wood rats, but at present its importance to human health is uncertain (Maupin et al., 1994; Eisen et al., 2003). Wood rats have also been found seropositive for hepatitis E virus in New Mexico (Favorov et al., 2000). Finally, wood rats have been reported to serve as hosts for the protozoon parasite *Trypanosoma cruzi* that causes Chagas disease. *Trypanosoma cruzi* is transmitted by a type of reduviid bug that can be found living within the large stick nests built by these animals (Peterson et al., 2002).

Four species of cotton rats occur in the temperate regions of North America, but only one of these, the hispid cotton rat, is likely to pose a significant threat to human health. Hispid cotton rats can occur near human habitations and are extremely abundant in much of the south-eastern and south-central United States, including some largely urban areas with appropriate habitat and other sites undergoing urbanization. In some regions they are the major hosts of the immature American dog tick, the primary vector of the RMSF agent (*R. rickettsii*) in the eastern United States (Gage, Burgdorfer & Hopla, 1990; Gage, Hopla & Schwan, 1992). Recently, Kosoy and colleagues (1997, 1999, 2004a,b) reported that cotton rats are hosts to four uncharacterized *Bartonella* genogroups. Hispid cotton rats also are known to be susceptible to infection with *B. burgdorferi* (Burgdorfer & Gage, 1987).

13.4.4.10. Rabbits and hares

Tularaemia, which is often referred to as rabbit fever, is frequently found in rabbits and hares, and these animals are common sources of human infection (Hopla, 1974; Jellison, 1974; Hopla & Hopla, 1994; Petersen & Schriefer, 2005). In the United States, rabbits are the source of tularaemia infection in 90% of human cases, 70% of which result from contact with the genus *Sylvilagus*. Jackrabbits also are an important source of infection in some areas of the United States, but are a minor factor nationally, and exposure to snowshoe hares (*Lepus americanus*) account for less than 1% of cases of human tularaemia in the United States.

In Europe, tularaemia has been detected in brown hares (*Lepus europaeus*), mountain hares (*Lepus timidus*), and rabbits (such as the European rabbit). High rates of mortality were usually observed among all these species, with death occurring 7–19 days after exposure. In Europe, die-offs of hares often followed epizootics in small rodents (Borg et al., 1969).

Rabbits have often been found to be naturally infected with *Y. pestis* in North America, and numerous human plague cases have been associated with handling these animals (Kartman, 1960; von Reyn et al., 1976). In most instances, these cases have occurred in hunters and others who have skinned infected carcasses. Cottontail rabbits, however, also occur frequently in urbanized environments, and rabbit carcasses positive for *Y. pestis* have been found in such areas, suggesting a possible risk for people who might come into contact with infected rabbits or their fleas in these areas.

Other bacterial or rickettsial disease agents reported from lagomorphs include those that cause pasteurellosis, brucellosis, yersiniosis, listeriosis, rickettsioses (*Rickettsia* spp., *C. burnetii*, *Anaplasma* spp.), salmonelloses and leptospirosis (Dunaeva, 1979; Dumler, 1994; Williams & Sanchez, 1994; Gage, Ostfeld & Olson, 1995; Goethert & Telford, 2003a), although no known cases of human infection with these agents have been traced to contact with wild rabbits or hares. Lagomorphs also carry endoparasites, such as helminths (*Trichostrongylus* spp., *Passalurus ambiguus* and *Graphidium strigosum* in hares), tapeworms (*Taenia pisiformis*, *Taenia serialis* and *Echinococcus granulosus* in rabbits; *Cittotaenia ctenoides* in rabbits; and *Protostrongylus* spp. in hares). Finally, rabbits and hares can carry protozoan parasites, including *T. gondii*, *Encephalitozoon cuniculi* and coccidia (*Eimeria* spp.) (Dunaeva, 1979; Marquardt, Demaree & Grieve, 2000). Although lagomorphs harbour viruses, such as myxoma virus, snowshoe hare virus, herpesvirus and the virus that causes European brown hare syndrome, these agents do not appear to be important causes of illness in people (Dunaeva, 1979).

13.5. Risk factors for rodent- and lagomorph-related diseases

In general, the risk of acquiring rodent- or lagomorph-related disease agents is linked to the likelihood of coming into close proximity to the rodent carriers of these diseases, their wastes or the disease-transmitting ectoparasites they carry. Even in the absence of changes induced by people, the risk rarely remains steady over time, but tends to fluctuate, depending on the status of various ecological factors, including climatic variables, quality of natural habitat and available food supplies (Gage & Kosoy, 2005; Mills, 2005), all of which can affect the survival and reproduction of rodents and lagomorphs through their impacts on the availability of food. Although the effects of various ecological factors on eruptions of rodent or lagomorph populations have been much debated (Begon, Harper & Townsend, 1996), any factor(s) that causes populations of these animals to increase above certain threshold levels is likely to lead to an increased spread of disease among these populations (Davis et al., 2004). Also, it is likely to lead to heightened risks of exposure for people, particularly when infected animals invade their home sites, recreational areas or workplaces where good rodent sanitation practices are lacking or inadequate (Childs et al., 1995; Zeitz et al., 1995; Enscore et al., 2002). The following paragraphs provide some examples of how the risk of disease in people can be affected by environmental factors that influence rodent population dynamics. These paragraphs also provide examples of the likelihood that these animals will move into close contact with people.

The 1993 HPS outbreak caused by SNV in the south-western United States occurred after a strong El Niño event that resulted in extremely high precipitation in this normally dry region. As a result of the large rainfalls, plant growth increased dramatically; in turn, this led to an increased availability of rodent food sources (such as vegetation, seeds, fruits and insects) and a tremendous eruption in the region's deer mouse population (Engelthaler et al., 1999; Hjelle & Glass, 2000; Glass et al., 2002), which was followed by an increase in hantavirus prevalence among these mice. As deer mouse populations increased, so did the frequency at which peridomestic environments, including homes, were invaded – a factor that greatly increased the risk of exposing people to hantavirus-infected animals or their excreta. Based on a study by Kuenzi and colleagues (2001), which demonstrated that the breeding season of deer mice in peridomestic sites was about two months longer than in naturally occurring habitats, Mills (2005) suggested that this factor could influence the risk of exposure of people living in these sites.

Human plague in the south-western United States largely occurs in peridomestic environments located in formerly rural areas that are rapidly undergoing urbanization. The risk of people acquiring plague in these areas is likely to be influenced by environmental factors, such as yearly variations in climatic variables, that affect rodent and vector abundance. Parmenter and colleagues (1999) also noted that increased cool season precipitation was associated with an increased risk of human plague in New Mexico. They proposed that the increased risk was linked to the positive effects of precipitation increases on the availability of rodent food sources; this, in turn, led to increased populations of the rodent hosts of plague and probably the flea vectors that feed on these animals. These ideas were further expanded by Enscore and colleagues (2002), who demonstrated that the frequencies of human plague cases in a given onset year in a region of the south-western United States were associated positively with late winter precipitation and relatively cool summers, findings believed to be related to rodent population dynamics and survival of vector fleas, respectively.

The incidence of LB also can be affected by changing ecological parameters, including the availability of food for rodents. Jones and colleagues (1998) proposed that increased oak masts (acorn crops) can lead to increased populations of rodents, particularly white-footed mice, that serve as hosts and sources of spirochetal infection for the immature stages of the deer tick, which is the primary vector of LB in the eastern United States and southern Canada.

13.6. The impacts of anthropogenic transformations

Although complete urbanization can result in the elimination of non-commensal species, the transition from rural to suburban environments often results in a mosaic of human development interspersed with largely natural areas that can harbour sizeable non-commensal rodent populations. Human activities often disturb landscapes, causing major alterations of rodent habitats, as well as significant changes in the composition of local rodent and lagomorph populations that serve as hosts for various disease agents (Barnes, 1982; Maupin et al., 1991; Gage, Ostfeld & Olsen, 1995; Gage et al., 2000; Randolf, 2001; Piesman, 2002). In some instances these landscape disturbances result in the loss of certain rodent species from the area affected, particularly when those species are highly specialized and are completely dependent on a very restricted range of environmental conditions for their survival. However, many other rodent species can be classified as generalists and are quite adaptable, being able to survive in isolated patches of remaining habitat or in disturbed areas. In still other instances, landscape disturbances might actually provide new habitats for opportunistic rodent species.

Such a process has been observed over the past 3–4 decades in plague-endemic regions of the south-western United States (Barnes, 1982; Gage, Ostfeld & Olson, 1995; Gage, 1998; Gage et al., 2000). Much of the development that has occurred in this area has involved the conversion of ranches and forests into semi-rural or suburban areas dotted with new home sites. Despite encroaching urbanization, these sites often retain significant amounts of native vegetation and thus remain attractive to a variety of non-commensal rodent species, including rock squirrels, which are important hosts and sources of *Y. pestis* infection for the flea species *O. montana* that is the primary vector of human plague in the United States.

Building rock walls around properties or piling rocks or other debris near new home sites further increases rock squirrel populations (Barnes, 1982). Problems with peridomestic rock squirrel populations are complicated further by residents who improperly dispose of garbage or allow these rodents to have access to spilled pet foods, seeds at bird feeders or other edible items. The risk of LB in peridomestic environments also can be affected by landscaping choices intended to make backyards or other sites more attractive (Maupin et al., 1991; Piesman, 2002).

Abandonment of agricultural lands in the north-eastern United States has resulted in landscapes characterized by various stages of succession. Initially, weeds and other invasive species colonize sites, but eventually other types of vegetation come to predominate, including the native tree species found in the climax forests of this region. Although abandoned agricultural lands might eventually return to a fully forested state, such sites are often converted to low-density housing, creating semi-rural or suburban environments that offer an excellent mosaic of habitats for non-commensal rodents and their ectoparasites. This abandonment and redevelopment of farm lands in the north-eastern United States has also been accompanied by recolonization of the region by white-tailed deer, which act as hosts for the adult stages of the deer tick vector of LB (Piesman, 2002). As tick-infested deer help to re-establish the deer tick populations in this region, local populations of white-footed mice and certain other species are able to act as competent reservoirs for infecting immature deer ticks with the agent that causes LB (*B. burgdorferi*). Although evidence suggests that *B. burgdorferi* has been present in this region for at least many decades, the landscape changes noted above have resulted, over the past 20 years, in the emergence of Lyme disease as the most commonly reported vector-borne disease in the United States. Recreational use of highly heterogeneous woodlands also is reported to be associated with an increased risk of LB in Europe (Gray, 1999). Many of these sites are secondary growth woodlands, and people have altered these environments extensively.

13.7. Public health impact

In Canada and the United States, local governments, physicians and others typically report cases of rodent-related illness to provincial or state public health officials who monitor the incidences of some of these diseases and report them periodically to agencies at the national level. In the United States, the CDC is responsible for the national surveillance of reportable diseases, which includes a few of the rodent- or rabbit-related diseases given in Table 13.2 (Hopkins et al., 2005). Unfortunately, incidence data for many of the remaining diseases associated with these animals are unavailable, making it difficult to assess their impact on human populations. Similar data also appear to be unavailable for Europe.

13.8. Costs, control and management of infestations

Although a few rodent- or lagomorph-related diseases, such as LB, are viewed as significant health problems, many others rarely come to the attention of people living in a particular area, which typically results in their believing that these diseases have little impact on their health or economic well-being. Such perceptions are often shared by policy-makers who see few reasons to use limited resources to prevent or control these often rare diseases or the rodent or lagomorph infestations associated with them. However, when the negative health and economic effects caused by non-commensal rodents and lagomorphs

Table 13.2. Incidence of nationally reportable rodent- or lagomorph-associated diseases in the United States

Disease	Incidence (1993–2003) per 100 000 population[a]	Years for which incidence data were available in the interval 1993–2003
HGA (HGE)	0.14	1998–2003
Encephalitis/meningitis: arboviral, California serogroup (primarily La Crosse virus) infections	0.04	1995–2003
Powassan viral encephalitis	< 0.01	2002–2003
Giardiasis	7.45	2002–2003
HPS	0.01	2000–2003
Leptospirosis	0.02	1993–1994
Lyme disease	5.86	1993–2003
Plague	< 0.01	1993–2003
RMSF	0.24	1993–2003
Tularaemia	0.05	1993–1994; 2000–2003

[a] Average for years in which data were available.

Note. HGA: human granulocytic anaplasmosis; HGE: human granulocytic ehrlichiosis; HPS: hantaviral pulmonary syndrome; RMSF: Rocky Mountain spotted fever.

Source: Hopkins et al. (2005).

are considered as a single, combined problem, their impact is likely to be significant and easier to recognize. Addressing rodent- or lagomorph-related issues in this way should be justifiable from a planning and policy perspective, because in many instances the measures used to prevent and control the various rodent- or lagomorph-associated pathogens are quite similar. The control measures recommended to reduce rodent or lagomorph infestations in homes or other sites also, are often nearly identical, although in some instances a specific technique must be employed for a particular species. Another advantage that is likely to result from treating rodent- or lagomorph-associated issues as a single problem is that programme duplication can be minimized, resulting in more successful, cost-effective programmes that can maximize the use of limited resources.

Unfortunately, estimating the costs of rodent-related diseases is difficult, and few publications have addressed this issue. An early evaluation of plague control in California campgrounds determined that control was economically beneficial compared with other options, including surveillance-only and no-programme alternatives (Kimsey et al., 1985). Similar estimates were not made, however, for benefits derived from plague control programmes targeted at peridomestic environments. More recently, a number of authors have attempted to estimate the costs associated with LB. Maes, Lecomte & Ray (1998) estimated that the total economic burden for this illness in the United States was US$2.5billion (1996 US$) over a five-year period. These costs are likely to have increased dramatically as a result of the general inflation of health care costs in the United States, as well as actual increases in the number of LB cases reported. In a later study, the mean cost per case of LB in the United States was estimated to be US$4466 (1996 US$) (Meltzer, Dennis & Orloski, 1999). A more recent study estimated the annual costs for LB in Scotland to be £331 000 (Joss et al., 2003).

Carabin and colleagues (2005) described methods for assessing the economic burden of echinococcosis in people, which they defined as involving infestations with either *E. multilocularis*, the agent of alveolar echinococcosis, or *E. granulosus*, which causes cystic echinococcosis. Although *E. granulosus* is not associated with rodents, *E. multilocularis* utilizes rodents as intermediate hosts (see section 13.4). According to the methods described by Carabin and colleagues, a case of echinococcosis in the United Kingdom costs about US$ 10 215 (2005 US$).

Although no estimate of costs for treating hantaviral cases was found, Hopkins and colleagues (2002) reported that inexpensive rodent proofing (less than US$ 500 per household in 2002) was effective in reducing the intensity of rodent intrusions in homes, a method that should prove cost effective and result in the decreased risk of SNV infection among the people living in these homes.

13.9. Control and management of non-commensal rodents and rodent-related diseases

A variety of approaches can be used to prevent rodent-related illnesses. Obviously, cost-effective and reliable surveillance methods must be developed, so that threatening situations can be identified quickly and appropriate preventive measures taken (Gage, Ostfeld & Olson, 1995; Glass et al., 1997; Trevejo et al., 1998; Gage, 1999; Hopkins et al., 2002; Douglas et al., 2003; Hayes & Piesman, 2003). Although the types of preventive measures used will vary from one situation to another, certain steps are likely to be appropriate in nearly all instances. Clearly, residents of the areas affected by threatening situations should be made aware of the risks and be educated about the best means of protecting themselves and their families. This includes taking certain precautions, such as avoiding sick or dead animals, using insect repellents, performing so-called tick checks, wearing protective clothing or gloves, or treating clothing with appropriate pesticides (Gage, Ostfeld & Olson, 1995; Hayes & Piesman, 2003). The intentional feeding of seeds or other foods to rodents and lagomorphs is to be discouraged, as it is likely to increase the risk of people coming in contact with infected animals or their ectoparasites. Residents of the areas affected also should be advised that pets can become infected after contact with non-commensal rodents or lagomorphs and can spread these infections to their owners. Pets also can carry infected vectors into homes. Applications of rodenticides, insecticides or acaricides in rodent habitats is sometimes warranted as a means of quickly reducing the risk of transmission, but such treatments are best done by pest control specialists. Although proper use of pesticides can often provide rapid reductions in the risk of disease, they should not be used solely for long-term management of problems posed by non-commensal rodents or lagomorphs.

Whenever possible, rodent-related disease problems should be addressed by a combination of personal protective measures, treatment and restraint of pets, limited applications of pesticides, reductions of rodent food and harbourage in or near people's homes, and modification of rodent habitats, so that people will be less likely to come in contact with infected rodents (Childs et al., 1995; Gage, Ostfeld & Olson, 1995; Gage et al., 2000; Mills et al., 2002). The last two measures (reductions of rodent food and harbourage and management of rodent habitats) are likely to represent the most sustainable and effective approach to reducing widespread threats posed by non-commensal rodents (Gage, Ostfeld & Olson, 1995; Levy & Gage, 1999; Glass et al., 1997; Hopkins et al., 2002; Douglass et al., 2003).

Temporary reductions in rodent numbers are of minimal use and might even be harmful. Douglass and colleagues (2003) reported that removal of deer mice from buildings actually resulted in an increased number of mice entering these structures, leading these authors to speculate that mouse removal in the absence of rodent-proofing is likely to increase the risk of human hantavirus.

The importation and ownership of exotic rodents, such as prairie dogs, as pets also should be discouraged, because this practice could result in direct human exposures to various pathogens. A remote possibility also exists that these animals might escape and spread pathogens to native rodent species, leading to the establishment of enzootic cycles of these agents in various rodent species. These issues were well illustrated by the recent identification of tularaemia bacteria in prairie dogs that were shipped from an exotic-pet facility in the United States to other sites in the United States and Europe (Avashia et al., 2004; Petersen et al., 2004). At least one animal handler at the exotic-pet facility was thought to have acquired tularaemia from handling these infected prairie dogs. In another incident in the United States, black-tailed prairie dogs became infected with monkeypox virus at an exotic-pet facility after being exposed to infected Gambian giant pouched rats imported from Africa. Following infection, the prairie dogs were shipped to pet stores in the Midwestern states, which resulted in numerous human monkeypox infections among people who handled these animals (Kile et al., 2005).

13.10. Conclusions

Non-commensal rodents and lagomorphs can be common in urbanized environments in Europe and North America, where they often threaten the health and economic well-being of residents who live in such environments. As the levels of contact between these animals and human beings increase, so do the risks that people will suffer illnesses caused by various rodent- or lagomorph-associated pathogens. Human contact with disease-carrying rodents and lagomorphs can be expected to increase on both continents as a result of ongoing human development activities, including urban sprawl. Prevention of rodent-borne diseases in urbanized environments will require adequate surveillance, public education and environmental management practices, as well as appropriate rodent- and vector control measures. The goal of these measures should be to reduce the degree of contact between the human residents of urbanized areas and local non-commensal rodent populations. Although such measures can be expensive, the costs are likely to be economically justifiable based on the medical expenses associated with treating human cases of rodent-related diseases and the property damage caused by some of these animals.

The following rules should be followed to reduce the risk of people acquiring a rodent- or lagomorph-related illness in areas undergoing urbanization.

Implement strict planning and regulatory processes intended to reduce the risk that urban sprawl will lead to increased exposures of people to non-commensal rodent- or lagomorph-related diseases.

Educate the public and government officials about the risks of rodent- or lagomorph-related diseases in sites undergoing urbanization, and provide these people with information about the steps that can be taken to protect individuals who are likely to be exposed to these diseases.

Minimize contact between people and disease-bearing non-commensal rodents or lagomorphs through implementation of effective and sustainable environmental-management programmes. Appropriate management strategies should include habitat modifications intended to reduce the attractiveness of sites for rodents, removal of sources of rodent food and shelter provided by people, and construction of barriers to exclude potentially infected animals from people's homes, recreation areas or work sites.

Provide, in emergency situations, for selective and limited use of pesticides to rapidly reduce the risk of disease or to protect property.

Develop a comprehensive plan for monitoring and controlling rodent- and lagomorph-related diseases. This plan should include provisions for increased disease surveillance that will provide a more comprehensive assessment of the risk of people being exposed to these diseases in sites undergoing urbanization. Such information is currently lacking for most rodent- or lagomorph-related diseases in Europe and North America, making it difficult to determine the true health and economic burdens posed by these diseases. The plan also should integrate all surveillance, prevention and control efforts for these diseases into a single programme for each region. A single integrated programme for addressing these diseases should be highly effective, because the techniques, equipment and trained personnel required to prevent or control many of them are quite similar. Combining all prevention and control activities into one programme also will reduce duplication of effort and will better utilize scarce resources and personnel, making these efforts more cost effective and easier to justify to programme managers, politicians and the general public.

References[2]

Alderton D (1996). *Rodents of the world*. New York, Facts on File Inc.

Amori G, Cantini M, Rota V (1995). Distribution and conservation of Italian dormice. *Hystrix (ns)*, 6:331–336.

Amori G, Gippoliti S (1995). Siberian chipmunk *Tamias sibiricus* in Italy. *Mammalia*, 59:288–289.

Anda P et al. (2001). Waterborne outbreak of tularemia associated with crayfish fishing. *Emerging Infectious Diseases*, 7(Suppl. 3):575–582.

Andrzejewski R et al. (1978). Synurbanization process in a population of *Apodemus agrarius*. I. Characteristics of population in urbanization gradient. *Acta Theriologica*, 23:341–358.

Astorga RJ et al. (1996). Pneumonic pasteurellosis associated with *Pasteurella haemolytica* in chipmunks (*Tamias sibiricus*). *Zentralblatt für Veterinärmedizin, Reihe B*, 43:59–62.

Avashia SB et al. (2004). First reported prairie-dog-to-human tularaemia transmission, Texas, 2002. *Emerging Infectious Diseases*, 10:483–486.

Babinska-Werka J, Gliwicz J, Goszczynski J (1981). Demographic processes in an urban population of the striped field mouse. *Acta Theriologica*, 26:275–283.

Banfield AWF (1974). *The mammals of Canada*. Toronto, University of Toronto Press.

Barnes AM (1982). Surveillance and control of bubonic plague in the United States. *Symposia of the Zoological Society of London*, 50:237–270.

Beaty BJ (2001). La Crosse virus. In: Service MW, ed. *The encyclopedia of arthropod-transmitted infections of man and domesticated animals*. Wallingford, CABI Publishing:261–264.

Begon M, Harper JL, Townsend CR (1996). *Ecology: individuals, populations and communities*, 3rd ed. Oxford, Blackwell Science:589–601.

Belk MC, Smith HD (1991). *Ammospermophilus leucurus*. *Mammalian Species*, 368:1–8.

Blanco JR, Oteo JA (2002). Human granulocytic ehrlichiosis in Europe. *Clinical Microbiology and Infection*, 8:763–772.

[2] Information for this chapter was obtained from Entrez-PubMed and Web of Science searches of peer-reviewed literature and from numerous books that dealt with various aspects of zoonotic and vector-borne diseases and their agents. Older citations that are not readily accessible via electronic search engines were discussed and listed in the literature cited when available and appropriate.

Beran G (1994). Rabies and infections by rabies-related viruses. In: Beran G, Steele JH eds. *Handbook of zoonoses, Section B: Viral*, 2nd ed. Boca Raton, CRC Press:307–357.

Berger JR, Weisman E, Weisman B (1997). Creutzfeldt-Jakob disease and eating squirrel brains. *Lancet*, 350:642.

Bonaccorso FJ, Brown JH (1972). House construction on the desert wood rat, *Neotoma lepida lepida*. *Journal of Mammalogy*, 53:283–288.

Borg K et al. (1969). On tularemia in the varying hare (*Lepus timidus* L.). *Nordisk Veterinaermedicin*, 21:95–104.

Bossi P et al. (2004). BICHAT guidelines for the clinical management of tularaemia and bioterrorism-related tularaemia. *Euro Surveillance*, 9:E9–10.

Boulouis HJ et al. (2005). Factors associated with the rapid emergence of zoonotic *Bartonella* infections. *Veterinary Research*, 36:383–410.

Burgdorfer W (1976). The epidemiology of relapsing fevers. In: Johnson RC, ed. *The biology of parasitic spirochetes*. New York, Academic Press:225–234.

Burgdorfer W, Friedhoff KT, Lancaster JL Jr (1966). Natural history of tick-borne spotted fever in the USA. Susceptibility of small mammals to virulent *Rickettsia rickettsii*. *Bulletin of the World Health Organization*, 35:149–153.

Burgdorfer W, Gage KL (1987). Susceptibility of the hispid cotton rat (*Sigmodon hispidus*) to the Lyme disease spirochete (*Borrelia burgdorferi*). *The American Journal of Tropical Medicine and Hygiene*, 37:624–628.

Burgdorfer W, Mavros AJ (1970). Susceptibility of various species of rodents to the relapsing fever spirochete, *Borrelia hermsii*. *Infection and Immunity*, 2:256–259.

Byers RE (1979). Meadow vole control using anticoagulant baits. *Horticultural Science*, 14:44–45.

Cahalane VH (1961). *Mammals of North America*. New York, The MacMillan Company.

Callisher C (2001). Colorado tick fever. In: Service MW, ed. *The encyclopedia of arthropod-transmitted infections of man and domesticated animals*. Wallingford, CABI Publishing:261–264.

Cameron GN, Spencer SR (1981). *Sigmodon hispidus*. *Mammalian Species*, 158:1–9.

Cappucci DT Jr, Emmons RW, Sampson WW (1972). Rabies in an eastern fox squirrel. *Journal of Wildlife Diseases*, 8:340–342.

Carabin H et al. (2005). Methods for assessing the burden of parasitic zoonoses: echinococcosis and cysticercosis. *Trends in Parasitology*, 21:327–333.

Carlton MD (1989). Systematics and evolution. In: Kirkland GL Jr, Layne JN, eds. *Advances in the study of Peromyscus (Rodentia)*. Lubbock, Texas Tech University Press:7–141.

CDC (1996). Prevention of plague: recommendations of the Advisory Committee on Immunization Practices (ACIP). *Morbidity and Mortality Weekly Report, Recommendations and Reports*, 45(RR-14):1–15.

CDC (2002). Tularemia – United States, 1990–2000. *Morbidity and Mortality Weekly Report*, 51:181–184.

CDC (2004). Lyme disease – United States, 2001–2002. *Morbidity and Mortality Weekly Report*, 53:365–369.

Ceballos GG, Wilson DE (1985). *Cynomys mexicanus. Mammalian Species*, 248:1–3.

Chantrey J et al. (1999). Cowpox: reservoir hosts and geographic range. *Epidemiology and Infection*, 122:455–460.

Chapman JA (1975). *Sylvilagus nuttallii. Mammalian Species*, 56:1–3.

Chapman JA, Hockman JG, Ojeda MM (1980). *Sylvilagus floridanus. Mammalian Species*, 136:1–8.

Chapman JA, Willner GR (1978). *Sylvilagus audubonii. Mammalian Species*, 106:1–4.

Chapuis JL (2005). Répartition en france d'un animal de compagnie naturalisé, le tamia de Sibérie (Tamias sibiricus). *Revue d'Ecologie de la Terre et la Vie*, 60:239–253.

Charrel RN et al. (2004). Tick-borne virus diseases of human interest in Europe. *Clinical Microbiology and Infection*, 10:1040–1055.

Chesnokova MV et al. (2003). [Evaluation of the polymerase chain reaction effectiveness in the investigation of outbreaks of pseudotuberculosis]. *Zhurnal Mikrobiologii Epidemiologii i Immunobiologii*, May–June (3):7–11 (in Russian).

Childs JE et al. (1994). Serologic and genetic identification of *Peromyscus maniculatus* as the primary rodent reservoir for a new hantavirus in the southwestern United States. *The Journal of Infectious Diseases*, 169:1271–1280.

Childs JE et al. (1995). A household based, case control study of environmental factors associated with hantavirus pulmonary syndrome in the southwestern United States. *The American Journal of Tropical Medicine and Hygiene*, 52:393–397.

Childs JE et al. (1997). Surveillance and spatiotemporal associations of rabies in rodents and lagomorphs in the United States, 1985–1994. *Journal of Wildlife Diseases*, 33:20–27.

Clark TW, Hoffmann RS, Nadler CF (1971). *Cynomys leucurus*. *Mammalian Species*, 7:1–4.

Clement J et al. (1997). The hantaviruses of Europe: from the bedside to the bench. *Emerging Infectious Diseases*, 3:205–211.

Corbet GB (1978). *The mammals of the Palaearctic Region: a taxonomic review.* London, British Museum (Natural History); Ithaca, New York, Cornell University Press.

Craine NG et al. (1997). Role of grey squirrels and pheasants in the transmission of *Borrelia burgdorferi* sensu lato, the Lyme disease spirochete, in the U.K. *Folia Parasitologica*, 44:155–160.

Cvetnic Z et al. (2003). A serological survey and isolation of leptospires from small rodents and wild boars in the Republic of Croatia. *Veterinární Medicina-Czech*, 48:321–329.

Daiter AB, Polotskii I, Tsareva G (1987). [Pathogenic properties of *Yersinia* and their role in yersiniosis pathology]. *Zhurnal Mikrobiologii Epidemiologiii i Immunobiologii*, (2):108–115 (in Russian).

Daiter AB et al. (1993). [*Leptospira* infection under the natural conditions of the environs of Saint Petersburg]. *Zhurnal Mikrobiologii Epidemiologii i Immunobiologii Immunobiologii*, (1):28–33 (in Russian).

Danell K (1977). Dispersal and distribution of the muskrat (*Ondatra zibethica*) in Sweden. *Viltrevy*, 10:1–26.

Davis S et al. (2004). Predictive thresholds for plague in Kazakhstan. *Science*, 304:736–738.

Douglass RJ et al. (2003). Removing deer mice from buildings and the risk for human exposure to Sin Nombre virus. *Emerging Infectious Diseases*, 9:390–392.

Dowda H, DiSalvo AF (1984). Naturally acquired rabies in an eastern chipmunk (*Tamias striatus*). *Journal of Clinical Microbiology*, 19:281–282.

Dubinsky P et al. (1995). Role of small mammals in the epidemiology of toxocariasis. *Parasitology* 110:187–193.

Dumler JS (1994). Rocky Mountain spotted fever. In: Beran GW, Steele JF, eds. *Handbook of zoonoses. Section A: Bacterial, rickettsial, chlamydial, and mycotic*, 2nd ed. Boca Raton, CRC Press:417–427.

Dunaeva T (1979). [Family *Leporidae*.] In: Sokolov VE, ed. *Meditsinskay Teriologia* [*Medical theriology*]. Moscow, Nauka Press:51–68 (in Russian).

Dunlap BG, Thies ML (2002). *Giardia* in beaver (*Castor canadensis*) and nutria (*Myocastor coypus*) from east Texas. *The Journal of Parasitology*, 88:1254–1258.

Eckert J, Deplazes P (2004). Biological, epidemiological and clinical aspects of echinococcosis, a zoonosis of increasing concern. *Clinical Microbiology Reviews*, 17:107–135.

Eisen L et al. (2003). Vector competence of *Ixodes pacificus* and *I. spinipalpis* (Acari:Ixodidae), and reservoir competence of the dusky-footed woodrat (*Neotoma fuscipes*) and the deer mouse (*Peromyscus maniculatus*), for *Borrelia bissettii*. *Journal of Medical Entomology*, 40:311–320.

Eisenberg JF (1981). *The mammalian radiations: an analysis of trends in evolution, adaption, and behavior*. Chicago, University of Chicago Press.

Ellis J et al. (2002). Tularemia. *Clinical Microbiology Reviews*, 15:631–646.

Elton C (1942). *Voles, mice and lemmings: problems in population dynamics*. Oxford, Oxford University Press.

Engelthaler DM et al. (1999). Climatic and environmental patterns associated with hantavirus pulmonary syndrome, Four Corners region, United States. *Emerging Infectious Diseases*, 5:87–94.

Enscore RE et al. (2002). Modeling relationships between climate and the frequency of human plague cases in the southwestern United States, 1960–1997. *The American Journal of Tropical Medicine and Hygiene*, 66:186–196.

Favorov MO et al. (2000). Prevalence of antibody to hepatitis E virus among rodents in the United States. *The Journal of Infectious Diseases*, 181:449–455.

Fenollar F, Sire S, Raoult D (2005). *Bartonella vinsonii* subsp. *arupensis* as an agent of blood culture-negative endocarditis in a human. *Journal of Clinical Microbiology*, 43:945–947.

Fishbein DB et al. (1986). Rabies in rodents and lagomorphs in the United States, 1971–1984: increased cases in the woodchuck (*Marmota monax*) in mid-Atlantic states. *Journal of Wildlife Diseases*, 22:151–155.

Fleming WJ, Caslick JW (1978). Rabies and cerebrospinal nematodosis in woodchucks (*Marmota monax*) from New York. *The Cornell Veterinarian*, 68:391–395.

Friedl A, Heinzer I, Fankhauser H (2005). Tularemia after a dormouse bite in Switzerland. *European Journal of Clinical Microbiology & Infectious Diseases*, 24:352–254.

Fulhorst CF et al. (1996). Isolation and characterization of Whitewater Arroyo virus, a novel North American arenavirus. *Virology*, 224:114–120.

Gage KL (1998). Plague. In: Collier L, Balows A, Sussman M, eds. *Topley and Wilson's microbiology and microbial infections. Vol. 3: Bacterial infections*, 9th ed. London, Edward Arnold Ltd:885–904.

Gage KL (1999). Plague surveillance. In: Dennis D et al., eds. *Plague manual: epidemiology, distribution, surveillance and control.* Geneva, World Health Organization:135–165 (document number:WHO/CDS/CSR/EDC/99.2; http://whqlibdoc.who.int/hq/1999/WHO_CDS_CSR_EDC_99.2.pdf, accessed 13 September 2006).

Gage KL, Burgdorfer W, Hopla CE (1990). Hispid cotton rats (*Sigmodon hispidus*) as a source for infecting immature *Dermacentor variabilis* (Acari:Ixodidae) with *Rickettsia rickettsii*. *Journal of Medical Entomology*, 27:615–619.

Gage KL, Hopla CE, Schwan TG (1992). Cotton rats and other small mammals as hosts for immature *Dermacentor variabilis* (Acari: Ixodidae) in central Oklahoma. *Journal of Medical Entomology*, 29:832–842.

Gage KL, Kosoy MY (2005). Natural history of plague: perspectives from more than a century of research. *Annual Review of Entomology*, 50:505–528.

Gage KL, Ostfeld RS, Olson JG (1995). Nonviral vector borne zoonoses associated with mammals in the United States. *Journal of Mammalogy*, 76:695–715.

Gage KL et al. (2000). Cases of cat-associated human plague in the Western US, 1977-1998. *Clinical Infectious Diseases*, 30:893–900.

Gage KL et al. (2001). Isolation and characterization of *Borrelia parkeri* in *Ornithodoros parkeri* Cooley (Ixodida: Argasidae) collected in Colorado. *Journal of Medical Entomology*, 38:665–674.

Gatei W et al. (2006). *Cryptosporidiosis*: prevalence, genotype analysis, and symptoms associated with infections in children in Kenya. *The American Journal of Tropical Medicine and Hygiene*, 75:78–82.

Gaus DP, Baumgardner DJ, Paretsky D (1996). Attempted isolation of *Blastomyces dermatitidis* from rectal cultures of beaver (*Castor canadensis*) from north central Wisconsin. *Wilderness and Environmental Medicine*, 7:192.

Glass GE et al. (1997). Experimental evaluation of rodent exclusion methods to reduce hantavirus transmission to humans in rural housing. *The American Journal of Tropical Medicine and Hygiene*, 56:359–364.

Glass GE et al. (2002). Satellite imagery characterizes local animal reservoir populations of Sin Nombre virus in the southwestern United States. *Proceedings of the National Academy of Sciences of the United States of America*, 99:16817–16822.

Goethert HK, Telford SR 3rd (2003a). Enzootic transmission of *Anaplasma bovis* in Nantucket cottontail rabbits. *Journal of Clinical Microbiology*, 41:3744–3777.

Goethert HK, Telford SR 3rd (2003b). What is *Babesia microti*? *Parasitology*, 127:301–309.

Gratz N (2006). *Vector- and rodent-borne diseases in Europe and North America: distribution, public health burden, and control.* Cambridge. Cambridge University Press.

Gray J (1999). Risk assessment in Lyme borreliosis. *Wiener Klinische Wochenschrift*, 111:990–993.

Grzimek B (1968). *Grzimek's animal encyclopedia. Vol. 11. Mammals II.* Zurich, Kindler Verlag A.G.

Grzimek B (1975). *Grzimek's animal life encyclopedia: mammals I–IV. Vol. 10–13.* New York, Van Nostrand Reinhold.

Hall ER (1981). *The mammals of North America*, 2nd ed. *Vol. I & II.* New York, John Wiley & Sons.

Handysides S (1999). WHO is collecting data on leptospirosis. *Eurosurveillance Weekly*, 3:2.

Hayes EB, Piesman J (2003). How can we prevent Lyme disease? *The New England Journal of Medicine*, 348:2424–2430.

Heymann DL (2004). *Control of communicable diseases manual*, 18th ed. Washington, DC, American Public Health Association.

Hjelle B, Glass GE (2000). Outbreak of hantavirus infection in the Four Corners region of the United States in the wake of the 1997–1998 El Niño-southern oscillation. *The Journal of Infectious Diseases*, 181:1569–1573.

Honlinger B et al. (2005). Generalized cowpox infection probably transmitted from a rat. *The British Journal of Dermatology*, 153:451–453.

Hoogland JL (1995). *The black-tailed prairie dog: social life of a burrowing mammal.* Chicago, University of Chicago Press.

Hopkins AS et al. (2002). Experimental evaluation of rodent exclusion methods to reduce hantavirus transmission to rodents in a Native American community in New Mexico. *Vector Borne and Zoonotic Diseases*, 2:61–68.

Hopkins RS et al. (2005). Summary of notifiable diseases – United States, 2003. *Morbidity and Mortality Weekly Report*, 52:1–85.

Hopla CE (1974). The ecology of tularemia. *Advances in Veterinary Science and Comparative Medicine*, 18:25–53.

Hopla CE, Hopla AK (1994). Tularemia. In: Beran GW, Steele JH, eds. *Handbook of zoonoses. Section A: Bacterial, rickettsial, chlamydial and mycotic*, 2nd ed. Boca Raton, CRC Press:113–126.

Hudson BW et al. (1971). Serological and bacteriological investigations of an outbreak of plague in an urban tree squirrel population. *The American Journal of Tropical Medicine and Hygiene*, 20:255–263.

Huegli D et al. (2002). *Apodemus* species mice are reservoir hosts of *Borrelia garinii* OspA serotype 4 in Switzerland. *Journal of Clinical Microbiology*, 40:4735–4737.

Humair P-F, Gern L (2000). The wild hidden face of Lyme borreliosis in Europe. *Microbes and Infection*, 2:915–922.

Hurkova L, Hadjusek O, Modry D (2003). Natural infection of *Cryptosporidium muris* (Apicomplexa: Cryptosporiidae) in Siberian chipmunks. *Journal of Wildlife Diseases*, 39:441–444.

Iushchenko GV (1970). [Some regularities of pseudotuberculosis epidemiology]. *Zhurnal Mikrobiologii, Epidemiologii, i Immunobiologii*, 47:54–57 (in Russian).

Jajosky RA et al. (2006). Summary of notifiable diseases – United States, 2004. *Morbidity and Mortality Weekly Report*, 53:1–79.

Jalava K et al. (2004). Multiple outbreaks of *Yersinia pseudotuberculosis* infections in Finland. *Journal of Clinical Microbiology*, 42:2789–2791.

Jameson EW Jr, Peeters HJ (2004). *Mammals of California*. Berkeley, University of California Press (California Natural History Guide No. 66).

Jardine C et al. (2005). Rodent-associated *Bartonella* in Saskatchewan, Canada. *Vector Borne and Zoonotic Diseases*, 5:402–409.

Jellison WL (1974). *Tularemia in North America, 1930–1974*. Missoula, University of Montana Foundation.

Jones CG et al. (1998). Chain reactions linking acorns to gypsy moth outbreaks and Lyme disease risk. *Science*, 279:1023–1026.

Jordan CN et al. (2005). Prevalence of agglutinating antibodies to *Toxoplasma gondii* and *Sarcocystis neurona* in beavers (*Castor Canadensis*) from Massachusetts. *The Journal of Parasitology*, 91:1228–1229.

Joss AW et al. (2003). Lyme disease – what is the cost for Scotland? *Public Health*, 117:264–273.

Junttila J et al. (1999). Prevalence of *Borrelia burgdorferi* in *Ixodes ricinus* ticks in urban recreational areas of Helsinki. *Journal of Clinical Microbiology*, 37:1361–1365.

Karaseva E (1963). [The role of wild mammals in natural foci of leptospirosis in USSR]. *Zoologicheskii Zhurnal*, 42:1699–1713 (in Russian).

Karaseva E (1971). [Ecological features of mammals – carriers of leptospires (*L. grippotyphosa*) and their role in natural foci of leptospirosis]. In: *Fauna and ecology of the rodents*, Series 10. Moscow, Moscow University Press:30–144 (in Russian).

Kartman L (1960). The role of rabbits in sylvatic plague epidemiology, with special attention to human cases in New Mexico and use of the fluorescent antibody technique for detection of *Pasteurella pestis* in field specimens. *Zoonoses Research*, 1:1–27.

Kays RW, Wilson DE (2002). *Mammals of North America*. Princeton, Princeton University Press.

Kerkhoff FT et al. (1999). Demonstration of *Bartonella grahamii* DNA in ocular fluids of a patient with neuroretinitis. *Journal of Clinical Microbiology*, 37:4034–4038.

Kile JC et al. (2005). Transmission of monkeypox among persons exposed to infected prairie dogs in Indiana in 2003. *Archives of Pediatrics & Adolescent Medicine*, 159:1022–1025.

Kimsey SW et al. (1985). Benefit-cost analysis of bubonic plague surveillance and control at two campgrounds in California, USA. *Journal of Medical Entomology*, 22:499–506.

King JA (1968). *Biology of* Peromyscus *(Rodentia)*, Stillwater, OK, American Society of Mammalogists (Special Publication No. 2).

Klein BS et al. (1986). Isolation of *Blastomyces dermatitidis* in soil associated with a large outbreak of blastomycosis in Wisconsin. *The New England Journal of Medicine*, 314:529–534.

Kosoy MY et al. (1996). Prevalence of antibodies to arenaviruses in rodents from the southern and western United States: evidence for an arenavirus associated with the genus *Neotoma*. *The American Journal of Tropical Medicine and Hygiene*, 54:570–576.

Kosoy MY et al. (1997). Distribution, diversity and host specificity of *Bartonella* in rodents from the Southeastern United States. *The American Journal of Tropical Medicine and Hygiene*, 57:578–588.

Kosoy MY et al. (1999). Experimental infection of cotton rats with three naturally occurring *Bartonella* species. *Journal of Wildlife Diseases*, 35:275–284.

Kosoy MY et al. (2003). *Bartonella* strains from ground squirrels are identical to *Bartonella washoensis* isolated from a human patient. *Journal of Clinical Microbiology*, 41:645–650.

Kosoy M et al. (2004a). Prospective studies of *Bartonella* of rodents. Part I. Demographic and temporal patterns in population dynamics. *Vector Borne and Zoonotic Diseases*, 4:285–295.

Kosoy M et al. (2004b). Prospective studies of *Bartonella* of rodents. Part II. Diverse infections in a single rodent community. *Vector Borne and Zoonotic Diseases*, 4:296–305.

Kramer V et al. (1999). Detection of the agents of human ehrlichiosis in ixodid ticks from California. *The American Journal of Tropical Medicine and Hygiene*, 6:62–65.

Krebs CJ, Myers JH (1974). Population cycles in small mammals. *Advances in Ecological Research*, 8:267–399.

Ksiazek TG et al. (1997). Isolation, genetic diversity, and geographic distribution of Bayou virus (Bunyaviridae: hantavirus). *The American Journal of Tropical Medicine and Hygiene*, 57:445–458.

Kuenzi AJ et al. (2001). Antibody to Sin Nombre virus in rodents within peridomestic habitats in west central Montana. *The Amercican Journal of Tropical Medicine and Hygiene*, 64:137–146.

Kurtenbach K et al. (2002). Host association of *Borrelia burgdorferi* sensu lato – the key role of host complement. *Trends in Microbiology*, 10:74–79.

Labuda M (2001). Tahyna virus. In: Service MW, ed. *The encyclopedia of arthropod-transmitted infections of man and domesticated animals*. Wallingford, CABI Publishing:482–483.

Lawson PA et al. (2005). *Streptococcus castoreus* sp. nov., isolated from a beaver (*Castor fiber*). *International Journal of Systematic and Evolutionary Microbiology*, 55:843–846.

Levett PN (2001). Leptospirosis. *Clinical Microbiology Reviews*, 14:296–326.

Levy CE, Gage KL (1999). Plague in the United States 1995–1997, with a brief review of the disease and its prevention. *Infections in Medicine*, 16:54–64.

Lloyd HG (1983). Past and present distribution of red and gray squirrels. *Mammal Review*, 13:69–80.

Long K (2002). *Prairie dogs: a wildlife handbook*. Boulder, CO, Johnson Books.

McDade JE et al. (1980). Evidence of *Rickettsia prowazekii* in the United States. *The*

American Journal of Tropical Medicine and Hygiene, 29:277–283.

McDade JE, Newhouse VF (1986). Natural history of *Rickettsia rickettsii*. *Annual Review of Microbiology*, 40:287–309.

Macdonald D (1984). *The encyclopedia of mammals.* New York, Facts on File Inc.

Macdonald D (2001). *The encyclopedia of mammals.* Oxfordshire, Andromeda Oxford Ltd:578–693.

Macdonald DW, Barrett P (1993). *Mammals of Europe.* Princeton, Princeton University Press.

Maes E, Lecomte P, Ray N (1998). A cost-of-illness study of Lyme disease in the United States. *Clinical Therapeutics*, 20:993–1008.

Magee JS et al. (1989). Tularemia transmitted by a squirrel bite. *The Pediatric Infectious Disease Journal*, 8:123–125.

Mahan CG, O'Connell TJ (2005). Small mammal use of suburban and urban parks in central Pennsylvania. *Northeastern Naturalist*, 12:307–314.

Mailles A et al. (2005). Larger than usual increase in cases of hantavirus infections in Belgium, France and Germany, June 2005. *Euro Surveillance*, 10:198–200.

Marquardt WC, Demaree RS, Grieve RB (2000). *Parasitology and vector biology*, 2nd ed. San Diego, CA, Harcourt/Academic Press.

Maupin GO et al. (1991). Landscape ecology of Lyme disease in a residential area of Westchester County, New York. *American Journal of Epidemiology*, 133:1105–1113.

Maupin GO et al. (1994). Discovery of an enzootic cycle of *Borrelia burgdorferi* in *Neotoma mexicana* and *Ixodes spinipalpis* from northern Colorado, an area where Lyme disease is nonendemic. *The Journal of Infectious Diseases*, 170:636–643.

Meer-Scherrer L et al. (2004). *Babesia microti* infection in Europe. *Current Microbiology*, 48:435–437.

Meltzer MI, Dennis DT, Orloski KA (1999). The cost effectiveness of vaccinating against Lyme disease. *Emerging Infectious Diseases*, 5:321–328.

Meylan A (1975). Fossorial forms of the water vole, *Arvicola terrestris* (L.), in Europe. *European Plant Protection Organization Bulletin*, 7:209–221.

Mills JN (2005). Regulation of rodent-borne viruses in the natural host: implications for human disease. *Archives of Virology, Supplementum*, 19:45–57.

Mills JN et al. (2002). Hantavirus pulmonary syndrome – United States: updated recommendations for risk reduction. *Morbidity and Mortality Weekly Report, Recommendations and Reports*, 51(RR-09):1–12.

Montgomery WI (1976). On the relationship between yellow-necked mouse (*Apodemus flavicollis*) and woodmouse (*A. sylvaticus*) in a Cotswold valley. *Journal of Zoology*, 179:229–233.

Montman CE, Barnes AM, Maupin GO (1986). An integrated approach to bubonic plague control in a southwestern plague focus. In: Salmon TP, ed. *Proceedings of the 12th Vertebrate Pest Conference, 4–6 March 1986, San Diego, California*. Davis, University of California at Davis:97–101.

Monzingo DL Jr, Hibler CP (1987). Prevalence of *Giardia* sp. in a beaver colony and the resulting environmental contamination. *Journal of Wildlife Diseases*, 23:576–585.

Morner T (1992). The ecology of tularaemia. *Revue scientifique et technique*, 11:1123–1130.

Moro MH et al. (1991). The epidemiology of rodent and lagomorph rabies in Maryland, 1981 to 1986. *Journal of Wildlife Diseases*, 27:452–456.

Murie JO, Michener GR (1984). *The biology of ground-dwelling squirrels: annual cycles, behavioral ecology and sociality*. Lincoln, NE, University of Nebraska Press.

Naktin J, Beavis KG (1999). *Yersinia enterocolitica* and *Yersinia pseudotuberculosis*. *Clinics in Laboratory Medicine*, 19:523–536.

Nayar PS, Crawshaw GJ, Neufeld JL (1979). Tularemia in a group of non-human primates. *Journal of the American Veterinary Medical Association*, 175:962–963.

Nechay G (2000). *Status of hamsters: Cricetus cricetus, Cricetus migratorius,* Mesocricetus newtoni *and other hamster species in Europe*. Strasbourg, Council of Europe Publishing (Nature and Environment Series 106).

Nelson BC (1980). Plague studies in California – the roles of various species of sylvatic rodents in plague ecology in California. In: Clark JP, ed. *Proceedings of the 9th Vertebrate Pest Conference, 4–6 March 1980, Fresno, California*. Davis, University of California at Davis:89–96.

Nowak RM (1991). *Walker's mammals of the world*, 5th ed. *Vol. II*. Baltimore, Johns Hopkins University Press.

Nowak RM (1999). *Walker's mammals of the world*, 6th ed. *Vol. II*. Baltimore, Johns Hopkins University Press.

Oaks EC et al. (1987). *Spermophilus variegatus*. *Mammalian Species*, 272:1–8.

Olsson GE et al. (2003). Human hantavirus infections, Sweden. *Emerging Infectious Diseases*. 9:1395–1401.

Olsson GE et al. (2005). Habitat factors associated with bank voles (*Clethrionomys glareolus*) and concomitant hantavirus in northern Sweden. *Vector Borne and Zoonotic Diseases*, 5:315–323.

Olsuf'ev N, Dunaeva T (1960). *[Epizootiology (natural focality) of tularemia]*. Moscow, Meditsina [Medical Press] (in Russian).

Panchola P et al. (1995). *Ixodes dammini* as a vector of human granulocytic ehrlichiosis. *The Journal of Infectious Diseases*, 172:1007–1012.

Parmenter RR et al. (1999). Incidence of plague associated with increased winter–spring precipitation in New Mexico. *The American Journal of Tropical Medicine and Hygiene*, 61:814–821.

Parola P (2004). Tick-borne rickettsial diseases: emerging risks in Europe. *Comparative Immunology, Microbiology and Infectious Diseases*, 27:297–304.

Parola P, Davoust B, Raoult D (2005). Tick- and flea-borne rickettsial emerging zoonoses. *Veterinary Research*, 36:469–492.

Parola P, Paddock CD, Raoult D (2005). Tick-borne rickettsioses around the world: emerging diseases challenging old concepts. *Clinical Microbiology Reviews*, 18:719–756.

Parola P, Raoult D (2001). Tick-borne bacterial diseases emerging in Europe. *Clinical Microbiology and Infection*, 7:80–83.

Perra A et al. (2002). Clustered cases of leptospirosis in Rochefort, France, June 2001. *Euro Surveillance*, 7:131–136.

Perz JF, Le Blancq SM (2001). *Cryptosporidium parvum* infection involving novel genotypes in wildlife from lower New York State. *Applied and Environmental Microbiology*, 67:1154–1162.

Pestryakova T et al. (1966). [About natural focality of toxoplasmosis in Vasyuganie. In: *Nature and economy of Pre-Vasyuganie Region*]. Tomsk, Tomsk University Press:221–223 (in Russian).

Petersen JM, Schriefer ME (2005). Tularemia: emergence/re-emergence. *Veterinary Research*, 36:455–467.

Petersen JM et al. (2004). Laboratory analysis of tularemia in wild-trapped, commercially traded prairie dogs, Texas, 2002. *Emerging Infectious Diseases*, 10:419–425.

Peterson AT et al. (2002). Ecologic niche modeling and potential reservoirs for Chagas disease, Mexico. *Emerging Infectious Diseases*, 8:662–667.

Piesman J (2002). Ecology of *Borrelia burgdorferi* sensu lato in North America. In: Gray J et al., eds. *Lyme borreliosis: biology, epidemiology and control*. Wallingford, CABI Publishing:223–249.

Pizzimenti JJ, Collier GD (1975). *Cynomys parvidens*. *Mammalian Species*, 52:1–3.

Pizzimenti JJ, Hoffmann RS (1973). *Cynomys gunnisoni*. *Mammalian Species*, 25:1–4.

Poljakov IY (1968). *Vrednije grizuni I borjba s nimi [Harmful rodents and their control]*, 2nd ed. Saint Petersburg, Kolos Press (in Russian).

Popov V, Fedorov Y (1958). [Chipmunk as a host of wood tick and a carrier of tick-borne encephalitis virus]. *Proceedings of Tomsk Institute of Vaccine*, 9:19–22 (in Russian).

Pucek Z (1989). A preliminary report on threatened rodents in Europe. In: Lidicker WZ Jr, ed. *Rodents: a world survey of species of conservation concern*. Gland, Switzerland, International Union for the Conservation of Nature/Species Survival Commission:26–32 (Occasional Papers of the International Union for the Conservation of Nature/Species Survival Commission, No. 4).

Randolph SE (2001). The shifting landscape of tick-borne zoonoses: tick-borne encephalitis and Lyme borreliosis in Europe. *Philosophical Transactions of the Royal Society of London, Series B, Biological Sciences*, 356:1045–1056.

Randolph SE et al. (1999). Incidence from coincidence: patterns of tick infestations on rodents facilitate transmission of tick-borne encephalitis virus. *Parasitology*, 118:177–186.

Reich LM (1981). *Microtus pennsylvanicus*. *Mammalian Species*, 159:1–8.

Reperant LA, Deplazes P (2005). Cluster of *Capillaria hepatica* infections in non-commensal rodents from the canton of Geneva, Switzerland. *Parasitology Research*, 96:340–342.

Reynolds JC (1985). Details of the geographic replacement of the red squirrel (*Sciurus vulgaris*) by the grey squirrel (*Sciurus carolinensis*) in eastern England. *Journal of Animal Ecology*, 54:149–162.

Richter D et al. (2004). Relationships of a novel Lyme disease spirochete, *Borrelia spielmani* sp. nov., with its hosts in Central Europe. *Applied and Environmental Microbiology*, 70:6414–6419.

Roher DP et al. (1981). Acute fatal toxoplasmosis in squirrels. *Journal of the American Veterinary Medical Association*, 179:1099–1101.

Rollin P et al. (1995). Isolation of Black Creek Canal virus, a new hantavirus from *Sigmodon hispidus* in Florida. *Journal of Medical Virology*, 46:35–39.

Rose AM et al. (2003). Patterns of Puumala virus infection in Finland. *Euro Surveillance*, 8:9–13

Rosicky B, Sebek Z (1974). To the evolution of natural foci of *Leptospira grippotyphosa* in Central Europe. *Folia Parasitologica*, 21:11–20.

Rothova A et al. (1998). *Bartonella* serology for patients with intraocular inflammatory disease. *Retina*, 18:348–355.

Sauvage F et al. (2002). *Puumala* hantavirus infection in humans and in the reservoir host, Ardennes region, France. *Emerging Infectious Diseases*, 8:1509–1511.

Schmaljohn C, Hjelle B (1997). Hantaviruses: a global disease problem. *Emerging Infectious Diseases*, 3:95–104.

Schneider F, Mossong J (2005). Increased hantavirus infections in Luxembourg, August 2005. *Euro Surveillance*, 10:204/E050825.1

Schwan TG et al. (2003). Tick-borne relapsing fever caused by *Borrelia hermsii*, Montana. *Emerging Infectious Diseases*, 9:1151–1154.

Settergren B (2000). Clinical aspects of nephropathia epidemica (Puumala virus infection) in Europe: a review. *Scandinavian Journal of Infectious Diseases*, 32:125–132.

Sexton DJ (2001). Rocky Mountain spotted fever. In: Service MW, ed. *The encyclopedia of arthropod-transmitted infections of man and domesticated animals*. Wallingford, CABI Publishing:437–442.

Song JW et al. (1994). Isolation of pathogenic hantavirus from white-footed mouse (*Peromyscus leucopus*). *Lancet*, 344:1637.

Soveri T et al. (2000). Disease patterns in field and bank vole populations during a cyclic decline in central Finland. *Comparative Immunology, Microbiology and Infectious Diseases*, 23:73–89.

Steere AC (1994). Lyme disease: a growing threat to urban populations. *Proceedings of the National Academy of Sciences of the United States of America*, 91:2378–2383.

Stevenson HL et al. (2003). Detection of novel *Bartonella* strains and *Yersinia pestis* in prairie dogs and their fleas (Siphonaptera: Ceratophyllidae and Pulicidae) using multiplex polymerase chain reaction. *Journal of Medical Entomology*, 40:329–337.

Strle F (2004). Human granulocytic ehrlichiosis in Europe. *International Journal of Medical Microbiology*, 293(Suppl. 37):27–35.

Telford SR 3rd, Goethert HK (2004). Emerging tick-borne infections: rediscovered and better characterized, or truly "new"? *Parasitology*, 129 (Suppl.):S301–S327.

Thomas K et al. (2003). A novel poxvirus lethal to red squirrels (*Sciurus vulgaris*). *The Journal of General Virology*, 84:3337–3341.

Thompson RS et al. (1969). Outbreak of tick-borne relapsing fever in Spokane County, Washington. *The Journal of the American Medical Association*, 210:1045–1050.

Tikhomirov E (1999). Epidemiology and distribution of plague. In: *Plague manual.* Geneva, World Health Organization:11–42 (document number: WHO/CDS/CSR/EDC/99.2; http://whqlibdoc.who.int/hq/1999/WHO_CDS_CSR_EDC_99.2.pdf, accessed 13 September 2006).

Tokarevich N et al. (2002). Wild small mammals and domestic dogs infected with zoonotic agents in Saint-Petersburg and its suburbs. *EpiNorth*, 3:12–15.

Trevejo RT et al. (1998). An interstate outbreak of tick-borne relapsing fever among vacationers at a Rocky Mountain cabin. *The American Journal of Tropical Medicine and Hygiene*, 58:743–747.

Trilar T, Radulovic S, Walker DH (1994). Identification of a natural cycle involving *Rickettsia typhi* infection of *Monopsyllus sciruorum sciurorum* fleas from the nests of the fat dormouse (*Glis glis*). *European Journal of Epidemiology*, 10:757–762.

Ulrich R et al. (2004). Verbreitung von Hantavirusinfektionen in Deutschland. *Bundesgesundheitsblatt, Gesundheitsforschung, Gesundheitsschutz*, 47:661–670.

Vapalahti O et al. (2003). Hantavirus infections in Europe. *The Lancet Infectious Diseases*, 3:653–661.

van den Brink FH (1968). *A field guide to the mammals of Britain and Europe*. Boston, Houghton Mifflin Company.

Veron G (1992). Histoire biographique du castor d'Europe, *Castor fiber* (Rodentia, Mammalia). *Mammalia*, 56:87–108.

Viel JF et al. (1999). Water vole (*Arvicola terrestris* Sherman) density as risk factor for human alveolar echinococcosis. *The American Journal of Tropical Medicine and Hygiene*, 61:559–565.

von Reyn CF et al. (1976). Bubonic plague from exposure to a rabbit: a documented case, and a review of rabbit-associated plague cases in the United States. *American Journal of Epidemiology*, 104:81–87.

Wallis PM et al. (1984). Reservoirs of *Giardia* spp. in southwestern Alberta. *Journal of Wildlife Diseases*, 20:279–283.

Weber K (2001). Aspects of Lyme borreliosis in Europe. *European Journal of Clinical Microbiology & Infectious Diseases*, 20:6–13.

Welch DF et al. (1999). Isolation of a new subspecies, *Bartonella vinsonii* subsp. *arupensis*, from a cattle rancher: identity with isolates found in conjunction with *Borrelia burgdorferi* and *Babesia microti* among naturally infected mice. *Journal of Clinical Microbiology*, 37:2598–2601.

Williams JC, Sanchez V (1994). Q fever and coxiellosis. In: Beran GW, Steele JH, eds. *Handbook of zoonoses. Section A: Bacterial, rickettsial, chlamydial, and mycotic*, 2nd ed., Boca Raton, CRC Press:429–446.

Zaikovskaia AV et al. (2005). [Rabies epizootic situation in the Novosibirsk region in 1997–2003]. *Zurnal Mikrobiologii, Epidemiologii i Immunobiologii*, (4):37–40 (in Russian).

Zeier M et al. (2005). New ecological aspects of hantavirus infection: a change of a paradigm and a challenge of prevention – a review. *Virus Genes*, 30:157–180.

Zeitz PS et al. (1995). A case-control study of hantavirus pulmonary syndrome during an outbreak in the southwestern United States. *The Journal of Infectious Diseases*, 171:864–870.

Zeveloff SI (1988). *Mammals of the Intermountain West*. Salt Lake City, University of Utah Press.

14. Pesticides: risks and hazards

Marco Maroni[1], Kevin J. Sweeney[2], Francesca Metruccio,
Angelo Moretto and Anna Clara Fanetti

Summary

This chapter examines human health risks from indoor pesticide exposure in residential settings. The primary focus is on risks to residential bystanders – not pesticide applicators – in home settings. Pesticides most frequently used for urban-pest management and pet treatments are discussed and evaluated in terms of hazard and exposure.

Due to the intrinsic toxicity of pesticides, their admission to the market and use is regulated. In Europe and North America, the European Chemicals Bureau, the Canadian Pest Management Regulatory Agency and the EPA are the principle pesticide regulatory agencies, with a legal regulatory framework also existing in most individual states and provinces. This framework and associated regulatory processes ensure a thorough review of pesticide effects, exposure and use patterns, to fully characterize risks to workers and the general public, with special consideration given to children, pregnant women and other sensitive subpopulations. In many regulatory agencies, the precautionary principle is applied where uncertainty exists and alternatives to more toxic pesticides are given priority in regulatory reviews and registration.

The process for assessing the risk to human health of pesticides, which objectively considers uncertainties and assumptions, entails four-steps: hazard identification, dose–response assessment, exposure assessment and risk characterization. In the section on "Toxicology", two main areas are covered: pesticide hazard identification/characterization and adverse effects of six groups of pesticides (organophosphates, pyrethroids, piperonyl butoxide, anticoagulant rodenticides, IGRs and neonicotinoid insecticides). In the sections that discuss exposure, the routes and magnitudes of pesticide exposure are considered and evaluated. In the home, the dermal and inhalation routes are the most common routes of exposure, with unintentional (incidental) oral exposure attributable primarily to toddlers putting their fingers in their mouth after crawling over treated surfaces or touching pets.

Based on these assessments and the weight of the scientific evidence, indoor applications of pesticides, which are regulated by a complex risk assessment before and after they are put on the market, do not pose a high level of risk to human health if the application of the product and the management of the application take place according to proper and adequate procedures. However, assessments of residential pesticides, such as chlorpyrifos, show that some pesticides are unsafe. These assessments, together with recent efforts to produce pesticides with a lower overall toxicity, are able to reasonably assure the absence of any unacceptable risk to human health and the environment.

[1] Deceased.

[2] Disclaimer: The findings and conclusions in this article are those of the author and do not necessarily reflect the views of the United States Environmental Protection Agency.

14.1. Introduction

Most people are exposed to pesticide[3] residues in four ways (Fig. 14.1), through:

1. the use of a pesticide (including mixing end-use solutions and applying them);
2. contact that results from such behaviour as touching pesticide-treated surfaces, inhaling vapours present in the air or putting contaminated hands or objects in one's mouth;
3. eating food that contains them; and
4. drinking water that contains them.

Pesticide production, sale, distribution and use are regulated to protect public health and the environment. The public relies on regulatory agencies assigned to enforce the laws that manage pesticide risks. The agencies do so through regulatory decisions founded on scientifically sound and evidence-based pesticide hazard and risk assessments. This reg-

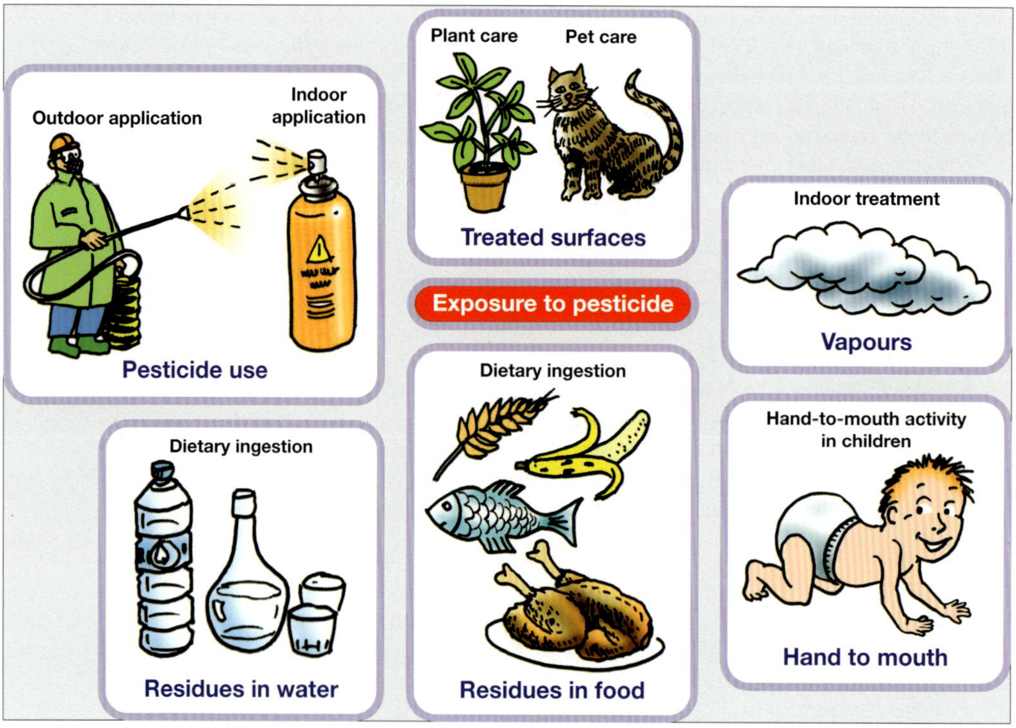

Fig. 14.1. Pesticide exposure pathways

[3] In this chapter, the terms "pesticide" and "biocide" have the same meaning and are used interchangeably, unless specified otherwise.

ulatory process and its associated framework must have built-in flexibility to evolve as our knowledge of the sciences of physiology, toxicology and risk assessment grows. Even though many effects on human health and the environment that result from pesticide exposure are well known, research continues to refine our knowledge of these effects. Today, much of what is known about risk assessment comes from studies conducted in occupational settings. Also, non-dietary residential exposure assessment has increased considerably in the past 10 years. Pesticide regulatory agencies are currently refining their assessments, to improve overall quality and achieve more realistic exposure estimates. Studies in residential and housing environments are relatively few, and recent efforts by public agencies have met with public opposition and scrutiny.

According to a recent survey, 75% of households in the United States used at least one pesticide product indoors during the past year. Products used most often are insecticides and disinfectants (EPA, 2001a, 2006a). Another study suggested that, for most people, 80% of exposure to pesticides occurs indoors and that measurable levels of up to a dozen pesticides have been found in the air inside homes (EPA, 2006a).

Yet the amount of pesticides found in homes appears to be greater than can be explained by recent pesticide use in those households. It should be noted that chemicals with biocidal activity are present in other categories of consumer products for personal or house environmental use, such as disinfectants, detergents, cleaners, wood preservatives and soaps. Although these biocidal products may contain some of the active ingredients present in the pesticide products used in agriculture for plant protection, they are often regulated in a different manner in some parts of the world. Other possible sources include contaminated soil or dust that floats or is tracked in from outside, stored pesticide containers, and household surfaces that collect and then release the pesticides. Pesticides used in and around the home include products to control insects (insecticides), termites (termiticides), rodents (rodenticides), fungi (fungicides) and microbes (disinfectants). Products that contain only a small percentage of the active ingredient (typically 0.5–1.0% wt/wt are sold directly to the public for nuisance pest control. They are sold as sprays, gels, liquids, sticks, dusts, crystals, baits and foggers. Consumers who use these pesticide products generally have no formal training in mixing and application techniques, but are provided label instructions for proper handling. Despite this condition, concentrated pesticide products are still sold in the United States, although ready-to-use products that can help to decrease hazard, user exposure and environmental contamination are becoming more popular.

Currently, there are no estimates and evaluations of pesticide risk from cumulative residential exposure to pesticides used in urban pest control applications. The risk and exposure estimates available are specific to certain chemicals and are often based on one or two use patterns – with most estimates performed recently. These risk evaluations are now required for reregistration of all chemicals with residential uses in the United States, and the same requirements are to be met under the Biocidal Products Directive in EU countries. Also, no cost estimates are available for the public health and environmental effects of pesticide use. Epidemiological data of this type are not collected routinely, but acute pesticide poisoning data are available from a variety of sources.

Consumer exposure to hazardous chemicals is a public concern fuelled by possible adverse effects. Most hazardous consumer pesticide exposures occur in residential settings, most likely at the time of their application or shortly following their application. Exposure occurs most commonly via dermal and inhalation routes, with incidental oral exposure attributable primarily to toddlers putting their fingers in their mouth after crawling over treated surfaces or touching pets.

The discussion in this chapter focuses on the pesticide risk (the relationship between hazard and exposure) in the indoor residential environment, where people (especially children) spend most of their time. The outcomes of the published pesticide risk assessments for chlorpyrifos, pyrethrins and permethrin are used as examples. Also, fundamental similarities and differences in regulatory philosophies between Europe and North America are discussed, especially as they relate to risk assessment outcomes. WHO assessments and advisory roles are also mentioned. The chapter, however, does not address issues of vector-borne disease or the topic of personal protection from vectors through the use of insect repellents. The WHO Pesticide Evaluation Scheme (WHOPES) has published many references on vector control, vector-borne diseases and insect-repellent evaluations (WHOPES, 2007). A comprehensive review of insect repellents is also available from Debboun, Frances & Strickman (2007), and the EPA performed a review of DEET-based insect repellents (EPA, 1998a).

The process for assessing the risk to human health of pesticides, as outlined in this chapter, entails four-steps:

1. hazard identification
2. dose–response assessment
3. exposure assessment
4. risk characterization.

Steps 1 and 2 comprise the toxicology evaluations for the pesticide and are discussed together.

14.2. Pesticide regulation in Europe and North America

14.2.1. Pesticide or biocide product regulations

This chapter refers to European and North American pesticide or biocide regulations. In particular, it refers to the United States Federal Insecticide, Fungicide, and Rodenticide Act of 1972 (FIFRA), administered by the EPA; to the Canadian Pest Control Products Act of 2006 (PCPA), administered by the Pest Management Regulatory Agency (PMRA); and to EU directives 98/8/EC (Biocidal Products Directive) and 91/414/EC (on placing plant protection products on the market). The responsible authority for the EU Directive 98/8/EC is the European Commission Environment Directorate-General (DG Environment), and technical support is provided by the European Chemicals Bureau (ECB). For EU Directive 91/414/EC, the European Commission Health and Consumer

Protection Directorate-General (DG Sanco) is the responsible authority. The first three of these four regulations pertain to the WHO urban pests and health project and are discussed separately in the sections below. The evaluation process of each focuses on whether or not the health and environmental risks posed by a pesticide or biocide product (when used as directed) are likely to be acceptable and whether the product offers a worthwhile contribution to pest management. The acceptability of the risk is a key element, regardless of whether the product is chemical, biological or biotechnological. The last directive, on plant protection products, is applicable to non-residential pesticides. More details on the legal and regulatory aspects of pesticide regulation appear in following subsections. A global guide to resources for Forum IV recommendations, published by the WHO Intergovernmental Forum on Chemical Safety (IFCS, 2006), provides information on pesticide regulation.

14.2.1.1. Health Canada and the PCPA

The PMRA of Health Canada administers the PCPA and manages the regulation of pesticides at the federal level in Canada. It also works closely with a committee of the Canadian provinces to consider provincial concerns and issues. A major component of this federal level regulatory system is the pre-market evaluation of new pesticides for potential health risks, environmental risks and value of products proposed for use in Canada. To fulfil this role, the PMRA requires that pesticide manufacturers provide extensive information on which to base these evaluations.

14.2.1.2. EPA and FIFRA

In the United States, the EPA has the national pesticide regulatory authority, as defined by FIFRA (FIFRA, 1972; EPA, 1996). The EPA is responsible for assessing and managing pesticide exposure and risk, for reviewing all data submitted in support of registration and reregistration actions and for performing exposure and risk assessments. Within the confines of FIFRA, the Agency develops new approaches to hazard and risk assessment, using input from experts and consideration of public comment (EPA, 1997b, 1999, 2002a).

14.2.1.3. EU regulatory framework for pesticides

EU Directive 98/8/EC (EC, 1998) provides a regulatory scheme for the registration of active ingredients in Member States. In general, under this Directive, each Member State can register pesticides, but the goal is to divide the work of registration review among the members capable of performing this work. Each country has a lead agency for pesticides, but the work is often divided among a number of federal agencies within a country, as is the case in Germany. Each country also reserves the right to be more restrictive than another Member State, though justification is required. The same applies to provinces or states within the country where such regulatory authority exists.

On 18 December 2006, the Council of Ministers adopted a new EU regulatory framework for Registration, Evaluation, Authorisation and Restriction of Chemicals, (REACH) – Directive 2006/121/EC. The law went into force on 1 June 2007. Implementation of REACH is the responsibility of the new European Chemicals Agency, located in Helsinki, Finland. The European Chemicals Agency will act as the hub in the

REACH system: it will run the databases necessary to operate the system, coordinate the in-depth evaluation of suspicious chemicals and run a public database that will provide consumers and professionals with information on chemical hazards. REACH aims to improve the protection of human health and the environment by improving the identification and understanding of chemical properties. Also, chemical manufacturers will be responsible for demonstrating the safety of chemicals produced and distributed and, in doing so, for substituting less hazardous chemicals for dangerous ones when suitable alternatives are identified. REACH affects the risk assessment and regulatory status of all chemicals.

14.2.1.4. EU community-level authorization of technical grade active ingredients

Under EU regulations, *pesticides* are not defined as described by the North American PCPA or FIFRA. Instead, they are divided into two major regulatory categories: plant protection products and biocidal products. Biocidal products, as defined by Directive 98/8/EC, provide protection against public health pests and include insecticides, repellents, attractants, rodenticides, wood preservatives, veterinary hygiene and pet protection products, food and feed disinfectants and most antimicrobial products (whether for public heath use or not), film preservatives, and molluscicides. Plant protection products, as defined under Directive 91/414/EEC, include insecticides, fungicides, herbicides and growth regulators. Under Directive 2001/18/EC, on the release into the environment of genetically modified organisms, crops can be insect or herbicide resistant. Pharmaceuticals and cosmetics are regulated under the authority of other directives not related to pesticides.

Directive 98/8/EC enables harmonization of legislation among EU Member States. The scope of the Directive is very broad, covering 23 different product types. Risk assessment of biocidal products is required before these can be placed on the European market. The EU has designated the ECB, together with experts from EU Member States, as the parties responsible for assessing whether or not the biocide data submitted fulfil the requirements of the Directive. These decisions are based on outcomes from technical meetings with all EU Member States. The Technical Guidance Document *Human exposure to biocidal products – guidance on exposure estimation* (EC, 2002) is used as a basis for conducting the risk assessment. The risk assessment methods for biocides and plant protection products are in accordance with the corresponding national legislation and (as much as possible) with the corresponding EU legislation.

Essentially, a new biocide active ingredient (AI) has to meet data requirements similar to those in the United States and Canada, and most test guidelines have been harmonized. Risk assessment and evaluation are similar, but the basis for accepting or rejecting a dossier can differ between EU and North American regulatory agencies. EU Member States may require additional data from that required for Annex I listing – that is, Annex I to the Directive is a list of active substances that have been successfully evaluated at the EC level and are considered to meet the requirements of Directive 91/414. The EU encourages work sharing in reviewing submissions and mutual recognition of approval.

14.2.1.5. Pesticide or biocide registration at various regulatory levels

In the United States, each of the 50 states registers pesticides at the state level. California and New York have their own pesticide evaluation and registration programmes that in some instances result in re-evaluation of data reviewed by the EPA. With regard to pesticide use in Europe, the United States and Canada, states or provinces can be more restrictive than the federal or EU registering authority.

14.2.1.6. Pesticide enforcement, applicator training and licensing activities

In North America, enforcement of pesticide regulations is generally delegated to state and local jurisdictions, even if the function is a federal mandate. The same is true of pesticide applicator training and licensing. The EU does not enforce pesticide regulations at the EU level, but instead enforcement takes place at the national level.

14.2.1.7. WHO and the Organisation for Economic Co-operation and Development

The WHO International Programme on Chemical Safety (IPCS), established in 1980, implements international activities related to chemical safety. WHO is the Executing Agency of the IPCS, whose main roles are to establish the scientific basis for safe use of chemicals and to strengthen national capabilities and capacities for chemical safety (Licari, Nemer & Tamburlini, 2005; WHO IPCS, 2005). Many developing countries are poorly equipped to respond to existing and emerging chemical safety issues. Strengthening the capacity of countries to soundly manage the chemicals they use is a theme that underpins most IPCS activities.

In the Organisation for Economic Co-operation and Development (OECD), the pesticide programme is one of 12 subprogrammes in the Environment, Health and Safety Programme. The goals of the programme are to help OECD countries share the work of developing pesticide risk assessments and to find new approaches to reducing the risk of pesticides. IPCS also works in the areas of international harmonization of pesticide risk and hazard assessments, not to mention labelling and classification.

The IPCS and OECD have developed a framework for cooperation in the field of risk assessment methods, which ensures mutual support and involvement in the projects conducted by each organization. IPCS is not a regulatory authority, but rather an advisory organization that provides expertise in the area of chemical safety.

WHOPES was set up in 1960. It promotes and coordinates the testing and evaluation of pesticides for public health use. Its objectives are: to facilitate the search for alternative pesticides and application methods that are safe and cost effective; to develop and promote policies, strategies and guidelines for the selective and judicious application of pesticides for public health use; and to assist and monitor their implementation by Member States.

14.2.2. All-embracing principles and approaches to pesticide regulation

Pesticide regulation involves a number of principles, doctrines and approaches. In particular, it involves the precautionary principle, the substitution doctrine, special consideration of children's health, and exposure assessments based on use patterns.

14.2.2.1. The precautionary principle

The *precautionary principle* is the all-embracing principle and so-called force behind the regulation of chemicals. The principle is the foundation of regulating chemicals in Europe, and although federal level regulatory authorities in the United States and Canada have never adopted this principle as a formal creed per se, it is gaining popularity with some local authorities in the United States. Expressed in another way, when scientific certainty is lacking, precaution should be applied to the breadth of regulatory concerns.

All regulatory agencies and authorities that register pesticides conduct thorough and extensive health and environmental reviews of pesticides before allowing their initial or continued use in the environment. Perhaps the best understood reason for applying precaution to pesticide regulation originates in the general abstinence from conducting or permitting tests with pesticides on human subjects. Therefore, toxicology data are *bridged* from animal testing to assess the risk of adverse effects on people. Also, computer models are used to refine exposure and risk assessment. Moreover, the special consideration now given to children's health in the pesticide risk assessment process enhances precautionary measures and is at the forefront of regulatory considerations.

14.2.2.2. The substitution doctrine

Simply defined, the *substitution doctrine* advocates, and in some regulatory scenarios requires, the substitution of a less hazardous pesticide for a more hazardous pesticide when an alternative is available. Supporters of this doctrine see its application to pesticide regulation as an opportunity to encourage research and development of less hazardous pesticides. Critics believe that application of this doctrine will lead to production, registration and use of less efficacious pesticides, especially when pest control is needed for public health and that this condition conflicts with *free market* forces, thus failing to meet consumer demands. Critics have also implied that such a condition places the public at risk.

14.2.2.3. Special considerations given to children's health

All major pesticide regulatory agencies in Europe and North America give special consideration to protecting children's health. Aside from the high political and social value given to protecting future generations, epidemiological evidence shows a relationship between certain children's ailments and chemicals. Disorders that result from exposure to heavy metals are well-known examples. Also, children have more life ahead of them than do adults, and this may be important for those health risks that result from the accumulation of doses or effects over a long period of time. Risk assessors in the United States, as required by law, presume that children are more susceptible to adverse effects from pesticides. Therefore, an additional margin of safety (usually tenfold) is applied to protect infants and children from risks posed by pesticide residues in food and to protect them when pesticides are used in and around homes and schools. In the EU, a special directive (known as the Baby Food Directive) has set at 0.001 mg/kg (taken as an enforceable *zero level* in practice) the maximum level of any pesticide residue admitted on food destined for infants. A global guide to resources on chemical safety and children's health has been published by IFCS (2005).

14.2.2.4. Basing risk and exposure assessments on use patterns

Pesticide data sets and requirements, including risk and exposure assessments, are based on the pattern of use, and the safety database of the technical grade AI manufacturer is large. Most regulatory agencies also refer to these data as *generic* or *core* safety data. Application formatting requirements for pesticides are detailed, but are not harmonized between different regulatory agencies, although efforts continue to harmonize them. Also, testing protocols must be approved and testing conducted in facilities that comply with Good Laboratory Practices – a system of management controls for laboratories and research organizations to ensure the consistency and reliability of results.

14.3. Toxicology[4]

This section covers two major areas: pesticide hazard identification and six groups of pesticides. In the WHO urban pests and health project, the primary focus is on exposure of residential bystanders and residential areas, not on professional pesticide applicators or food handling establishments. Hazards related only to the pesticides used most frequently for urban-pest management and pet treatments are considered.

14.3.1. Pesticide hazard identification

Pesticides are unique among chemical products, since a lot of toxicological information is available before marketing. Available data are mainly from tests on animals and this poses some problems in extrapolating and applying these data to people. These tests may not always be the most sensitive indicators of human response, but test data on people are limited, given that pesticides are toxicants. Bridging toxicology test results from one species to another is not foolproof. However, procedures have been devised to take this into account when assessing the risk to people of pesticide exposures. This section deals mostly with existing general toxicological information on pesticide poisoning and known toxic effects. Emphasis is on data that relate to people; when such data are absent or minimal, a summary of animal toxicological data is provided.

14.3.1.1. Acute toxicity

WHO classifies pesticides according to the acute risk to health – that is, the risk of single or multiple exposures over a relatively short period of time – that might be encountered accidentally by any person handling the product, in accordance with the manufacturer's directions for handling or in accordance with the rules laid down for storage and transportation by competent international bodies. The classification distinguishes between the more and the less hazardous forms of each pesticide, in that it is based on the toxicity of the technical compound and on its formulations. In particular, allowance is made for the lesser hazards presented by solids, compared with liquids. The classification

[4] Data on acceptable daily intake and acute reference dose reported below have been established by the FAO/WHO Joint Meeting on Pesticide Residues (JMPR) and can be found in the *Inventory of IPCS and other WHO pesticide evaluations and summary of toxicological evaluations performed by the Joint Meeting on Pesticide Residues* (JMPR) (WHO IPCS, 2006b).

Table 14.1. WHO classification of pesticides according to hazard

Class	Rat LD_{50} (mg/kg BW)			
	Oral		Dermal	
	Solids	Liquids	Solids	Liquids
1a. Extremely hazardous	≤5	≤20	≤10	≤40
1b. Highly hazardous	5–50	20–200	10–100	40–400
II. Moderately hazardous	50–500	200–2000	100–1000	400–4000
III. Slightly hazardous	>500	>2000	>1000	>4000

Note. The terms solids and liquids refer to the physical state of products and formulations.

Source: WHO IPCS (2006a).

is based primarily on acute oral and dermal toxicity (lethal dose, 50% [LD_{50}]) to rats used in experiments, since these determinations are standard procedures in toxicology. The LD_{50} value is a statistical estimate of the number of milligrams of toxicant per kilogram of body weight required to kill 50% of test animals. Provision is made for adjusting the classification of a particular compound if, for any reason, the acute hazard to people differs from that indicated by LD_{50} assessments alone. Among other considerations, classification adjustments are made for the following reasons.

- If it is shown that the rat is not the most suitable test animal, then data from species other than the rat are used.

- If the AI produces irreversible damage to vital organs, is highly volatile, is markedly cumulative in its effect, or is found after direct observations to be particularly hazardous or significantly allergenic to human beings, then it can be classified in a higher hazard category.

Table 14.1 shows the criteria used to classify solid and liquid pesticides, according to hazard from oral or dermal exposure. It also shows the criteria for a group or class made up of compounds unlikely to present an acute hazard in normal use. The updated WHO classification of pesticides by hazard can be found at a dedicated web site (WHO IPCS, 2006a).

Presently, WHO has classified 28 extremely hazardous compounds (Ia), 56 highly hazardous compounds (Ib), 117 moderately hazardous compounds (II), 119 slightly hazardous compounds (III) and 248 compounds unlikely to

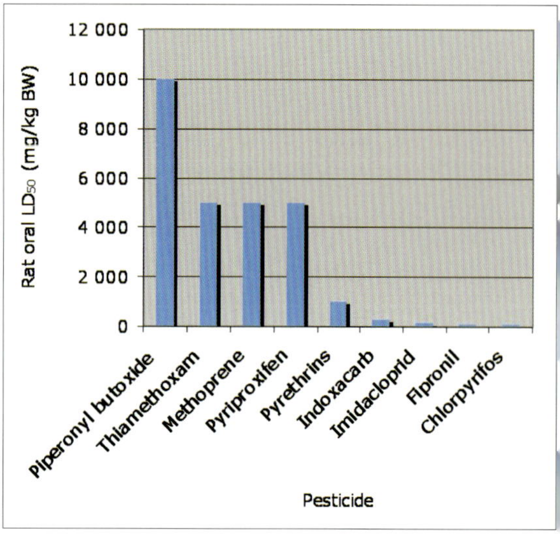

Fig. 14.2. Acute toxicity of commonly used pesticides

Source: WHO IPCS (2002), EPA (2000a, b).

present an acute hazard in normal use (no category). Most of the recently registered compounds are classified as presenting slight or no acute hazard. For our discussion on pesticide hazards in residential settings, Fig. 14.2 shows the oral LD_{50} values (for rats) for some of the most commonly used household biocides.

14.3.1.1.1. Epidemiology of acute pesticide poisoning

Our knowledge of the effects of pesticides on people generally comes from reports of acute poisoning that occur worldwide. Acute pesticide poisoning can result from intentional, occupational or accidental exposure to pesticides, but worldwide figures of pesticide poisonings are not available. WHO & the United Nations Environment Programme (WHO, 1990) estimated an annual incidence of unintentional acute poisoning of about 1 million people, with an overall mortality rate of about 1% (of which only 1% is in middle- and high-income countries). The majority of unintentional pesticide poisonings are occupational, although cases occur in the general population due to improper use or storage of pesticides intended for amateur uses or in-house pest control. The most hazardous compounds cause the majority of these poisonings. Population-based studies in 17 countries gave annual incidence rates of unintentional pesticide poisoning of 0.3–18.0 cases per 100 000 population (Jeyaratnam, 1990).

Pesticides are estimated to be responsible for less than 4% of deaths from all types of accidental poisoning, based on reports from poison control centres: the apparent increase in the number of cases in recent years may reflect increased use, but it may also reflect the availability of better statistics. The estimated annual incidence of intentional acute poisoning is about 2 million people, with a 5.7% mortality rate, which appears to be higher in low-income countries (up to 23%) (WHO, 1990). Suicide attempts (usually, but not exclusively, with organophosphorous compounds) represent 44–91% and 26–60% of acute pesticide poisonings in South-East Asia and in Central America, respectively, whereas in California all non-occupational pesticide poisonings represent about 5% of the total (Jeyaratnam, 1990). An unknown fraction of these attempts involves pesticides intended for uses other than agriculture. WHO (Peden, McGee & Krug, 2002) reports accidental pesticide poisonings as the ninth leading cause of death.

14.3.1.2. Toxicity end-points

Standardized sets of toxicological tests are used to screen a pesticide for toxicity, and the results are used to construct the pesticide's toxicology profile. Ultimately, this results in the identification of the dose–response relationship for toxic end-points of concern to human health. A dose level that causes no observed adverse effect (NOAEL) is generally used to identify safe levels of intake (reference dose), which are then compared with the expected or measured exposure. Generally, the reference dose (acute, short-term, or long-term) is derived from the NOAEL for the most relevant effect, by applying to its value, a factor (called safety or uncertainty factor) to take into account such issues as species extrapolation, individual variability and quality of the toxicological database. The NOAEL values selected for the reference dose vary according to the exposure scenario that needs to be assessed.

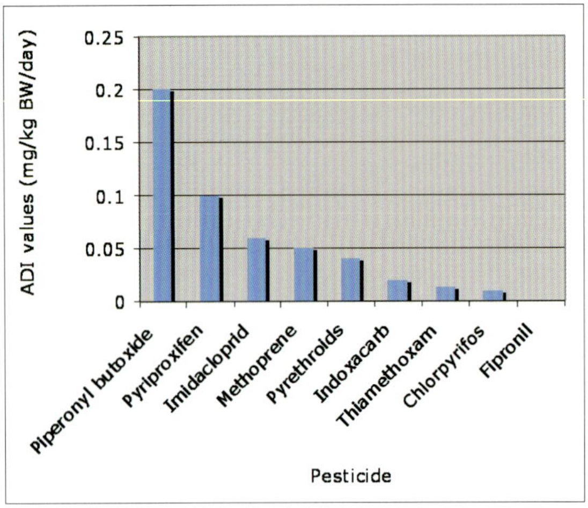

Fig. 14.3. ADI values for commonly used pesticides

Source: WHO IPCS (2002), EPA (2000a, b).

Therefore, there might be toxicity end-points that are specific to a specific route and duration of exposure. For instance, a dietary risk evaluation that is usually defined as the acceptable daily intake (ADI) will (usually) use an end-point from a long-term oral feeding study, while a short-term dermal exposure risk evaluation might use an acute dermal toxicity study.

Toxicity end-points discussed in this chapter were chosen from chemical-specific FAO/WHO Joint Meeting on Pesticide Residues (JMPR) assessments (WHO IPCS, 2002) and EPA regulatory end-points (EPA, 2000a, b). Fig. 14.3 shows the ADI values for some commonly used pesticides.

14.3.1.3. Long-term effects

Long-term exposures occur to farmers and professional pesticide users. To a much lesser extent, they also occur to the general population via residues in food and water and by environmental exposure to pesticides from indoor and outdoor use for pest control. Identifying subjects who have been occupationally or non-occupationally chronically exposed to pesticides is relatively easy, but toxicological evidence of exposure is seldom available. Moreover, extrapolation from current data to assess past exposures as well as the risk associated with a given pesticide is difficult, since AIs and application practices differ and change with time. This is particularly true in the general population where data on exposure and biological monitoring are scanty or non-existent. Attention has been focused on the carcinogenicity, allergenicity and teratogenicity of pesticides, and most recently on their effect on endocrine disruption and neurological development. However, the long amount of time it takes for these effects to develop and show clinically detectable signs hampers their identification in population studies. Evaluations of the carcinogenic potential of relatively few pesticides have been performed by the International Agency for Research on Cancer (IARC), even though carcinogenicity studies in animals are available for all pesticides. However, most of these studies have not been published in the open literature. Several criteria are used by the IARC for choosing the compounds to be evaluated and include: evidence of human exposure and some experimental evidence of carcinogenicity or some evidence (or suspicion) of a risk to people, or both.

The assessment of a pesticide's potential to cause cancer requires a different kind of assessment and expression of risk. Assessing the risk of cancer from exposure to pesticides is based on evidence from cancer studies in at least two species, usually the rat and the mouse, together with evidence from in vitro and in vivo genotoxicity studies. Dose levels in these studies are much higher than expected for human exposures. Studies that shed light on the mechanism by which the pesticide causes the carcinogenic effect often accompany these studies. Based on study outcomes, a weight-of-evidence approach is used to decide if a pesticide is likely to pose a cancer risk to people (PMRA, 2000).

The results obtained so far by the IARC include more than 60 pesticide AIs, most of which are no longer in common use. The AIs used also for urban pest control include: the pyrethroids deltamethrin, fenvalerate and permethrin; the organophosphates dichlorvos, malathion, methyl parathion, parathion, tetrachlorvinphos and trichlorfon; and the synergist piperonyl butoxide. These compounds were included in IARC Carcinogen Classifications[5] Group 3 (agents not classifiable as to their carcinogenicity to people), except dichlorvos, which was classified as Group 2B (an agent possibly carcinogenic to people). Among other things, dichlorvos has been criticized on the basis of biochemical and toxicological considerations (FAO/WHO, 1994). More recently, the Scientific Panel on Plant Health, Plant Protection Products and their Residues (PPR Panel) of the European Food Safety Authority considered the carcinogenic potential of dichlorvos in animals to be not relevant to human beings, on the basis of mechanistic considerations (PPR Panel, 2006). Also, among the compounds used in urban pest control, dichlorvos and pyrethrum have been reported to cause allergic contact dermatitis (Moretto, 2002).

Children (especially in the first 6–12 months after birth) are considered by some at higher risk of toxic effects from pesticide exposure, as their metabolic processes are immature and they are less able to detoxify chemicals. In some instances, however, metabolic immaturity may be beneficial, because the metabolic pathways that activate their toxic metabolites are not yet developed. Infants and children are growing and developing, and their delicate developmental process can be disrupted. However, the data available suggests that this possible increased susceptibility is evident at high doses, whereas young animals do not appear to be more susceptible to low doses that cause no toxic effects in adults (see the discussion in subsection 14.3.2.2 on pyrethroids).

Exposure to pesticides during pregnancy can have potentially adverse effects on fœtal growth and child neurodevelopment (Landrigan et al., 1999). However, when specifically designed studies of developmental neurotoxicity were available for approved pesticides, these effects were not detected at levels lower than those observed in the usual studies of developmental toxicity, multigeneration reproductive toxicity, and acute and short-term neurotoxicity. In fact, when studies with 14 pesticides evaluated by the EPA were reviewed by JMPR, among others, the comparison of the toxicity end-points and dose levels without toxic effects (that is, NOAEL) or the minimal dose causing toxic effects (that is, the lowest observed adverse effect level or LOAEL) of each study (and of four related studies that had been performed with each chemical) showed that, in general, the NOAELs and LOAELs did not differ significantly (FAO/WHO, 2005).

Exposure to disrupting substances during fœtal life can also contribute to the development of a number of diseases in adult life, including cancer (Birnbaum & Fenton, 2003).

[5] IARC Carcinogen Classifications: Group 1: known human carcinogen; Group 2A: probable human carcinogen; Group 2B: possible human carcinogen; Group 3: not classifiable for human carcinogenicity; and Group 4: probably not carcinogenic to humans.

14.3.2. Pesticides

Toxicological data reported in this section are derived from published scientific literature that is quoted and reported in the reference list. When such quotations are missing, the information is derived from proprietary data reviewed yearly by JMPR. Data on the individual compounds can be found in a dedicated web site (WHO IPCS, 2002).

14.3.2.1. Organophosphates

Thousands of cases of acute poisoning by organophosphates (OPs) have been reported. The majority of cases were suicides or accidents, and several, including some fatal ones, were occupational in origin, due mainly to dermal exposure. OP poisoning generally represents a high percentage of total systemic pesticide poisonings, varying from about 30% in California (CDC, 1999) to 77% in China (WHO, 1990).

OP insecticides are derivatives of phosphoric acid. Depending on the atom (oxygen or sulfur) double-bonded to the phosphorus, OPs are called *phosphates* or *phosphorothioates*, respectively. OPs exert their lethal toxic effect on both insects and mammals through inhibition (phosphorylation) of acetylcholine esterase (AChE) activity at nerve endings (Lotti, 1991). Inhibition of AChE leads to accumulation of acetylcholine, which is responsible for cholinergic syndrome. Signs and symptoms may include salivation, lachrymation, bronchoconstriction, increased bronchial secretions, nausea, vomiting, diarrhoea, bradycardia, headache, dizziness, meiosis (unreactive to light), urinary and faecal incontinence, muscle fasciculation (twitching), dysarthria, ataxia and, in severe cases, convulsions and coma. Symptoms and signs may occur in various combinations and at different times after exposure. Currently available data from animals treated with organophosphorous pesticides indicate that the functional and pathological effects of these pesticides are not seen at doses lower than those at which cholinesterase inhibition is observed (Lotti, 2000).

Long-term effects, such as reduced vibrotactile sensitivity and memory deficits on test batteries, have been reported in subjects poisoned by OPs years before (Rosenstock et al., 1991; Steenland et al., 1994). However, the neurological examinations of these patients were normal. Unless the patients had prolonged brain hypoxia, seizures or both, the significance of these findings is unclear.

A so-called intermediate syndrome, apparently non-cholinergic, has been described in some patients acutely intoxicated with OPs (Senanayake & Karalliedde, 1987). After the cholinergic phase, the patients developed paralysis of the proximal limb muscles, neck flexors, motor cranial nerves and respiratory muscles. This syndrome occurs in patients with high and prolonged inhibition of AChE, probably due to the slow elimination of the compounds involved (De Bleeker, van den Neucker & Colardyn, 1993) and may lead to death, if artificial respiration is not provided.

Some OP insecticides (such as chlorpyrifos, dichlorvos, methamidophos, trichlorfon, trichlornat and isofenphos) cause a sensory–motor polyneuropathy, known as organophosphate-induced delayed polyneuropathy (OPIDP) (Lotti & Moretto, 2005).

Most cases of OPIDP follow a massive ingestion of an OP by suicidal people, and only a few cases involve careless occupational exposures to methamidophos. No human case was reported after residential exposure (Moretto & Lotti, 1998; Lotti & Moretto, 2005). OPIDP is characterized by flaccid paralysis of the lower limbs, but the upper limbs might also be affected in severe cases. The sensory peripheral nervous system is affected to a lesser degree (Moretto & Lotti, 1998). The histopathology of OPIDP shows degeneration of long and large-diameter axons in peripheral nerves and the spinal cord. OPIDP development is unrelated to inhibition of AChE, and the putative molecular target is a nervous system protein called neuropathy target esterase (Lotti, 1991). Since all commercial OP insecticides display high potency for AChE, OPIDP always develops after doses that cause severe cholinergic syndrome.

Repeated exposures to OPs at doses that do not cause AChE inhibition do not cause either neuropsychiatric disorders or behavioural disturbances (Lotti, 1991). Also, persistent electroencephalogram (EEG) changes have been reported in industrial workers who had repeated accidental exposures to sarin (a nerve agent similar in structure and biological activity to some commonly used OP insecticides). The exposures caused symptoms and significant inhibition of red blood cell AChE. The toxicological significance of these EEG changes has, however, been questioned (Lotti, 1991).

Observational studies aimed at detecting mild peripheral neuropathy or changes in peripheral nerve functions have been performed on individuals with varying long-term, low-level exposures to OPs. These studies include different occupational exposures, such as those that occur in sheep dip farmers, and exposures of military personnel during the first Gulf War in 1990/1991, and they have been reviewed recently (Lotti, 2002). It was concluded that these studies suffered from a number of limitations. For instance, they did not accurately assess exposure and reported changes in peripheral nerves were usually mild and inconsistent, sometimes reversible and sometimes apparently irreversible, because they were observed a long time after cessation of exposure. Understanding these changes is difficult, because of the lack of histopathological studies of tissues, follow-up data and an experimental model for such peripheral nerve changes that seem different from classic OPIDP. In addition, electrophysiological results were usually examined together as a group, and a correlation with clinical data was almost always missing. Finally, since these pesticides are far better inhibitors of AChE than neuropathy target esterase, they are expected to cause peripheral neuropathy at doses that inevitably cause cholinergic toxicity, irrespective of type of exposure.

ADIs of OPs vary between 0–0.0003 and 0–0.3 milligrams per kilogram body weight (mg/kg BW) per day, depending on the AI considered, and acute reference doses (ARfDs) (measured in mg/kg BW) are roughly an order of magnitude higher (WHO IPCS, 2006b).

14.3.2.2. Pyrethroids

Pyrethroids are synthetic derivatives of natural pyrethrins from *Chrysanthemum cinerariaefolium*. Their characteristics include high stability in the field (greater than OPs and carbamates), persistence in soil lower than that of organochlorines, greater insecticidal potency and low mammalian toxicity. The latter quality is reflected in ratios of rat LD_{50} to insect LD_{50} being generally higher than 1000, whereas for other pesticides they are in the range 1–50 (Elliot, 1976).

Acute poisoning from pyrethroids is characterized by dizziness, headache, nausea, muscular fasciculation, convulsive attacks and coma (He et al., 1989; Chen et al., 1991). Two patterns of symptoms are described in rats after acute intoxication, depending on the presence or the absence of an (α)cyano-group substituent: the so-called T-syndrome (aggressive sparring, sensitivity to external stimuli and tremors) and the so-called CS-syndrome (choreathetosis (a disorder that causes involuntary movement or spasms), salivation and seizures), respectively. Sometimes, however, the two syndromes may combine to give a more complex one (Aldridge, 1990). Cases of acute pyrethroid poisoning in China have been reviewed, but it was not possible to differentiate the two syndromes in people (He et al., 1988).

Occupational exposures often result in abnormal skin sensations – mainly of the face – described as burning and tingling (He et al., 1989; Chen et al., 1991; Moretto, 1991; Zhang et al., 1991). Symptoms appear shortly after beginning work and disappear within 24 hours. Most pyrethroids cause these sensations, with the following order of decreasing potency: deltamethrin, flucythrinate, cypermethrin = fenvalerate, permethrin (Aldridge, 1990). This neurotoxicity is due to a local effect, since only unprotected parts of the skin are affected. Pyrethroids are known to act on sodium channels, thereby causing repetitive firing of the sensory nerve endings of the skin (Aldridge, 1990). However, electrophysiological studies performed on the arms and legs of exposed subjects that complained of cutaneous sensations were negative (Le Quesne, Maxwell & Butterworth, 1981).

Neonatal rats are known to be 4–17 times more vulnerable to acutely toxic doses of pyrethroids than adults. This probably can be attributed wholly to their smaller capacity for metabolic detoxification. In contrast, there is no evidence that shows them to be more susceptible to low doses that cause no toxic effects in adults (Ray, 2001). In several studies performed in one laboratory, the pyrethroids permethrin and deltamethrin were reported to induce changes in behaviour and in some neurochemical parameters in adult mice that were administered the pyrethroids, during their neonatal life, at doses not causing overt toxicity. However, others did not duplicate these findings, and the findings are not consistent with the results of several multigeneration, regulatory studies (Ray, 2001).

ADIs for pyrethroids vary between 0–0.07 mg/kg BW per day and 0–0.002 mg/kg BW per day, depending on the AI considered ARfDs are roughly an order of magnitude higher.

14.3.2.3. Insecticide synergists

Insecticide synergists are chemicals that enhance the insecticidal activity of other chemicals, such as pyrethrins and synthetic pyrethroids. These synergists are mixed with insecticides in end-use product formulations. They are not used alone. These compounds have very low or negligible levels of toxicity. Piperonyl butoxide is the most commonly used insecticide synergist.

14.3.2.3.1. Piperonyl butoxide

Piperonyl butoxide has negligible acute toxicity, and WHO has classified it as unlikely to present an acute hazard in normal use. Both short-term and long-term studies, however, show that it is responsible for hepatic toxicity, which is characterized by liver enlargement with associated hypertrophic hepatocytes, focal necrosis and (at times) alteration of some clinical chemical parameters. Piperonyl butoxide was shown to be carcinogenic at doses toxic to the liver, and it caused general toxicity. Also, piperonyl butoxide was not shown to be genotoxic, and therefore JMPR did not consider the cocarcinogenic effect observed in animals to be relevant to people (WHO IPCS, 2002). These effects were considered to be secondary to the ability of piperonyl butoxide to induce hepatic cytochrome P450 enzymes, a large group of monooxygenase enzymes responsible for the metabolism of toxic hydrocarbons. Moreover, piperonyl butoxide was not embryotoxic or teratogenic in rats or rabbits. Furthermore, it was a mild dermal and ocular irritant, but not a dermal sensitizer, in rabbits.

Some experiments with piperonyl butoxide have been performed on people. A study reported by JMPR (WHO IPCS, 2002), in which a formulation containing 3% piperonyl butoxide was spread onto the ventral forearm of adult male volunteers, indicated that about 8% of the applied dose was absorbed through the skin. The percutaneous absorption of pyrethrins and piperonyl butoxide from the scalp was calculated to be 7.5% of the applied dose for pyrethrins and 8.3% for piperonyl butoxide. With a 7-day urinary collection, 1.9 ± 1.2% (standard deviation) of the dose of pyrethrins and 2.1 ± 0.6% of the dose of piperonyl butoxide applied, were absorbed through the forearm skin. An hour after application, blood samples contained no detectable radioactivity (Wester, Bucks & Maibach, 1994). The ADI was 0–0.02 mg/kg BW per day, but an ARfD value was not established.

14.3.2.4. Anticoagulant rodenticides

Anticoagulants used as rodenticides are antimetabolites of vitamin K and inhibit the synthesis of prothrombin (Ecobichon, 1996). Warfarin was the first compound to be introduced and is the prototypical short-acting rodenticide. Due to the emergence of warfarin-resistant rats, so-called superwarfarins or long-acting anticoagulants were developed. These include diphacinone, brodifacoum, bromadiolone and chlorophacinone. They differ from warfarin in a number of respects: a longer polycyclic hydrocarbon side chain, higher lipid solubility, accumulation in the liver, higher potency (on a molar basis) and prolonged action. Accidental or intentional exposure to either type of anticoagulant caused prolonged clotting failure (coagulopathy) and, in severe cases, death. Short-acting anticoagulants have caused death at an estimated dose of 50 mg/kg per day for 8 days (Lange & Terveer, 1954).

In people, accidental percutaneous exposure to talcum powder contaminated with 1.7–6.5% warfarin caused toxicity and death in Vietnamese infants (Martin-Bouyer et al., 1983). In adults, a single oral dose of 1 mg/kg is considered to cause therapeutic prothrombin values. Less information is available for superwarfarins, which are expected to cause longer-lasting coagulation defects. Data collected from poison control centres indicate that a single acute unintentional rodenticide anticoagulant ingestion rarely causes laboratory or clinical evidence of excessive anticoagulation (Wedin & Benson, 2000). Therefore, although the number of cases of accidental ingestion is relatively high, the outcome is generally without consequences.

It is known that warfarin anticoagulation therapy during the first trimester of gestation may cause developmental disorders in embryos (such as nasal cartilage hypoplasia and skeletal abnormalities), whereas in the late third trimester it may result in prenatal, perinatal or postnatal haemorrhages. Ocular and neurological abnormalities have also been observed after warfarin treatment during pregnancy. Known adverse effects that occur during anticoagulant treatment, such as cutaneous and subcutaneous tissue necrosis, purple toes syndrome and dermatitis medicamentosa, do not occur after massive accidental exposure or after prolonged environmental exposure. The laboratory parameter to be assessed after anticoagulant ingestion is PT/PTT (prothrombin/partial thromboplastin time) or INR (international normalized ratio), to be measured 24–48 hours after poisoning. No ADI or ARfD values have been established.

14.3.2.5. IGRs

This group of AIs is generally devoid of significant acute and chronic toxic effects in animals. Due to their relatively recent introduction in the market, little or no information is available on effects on people. Methoprene and pyriproxyfen are discussed in some detail because they appear to be the AIs most used.

14.3.2.5.1. Methoprene

No information is available on the toxicity to people of methoprene. Methoprene showed little acute toxicity in animals, the LD_{50} being more than 5000 mg/kg BW when taken orally by rats and more than 2000 mg/kg BW when applied dermally to rabbits. WHO has classified methoprene as "unlikely to present acute hazard in normal use". On the rabbits tested, methoprene was neither irritating to the eye or skin nor sensitizing. Short- and long-term studies in mice, rats and dogs showed little toxic potential. The main effect was an increase in the weight of the liver relative to BW at very high and repeated doses, and this effect was not always associated with histopathological changes. No increase in the incidence of tumours at any site was seen in mice or rats, and no effects on reproduction were observed. The ADI for S-methoprene was 0–0.05 mg/kg BW per day, and for methoprene racemate it was 0.09 mg/kg BW per day; no ARfD value has been established.

14.3.2.5.2. Pyriproxyfen

No data are available on the toxicity to people of pyriproxyfen. The acute toxicity of pyriproxyfen in animals is low, with oral LD_{50} values more than 5000 mg/kg BW in mice, rats and dogs, dermal LD_{50} values more than 2000 mg/kg BW in mice and rats, and an LC_{50} value (for inhalation experiments, it is the concentration of a chemical in air that kills 50% of the test animals in a given time) more than 1.3 mg/l air in mice and rats. WHO has classified pyriproxyfen as "unlikely to present acute hazard in normal use". Pyriproxyfen was found to be mildly irritating to rabbit eyes, but not to their skin. Also, it did not sensitize the skin of guinea-pigs. In short- and long-term studies reviewed by JMPR, the liver was the main toxicological target, with increases in liver weight and changes in plasma lipid concentrations, particularly cholesterol, at doses of 120 mg/kg

BW per day and above in rats. Also, some evidence showed that the compound might cause modest anaemia in mice, rats and dogs at high doses. In long-term studies of toxicity in mice, pyriproxyfen also caused a dose-dependent increase in the occurrence of systemic amyloidosis, which was associated with increased mortality rates at doses greater than 16 mg/kg BW per day. Pyriproxyfen was not carcinogenic in mice or rats (WHO IPCS, 2002), and it showed no evidence of carcinogenicity in a one-year study in dogs at doses up to 1000 mg/kg BW per day. Pyriproxyfen was not genotoxic in a range of tests for mutagenicity and cytogenicity in vitro and in vivo. Moreover, it caused little developmental toxicity and was not teratogenic. The ADI was 0–0.1 mg/kg BW per day; and no ARfD value has been established.

14.3.2.6. Neonicotinoid insecticides

These insecticides have very low toxicity in mammals and, being of relatively recent introduction, have a very small, if any, record of human poisoning or excessive exposure.

14.3.2.6.1. Imidachloprid

JMPR reports that a 4-year-old child who ingested about 10 mg/kg BW of a veterinary preparation of imidacloprid showed no signs of poisoning or adverse health effects. In two fatal cases with this compound, blood concentrations were 12.5 µg/ml and 2.05 µg/ml (Proenca et al., 2005). In a case of acute ingestion of a formulation containing 9.7% imidacloprid (<2% surfactant) and the balance as solvent (N-methyl pyrrolidone), clinical manifestations included drowsiness, disorientation, dizziness, oral and gastroesophageal erosions, haemorrhagic gastritis, productive cough, fever, leukocytosis and hyperglycaemia. The patient recovered without complication with supportive treatment, was discharged four days after ingestion, and the follow-up barium upper gastrointestinal examination a month later was normal. Because a moderate to high dose of imidacloprid in animals causes central nervous system activation similar to nicotine, including tremors, impaired pupillary function and hypothermia, it is unclear whether imidacloprid had a causal role in the patient's initial drowsiness and dizziness. It is more likely that the formulation ingredients, particularly N-methyl pyrrolidone, caused most of the clinical symptoms, including minor central nervous system depression, gastrointestinal irritation and hyperglycaemia (Wu, Lin & Gheng, 2001). As reported by JMPR, periodic examinations of employees exposed to imidacloprid showed no adverse health effects.

Imidacloprid is moderately toxic to rats (oral LD_{50}: 380–650 mg/kg BW) and mice (oral LD_{50}: 130–170 mg/kg BW). Behavioural and respiratory signs, disturbances of motility, narrowed palpebral fissures, transient trembling and spasms were seen in rats and mice treated orally at doses greater than or equal to 200 mg/kg BW and greater than or equal to 71 mg/kg BW, respectively. The clinical signs were reversed within six days. The LC_{50} for acute exposure to an aerosol could not be determined exactly, as rats tolerated inhalation for four hours of the maximum concentration of dust that could be produced technically (0.069 mg/l of air) without signs or deaths.

Imidacloprid did not irritate the skin or eyes of rabbits and did not sensitize the skin of guinea-pigs in a maximization test. A reduced gain in BW was the most sensitive toxicological end-point in animals, occurring at doses greater than or equal to 22 mg/kg BW per day. Effects on the liver occurred at higher doses, but no evidence of a carcinogenic effect of imidacloprid was found in either mice or rats in long-term studies of dietary administration. Imidacloprid gave negative results in in vitro and in vivo genotoxicity studies and was not teratogenic. The ADI value was 0–0.06 mg/kg BW per day; the ARfD value was 0.4 mg/kg BW.

14.3.2.6.2. Indoxacarb

Indoxacarb exerts its insecticidal activity by interfering with the normal transmission of nerve impulses along neurons, by blocking sodium channels and binding to neuronal nicotinic acetylcholine receptors (Zhao et al., 1999; Lapied, Grolleau & Sattelle, 2001). No toxicity data on indoxacarb are available for people.

Indoxacarb administered by gavage at low doses (5 mg/kg BW) is extensively, but slowly, absorbed (69–81%), but at higher doses (150 mg/kg BW) saturation becomes evident (8–14% absorption). Absorption plateaued between 5 and 8 hours at low doses and between 3 and 27 hours at high doses. Elimination due to preferential accumulation of metabolites in fat and erythrocytes was slow, with the half-life in plasma that was 92 hours (in males) and 114 hours (in females).

Indoxacarb has low acute oral toxicity (LD_{50}: 1,730 mg/kg BW) in male rats and moderate oral toxicity (LD_{50}: 268 mg/kg BW) in female rats, and low dermal and inhalation toxicity in rats. The difference in oral toxicity between the sexes is thought to arise from the more efficient biotransformation of indoxacarb to an acutely toxic metabolite in females and its subsequent disposition to accumulate in fat. It caused moderate eye, but not skin, irritation in rabbits, and it was a skin sensitizing agent in the guinea-pig maximization test.

Although indoxacarb inhibits neuronal sodium channels and nicotinic acetylcholine receptors, clear evidence of neurotoxicity occurred only at high acute doses (in excess of 100 mg/kg BW), where ataxia, reduced motor activity, forelimb grip strength and foot splay were observed in rats. Clinical signs suggestive of neurotoxicity were noted in short-term repeat-dose mouse dietary studies and included abnormal gait or mobility and abnormal head tilt at high doses (30 mg/kg BW per day). In contrast, repeat-dose rat studies showed no effects on motor activity or functional observational battery assessments, and no histological evidence of neurotoxicity.

In mice, rats and dogs, the major toxicological finding after repeated dosing with indoxacarb was mild haemolytic anaemia. Also, indoxacarb and its major metabolites were not genotoxic, either in vitro or in vivo. Moreover, in developmental studies in rats and rabbits, indoxacarb was not teratogenic and did not show effects on reproductive performance.

In a one-year dietary study in dogs, the 2005 JMPR established an ADI of 0–0.01 mg/kg BW per day, based on a NOAEL of 1.1 mg/kg BW per day for erythrocyte damage and the secondary increase in haematopoiesis in the spleen and liver. After a single administration of indoxacarb in rats, an ARfD of 0.1 mg/kg BW was also established, based on the NOAEL of 12.5 mg/kg BW for reduction in BW gain and food intake.

14.3.2.6.3. Thiamethoxam

Thiamethoxam has low acute toxicity since the NOAEL in a single dose neurotoxicity study in the rat was 100 mg/kg BW, based on neurobehavioural effects (such as dropped palpebral closure, decreased rectal temperature and locomotor activity, and increased forelimb grip strength). No toxicity data on thiamethoxam are available for people.

Thiamethoxam is not genotoxic; however, it caused species-specific liver damage and associated increased incidence of liver tumours in mice. This is due to the formation of a toxic metabolite in mice, but this metabolite was not formed to a relevant extent in rats and people, as shown in in-vitro studies with liver microsomes. Consequently, liver tumours caused by thiamethoxam in mice are not considered relevant for human risk assessment, because formation of that toxic metabolite in people at environmentally significant exposures is unlikely to occur (Green et al., 2005; Pastoor et al., 2005).

Other toxic effects involve the kidneys (tubular lesions and inflammatory cell infiltration at higher than 20 mg/kg BW per day) and testes (increased incidence and severity of tubular atrophy at 1.8 mg/kg BW per day, with a NOAEL of 0.6 mg/kg per day). These effects were the basis for the chronic reference dose established by the EPA (2005). The EPA also published toxicity end-points for thiamethoxam (EPA, 2000a).

14.3.2.6.4. Fipronil

Fipronil is a potent disrupter of the insect central nervous system, interfering with the gamma-aminobutyric acid (GABA)-regulated chloride channel. Acute fipronil poisoning, after a suicide attempt, is characterized by vomiting, agitation and seizures, and it normally has a favourable outcome. Seizures were associated with peak plasma fipronil concentrations of 1600 mg/l and 3744 mg/l in two patients (Mohamed et al., 2004).

In a single-dose study of neurotoxicity in rats, decreased hind-foot splay was observed in males at 7.5 mg/kg BW and 25 mg/kg BW, seven hours after dosing. Females at these doses showed decreased grooming and decreased BW gain, food consumption and food efficiency. No effect was observed at 2.5 mg/kg BW. In a repeated dosing study, fipronil was administered in the diet of mice. Overactivity and irritability were observed consistently in males and females at about 5 mg/kg BW per day, whereas general toxicity (reduced BW and food consumption, increased absolute or relative weights (or both) of the liver) was observed at less than 1 mg/kg BW per day. In rats, dermal absorption was less than 1% of the applied dose. In vitro, at the lowest concentration tested (0.2 g/l), the percentage of the dose absorbed across human and rat membranes was similar. At higher concentrations (4 g/l and 200 g/l), penetration was greater through rat and rabbit skin than through human skin. Fipronil was moderately hazardous to rats (LD_{50}: 92 mg/kg BW) and mice (LD_{50}: 91 mg/kg BW) after oral administration of single doses and to rats after single exposure by inhalation (LC_{50}: 0.36 mg/l). After a single dermal exposure, fipronil was relatively non-hazardous to rats (LD_{50}: > 2000 mg/kg BW). At all doses and times up to 24 hours, the quantity of fipronil absorbed via rat skin was less than 1% of the applied dose.

In dogs treated for a year, convulsions, twitching, tremors, ataxia, unsteady gait, rigidity of limbs, nervous behaviour, hyper- or hypoactivity, vocalization, nodding, aggression, resistance to dosing, lack of appetite and abnormal neurological responses were observed at 2 mg/kg BW per day and higher. No effects were observed at 0.2 mg/kg BW per day.

Fipronil and its metabolites are not genotoxic. Thyroid tumours observed in rats are not relevant to human beings. Fipronil is a slight skin irritant and a weak skin sensitizer in animals. In a study of developmental neurotoxicity, higher sensitivity of young animals to the neurotoxic effects of fipronil was not seen. The ADI was 0–0.0002 mg/kg BW per day, and the ARfD was 0.003 mg/kg BW.

14.4. Pesticide use patterns: application scenarios

Information on the tasks involved in using a product, the method used to apply it, and the training and choice of the person applying it are all essential in ascertaining how exposure will arise. This section covers applications scenarios for spray products, baits, rodent tracking powders, and cat and dog spot-on formulations.

14.4.1. Spray products

Broadcast spray pest control products are available on the market in many formulation types and delivery devices. The target organisms for these pest control products are arthropods, mainly such pests as fleas, mosquitoes and ticks.

For these applications, two main aspects characterize exposure: (a) the processing needs of the product before application, such as diluting or mixing a concentration then loading the end-use dilution into a sprayer, and (b) the target site of the application. Mixing and loading differ for liquids and powders: concentrated liquid products, which are diluted and dispensed from a sprayer, evaporate during dilution; powdered and granular products, which are dissolved in water and dispensed from a sprayer, disperse the dissolved powder into the air.

14.4.2. Liquid spray application types

With regard to the target, one can distinguish between the following four types of application: spot applications, crack-and-crevice applications, general surface applications and air space applications.

14.4.2.1. Spot applications

Spot applications refer to spraying hiding places of crawling insects and spraying ant tunnels. Although a relatively small surface area is sprayed, sometimes it is difficult to access the area, both for the applicator and for the bystander. Such an area may be behind a refrigerator or radiator, or in or under kitchen cabinets. When considering the method and extent of exposure, spot applications can be compared with the spraying of indoor plant leaf surfaces against red spider mites. The EPA defines a *spot application* as 20% or less of the infested area of a home.

14.4.2.2. Crack-and-crevice applications

Crack-and-crevice applications involve spraying these building features to control such pests as silver fish, cockroaches and ants – for example, on baseboards in living and accommodation areas, in cracks and holes in wooden floors, and areas such as door frames.

14.4.2.3. General surface applications

General surface applications involve spraying such large surfaces as carpets or couches – for example, to control dust mites or fleas.

14.4.2.4. Air space applications

Air space applications involve spraying of living, working or accommodation areas against flying insects. In these applications, the user stands in the middle of the room and sprays all four of its upper corners.

14.4.2.5. Differences between application types

These spray applications differ from each other in the manner and extent to which the person applying the pesticide and bystanders are exposed. For example, the exposures for a crack-and-crevice application and for a general surface spray are expected to be different, due to the longer application time of the latter treatment. The height and angle at which the spraying occurs also account for a difference in exposure – for example, above the head, which is common during an air space application, or aimed at the floor, which is common during a general surface spray. Among other things, after applying these sprays, the size of the wipeable surfaces differs. In the worst case, the entire sprayed surface is assumed to be within the reach of crawling children.

14.4.3. Baits

Baits are used to kill mice, rats, ants and cockroaches and are very specific for the target pest. Some products deliver enough toxicant for a lethal dose in one feeding, while others are fed upon a number of times before the pest dies. The pesticide products used against rats and mice are composed mainly of grains to which the AI has been added. Because it is a poison, the product is always dyed. To prevent children and non-target species and pets from being poisoned, rodenticides usually require a bittering agent as well. Most bait products are sold in child-resistant packaging or locked in boxes, to prevent access to all but the target pest.

For consumer use, the net contents of a single packet may not be higher than 200g, and bait stations must be included. For use in rooms, the bait must be placed in feeding boxes that are closed at the top. For outdoor use, it must be placed in specially designed feeding stations, in such a way that the bait is not within the reach of children, cattle, pets or birds.

14.4.4. Rodent tracking powders

Rodenticides are used for rodent control and, in most cases, are formulated as ready-for-use products. For special purposes, some concentrates are available and some rodenticides are formulated as tracking powders. It is a general rule that rodenticides are formulated and kept in such a way that people and non-target animals are not exposed. Tracking powders, which are rodent poisons in the traditional sense that they must be eaten to kill the pest, are placed along rodent runways, in and around buildings. When the animal passes by, its fur picks up the powder, which is then ingested during grooming. Consequently, the concentration of rodenticide in contact powders is much larger than in food baits. In view of the possible exposure of people and other non-target species, the treated areas should be covered (EC, 2002).

14.4.5. Cat and dog spot-on formulations

Spot-on insecticides are becoming a popular type of flea control for pets. The products may be applied monthly for flea control or every two weeks for tick control. People are exposed to the insecticide when residues are transferred from the treated fur to the hand or body of an adult or child. Children are most likely to be exposed in this manner, because they are often in direct contact with the animal. Residues on hands can be transferred to the mouth, especially by toddlers.

14.5. Residential exposure

14.5.1. The nature of residential exposure

The term *residential* refers to the generic conditions of non-occupational exposures, regardless of where they occur. The term *general population exposure* could be easily substituted. If exposures occur as a result of activity directly related to an application, they are referred to as *handler* exposures – for example, the handler can be someone who mixes or applies a pesticide product. On the other hand, if exposures occur as a result of activities in a previously treated area, they are referred to as *post-application* exposures. The other distinction that is made by the EPA is the one between the terms *residential* and *homeowner*.

- **Homeowner, handler exposures** result from an individual, not as a condition of his/her employment, applying a pesticide.

- **Residential, post-application exposures** result from entry and activity in an environment previously treated with a pesticide. These exposures may result from both occupational or homeowner applications and may occur in a variety of settings, including homes, schools, day-care facilities and other public places, such as parklands.

The term *home* is not limited to the inside of the actual building. Use of a pesticide on the property outside a home (such as use of a lawn pesticide product) is also considered a residential use; examples include products that are used on pets, lawns or gardens.

For people, residential exposure to biocides can be categorized as primary and secondary exposures (Fig. 14.4). Primary exposure to biocidal products occurs to the individual who actively uses the products that contain biocides (the user). Secondary exposure occurs to non-users or bystanders; these are individuals who do not actively use the biocidal products, but are indirectly exposed to biocides released during or after product use by another person (the user).

Fig. 14.4. Primary and secondary residential exposure

Primary exposures are invariably higher than secondary exposures; however, some specific subgroups of the population may experience higher secondary exposures because of their specific behaviour – for example, children crawling on the floor. In addition, secondary exposure can be experienced over a much longer period of time than primary exposure, particularly for persistent products.

14.5.1.1. Primary exposure of non-professional users and secondary exposures

Non-professional users are usually consumers – who may or may not read a product label. Although they are expected to comply with instructions for use of a product, there is little guarantee of this. Also, they have no access to controls or formal personal protective equipment (PPE), though they may use household protective equipment, such as gardening or kitchen gloves.

The groups (residential bystanders, as defined by the EPA) at risk through secondary exposure are less easy to identify. However, the intended location of use – for example, indoors, outdoors, residential or recreational – will provide useful indicators. Secondary exposure results from contamination of the indoor environment caused by residential applications. The following two post-application scenarios for the residential environment are examples of secondary exposure. The first applies to toddlers (less than 8 kg), while the second is a bystander exposure applicable to all age groups.

1. **Hand-to-mouth ingestion** includes children playing on the floor where biocides have been applied. In this scenario, children transfer the biocide to their skin by contact with contaminated surfaces, such as floors and walls. Oral contact may take place via hand-to-mouth transfer and toy-to-mouth transfer.

2. **Residential bystander** covers people, present in the house after application, who are exposed to the residues in the air and on surfaces.

There is a paucity of data on exposure to biocides. Currently, various national approaches or models are used to estimate human exposure to biocides. Therefore, the EU funded a project to fill this knowledge gap and establish a harmonized approach to assessing human exposure to biocides. The report *Technical notes for guidance: human exposure to biocidal products – guidance on exposure estimation* (EC, 2002) was produced as a result of this exercise and is available on the ECB's web site.

14.5.2. Pattern of use considerations

The *use pattern*, a fundamental aspect of exposure, must not be overlooked, but adequate data are sparse. This pattern contains information about the frequency and duration of elements of tasks that comprise the scenario, information about the ancillary operations and information about those who may be exposed as a result of a product having been used. Many of the elements in the pattern of use will result in distributions – for example, the indoor air concentration at time of use and the application time. The pattern of use is not universal and is likely to show considerable variability between use sectors and between nations or regions.

The information on the pattern of use is used to develop exposure scenarios, which are then evaluated to derive quantitative exposure estimates. The essential information on the pattern of use required for deriving exposure scenarios includes:

- the product (such as physical state, concentration and vapour pressure);
- where and how the product will be used (such as location and method of application);
- who will use or apply the product (primary exposure);
- tasks (such as spraying, or mixing or loading, frequency of application or mixing or loading (or both application and mixing or loading) with time, and duration of each stage of use);
- expected exposure controls (such as PPE); and
- others who may be exposed – bystanders (secondary exposure).

Information on the pattern of use can only be gathered through surveys or by conducting actual observational studies in the home itself, in a simulated laboratory environment or at the site of application. Such information is rarely available in scientific or published literature, because of concerns about human testing and exposure. Biocidal product manufacturers will need to conduct research into the pattern of use – directly with the users, if actual or surrogate data are not available. While there may be common tasks – for example, coupling a reservoir of biocide (product container or spray tank) to a dilution system – and the ranges of duration of these tasks may be known, the number of times a day such tasks could be undertaken depend on the product type and location of use, as directed by the product label.

The pattern of use may be a seasonal, regional or local issue, and regulators need to assure the relevance of a stated pattern of use in product authorization or registration. For example, ground injection of insecticides for termite control occurs only in countries where termites thrive.

14.5.3. Consumer safety considerations

Regulatory agencies usually register a pesticide for specific uses and require manufacturers to put information on the label about when and how to use it. The information on the product label describes the approved uses and usually warns against mistakes that may endanger the user. While it is reasonable to assume that training and experience will guide professional users to follow the directions of use, there is less certainty that the general public and consumers in general will do the same – for example, some of them may be illiterate or generally unaware of chemical risks. For this reason, the product authorization process usually applies a margin of safety and, when the toxicity of the ingredients is acute, special provisions are recommended for packaging, stinking warning additives and safety-proof containers.

14.5.3.1. Safety considerations for sensitive individuals and children

Some individuals may be exposed to higher concentrations than others, because of differences in their behaviour and physiological variables. Young children, for instance, may be exposed to higher levels than adults, due to their distinct (hand-to-mouth or crawling) behaviour and relatively lower BW. Also, a child's greater ratio of surface area to BW might increase the risk of hazardous dermal exposure (Wolff & Schecter, 1991; Snodgrass, 1992). Children are also at higher risk because their metabolic processes are immature and they are less able to detoxify chemicals. Moreover, infants and children are growing and developing, and their delicate developmental processes can be more easily disrupted. As children have more future years of life than most adults, early exposure can result in the development of chronic diseases (Licari, Nemer & Tamburlini, 2005).

Other population groups that share a higher risk include pregnant women and immunocompromised individuals. Exposure to pesticides during pregnancy can result in potentially adverse effects on fœtal growth and child neurodevelopment. Exposure to disrupting substances during fœtal life can also contribute to the development of a number of diseases in adult life, such as cardiovascular diseases, diabetes and cancer. Furthermore, there is a large amount of experimental data that describes pesticide-induced immunosuppression, and exposure to pesticides may thus impair immune functions – particularly of people already affected by immunological disorders.

14.5.4. Assessing residential exposure

Addressing exposure of the general population and (particularly) of children is a complex task. Because of the many ways in which non-dietary residential exposure can occur, the EPA has developed standard operating procedures (SOPs) for residential exposure assessment, to add consistency and transparency to the risk assessment and regulatory process and to provide guidance to scientists. The SOPs address over 40 different exposure scenarios, were developed using the most recent data available and provide a so-called handbook approach, by presenting a description of each scenario along with recommended algorithms, sample calculations, a discussion of uncertainties and references.

All exposure scenarios addressed in the EPA's SOPs (EPA, 1997a) are non-occupational. Exposure of bystanders from occupational applications or from bring-home events to children (such as spray drift and residue track-in) are also being considered by the EPA Office of Pesticide Programs, as they may expose individuals not involved in the occupational activity (such as children of farmers or pest control operators).

The EPA's residential exposure assessments are designed to be as realistic as possible. They are, however, generally conservative, which adds an extra measure of safety to regulating pesticides. More recently, the EC has developed guidance (EC, 2002) that addresses a variety of exposure scenarios.

14.6. Routes of exposure

This section covers three routes of exposure: inhalation, dermal exposure, and incidental oral exposure.

14.6.1. Inhalation

Exposure from inhaled pesticides is sometimes a small component of total exposure to biocides, but it can (in some cases) become the predominant route of exposure. Conditions where exposure through inhalation becomes important, usually involve the use of volatile biocides or of dusts, fumigants and sprays, especially in enclosed spaces. It should also be borne in mind that a higher proportion, (up to 100%), of the inhaled dose may be bioavailable, compared with a lower percentage absorbed by dermal exposure.

The assessment of inhalation exposure is well characterized by standard metrics and sampling methods. Because there is a large body of national guidance and scientific literature on conducting surveys to determine exposure to vapours and aerosols by inhalation, this matter is not developed further in the present chapter. It is important, however, to have some knowledge of the likely distribution of particle sizes of an aerosol generated from a solid product. Also, because some biocides have a low, but nonetheless significant vapour pressure, and because deposits on air sampling filters can evaporate into the sampled air stream, special sampling techniques are required.

The airborne concentration of a biocidal product sprayed in an enclosed space, varies according to the ventilation of the room during and after the application, the concentration decay over time being directly related to the rate the air in the room is changed

(Fenske et al., 1990). After application, the airborne chemical may diffuse onto surfaces (such as walls, furniture and floors) and into sorptive media (such as textile materials, curtains, carpets and plush toys), from which it can be subsequently re-emitted for quite a long time. For example, Gurunathan and colleagues (1998) showed that after application indoors, chlorpyrifos residues increased on the surface of plastic toys and peaked at one week after application.

The user's exposure via inhalation (primary exposure) should be measured, using personal monitoring, to assess the airborne concentration in the breathing zone (by convention within 30 cm of the nose and mouth). This measurement is incorporated into inhalation exposure risk assessments.

Exposure of others, as a consequence of use (secondary exposure), is often evaluated either by using static (background) monitoring or mathematical models, which are used more frequently. Secondary exposure by inhalation is generally expressed by the time-weighted average of a particular substance (in mg/m^3), over a defined period of time.

Few validated methods relate exclusively to monitoring air and the determination of biocidal agents. Less than 10% of the substances listed by the EC in its provisional list of existing biocidal substances have been found to have specific workplace measurement methods (as vapour or aerosol).

The important criteria and appropriate selection of sampling devices are outlined in a review by Findlay (1995). Other relevant texts include the European Committee for Standardization (CEN) standard on workplace atmospheres (CEN, 1995), the Deutsche Forschungsgemeinschaft (DFG) publication on analysing hazardous substances in air (DFG, 1993) and the Health & Safety Executive (HSE) publications on methods for determining hazardous substances (HSE, 2007).

14.6.2. Dermal exposure

Exposure of (and via) the skin is usually a significant aspect of human exposure to biocides. While this has been commonly considered for risk assessments of plant protection products, it is not so for biocides. Exposure data for deposition of biocides on work clothing and exposed skin have only recently been established. The pattern of distribution over the body differs with the task – for example, sometimes only the hands are exposed.

The concepts of *potential dermal exposure, estimated exposure* and *actual dermal exposure* are used to gauge exposure.

Potential dermal exposure. This is the amount of biocide that may deposit on the clothes and on exposed skin over some defined period of time. Common metrics include mg AI deposited/kg AI handled (mg/kg AI). However, in numerous biocide exposure scenarios, the amount of biocide handled simply cannot be estimated – for example, drilling mud. Another common metric is the amount of in-use biocide that deposits per unit of time or per task (in mg/min and mg/cycle, respectively). Practical evidence from field studies indicates that metrics for potential dermal exposure, such as mg/min or mg/cycle, are useful.

Estimated exposure. In the absence of measured exposure data or representative data on analogous substances, exposure must be estimated using recommended modelling approaches. To ensure that the predictions are realistic, all relevant exposure-related information on the substance should be used iteratively.

Actual dermal exposure. This is the amount of compound that actually reaches the skin. It is affected by the efficiency and effectiveness of clothing in acting as a barrier and is often expressed simply as the weight of biocide product on the skin (mg on skin). Another metric of importance is the concentration of a substance (AI) on the skin (mg AI/cm^2 of skin), which in turn depends on the amount deposited and the surface of contaminated skin. By knowing the skin penetration rate of a substance, its concentration on the skin enables the actual intake through the skin to be assessed. It is worth noting that the percentage for the skin penetration rate is generally inversely proportional to the concentration on the skin, being highest for the low concentration values. Moreover, damages to skin (such as small wounds and fissures), conditions that alter skin permeability (such as inflammation and eczema), or the presence of certain solvents may increase the skin penetration rate.

Although not a major route of exposure, the potential of exposure to the eyes will also need to be considered, particularly when handling irritants or corrosive substances.

Potential dermal exposure normally has to be measured, or estimates may be obtained from database models. Actual dermal exposure, however, arises through:

- direct deposition on exposed skin, such as the face;
- permeation through clothing and penetration of clothing around fastenings, openings, and along seams;
- incidental contact with residues on surfaces; and
- putting on and taking off contaminated clothing (including protective gloves).

It is often impossible to know the actual amount of dermal exposure – that is, the sum of the total direct and indirect exposures of the skin. Studies that use fluorescent dyes can provide a useful indicator of actual dermal exposure, but there are few data available. The quantity of a substance deposited on the skin can be expressed in terms of mg/cm^2, with the amount of skin exposed expressed in cm^2. However, the quantity on the skin is more likely to be simply expressed as a weight (mg on skin). Such metrics, however, do not account for the rate of accumulation on the skin.

Dermal exposure data are difficult to acquire and interpret. However, methods are available for the sampling process, such as the WHO recommended methods developed for occupational exposure assessment (WHO IPCS, 1999) and dermal absorption (Kielhorn, Melching-Kollmuss & Mangelsdorf, 2006). These methods make use, as in other methods, of a set of gauze pads applied to the skin of different parts of the body; the concentration measured in the pads is used to calculate the dermal exposure of each part of the body.

Exposure of the hands is often highly significant. Although residential labels of several products recommend gloves as best management practice, homeowners can hardly be assumed to typically wear protective gloves when applying pesticides in the home. Moreover, poor procedures in putting on and taking off gloves can lead to significant exposure to the hands, regardless of the barrier properties of the protective glove material.

Dermal exposure may also result from contacting treated areas. Even where a pesticide has been applied to cracks and crevices, its residues can be detected throughout the house, and contact may occur through everyday activities. For example, in a case where chlorpyrifos applications had been made to the cracks and crevices of the homes and along the perimeters of the walls behind appliances and furniture, surface wipe samples collected from non-targeted surfaces (such as play areas, bedrooms and plush toys) showed the presence of chlorpyrifos (Hore et al., 2005).

Sampling gloves provide a measure of potential dermal exposure when coming directly into contact with solids, fluids and aerosols; these gloves may over-sample, but they can reflect actual dermal exposure. Thin cotton sampling gloves worn beneath protective gloves demonstrate actual hand exposure but again may overestimate exposure. Sampling protocols, however, need to recognize that sampling gloves will collect pre-existing contamination inside protective gloves. Another technique for assessing hand exposure uses hand washing with solvent–water solutions and wiping the skin on the hand at the end of use (see also Kielhorn, Melching-Kollmuss & Mangelsdorf, 2006).

Using surrogate biocide products, further research in progress will indicate the likely percentages and spatial distributions of typical work clothing penetration. Finally, it should be mentioned that washing contaminated clothing (clothing used while applying biocides) with everyday laundry might contaminate the household laundry or clothing.

14.6.3. Incidental oral exposure

Unintentional oral exposure is the amount of a biocide entering the mouth. At present, it can only be inferred from biomonitoring or (worst case) modelling that uses EPA SOPs (EPA, 1997a). Exposure is expressed in mg/event or mg/day. Biomonitoring for biocides requires expert advice and appraisal of the results. As biomonitoring measures total exposure, the percentage of the exposure that results from oral unintentional exposure can only be differentiated from other routes of exposure when they are negligible or have been assessed separately.

As an exposure route, ingestion is currently weakly defined, though it may be the most important route in some circumstances. This is particularly true where inhalation and dermal uptake are low, where contaminated hands are not properly washed at the end of an application (poor hygienic practice) or where children have secondary exposure. A special case occurs when foodstuffs or food containers or plates and glasses are inadvertently contaminated by spraying or direct contamination.

Children spend a substantial part of their time walking and toddling indoors. Ingestion may occur by hand–mouth contact or by object–mouth contact. Such contact may be most prominent in young children, who show extensive mouthing behaviour around 1 year of age, when teeth erupt. Also, such contact transfers, to the mouth, pesticide residues, that have accumulated on hands or on objects. These may have come in contact with pesticide residues on such indoor surfaces as carpets, countertops or hard flooring or on lawns or other outdoor surfaces, such as soil. Oral non-dietary ingestion may be particularly important for infants and children exposed to lawn chemicals and household pesticide products in residential settings, because of the incidence of hand-to-mouth activity (Hawley, 1985) or object-to-mouth activity (Reed et al., 1999) and because of the activities performed by children (such as crawling) that put them in close proximity to treated surfaces (Cohen Hubal et al., 2000). Some assumptions and observations have been made for hand-to-mouth activity in children (Reed, 1998; Zartarian, Ferguson & Leckie, 1997, 1998; Zartarian et al., 2000; Tulve et al., 2002).

Exposure from non-dietary ingestion may be estimated using residue data for the media of interest (such as turf, soil or indoor surfaces) and standard ingestion rates, based on the age group or activity of interest. Non-dietary ingestion can be assessed by using data on transferable residues for the surface area of the hands of children and also using standard assumptions about the frequency of hand-to-mouth and object-to-mouth activity. The mass of residue ingested may be calculated for various age groups as the product of the transferable residue concentration (in mg/cm^2), frequency of hand-to-mouth or object-to-mouth activity (in events/day), and contacted surface area from which residues are ingested (in cm^2/event) to yield the potential dose (in mg/day) from hand-to-mouth transfer. Depending on the scenario of interest – that is, indoor or outdoor – dislodgeable residues should be assessed (PMRA, 1998).

Quantifying the intake rate from hand-to-mouth exposure is very difficult. Intake values have only been established for soil uptake, based on data collected in soil contamination study assessments that used chemical markers, although data specific to biocides have not been reported (EC, 2002).

14.7. Pesticide exposure and risk assessment

Risk assessment is a formal means of evaluating risk in an objective manner that takes into consideration uncertainties and assumptions. Quantitative terms, such as fractions or percentages, can be used to express risk as a function of hazard and exposure. Risk assessments usually apply tiered modelling approaches that range from deterministic modelling based primarily on conservative assumptions (Tier 1) to probabilistic models that use refined assumptions based on real data (Tier 3). Pesticide risk assessments apply conservative assumptions that overestimate exposure and hazard, resulting in a quantitative assessment that has a large margin of safety. The conservatism of these approaches is supported by surrogate data; for example, exposure often can be estimated by using surrogate data developed previously for other chemicals or an exposure database can be used for a surrogate estimate of inhalation and dermal exposure for many exposure scenarios.

Relatively few studies in residential environments have examined the exposure of occupants to pesticides (EPA, 1980). Most risk assessments are based on surrogate data and conservative assumptions (EPA, 1998b, 1999, 2002a). To our knowledge, no study has compared the risk of disease from exposure to urban pests with the risk of pesticide exposure. These risk–risk and risk–benefit analyses would benefit greatly from more exposure studies on both of these topics.

14.7.1. Steps of pre-market risk assessment of pesticides

The first step in the risk assessment process identifies the uses of a pesticide from its proposed label (such as use on lawns, use on carpets and crack-and-crevice treatment) and how it is applied (such as powder, spray or fogger). In the second step, the pesticide's toxicity is studied. The regulatory agency determines the most sensitive effect – the so-called critical effect – for each exposure route and for several exposure durations (such as acute and chronic). Next, all appropriate exposure scenarios are determined. Examples of scenarios include:

- person spraying a liquid pesticide
- person living in a house treated for insects.

A registrant may be required to conduct exposure studies that address specific exposure scenarios for a specific chemical. If an exposure study that addresses a specific exposure scenario has not been conducted and a bridge to the existing use pattern is found unacceptable, scientists will use a model to evaluate available data, published literature, or some other source of information to develop a risk assessment for that scenario.

A different modelling approach focuses on those exposed to pesticides who have not directly used them – that is, secondary exposures of bystanders. The premise of this model is that exposure levels are related to one's activity while in a previously treated area and to the amount of pesticide available to the individual in that environment.

14.7.1.1. Exposure potential

Potential exposures may be measured or modelled, and each approach has advantages and disadvantages. Exposure measurements represent precise observations for a limited number of cases: they can be carried out in the workplace, in the residential environment or through laboratory or workshop studies. Modelled exposures can be estimated from underlying physical processes, the physicochemical properties of the product, characteristics of the formulation and an understanding of the nature of contact with the chemical.

14.7.1.2. Tiered approaches to exposure estimation: a basis for risk assessment

The tiered approach is a logical stepwise approach to risk assessment that uses the available information to the optimum extent, while reducing unnecessary requirements for human exposure surveys or studies. Alternatively, the need for an exposure study can be justified through elimination of all other possibilities. Tiered approaches use increasingly sophisticated analyses, exposure controls and parameter sets. Initial tiers should provide conservative assessments of exposure that are refined in subsequent tiers. The tiered process explained below is a standardized approach to the evaluation of risk.

In residential settings, most risk assessments can be characterized as Tier 1 assessments. If a Tier 1 assessment does not adequately address the margin of safety for a pesticide use pattern, a Tier 3 level approach may be applied to refine the assessment, to eliminate uncertainty. Tier 2 assessments apply to professional applicators and occupational exposures only.

In Tier 1, the assessor selects an indicative exposure value from an empirical (database) or mathematical model, or a reasoned worst case, or by selecting validated data from tasks likely to produce similar exposure distributions. For example, Tier 1 estimates must not take PPE into account. When the result of a Tier 1 exposure assessment produces an unacceptable outcome in risk assessment, a Tier 2 estimate is required.

In Tier 2, the exposure estimate needs to state the default values – these are the assumptions used in the absence of scientific data, and they are set on the basis of scientific information and for conservative purposes – selected and also all assumptions; the assessments may combine some chemical specific data with standard default values or generic data. Tier 2 estimates are appropriate for a detailed exposure assessment of specialized professional users – for example, protective measures are supposed to be carefully observed. If the resulting exposure estimate produces an unacceptable outcome in the risk assessment, the exposure abatement measures may be successively refined and the exposure estimate revised, until the options for exposure reduction are exhausted. If after this remodelling the predicted exposure is still unacceptable, then a third iteration of the exposure assessment will be required.

In Tier 3, the final tier of the assessment, valid estimates of human exposure are produced through surveys or studies with the actual product or with a surrogate. Studies may need to cover an entire scenario and may include biomonitoring to show systemic uptake. The information is particularly useful in the case of a workforce that has been studied over a period of time and at a known (fairly continuous) level of exposure. For residential settings, these types of studies are the exception and usually involve environmental monitoring for pesticide residue or air concentration levels instead of biomonitoring. These data often confirm previous findings and are used to put a more realistic perspective on exposure patterns. Fig. 14.5–14.7 give tiered approach schemes on the three different routes of exposure: inhalation exposure, dermal exposure and oral exposure.

14.7.2. Modelling residential exposures

In the absence of measured data on exposure or representative data on analogous substances, exposure must be estimated using recommended modelling approaches. To ensure that the predictions are realistic, all relevant exposure-related information on the substance should be used iteratively.

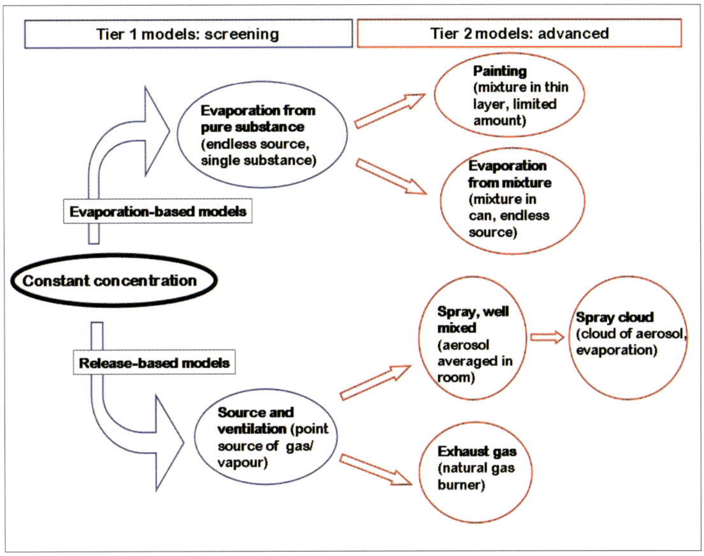

Fig. 14.5. Tiered approach to estimation of inhalation exposure

Source: Modified from EC (2002).

General predictive models are available for generic substances and for specific scenarios. These models take into consideration the physical properties of active substances, such as the particle size of aerosols and volatility of liquids. Mathematical and empirical (database) models exist for a number of scenarios and tasks. Mathematical models for physical evaporation processes are normally for a specific substance and require data on such physical properties as saturated vapour pressure. Models for dispersive processes, such as spraying, typically apply to the product in use, emerging from the spray nozzle. Database models may be highly specific (for example, an active substance discharged from a handheld aerosol can) or generic (for example, a product in use, including propellant, held within the can).

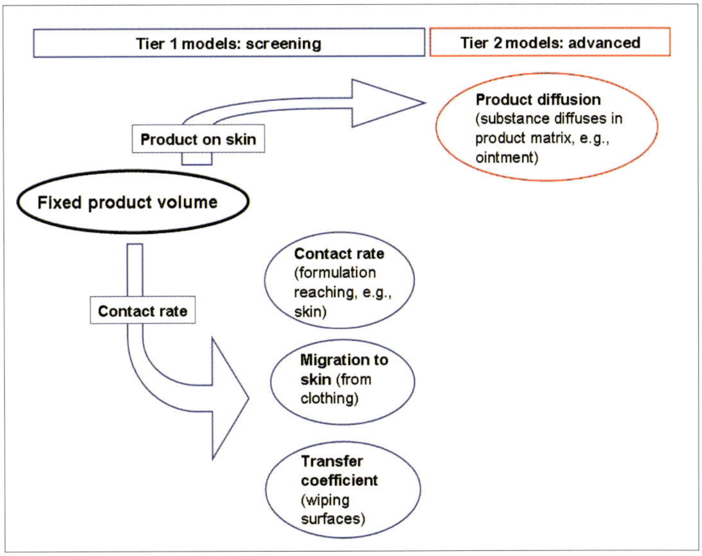

Fig. 14.6. Tiered approach to estimation of dermal exposure

Source: Modified from EC (2002).

The use of exposure models requires the selection of various input parameters. Insufficiently detailed information on exposure scenarios or lack of sufficient data may require the use of default values. Input data or default values used for the calculations must be clearly documented.

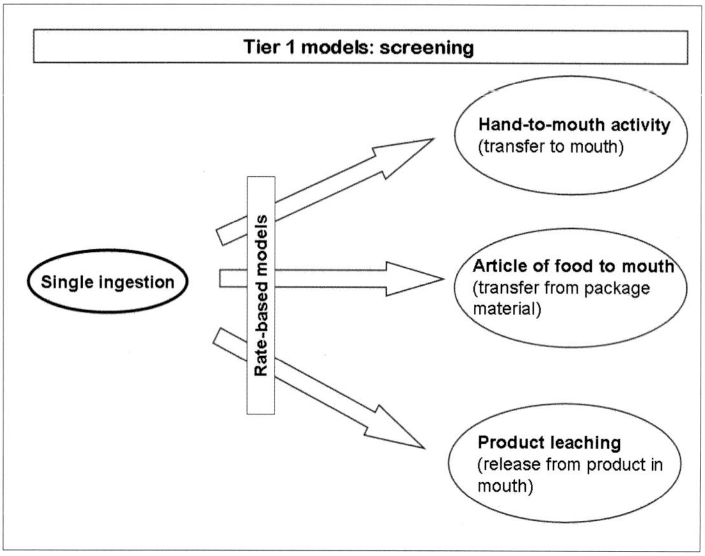

Fig. 14.7. Tiered approach to estimation of oral exposure

Source: Modified from EC (2002).

Computer programs have been developed to implement mathematical predictive models and empirical models. Statistical models have been developed using available data and appropriate statistical methods. Model choice should be justified by showing that the model uses the appropriate exposure scenario – for example, as judged from the underlying assumptions of the model. Expert judgement may be required to check the realism of the exposure value derived from a model, particularly if default or so-called reasonable worst-case values have been used. Modelling exposure can be performed either by taking discrete values (point estimates) or distributions for the model variables (probabilistic modelling).

Generally, exposure models fall into one of three types: mathematical mechanistic models, empirical or knowledge-based models, and statistical mathematical models. These models predict exposure levels from a mechanistic description of a process, an empirical database or statistical relationships.

14.7.2.1. Mathematical mechanistic models

Normally, mathematical models are based on mass balance equations. These models can incorporate the physical and chemical properties of the substance, together with patterns of use. They are used to characterize the rate of release of the product into a space and its subsequent behaviour. Mathematical models should cover all relevant processes or tasks that contribute to exposure in a scenario. For many tasks, a number of different models may be appropriate.

14.7.2.2. Empirical models

Empirical models are probably best described as models based on exposure measurements obtained from real situations. This type of model can be used to predict the likely exposures in other comparable situations. If sufficient and high-quality data are used in empirical models, they are likely to account for the many variables that influence exposure. Currently, no empirical models exist for predicting consumer exposures, since the available databases on exposure measurements are not sufficiently large.

The two model types described above have different attributes and are strictly applicable only within their defined realms of use. In general, mathematical mechanistic models are concerned with specific circumstances or processes (such as paint spraying and drum filling) and usually cannot be used for more general applications. They are likely to give a lot of detailed information on the exposure to be expected. Empirical or knowledge-based models, on the other hand, tend to be applicable to a wider range of circumstances and are based on many years of accumulated experience. While they can give a broad idea of likely exposure, they are imprecise. They are, however, able to predict exposure to new and existing substances and to assess likely exposure across a range of uses of a substance. Model estimates are more widely used for exposure assessment of new substances, because the data available are often poor or non-existent (EC, 1996).

14.7.2.3. Statistical mathematical models

Statistical models have not been used yet for exposure estimations in the EU. Such models use empirical relationships to predict exposures from statistical distributions. In principle, they reflect a combination of empirical and mechanistic models, together with consideration of the distribution of the input parameters. One of the most important steps in the procedure is represented by the implementation of the probabilistic approach, which allows the use of distributions in the calculation. While these models are in use in other related fields, such as the dietary exposure from pesticide residues, their application to exposure to biocidal products is still experimental.

One of the reasons for developing statistical models and replacing deterministic models with probabilistic models is the recognition that deterministic models, by nature, tend to be overly conservative. In fact, they tend to introduce several conservative assumptions in a serial way, and the resulting so-called point estimate reflects worst-case scenarios so extreme as to become clearly unrealistic. Unlike deterministic models, probabilistic models integrate the distribution of discrete input variables and generate a final risk distribution as a continuous variable, allowing the assessor to select the extent of protection (or uncertainty) desired. A recent paper by Lunchick (2001) has an interesting discussion of the pros and cons of deterministic exposure assessments versus probabilistic exposure assessments. He also provides an example of operator exposure where the deterministic exposure estimate corresponds to the 96th percentile of the probabilistic exposure distribution.

How regulators will decide to interpret the results of probabilistic assessments is still open. These models demand the regulator to set the level of desired conservatism and in some cases determine the extent of protection of the population at risk.

Current exposure models are limited to considering a single source and a single agent at a time. However, aggregate and cumulative exposure analyses are now a requirement of registration or regulation in the EU and North America (EPA, 2001b, 2002a–c; PMRA, 2003). Definitions of aggregate and cumulative risk are as follows.

- **An aggregate risk** involves the likelihood of the occurrence of an adverse health effect from all routes of exposure to a single substance.

- **A cumulative risk** involves the likelihood of the occurrence of an adverse health effect from all routes of exposure to a group of substances sharing a common mechanism of toxicity.

14.7.2.4. Some existing models

Newly emerging exposure models are set up to accommodate aggregated residential exposure scenarios that contain multiple sources of a chemical. These models are mostly initiated in response to the demands of the Food Quality Protection Act in the United States. They aggregate exposure from multiple sources, at the cost of requiring good input data for each source. Some examples of these newly emerging exposure models are listed below.

A modelling effort by the Hampshire Research Institute, funded by the EPA, developed a dietary and non-dietary residential exposure model that helps estimate aggregate exposure over a person's lifetime (Price, Young & Chaisson, 2001).

Another model can estimate daily annual non-dietary residential exposure and works in a probabilistic environment. The outputs from this model can be linked with the dietary exposure estimation model to produce an aggregate model.

Exposure to pest control products applied by consumers can be assessed with the ConsExpo (Consumer Exposure) model (Delmaar, Park & van Engelen, 2005; Bremmer et al., 2006; RIVM, 2006). It is relevant to the Biocides Directive (EC 1998, 2002).

In the SHEDS (stochastic human exposure and dose simulation) model, sequential dermal and non-dietary ingestion exposure and dose–time profiles are simulated by combining measured surface residues and residue transfer efficiency with actual micro-level activity data quantified from videotapes (Zartarian et al., 2000). This model has been used for some specialized residential exposure assessments (EPA, 2002d).

The REx (residential exposure) model is another model for aggregated exposure assessment. It is structured according to EPA SOPs for pesticidal residential exposure assessment (EPA, 2000c).

All these models can be used and refined by regulatory agencies as needed.

14.7.3. SOPs and exposure scenario types

SOPs for residential exposure assessments have been developed by the EPA (1997a) and EU (EC, 2002). In both cases, the objective of the SOPs is to provide standard default methods for developing residential exposure assessments for both application and post-application exposures, when applicable monitoring data are limited or not available. The SOPs cover calculation algorithms for estimating dermal, inhalation or incidental ingestion doses for a total of 13 major residential exposure scenarios: (a) lawns; (b) garden plants; (c) trees; (d) swimming pools; (e) painting with preservatives; (f) fogging; (g) crack-and-crevice treatments; (h) pet treatments; (i) detergents; (j) impregnated materials; (k) termiticides; (l) inhalation of residues from indoor treatments; and (m) rodenticides. Default values for the underlying exposure factors, such as amount used or dermal transfer factors, are specified. These default values represent (reasonable) worst-case values. While the SOPs provide methods and default assumptions for conducting screening-level residential exposure assessments for indoor and outdoor settings, they do not preclude the use of more sophisticated methods (including stochastic analyses) and the replacement of default values for exposure parameters with new data. WHOPES has published an easy to understand quantitative exposure and risk assessment for insecticide-treated mosquito nets, which provides a stepwise approach to assessing toxicity, exposure and risk (WHOPES, 2004).

14.7.3.1. Comparing pesticide risks from residential and dietary exposures

When discussing exposure of the general population to pesticides, as a result of home use and application, it has to be kept in mind that dietary exposure to pesticide residues is known to occur in most parts of the world. This dietary exposure is regulated by law in most countries and is legally permitted in so far as it is considered to be *safe* or *acceptable* when the use of pesticides on specific agricultural products is authorized. Since such exposure involves millions of people, the regulating authorities in the various countries of the world have set up complex monitoring systems aimed at checking regularly the residues present on a large variety of food items sampled from the market. The results of this monitoring are collected and analysed periodically to ensure that the foods on the market comply with the legally established levels of residues and that the health risk for consumers remains within the limits set up by legislation.

A general overview of the results produced by the residue monitoring systems indicates that the proportion of samples (of food items) found to be *irregular* is generally rather low and decreasing over time; for example, in the EU, from a frequency of detection of irregular samples of 5–6% typically observed in the mid-1990s, the actual frequency of detection has decreased to 2–3% and in some countries to 1%. Two major reasons account for a sample being defined as irregular.

- A sample contains a pesticide residue permitted, but in a concentration slightly exceeding the maximum residue limit (MRL).

- A food item contains the residue of a pesticide the use of which is not legally permitted for that crop.

A toxicological assessment of such irregularities in pesticide residues suggests that the overall risk for the population is minimum, if any, given the margin of safety adopted in setting MRLs and the modestly excessive levels found in the samples.

These population-level data, though reassuring in terms of toxicological risk, do indicate that consumers normally ingest pesticide residues with their diet, that only about 50% of the food items on the market do not have (measurable) residues in or on them and that a fair percentage of food items (10–20%) bear more than one residue (up to six) at the same time. These analytical observations are in agreement with the results obtained by large biomonitoring surveys of general populations, such as the survey conducted by the National Center for Environmental Health (CDC, 2005) in the United States, that revealed a widespread presence of small amounts of measurable pesticide residues in the urine of subjects who were not employed in agriculture nor were known to be professional users of pesticides.

It is difficult to say how much the home use of pesticides and biocidal products contributes to the pesticide residues detected in general population surveys. In principle, the possibility that home use may be responsible for the urinary occurrence of pesticide residues in some subjects and in others may add considerably to dietary exposure, cannot be excluded. However, the frequency and continuity (basically daily) of dietary exposure, when compared with the infrequent and discontinuous and relatively low level of residential exposure for home use of biocides, make dietary exposure more likely blameworthy than home exposure for the widespread pesticide residue contamination at the general population level.

14.8. Examples of residential risk assessments

This section covers understanding and comparing risks from routes of residential exposure, chlorpyrifos and pyrethrin exposures and risks, and a summary of examples of residential risk assessments.

14.8.1. Understanding and comparing risks from routes of residential exposure

Housing residents are likely to be exposed to pesticides used in the home. The most common routes of exposure are the dermal and inhalation routes, with unintentional (incidental) oral exposure being attributable primarily to toddlers putting their fingers in their mouth after crawling over treated surfaces or touching pets. To better understand the magnitude of exposures by the principal routes and their significance in the light of patterns of application use, we provided (below) examples from EPA exposure evaluations and the public literature. In these evaluations, residents who apply pesticides are called *handlers* or *applicators*, while non-applicator housing residents are called *residential bystanders*.

Quantifying exposure to pesticides from the oral, inhalation, and dermal routes is essential to risk characterization and management. The margin of safety (MOS) is defined as the ratio of the NOAEL, derived from relevant toxicity studies, to the actual (estimated, calculated or measured) exposure. An acceptable exposure level (AEL) is a MOS derived by dividing NOAELs by factors, called uncertainty factors or safety factors by different bodies, that are applied to overcome uncertainty. This chapter uses the term *uncertainty factor* (UF). AELs are identified as such because the uncertainty factor has been applied and represents a benchmark regulation based on the level of human health concern. A MOS calculated from actual data is referred to as *MOS* in this chapter.

Note that the EPA uses the term *margin of exposure* (MOE) instead of MOS. Also, it uses MOE (UF) instead of AEL.

When performing risk assessments, uncertainties arise from the possibility of differences in toxicological responses between people and test animals (interspecies responses), differences within the human population (intraspecies responses) and, where applicable, within subgroups of human populations. UFs are applied to account for these differences, in the absence of evidence to the contrary, as follows.

- People are assumed to be 10 times (10X UF) more sensitive to the adverse effects of a pesticide than tested animals.

- An additional 10X UF is added to account for human intraspecies responses.

In the United States, an additional factor (known in the United States as the Food Quality Protection Act (FQPA) factor) is added by default, assuming that (if not otherwise demonstrated) children are 10 times (10X UF) more sensitive than adults to the adverse effects of pesticides.

If all three UFs are applied, multiplying them together yields an AEL for residential settings equal to 1000, and this AEL is used often in the United States for Tier 1 risk assessments. AEL values less than 1000 may be used if data exist to refine the assessment. An AEL of 1000 means that it is only acceptable to allow a person to be exposed daily to a pesticide amount that is 1000X less than the level at which no adverse effects were observed in laboratory animals – that is, the AEL is 1000 or more. Therefore, an acceptable AEL may range in value from 100 to 1000 or more, depending on the amount and quality of toxicity and exposure data available to refine the risk assessment.

Chlorpyrifos and pyrethrin exposures and risks, as assessed by the EPA, are described below (EPA, 2000d, 2006b). For these compounds, the acute and sub-chronic toxicity end-points are described in Table 14.2.

Table 14.2. Regulatory end-points and toxic effects of chlorpyrifos and pyrethrins

	Acute		Sub-chronic	
Pesticide	**Regulatory end-point (NOAEL)**	**Toxic effect**	**Regulatory end-point (NOAEL)**	**Toxic effect**
Chlorpyrifos	Incidental oral route = 0.5 mg/kg	Plasma and RBC[a] cholinesterase inhibition	—	Cholinesterase inhibition
	Dermal route = 0.15 mg/kg/day		Dermal route = 0.3 mg/kg/day	
	Inhalation route[b] = 0.1 mg/kg/day		Inhalation route[b] = 0.1 mg/kg/day	
Pyrethrins	Inhalation route = 7.67 mg/kg/day	Neurotoxicity	Inhalation route LOAEL = 2.57 mg/kg/day	Neurotoxicity
	Incidental oral route = 20 mg/kg/day		Incidental oral route = 6.4 mg/kg/day	

[a] RBC = red blood cell.

[b] As a conservative assumption, 100% lung absorption was assumed.

Source: EPA (2000d, 2006a).

14.8.2. Risk assessment of residential exposures to chlorpyrifos

Chlorpyrifos is an organophosphorous insecticide that causes toxicity by inhibiting AChE in both insect and mammalian nervous systems. The EPA completed a comprehensive chlorpyrifos exposure and risk assessment in 2000 (EPA, 2000d). The patterns of use most common in indoor environments for control of pests were crack-and-crevice or spot applications and general surface treatments. Chlorpyrifos applied in homes by these methods vaporized after application, contaminated indoor air and led to secondary inhalation exposure. Secondary dermal exposure was most likely from general surface treatments, leading to incidental exposures, especially among children. The effect that was considered adverse by the EPA, which was common to all exposure routes, was inhibition of cholinesterases (not only AChE).

An AEL equal to 1000 was considered acceptable, because it took into account what was considered to be young animals' greater sensitivity to the toxic effects of chlorpyrifos. It should be noted that most residential exposure assessments done for chlorpyrifos were Tier 1 assessments. SOPs and Pesticide Handlers Exposure Database (PHED) data were used in these evaluations. For a few scenarios, studies were available on environmental and biomonitoring, enabling the EPA to refine some of the exposure values.

14.8.2.1. Residential applicator exposure assessment

Three residential applicator exposure assessment scenarios were evaluated:

1. paint-brush applications to surfaces
2. spot treatments with low-pressure hand sprayers
3. outdoor hose-end applications to lawns and shrubs.

Residential applicator data were unavailable, and the following assumptions were taken into consideration in performing the assessment.

- Dermal and inhalation exposures were combined, because cholinesterase inhibition was the toxicological end-point of concern in both cases.

- The average BW of an adult is 70 kg.

- The application of chlorpyrifos is made according to the label directions.

- Applicators wore short-sleeved shirts and no gloves.

- Exposure is short-term (one day to one month after application).

- When data were unavailable, surrogate unit exposures were based on PHED data.

- The AEL is 1000.

The daily dermal dose, inhalation dose and total MOS are defined as follows.

- The daily dermal dose (in µg/kg/day) is equal to the unit exposure (in µg/kg/AI) times kg/AI handled times the dermal absorption factor/70 kg BW.

- The inhalation dose (in µg/kg/day) is equal to:

$$(BZC \times BR \times ED)/BW,$$

where BW is the body weight; BZC is the breathing zone concentration; BR is the breathing rate; and ED is the estimated duration – and is also equal to the unit exposure (µg/m^3 x 13.3 m^3/day x hours/day)/70 kg.

- The total MOS is equal to:

$$1/(\text{Inhalation MOS} + \text{Dermal MOS}).$$

- The MOSs ranged from 6 to 150, and thus were lower than the AEL of 1000.

This EPA assessment concluded that when dermal and inhalation exposures were combined, the risk to residential applicators was low, due to an adequate margin of safety.

14.8.2.2. Residential bystander exposure assessment

As exposure data were unavailable, this assessment is based only on EPA SOPs. The general assumptions for this assessment were as follows.

- The average BW of an adult is 70 kg.
- The average BW of a child (toddler) is 15 kg.
- The application of chlorpyrifos is made according to label directions.
- Exposure is short-term (one day to one month following application).
- Only post-application exposure occurs.
- The method of application was crack-and-crevice or spot treatment.
- The chlorpyrifos dilution applied was 0.5%.
- The application rate of dilution to surfaces was 0.08 litre/m^2.
- AEL is 1000.

14.8.2.3. Inhalation exposure

The general assumptions for assessing this route of exposure were as follows.

- The BRs were assumed to be 1.0 m^3/hour for adults and 0.7 m^3/hour for children.

- The exposure occurred 1–10 days after treatments.

- The inhalation dose is expressed as:

$$(BZC \times BR \times ED)/BW.$$

- The MOS is expressed as: NOAEL/Inhalation Dose.

The AEL is 1000.

14.8.2.4. Incidental (hand-to-mouth) oral exposure

For this route of exposure, the following assumptions were made.

- The saliva extraction factor (SEF: the percentage of the dislodged chlorpyrifos residue that is extracted from the fingers by the saliva of the child) is 50%.
- The surface area (SA) of a hand put in a mouth is 20 cm^2.
- The frequency (Freq) of hand-to-mouth events is one per day (adjusted).
- On surfaces, 5% of dislodgeable residues (indoor surface residue, ISR) are available for transfer to skin coming in contact with these treated surfaces.

The daily oral dose (in mg/day) is given by the expression:

(ISR x HTE x SEF x SA x Freq x ED)/BW,

where AEL equals 1000 and HTE is the hand transfer efficiency.

14.8.2.5. Dermal exposure assumptions

For this route of exposure, the following assumptions were made.

- The body surface available for transfer of residue is 16 700 cm^2 for adults and 6600 cm^2 for children.
- On surfaces, 5% of dislodgeable residues are available for transfer to skin coming in contact with these treated surfaces.

The daily dermal dose (in µg/kg/day) is given by the following expression:

The Unit Exposure (in µg/kg/AI) x kg/AI Handled x Dermal Absorption Factor/*x* BW,

where *x* is 70 kg for adults and 15 kg for children, and the AEL is 1000.

14.8.2.6. Comparison of evaluation of MOSs

For the inhalation route of exposure in treated rooms (kitchen and toilet), the MOSs were less than the AEL of 1000. For crack-and-crevice applications, the dermal and oral combined MOSs were greater than the AEL of 1000. When all exposures were combined, the MOSs for adults and children were less than the AEL of 1000 (for details and the specific range of values, see EPA, 2000d)

Therefore, residential exposures that result from consumer applications of pesticides and post-application exposure in pesticide-treated and untreated rooms were found to be unacceptable because, according to EPA calculations, the MOSs exceeded the level of concern indicated by the required AEL. Additional data were not collected to support the use pattern, and residential use of chlorpyrifos was discontinued in the United States.

14.8.3. Residential exposures to pyrethrins

Pyrethrins are insecticides that interact with sodium channels, disrupting the transmission of impulses along the axons. In 2006, the EPA completed a comprehensive assessment of exposure to pyrethrins and the risk they posed (EPA, 2006b). The use patterns most common in indoor environments for control of pests were crack-and-crevice or spot applications, general surface treatments, and indoor fogging. Pyrethrins applied in homes have a very short residual life and secondary exposure scenarios were unlikely. Inhalation is the primary route of exposure and neurotoxicity is the adverse effect common to all exposure routes.

EPA SOPs (EPA, 1997a) and PHED data were used in these evaluations. However, uncertainties in the assessment arose due to a lack of data on some consumer applicator habits and children's exposure to pesticides. Registrant task forces performed residential studies that measured these exposures, thus enabling the EPA to perform more refined exposure and risk assessments. Due to the availability of these data, an AEL below 1000 could be used. AELs equal to 100 for residential exposures and to 300 for incidental oral exposure were adopted. Therefore, if the MOS calculated as MOS = NOAEL/Daily Dose was greater than 100 or 300, respectively, the exposure was acceptable.

14.8.3.1. Toxicity end-points and MOS

Due to the short life of pyrethrins, only short-term exposures were evaluated. Dermal exposure was not considered a significant exposure route, but toxicity end-points were identified for the inhalation and oral routes of exposure. The oral NOAEL was 20mg/kg/day, while the inhalation NOAEL was 7.67mg/kg/day. As already noted, neurotoxicity is the toxic effect. An AEL of 300 (10 x 10 x 3) for oral incidental exposure was selected because a developmental neurotoxicity study was not available to fully characterize exposure to children. An AEL of 100 was selected for inhalation exposures, because toxicity and exposure were well characterized from the data available.

14.8.3.2. Residential applicator exposure

The assessment of residential applicator exposure evaluated the following application methods: aerosol cans, dusts, handheld wand sprayers and handheld trigger sprayers. The general assumptions made were as follows.

- The maximum application rate was 0.1 kg pyrethrins per 92.90 m^2.
- Only inhalation exposures were assessed, because dermal and oral exposures were assumed to be insignificant. Exposure was assumed to occur at the time of application.

The inhalation MOS was calculated as NOAEL/Inhalation Dose (in mg/kg/day), and the AEL equalled 100.

The calculated MOSs exceeded the AEL of 100. Thus, consumer applicator exposure to pyrethrins did not exceed the level of concern and were considered to be acceptable (for details on the range of MOS values, see EPA, 2006b).

14.8.3.3. Residential bystander exposures

The general assumptions made for residential bystander exposures are as follows.

- The average BW of an adult is 70 kg.
- The average BW of a toddler is 15 kg.
- The highest label rate for each use pattern was applied.
- Exposures were secondary, except for indoor foggers.

For residential bystander exposure, four scenarios were assessed to take into account possible exposures of adults and children after applying pyrethrins:

1. dermal transfer and incidental oral ingestion after pet treatments
2. inhalation exposure
3. incidental oral exposure during and after indoor space-spray or fogging treatments
4. dermal exposure for all use patterns.

14.8.3.4. Scenarios

14.8.3.4.1. Scenario 1: child exposure to pet treatments

The assumptions made for post-application spray treatments of pets were as follows.

- One half of a 473.5-ml spray container is used to treat each animal.

- The transferable residue (TR) from a treated pet is assumed to be 20% of the maximum application rate for sprays.

- The SA of a treated (15 kg) dog is 6000 cm^2 (EPA, 1993).

- The SEF is 50% (50% of the dislodged residue of pyrethrins extracted from the fingers by saliva).

- The SA of a hand put in a mouth is 20 cm^2.

- The Freq of hand-to-mouth events is one per day.

The daily incidental oral dose was given by the following expression:

$$TR \times SEF \times SA_{hands} \times Freq.$$

The MOS is given by the following expression:

Short-term Oral NOAEL (20 mg/kg/day)/Daily Oral Dose (mg/kg/day).

An AEL equal to 100 was used.

The MOS calculated for this route of exposure was greater than 300, which meant this route of exposure was not considered to be of concern.

14.8.3.4.2. Scenario 2: inhalation exposure from space sprays
The assumptions made for inhalation exposure from space sprays, including indoor foggers, were as follows.

- The label rate of 0.15-g AI/28 m^3 is applied (volume of an average size room).
- The occupant inhales aerosol space spray during and after an application.
- The exposure duration (ED) is two hours (based on label directions for use).
- The BRs are 1.0 m^3/hour for adults and 0.7 m^3/hour for children (EPA, 1997a). The actual concentration of pyrethrins (in mg/m^3) in the breathing zones (BZC) was based on data that are proprietary.
- The 9.31 g of product, which contains 0.5% pyrethrins by weight, is applied to a room with an internal volume of 57.99 m^3.

The inhalation dose is given by the expression: (BZC x BR x ED)/BW; the MOS is found from: NOAEL/Inhalation Dose; and an AEL of 100 was used.

For scenario 2, it was concluded that this type of short-term exposure did not exceed the level of concern, because the MOSs were greater than the AEL of 100. This assessment was based on actual data collected from studies conducted indoors when the product was applied as directed on the label.

14.8.3.4.3. Scenario 3: incidental oral exposures after indoor fogging
The assumptions made for incidental oral exposures after indoor fogging were as follows.

- ISR values (reported in µg/cm^2) used in the assessment are based on study data and a maximum label application rate of 0.015-g AI/28.32 m^3 for indoor foggers.
- HTE is measured for carpet and vinyl tile in a study of simulated use. These values are less than the EPA SOP assumption of 20% residue transfer from the surface treated to the hand of a toddler.
- The SA of two fingers is 20 cm^2.
- The SEF is 50%.
- The Freq of hand-to-mouth events is 20 per hour.
- ED is 2 hours/day.
- The BW is 15 kg for toddlers.

The daily oral dose (in mg/day) is given by the expression:

(ISR x HTE x SEF x SA x Freq x ED)/BW.

The MOS is given by the expression:

Short-term Oral NOAEL (20 mg/kg/day)/Daily Oral Dose.

An AEL of 300 was used.

In this assessment, hand-to-mouth transfer from treated surfaces did not present a risk of concern as the MOS was greater than 300.

14.8.3.4.4. Scenario 4: dermal exposure from applications of pyrethrins
Dermal risk assessments were not required for pyrethrins, due to negligible dermal absorption and dermal toxicity. A 21-day dermal toxicity study in rabbits showed no systemic or dermal toxicity at the limit dose of 1000 mg/kg/day. Also, a human dermal penetration study showed absorption was less than 0.22%.

For pyrethrins, the comparative risk from the use patterns can be described as follows:

Inhalation > Incidental Oral Ingestion > Dermal Exposures.

The dermal exposure presents the lowest risk and inhalation exposure the highest. However, inhalation risks are higher for children than for adults. Also, residues on pets treated with pyrethrins for control of pests present little risk to children and adults.

14.8.4. Summary of examples of residential risk assessments

In summary, chlorpyrifos and pyrethrins were used to provide overviews of residential exposures and risk assessments performed by the EPA. These examples show that residential assessments are based on conservative assumptions and surrogate data, together with chemical-specific exposure data, to reduce uncertainties.

14.9. Benchmarks

The approach described above relies on sophisticated models and direct detection (when possible) of pesticide residues to determine risks from exposure to pesticides. The MOS, also called MOE by the EPA, is the benchmark used to describe most pesticide risks. Generally, if the pesticide's toxicology database and characterization of exposure are adequate, the acceptable AEL = 100. On the other hand, if the toxicology database or exposure characterization (or both) is inadequate, the benchmark is usually AEL = 1000. This framework ensures an adequate MOS for residential occupants, including children and pregnant women.

Assessing the risk of cancer requires a much different kind of assessment and expression of risk (benchmark). Assessing the risk of cancer from pesticides is based on evidence from cancer studies in at least two animal species, together with evidence from genotoxicity studies. Pesticide exposure is always much higher in these studies than any expected exposure in people. Where possible, studies are conducted that determine (or can provide a hint about) the cancer-causing mechanism. The outcome of these animal studies is used in a so-called weight-of-the-evidence approach, to decide if a pesticide is likely to have carcinogenic effects in people. Risk management is applied to further mitigate any risks.

The quantitative assessment of the risk of cancer requires the use of sophisticated statistical models to estimate its potential at the lower levels of exposure seen in people. A model used widely for regulatory purposes is the linearized multistage model. This model results in an expression of a unit risk of cancer, Q1* (known as a Q-star), that allows the calculation of the likelihood or probability of cancer (lifetime cancer risk) for an average daily lifetime exposure. For example, a 1×10^{-6} risk of cancer means that an individual has a one in a million chance of developing cancer from an average daily lifetime exposure to a particular pesticide.

The acceptability of the risk of cancer is a risk management decision that cannot rely exclusively on a numerical standard, but needs to take into consideration all the factors that influence the risk. From the historical actions of regulatory agencies, it is recognized that areas of regulatory concern for the lifetime risk of cancer are in the neighbourhood of 10^{-4} to 10^{-6} (one in ten thousand to one in a million). A lifetime risk of cancer that is below one in a million usually does not indicate an unacceptable risk for the general population – and to otherwise unintentionally exposed people, such as housing residents – when exposure occurs through pesticide residues in or on food. In some instances, cancer risks in the range of 1×10^{-5} to 1×10^{-6} (one in one-hundred thousand to one in a million) have been tolerated for industrial workers exposed occupationally to carcinogenic chemicals. These risk ranges are used by pesticide regulators as benchmarks to guide them through decisions about the acceptability of the lifetime risk of cancer.

Both types of pesticide risk benchmarks, the MOS approach and the quantitative cancer risk assessment, provide estimates of risk that arise from defined exposures. Usually the estimate reflects a so-called typical exposure and use situation, taking into consideration whether the exposure is occasional or frequent and of short or long (lifetime) duration. The estimate is kept conservative by generally overestimating exposure and risk and by using many so-called worst case assumptions, such as assuming that 100% of the airborne pesticide exposure would be inhaled and absorbed or that all applicators will make applications at the maximum application rate, or that 100% of the pesticide deposited on the skin would penetrate through the skin. These benchmarks serve as a basis for regulatory decision-making and aid in residential risk-management approaches.

14.10. Review of main points

Risk assessment is a formal means of evaluating risk in an objective manner by considering uncertainties and assumptions. Risk, as a function of hazard and exposure, can be expressed in quantitative terms. Risk assessments usually apply tiered modelling approaches that range from deterministic modelling based primarily on conservative assumptions (Tier 1) to probabilistic modelling that uses refined assumptions based on real data (Tier 3). Pesticide risk assessments apply conservative assumptions that overestimate exposure and hazard, resulting in a quantitative assessment that has a large MOS. The conservatism of these approaches is supported by surrogate data, because there have been relatively few studies in residential environments that have examined the exposure of occupants to pesticides. The objective of this chapter is to examine human health risks from indoor pesticide exposure in residential settings.

Due to the intrinsic toxicity of pesticides, the need for strict regulation of their admission to the market and their use has been recognized for a long time. Also, in most countries specific and complex legislation prescribes a thorough risk assessment process prior to their entrance in the market, to guarantee health safety for consumers and protection of the environment.

Pesticides are regulated nationally and internationally as toxicants. In Europe and North America, the ECB, EPA, and PMRA are the principle pesticide regulatory agencies. A legal regulatory framework also exists in most individual states. This framework includes standardized approaches to risk characterization, and many decisions are made in a forum that includes the general public and other stakeholders. This framework and associated regulatory processes ensure a thorough review of pesticidal effects, and exposure and use patterns, to fully characterize risks to workers and the general public. Special consideration is given to children, pregnant women and other sensitive sub-populations. In many regulatory agencies, the precautionary principle is applied where uncertainty exists, and alternatives to more toxic pesticides are given priority in regulatory reviews and registration.

Post-marketing risk assessment takes place after a pesticide has been put on the market and is aimed at evaluating the safety of the actual conditions and pattern of use. That some pesticides are being removed from the market due to unacceptable intrinsic toxicity does not have to be regarded with concern. On the contrary, it represents a part of a complex approach to consumer protection and has to be addressed as part of the larger pesticide regulatory framework to prevent risks to human health.

The first two steps in the risk assessment process are identifying the hazard and describing the dose–response relationship. These steps establish the toxicological profile of the pesticide, based primarily on standardized toxicity testing in animal models. Acute, subchronic and chronic testing describes the short-term and long-term effects from exposure by the oral, dermal and inhalation routes. However, to date, standardized testing in animal models has not been effective in predicting human responses to pesticide exposure that cause allergenic or neuropathic effects. Furthermore, the development of animal and computer models to detect and understand possible, if any, relationships between pesticide exposures and endocrine disruption is still in its infancy. Another area of uncertainty concerns the interaction of pesticides with inert ingredients in formulations. Currently, these effects are quantified for acute exposures, but toxicity evaluations by regulatory agencies continue on inert ingredients.

Quantifying exposure to pesticides is essential to characterizing and managing risk. Because risk generally increases with exposure, exposures from the oral, inhalation, and dermal routes must be determined. To evaluate risk, a risk assessment integrates hazard and exposure. An MOS is calculated as a function of hazard and exposure and AELs are applied as precautionary measures to ensure an adequate MOS, to minimize risks from pesticide exposures. For most assessments of the residential risk of pesticides, the AEL equals 1000. This exposure is 1000 less than the NOAEL derived from animal testing.

For residents exposed to pesticides used in the home, the dermal and inhalation routes are the most common routes of exposure, with unintentional (incidental) oral exposure attributable primarily to toddlers putting their fingers in their mouth after crawling over treated surfaces or touching pets. To better understand the magnitude of exposures by the principal routes, together with their significance in light of application use patterns, we provided examples from EPA exposure evaluations and cited public literature.

Risks from residential pesticide use and exposure are affected by a number of factors, including: the hazard and the residual quality of the pesticide applied; the amount of AI applied per unit area; the application method; the formulation type; and the route or routes of exposure. Exposures to chlorpyrifos and pyrethrins and the corresponding risks were described to illustrate the evaluation process. For chlorpyrifos, exposure took place at the time of application and for weeks thereafter, even with crack-and-crevice treatments. On the other hand, due to the low application rates and to pyrethrins being short lived, exposure was minimal by the inhalation route, and dermal exposure was not a concern.

Based on these assessments and the weight of the scientific evidence, the risk of pesticide use and pesticide application by consumers in residential environments usually does not exceed a level of concern. Compared with agricultural applications, residential applications of pesticides are made at much lower rates and use application methods designed to minimize direct or incidental exposures. However, assessments of residential pesticides, such as chlorpyrifos, show that some pesticides are unsafe.

The chemical nature of a pesticide and its patterns of use can result in significant exposure concerns. Such concerns have fuelled the regulatory review of all pesticide chemicals and led to extensive international efforts to protect sensitive sub-populations, such as children, pregnant women and people sensitive to chemicals. This concern has also led to the application of the precautionary principle to the regulatory decision-making process. Where uncertainty exists, caution is used as a regulatory measure to ensure that exposures to pesticides do not cause adverse effects. IPCS is coordinating efforts to harmonize international approaches to hazard, exposure and risk assessment.

When discussing exposure of the general population to pesticides, as a result of home use and application, it has to be borne in mind that dietary exposure to pesticide residues is known to occur in most parts of the world. This dietary exposure is regulated by law in most countries and is legally permitted, in so far as it is considered to be *safe* or *acceptable* when the use of pesticides on specific agricultural products is authorized. Since such exposure concerns millions of people, regulating authorities throughout the world have set up complex monitoring systems aimed at regularly checking the residues present on a large variety of food items sampled from the market. The results of this monitoring are collected and analysed periodically, to ensure that the foods on the market comply with the legally established levels of residues and that the health risk for consumers remains within the limits set up by legislation.

A general overview of the findings of residue monitoring systems indicates that the proportion of samples (food items) found to be *irregular* is generally rather low and is decreasing over time: for example, in the EU, from a frequency of detection of irregular samples of 5–6% typically observed in the mid-1990s, the actual frequency of detection has decreased to 2–3%, and in some countries to 1%. Among the reasons for a sample to be defined irregular, two major circumstances may occur that result in this classification: (a) a sample may contain a permitted pesticide residue, but in a concentration slightly exceeding the respective MRL; or (b) a food item may contain the residue of a pesticide the use of which is not legally permitted for that crop. A toxicological assessment of such irregularities in pesticide residues suggests that the overall risk for the population is minimum, if any, given the MOS adopted in setting the MRLs and given the modestly exceeded levels found in the samples.

These data at the population level, though reassuring in terms of toxicological risk, do indicate that consumers normally ingest pesticide residues with their diet, that only about 50% of the food items on the market do not have (measurable) residues in or on them and that a fair percentage of food items (10–20%) bear more than one residue (up to six) at the same time. These analytical observations are in agreement with the results obtained by large biomonitoring surveys of the general population, which have revealed a widespread presence of small amounts of measurable pesticide residue in the urine of subjects who were not employed in agriculture nor were known to be professional users of pesticides.

It is difficult to say how much the home use of pesticides and biocidal products contributes to the pesticide residues detected in general population surveys. In principle, the possibility that home use may be responsible for the urinary occurrence of pesticide residues in some subjects and in others may add considerably to dietary exposure cannot be excluded. However, both the relative frequency and continuity of dietary exposure (basically daily), when compared with the infrequent and discontinuous home use of biocides, and the relatively low level of residential exposure, when compared with dietary exposure, make dietary exposure more likely blameworthy than home exposure for the widespread pesticide residue contamination at the general population level.

Cancer from exposure to pesticides has long been a concern of regulators and public health professionals. Current risk assessment methods quantify cancer risk and probability at the population level, usually expressed as a value, such as the chance of one cancer occurring in one million people. Although most pesticides present little risk of cancer, people in occupations that handle them are likely to have significant exposures and must use PPE, as directed by the pesticide label, to mitigate such risks.

To avoid an unacceptable risk to the consumer, the application of a pesticide must take place according to good application practices. The information necessary for good practices are reported by the producer on the label: a careful reading provides the consumer with all the preventive measures that should be applied to avoid a risk to health. The use of PPE is one such measure. Moreover, the label reports data about pesticide toxicity, symptoms of poisoning and first aid.

The application of a pesticide has to be rationally planned: application on surfaces that can represent sources of secondary exposure must be carefully evaluated, and subsequent precautionary measures must be adopted. For instance, before applying a pesticide in a room where food is consumed or cooked, all food, utensils, and table crockery must be removed.

Floors and other touchable surfaces represent another source of secondary exposure. These contaminated areas should receive attention. Exposing children to pesticides is of special concern. As they spend most of their time toddling, crawling and playing on the floor, children have frequent hand-to-mouth or object-to-mouth activity. Because of this, they should not be allowed to move freely shortly after a pesticide application (depending on the specific substance applied). Also, pets can be a source of secondary exposure when treated for fleas and ticks.

Several steps can be identified to reduce exposure and manage consumer risks. Among them are the following.

- Read the label and follow the directions when working with a pesticide. It is illegal to use any pesticide in any manner inconsistent with the directions on its label.

- Mix or dilute pesticides outdoors.

- Apply the pesticide only in recommended quantities.

- Increase ventilation when using pesticides indoors. Take plants or pets outdoors when applying pesticides or flea-and-tick treatments.

- If you use a pest control company, select a company licensed to perform this service in your locality.

- Do not store unneeded pesticides inside the home; dispose of unwanted containers safely.

- Store clothes with moth repellents in separately ventilated areas, if possible.

- Keep indoor spaces clean, dry and well ventilated to avoid pest and odour problems.

Unless you have had special training and are certified, never use a pesticide that is restricted to use by state-certified pest control operators.

In conclusion, indoor applications of pesticides, which are regulated by a complex risk assessment before and after they are put on the market, do not pose a high level of risk to human health if the application of the product and the management of the application take place according to proper and adequate procedures. This adherence to proper procedures, together with recent efforts to produce pesticides with a lower overall toxicity, is able to reasonably assure the absence of any unacceptable risk to human health and the environment.

To our knowledge, no studies have considered which risk is worse: the risk of disease from exposure to urban pests or the risk of pesticide exposure. These risk–risk and risk–benefit analyses would benefit greatly from more exposure studies on both of these topics.

14.11. Future actions and data development

A number of future actions would help reduce the risk posed by pesticides to consumers and the environment. Ten such actions follow.

1. Consumer applicators should be educated in pesticide use, including reading and following the product label. Labels should be short and easy to read, with clear and concise first-aid instructions. Also, labels should always be written in the language of the target user population.

2. International harmonization of all pesticide labelling should be implemented. The use of international hazard, risk and use symbols should be employed whenever practical.

3. All pesticide products for consumer use should be packaged in child-resistant containers, to avoid accidental poisoning.

4. Hazardous pesticides should not be sold to consumers. Only WHO Category IV or equivalent category products formulated as ready-to-use products should be sold for consumer use.

5. Sales of pesticide concentrates to consumers should be forbidden, to avoid exposing mixers or loaders and to avoid over-application of AIs to the environment.

6. More research on pesticide use in residential settings should be conducted, and this research should better quantify environmental concentrations of pesticides in residential

environments, to increase the level of certainty associated with residential pesticide exposure assessment and risk characterization.

7. The precautionary principle should always be applied where uncertainty exists; however, regulatory decision-makers should rely on the results of quantitative risk assessments. Also, international harmonization of risk and exposure assessment efforts should continue to increase the uniformity and scope of public health protection.

8. In residential environments, only substances with the least potential for causing cancer should be registered (authorized) and used.

9. Data should be developed on the public health benefits of pesticide use, such as the role of pesticides in preventing disease transmission and pest or vector infestation.

10. The risk of disease from exposure to urban pests must be weighed against the risk of pesticide exposure. A comparison of these two risks must be developed and considered. These risk–risk and risk–benefit analyses would benefit greatly from more exposure studies on urban pest and pesticide exposure to human beings.

References[6]

Aldridge WN (1990). An assessment of the toxicological properties of pyrethroids and their neurotoxicity. *Critical Reviews in Toxicology*, 21:89–104.

Birnbaum LS, Fenton SE (2003). Cancer and developmental exposure to endocrine disruptors. *Environmental Health Perspectives*, 111:389–394.

Bremmer HJ et al. (2006). *Pest control products fact sheet*. Bilthoven, National Institute for Public Health and the Environment (RIVM), (RIVM report 320005002/2006; http://www.rivm.nl/bibliotheek/rapporten/320005002.pdf; accessed 3 September 2007).

CDC (1999). Illnesses associated with occupational use of flea-control products – California, Texas, and Washington, 1989–1997. *Morbidity and Mortality Weekly Report*, 48:443–447.

CDC (2005). *Third national report on human exposure to environmental chemicals*. Atlanta, GA, National Center for Environmental Health, Centers for Disease Control and Prevention (NCEH Publication No. 05-0570; http://www.cdc.gov/exposurereport/3rd/pdf/thirdreport.pdf, accessed 28 December 2006).

CEN (1995). *Workplace atmospheres – guidance for the assessment of exposure by inhalation to chemical agents for comparison with limit values and measurement strategy*. Brussels, European Committee for Standardization (Standard reference: EN 689:1995).

Chen SY et al. (1991). An epidemiological study on occupational acute pyrethroid poisoning in cotton farmers. *British Journal of Industrial Medicine*, 48:77–81.

Cohen Hubal EA et al. (2000). Children's exposure assessment: a review of factors influencing children's exposure, and the data available to characterize and assess that exposure. *Environmental Health Perspectives*, 108:475–486.

De Bleeker J, van den Neucker K, Colardyn F (1993). Intermediate syndrome in organophosphate poisoning: a prospective study. *Critical Care Medicine*, 21:1706–1711.

Debboun M, Frances SP, Strickman DA (2007). *Insect repellents: principles, methods and uses*. Boca Raton, FL, CRC Press, Taylor & Francis Group.

[6] Information on pesticide risk assessment and risk management was obtained from sites of governmental agencies and outstanding international institutions involved in pesticide regulation and consumer protection, as well as sites of pesticide producers' associations. The sites visited included: EPA, European Commission, Joint Research Centre, WHO, FAO and the American Crop Protection Agency. Data on risk characterization was represented by the scientific literature obtained from PubMed and TOXNET. The criteria used for selection included: scientific relevance of the subject, impact factor of the journal and date of publication. Information was also obtained from the most recently published books in the field of pesticide toxicology. Information from sites of outstanding universities was also taken into consideration.

Delmaar JE, Park MVDZ, van Engelen JGM (2005). *ConsExpo 4.0: consumer products and uptake manual*. Bilthoven, National Institute for Public Health and the Environment (RIVM) (RIVM report 320104004/2005; http://rivm.openrepository.com/rivm/bitstream/10029/7307/1/320104004.pdf, accessed 23 September 2007).

DFG (1993). *Analysis of hazardous substances in air, Vol. 1–2*. Weinheim, Wiley-VCH Verlag.

EC (1996). *Technical Guidance Document in support of Commission Directive 93/67/EEC on Risk Assessment for new notified substances. Commission Regulation (EC) No 1488/94 on Risk Assessment for existing substances. Part IV*. Brussels and Luxembourg, European Commission (http://ecb.jrc.cec.eu.int/documents/TECHNICAL_GUIDANCE_DOCUMENT/EDITION_1/tgdpart4.pdf, accessed 25 May 2007).

EC (1998). Directive 98/8/EC of the European Parliament and of the Council concerning the placing of biocidal products on the market. *Official Journal of the European Communities*, L123:1–63.

EC (2002). *Technical notes for guidance: human exposure to biocidal products – guidance on exposure estimation*. Brussels, European Commission (http://ecb.jrc.it/documents/Biocides/TECHNICAL_NOTES_FOR_GUIDANCE/TNsG_ON_HUMAN_EXPOSURE, accessed 12 March 2007).

Ecobichon DJ (1996). Toxic effects of pesticides. In: Klaassen CD, Doull J, eds. *Casarett and Doull's toxicology: the basic science of poisons*, 5th ed. New York, McGraw-Hill:643–689.

Elliot M (1976). Properties and application of pyrethroids. *Environmental Health Perspectives*, 14:1–2.

EPA (1980). *National household pesticide usage study, 1976–1977*. Washington, DC, Office of Pesticide Programs, United States Environmental Protection Agency (EPA 540/9-80-002).

EPA (1993). *Wildlife exposure factors handbook*. Washington, DC, National Center for Environmental Assessment, United States Environmental Protection Agency (EPA/600/R-93/187; http://cfpub.epa.gov/ncea/cfm/recordisplay.cfm_deid=2799, accessed 28 December 2006).

EPA (1996). *The Food Quality Protection Act of 1996. U.S. Public Law 104–170 to amend the Federal Insecticide, Fungicide and Rodenticide Act*. Washington, DC, United States Environmental Protection Agency (http://www.epa.gov/pesticides/regulating/laws/fqpa/, accessed 29 December 2006).

EPA (1997a). *Standard operating procedures (SOPs) for residential exposure assessments.* Washington, DC, United States Environmental Protection Agency (http://www.epa.gov/oscpmont/sap/meetings/1997/september/sopindex.htm, accessed 12 May 2007).

EPA (1997b). *Pesticide Registration (PR) Notice 97-1*. Washington, DC, Office of Pesticide Programs, United States Environmental Protection Agency (www.epa.gov/opppmsd1/PR_Notices/pr97-1.html, accessed 28 December 2006).

EPA (1998a). *Reregistration eligibility decision (RED) DEET*. Washington, DC, United States Environmental Protection Agency (EPA738-R-98-010; http://www.epa.gov/oppsrrd1/REDs/0002red.pdf, accessed 3 September 2007).

EPA (1998b). *Series 875 occupational and residential exposure test guidelines – final*. Washington, DC, United States Environmental Protection Agency (http://www.epa.gov/opptsfrs/publications/OPPTS_Harmonized/875_Occupational_and_Residential_Exposure_Test_Guidelines/Series, accessed 19 March 2007).

EPA (1999). *Framework for assessing non-occupational, non-dietary (residential) exposure to pesticides (draft 12/22/98). Executive Summary*. Washington, DC, United States Environmental Protection Agency (www.epa.gov/fedrgstr/EPA-PEST/1999/January/Day-04/6030.pdf, accessed 28 December 2006).

EPA (2000a). Thiamethoxam: pesticide tolerance. *Federal Register Environmental Documents*, 65(246):80343–80353 (DOCID:fr21de00-17; http://www.epa.gov/fedrgstr/EPA-PEST/2000/December/Day-21/p32570.htm, accessed 28 December 2006).

EPA (2000b). *Indoxacarb pesticide fact sheet*. Washington, DC, Office of Prevention, Pesticides and Toxic Substances, United States Environmental Protection Agency (http://www.epa.gov/opprd001/factsheets/indoxacarb.pdf, accessed 28 December 2006).

EPA (2000c). Intro and Background: Session III: models – residential exposures – REx model. In: *Scientific Advisory Panel Meeting, 26–29 September 2000, Arlington, Virginia*. Washington, DC, United States Environmental Protection Agency (http://www.epa.gov/scipoly/sap/meetings/2000/september/rex_background_memo2.pdf, accessed 12 March 2007).

EPA (2000d). *Reregistration eligibility decision for chlorpyrifos*. Washington, DC, United States Environmental Protection Agency (http://www.epa.gov/oppsrrd1/REDs/chlorpyrifos_ired.pdf, accessed 29 December 2006).

EPA (2001a). *National home and garden pesticide use survey*. Washington, DC, United States Environmental Protection Agency (No. RTI/5100/17-01f).

EPA (2001b). *General principles for performing aggregate exposure and risk assessments*. Washington, DC, Office of Pesticide Programs, United States Environmental Protection

Agency (www.epa.gov/pesticides/trac/science/aggregate.pdf, accessed 29 December 2006).

EPA (2002a). *Child-specific exposure factors handbook (Interim Report) 2002.* Washington, DC, Office of Research and Development, National Center for Environmental Assessment, United States Environmental Protection Agency (EPA-600-P-00-002B; http://cfpub.epa.gov/ncea/cfm/recordisplay.cfm_deid=55145, accessed 29 December 2006).

EPA (2002b). *Organophosphate pesticides: revised cumulative risk assessment.* Washington, DC, United States Environmental Protection Agency (http://www.epa.gov/pesticides/cumulative/rra-op, accessed 29 December 2006).

EPA (2002c). Guidance on cumulative risk assessment of pesticide chemicals that have a common mechanism of toxicity. *Federal Register Environmental Documents*, 67(11): 2210–2214 (http://www.epa.gov/fedrgstr/EPA-PEST/2002/January/Day-16/p959.htm, accessed 29 December 2006).

EPA (2002d). Stochastic human exposure and dose simulation Model (SHEDS): system operation review of a scenario specific model (SHEDSWood) to estimate children's exposure and dose to wood preservatives from treated playsets and residential decks using EPA's SHEDS probabilistic model. In: *FIFRA Scientific Advisory Panel Meeting, 30 August 2002, Arlington, Virginia*. Washington, DC, United States Environmental Protection Agency (http://www.epa.gov/scipoly/sap/meetings/2002/august30/minutes.pdf, accessed 29 December 2006).

EPA (2005). Thiamethoxam: pesticide tolerance. *Federal Register Environmental Documents*, 70:708–720 (http://www.epa.gov/EPA-PEST/2005/January/Day-05/p089.htm, accessed 12 March 2007).

EPA (2006a). *Pesticides and child safety.* Washington, DC, United States Environmental Protection Agency (http://www.epa.gov/pesticides/factsheets/childsaf.htm, accessed 12 March 2007).

EPA (2006b). *Registration eligibility decision for pyrethrins*. Washington, DC, United States Environmental Protection Agency (http://www.epa.gov/oppsrrd1/REDs/pyrethrins_red.pdf, accessed 29 December 2006).

FAO/WHO (1994). *Pesticide residues in food – 1994. Report of the Joint Meeting of the FAO Panel of Experts on Pesticide Residues in Food and the Environment and the WHO Expert Group on Pesticide Residues, Rome, 19–28 September 1994.* Rome, Food and Agriculture Organization of the United Nations and World Health Organization (FAO Plant Production and Protection Paper 127).

FAO/WHO (2005). *Pesticide residues in food – 2005. Report of the Joint Meeting of the FAO Panel of Experts on Pesticide Residues in Food and the Environment and the WHO Core*

Assessment Group on Pesticide Residues. Geneva, Switzerland, 20–29 September 2005. Rome, Food and Agriculture Organization of the United Nations and World Health Organization (FAO Plant Product and Production Paper 183; http://www.fao.org/ag/AGP/AGPP/Pesticid/JMPR/Download/2005_rep/report2005jmpr.pdf, accessed 28 December 2006).

Fenske RA et al. (1990). Potential exposure and health risks of infants following indoor residential pesticide applications. *American Journal of Public Health*, 80:689–693.

FIFRA (1972). *The Federal Insecticide, Fungicide and Rodenticide Act of 1972; 7 U.S.C. s/s 136 et seq.* Washington, DC, United States Government Printing Office (http://www.access.gpo.gov/uscode/title7/chapter6_.html, accessed 29 December 2006).

Findlay A (1995). *The assessment of respiratory and dermal exposure to pesticides: a review of current practice.* Canberra, Worksafe Australia, Australian Government Publishing Service.

Green T et al. (2005). Thiamethoxam induced mouse liver tumors and their relevance to humans. *Toxicological Sciences*, 86:36–47.

Gurunathan S et al. (1998). Accumulation of chlorpyrifos on residential surfaces and toys accessible to children. *Environmental Health Perspectives*, 106:9–16.

Hawley JK (1985). Assessment of health risk from exposure to contaminated soil. *Risk Analysis*, 5:289 302.

He F et al. (1988). Effects of pyrethroid insecticides on subjects engaged in packaging pyrethroids. *British Journal of Industrial Medicine*, 45:548–851.

He F et al. (1989). Clinical manifestations and diagnosis of acute pyrethroid poisoning. *Archives of Toxicology*, 63:54–58.

Hore P et al. (2005). Chlorpyrifos accumulation patterns for child-accessible surfaces and objects and urinary metabolite excretion by children for 2 weeks after crack-and-crevice application. *Environmental Health Perspectives*, 113:211–219.

HSE (2007). Methods for the determination of hazardous substances [web site]. London, Health & Safety Executive (MDHS Series; (http://www.hse.gov.uk/pubns/mdhs/index.htm, accessed 12 May 2007).

IFCS (2005). *Chemical safety and children's health: protecting the world's children from harmful chemical exposures.* Geneva, WHO Intergovernmental Forum on Chemical Safety, Children and Chemical Safety Working Group (http://www.who.int/ifcs/champions/booklet_web_en.pdf, accessed 7 August 2007).

IFCS (2006). *Acutely toxic pesticides – a global guide to resources.* Geneva, WHO

Intergovernmental Forum on Chemical Safety (http://www.who.int/ifcs/documents/champions/atp_web_prel.doc, accessed 12 September 2007).

Jeyaratnam J (1990). Acute pesticide poisoning: a major global health problem. *World Health Statistics Quarterly*, 43:139–144.

Kielhorn J, Melching-Kollmuss S, Mangelsdorf I (2006). *Dermal absorption*. Geneva, World Health Organization (Environmental Health Criteria 235; http://www.who.int/ipcs/publications/ehc/ehc235.pdf, accessed 12 March 2007).

Landrigan PJ et al. (1999). Pesticides and inner-city children: exposures, risks, and prevention. *Environmental Health Perspectives*, 107(Suppl. 3):431–437.

Lange PF, Terveer J (1954). Warfarin poisoning: report of fourteen cases. *United States Armed Forces Medical Journal*, 5:872–877.

Lapied B, Grolleau F, Sattelle DB (2001). Indoxacarb, an oxadiazine insecticide, blocks insect neuronal sodium channels. *British Journal of Pharmacology*, 132:587–595.

Le Quesne PM, Maxwell IC, Butterworth ST (1981). Transient facial sensory symptoms following exposure to synthetic pyrethroids: a clinical and electrophysiological assessment. *Neurotoxicology*, 2:1–11.

Licari L, Nemer L, Tamburlini G (2005). *Children's health and environment: developing action plans*. Copenhagen, WHO Regional Office for Europe (http://www.euro.who.int/document/E86888.pdf, accessed 29 December 2006).

Lotti M (1991). The pathogenesis of organophosphate polyneuropathy. *Critical Reviews in Toxicology*, 21:465–487.

Lotti M (2000). Organophosphorus compounds. In: Spencer PS, Schaumburg HH, Ludolph AC, eds. *Experimental and clinical neurotoxicology*, 2nd ed. New York, Oxford University Press:898–925.

Lotti M (2002). Low-level exposures to organophosphorus esters and peripheral nerve function. *Muscle & Nerve*, 25:492–504.

Lotti M, Moretto A (2005). Organophosphate-induced delayed polyneuropathy *Toxicological Reviews*, 24:37–49.

Lunchick C (2001). Probabilistic exposure assessment of operator and residential non-dietary exposure. *Annals of Occupational Hygiene*, 45(Suppl. 1):S29–S42.

Martin-Bouyer G et al. (1983). Epidemic of haemorrhagic disease in Vietnamese infants caused by warfarin-contaminated talcs. *Lancet*, 1:230–232.

Mohamed F et al. (2004). Acute human self-poisoning with the *N*-phenylpyrazole insecticide fipronil – a GABAA-gated chloride channel blocker. *Journal of Toxicology: Clinical Toxicology*, 42:955–963.

Moretto A (1991). Indoor spraying with the pyrethroid insecticide lambda-cyhalothrin: effects on spraymen and inhabitants of sprayed houses. *Bulletin of the World Health Organization*, 69:591–594.

Moretto A (2002). Occupational aspects of pesticide toxicity in humans. In: Marrs TC, Ballantyne B. eds. *Pesticide toxicology and international regulation*. Chichester, John Wiley & Sons Ltd:431–458.

Moretto A, Lotti M (1998). Poisoning by organophosphorus insecticides and sensory neuropathy. *Journal of Neurology, Neurosurgery and Psychiatry*, 64:463–468.

Pastoor T et al. (2005). Case study: weight of evidence evaluation of the human health relevance of thiamethoxam-related mouse liver tumors. *Toxicological Sciences*, 86:56–60.

Peden M, McGee KS, Krug EG, eds (2002). *Injury: a leading cause of the global burden of disease, 2000.* Geneva, World Health Organization (http://whqlibdoc.who.int/publications/2002/9241562323.pdf, accessed 29 December 2006).

PMRA (1998). Nondietary ingestion exposure assessment. In: *Postapplication exposure monitoring test guidelines (PRO98 04).* Ottawa, Canada, Pest Management Regulatory Agency:B9-i–B9-7 (http://www.pmra-arla.gc.ca/english/pdf/pro/pro98-04/ChapterB09-e.pdf, accessed 12 March 2007).

PMRA (2000). *A decision framework for risk assessment and risk management in the Pest Management Regulatory Agency.* Ottawa, Canada, Pest Management Regulatory Agency (Technical Paper SPN 2000-01; http://www.pmra-arla.gc.ca/english/pdf/spn/spn2000-01-e.pdf, accessed 28 December 2006).

PMRA (2003). *General principles for performing aggregate exposure and risk assessments.* Ottawa, Canada, Pest Management Regulatory Agency (www.pmra-arla.gc.ca/english/pdf/spn/spn2003-04-e.pdf, accessed 28 December 2006).

PPR Panel (2006). Opinion of the Scientific Panel on Plant Health, Plant Protection Products and their Residues on a request from EFSA related to the evaluation of dichlorvos in the context of Council Directive 91/414/EEC. *European Food Safety Authority Journal*, 346:1–13 (http://www.efsa.europa.eu/etc/medialib/efsa/science/ppr/ppr_opinions/1451.Par.0001.File.dat/ppr_op_ej346_summary-tox_en1.pdf, accessed 28 December 2006).

Price PS, Young JS, Chaisson CF (2001). Assessing aggregate and cumulative pesticide risks using a probabilistic model. *Annals of Occupational Hygiene*, 45(Suppl. 1):S131–S142.

Proenca P et al. (2005). Two fatal intoxication cases with imidacloprid: LC/MS analysis. *Forensic Science International*, 153:75–80.

Ray DE (2001). Pyrethroid insecticides: mechanisms of toxicity, systemic poisoning syndromes, paresthesia, and therapy. In: Krieger R, Doull J, Ecobichon D, eds. *Handbook of pesticide toxicology. Vol. 2. Agents.* San Diego, CA, Academic Press:1289–1303.

Reed KJ (1998). *Quantification of children's hand and mouthing activities through videotaping methodology* [PhD thesis]. Piscataway, NJ, Rutgers University.

Reed K et al. (1999). Quantification of children's hand and mouthing activities through a videotaping methodology. *Journal of Exposure Analysis and Environmental Epidemiology*, 9:513–520.

RIVM (2006). *Human exposure to consumer products.* Bilthoven, National Institute for Public Health and the Environment (RIVM); http://www.rivm.nl/en/healthanddisease/productsafety/Main.jsp, accessed 3 September 2007)

Rosenstock L et al. (1991). Chronic central nervous system effects of acute organophosphate pesticide intoxication. *Lancet*, 338:223.

Senanayake N, Karalliedde L (1987). Neurotoxic effects of organophosphorus insecticides. An intermediate syndrome. *The New England Journal of Medicine*, 316:761–763.

Snodgrass WR (1992). Physiological and biological differences between children and adults as determinants of toxic response to environmental pollutants. In: Guzelian PS, Henry CJ, Olin SS, eds. *Similarities and differences between children and adults: implications for risk assessment.* Washington, DC, ILSI Press:35–42.

Steenland K et al. (1994). Chronic neurological sequelae to organophosphate pesticide poisoning. *American Journal of Public Health*, 84:731–736.

Tulve NS et al. (2002). Frequency of mouthing behavior in young children. *Journal of Exposure Analysis and Environmental Epidemiology*, 12:259–264.

Wedin GP, Benson BE (2000). Treatment of pesticide poisoning. In: Marrs TC, Ballantyne B, eds. *Pesticide toxicology and international regulation*. Chichester, John Wiley & Sons Ltd:472–498.

Wester RC, Bucks DA, Maibach HI (1994). Human in vivo percutaneous absorption of pyrethrin and piperonyl butoxide. *Food and Chemical Toxicology*, 32:51–53.

WHO (1990). *Public health impact of pesticides used in agriculture*. Geneva, World Health Organization (http://whqlibdoc.who.int/publications/1990/9241561394.pdf, accessed 29 December 2006).

WHO IPCS (1999). *Principles for the assessment of risks to human health from exposure to chemicals.* Geneva, World Health Organization (Environmental Health Criteria 210; http://www.inchem.org/documents/ehc/ehc/ehc210.htm, accessed 29 December 2006).

WHO IPCS (2002). *International Programme on Chemical Safety: inventory of IPCS and other WHO pesticide evaluations and summary of toxicological evaluations performed by the Joint Meeting on Pesticide Residues (JMPR) – evaluations through 2002,* 7th ed. Geneva, United Nations Environment Programme, International Labour Organisation, and World Health Organization (WHO/IPCS/02.3; http://www.inchem.org/documents/jmpr/jmpeval/jmpr2002.htm, accessed 28 December 2006).

WHO IPCS (2005). *IPCS pesticide activities.* Geneva, World Health Organization (http://www.who.int/ipcs/publications/jmpr/pesticide_2005.pdf, accessed 29 December 2006).

WHO IPCS (2006a). *The WHO recommended classification of pesticides by hazard.* Geneva, WHO International Programme on Chemical Safety (http://www.who.int/ipcs/publications/pesticides_hazard/en/index.html, accessed 7 January 2007).

WHO IPCS (2006b). *Inventory of IPCS and other WHO pesticide evaluations and summary of toxicological evaluations performed by the Joint Meeting on Pesticide Residues (JMPR).* Geneva, WHO International Programme on Chemical Safety (http://www.who.int/ipcs/publications/en/inventory2.pdf, accessed 7 January 2007).

WHOPES (2004). *A generic risk assessment model for insecticide treatment and subsequent use of mosquito nets.* Geneva, World Health Organization, Department of Communicable Disease Control, Prevention and Eradication (http://whqlibdoc.who.int/hq/2004/WHO_PCS_04.1.pdf, accessed 7 August 2007).

WHOPES (2007). WHO pesticide evaluation scheme: links and resources [web site]. Geneva, World Health Organization, Department of Communicable Disease Control, Prevention and Eradication (http://www.who.int/whopes/resources/en/, accessed 3 September 2007).

Wolff MS, Schecter A (1991). Accidental exposure of children to polychlorinated biphenyls. *Archives of Environmental Contamination and Toxicology*, 20:449–453.

Wu IW, Lin JL, Gheng ET (2001). Acute poisoning with the neonicotinoid insecticide imidacloprid in N-methyl pyrrolidone. *Clinical Toxicology*, 39:617–621.

Zartarian VG, Ferguson AC, Leckie JO (1997). Quantified dermal activity data from a four-child pilot field study. *Journal of Exposure Analysis and Environmental Epidemiology*, 7:543–552.

Zartarian VG, Ferguson AC, Leckie JO (1998). Quantified mouthing activity data from a four-child pilot field study. *Journal of Exposure Analysis and Environmental Epidemiology*, 8:543–553.

Zartarian VG et al. (2000). A modeling framework for estimating children's residential exposure and dose to chlorpyrifos via dermal residue contact and nondietary ingestion. *Environmental Health Perspectives*, 108:505–514.

Zhang ZW et al. (1991). Levels of exposure and biological monitoring of pyrethroids in spraymen. *British Journal of Industrial Medicine*, 48:82–86.

Zhao X et al. (1999). Effects of the oxadiazine insecticide indoxacarb, DPX-MP062, on neuronal nicotinic acetylcholine receptors in mammalian neurons. *Neurotoxicology*, 20:561–570.

15. Integrated pest management

John P. Sarisky[1], Randall B. Hirschhorn and Gregory J. Baumann

Summary

Vector-borne diseases have been responsible for much suffering and death throughout human history. Illnesses caused by arthropods, rodents and other pests affect all races, ethnicities, ages and cultures. Even in the most modern societies of the world today, vector-borne diseases are a continuing threat, and efforts to prevent these illnesses must be undertaken to protect public health.

The principles of integrated pest management (IPM) have been practised successfully throughout the world for half a century. As a continuously evolving practice, IPM has incorporated and will continue to incorporate new knowledge and technologies in the field of pest management and vector control. However, the foundation of IPM will remain the same: control pests' access to food, water and shelter and you will control the pest.

IPM is a common-sense approach to pest management. By using a hierarchy of control practices – including public education, sanitation, pest exclusion, and other biological and mechanical control methods, while limiting pesticide application – long-term pest management can be achieved while minimizing environmental and public health hazards.

While establishing an IPM programme may prove more costly and time intensive at its onset, the success of such programmes, when compared with pest management programmes that use a non-integrated approach, are well established. It is also likely that the long-term costs of using a proactive integrated approach to pest management will be far less than those of continuing reactive non-integrated programmes that rely on chemical control.

This chapter provides a basic introduction to the science of IPM. It is expected that by using the principles described in this chapter, effective control of pests can be achieved while minimizing the impact of pest-control measures on the environment.

[1] Disclaimer: The findings and conclusions in this article are those of the author and do not necessarily reflect the views of the United States Department of Health and Human Services.

15.1. An integrated approach to managing urban insects and rodents

15.1.1. A sustainable, holistic and integrated approach to managing urban pests

The control of insect- and rodent-vectored disease depends on understanding the agent and host and on how the agent establishes itself in an ecosystem, survives and moves through the environment to reach and infect a susceptible host. The primary objective of a vector-borne disease control programme is to interrupt the disease transmission cycle. To break this cycle, public health authorities and pest management organizations use a science-based outcome-driven decision-making process to identify and reduce health risks associated with pests. This decision-making process also considers the human health risks associated with implementing a pest management strategy.

This integrated decision-making process, known as IPM, was first used to control agricultural pests (Jacobsen, 1997; Kelly, 2005; Robson & Hamilton, 2005). IPM is a system of principles, practices and procedures applied to improve pest-control outcomes. Pest-control programmes that are not integrated often focus on killing pests instead of on why pests are present. The focus of IPM is to eliminate the source of pest problems – that is, the conditions conducive to the establishment, survival and reproduction of pests. In doing so, IPM controls pest infestations and pest access to people and their dwellings.

Like people, all insects and rodents need food, water and shelter to survive. By removing elements needed to sustain insect and rodent populations, infestations can be controlled. Table 15.1 describes how IPM differs from non-integrated pest control.

Table 15.1. Differences between IPM and non-integrated pest control

Pest management programme components	Non-integrated pest control	IPM
Programme strategy	Reactive	Preventive
Customer education	Minimal	Extensive
Potential liability	High	Low
Emphasis	Routine pesticide application	Pesticides used when exclusion, sanitation and other means are inadequate
Inspection and monitoring	Minimal	Extensive
Use of non-chemical controls	Minimal	Extensive
Positive identification of pests	Sometimes	Required
Use of pest thresholds	Minimal	Extensive
Outcome evaluation	Sometimes	Required

Source: Adapted from Earth Tech, Inc. (1996).

Successful outcome-driven IPM programmes emphasize management over eradication and reduce the amount and frequency of pesticide applications by using other available interventions (Robinson, 1996). Implementers of IPM programmes do three things (Howard, 1988):

1. consider pest biology and behaviour
2. understand the environment in which pests thrive
3. use interventions that prevent unacceptable levels of pest and human interaction.

This holistic approach is more effective and economical and presents fewer risks to people and non-targets, such as pets, birds, livestock and the environment (AFPMB, 1997; UNEP Chemicals, 2000; CDC, 2006).

The United States Fish and Wildlife Service (2004) uses IPM to:

- reduce human exposure to rodent and insect disease vectors;
- protect property and the environment from pest damage;
- develop evidence-based cost-effective interventions;
- improve the scientific management of disease vectors;
- shift pest management from a reactive mode to a preventive mode;
- monitor vector populations and document management actions;
- identify inadequate management practices, along with successful practices, for reducing insect and rodent populations;
- increase coordination and partnerships necessary to control vectors;
- decrease pesticide use and associated human health issues;
- eliminate or reduce the release of pesticides into the environment;
- decrease vector resistance from repetitive applications of pesticides; and
- increase safer use of pesticides and other management tools.

Public health authorities and pest-control organizations throughout the world use the IPM process. With the adoption in 1992 of Agenda 21, a global action plan for sustainable development into the 21st century, the United Nations established its support of IPM. The IPM process fosters informed decision-making at the local level and results in the development of more sustainable pest management strategies (United Nations Department of Economic and Social Affairs, Division for Sustainable Development,

1992; United Nations Educational, Scientific and Cultural Organization, 1992; Mörner, Bos & Fredrix, 2002; Dreyer et al., 2005).

WHO, through the adoption of World Health Assembly resolution 50.13 in 1997, called on Member States to "control vector-borne diseases through the promotion of integrated pest management". The concepts of IPM are part of the *Global strategic framework for integrated vector management* (WHO Strategy Development and Monitoring for Parasitic Diseases and Vector Control Team, 2004).

15.1.2. Components and tools of IPM

Pest control guidance developed by the United States Fish and Wildlife Service (2004) and the National Park Service (2004) describe IPM as "a science-based decision-making process that coordinates knowledge of pest biology, the environment and available technology to prevent unacceptable levels of pest damage through cost-effective means, while posing the least possible risk to people, resources and the environment" (also see Kogan, 1998; AFPMB, 2003; Massachusetts Department of Food and Agriculture Pesticide Bureau, 2006).

The NPMA operationalizes IPM – that is, defines it so that it can be measured or expressed quantitatively – in five steps (NPMA, 2006):

1. inspection
2. identification
3. establishment of threshold levels
4. employment of two or more appropriate control measures
5. evaluation of effectiveness.

15.1.2.1. Inspection

The purpose of inspection is to determine whether a current or potential insect or rodent infestation exists at a specific location.

The inspection process is as follows. The interior space and exterior envelope of buildings and all outdoor areas, including adjacent properties, are inspected to determine whether an infestation or a potential for infestation exists. The location and scope of the infestation are determined in this step. Arthropod (including insects)- or rodent-related structural damage (or damage related to both groups) and damage to equipment, supplies, stored items and food products are documented. Conditions that allow and support the presence of insect, rodent, or both types of pests, as well as the potential for negative effects of pests, including public health considerations, are also documented. These conditions may include a leaking water supply and wastewater drain lines, damp and wet areas, domestic animals, harbourage and coverage, avenues of entry, unsanitary conditions, and access to food and water. The location, number and frequency of insect and rodent sightings by occupants and pest-control professionals are also documented. This information is used to guide pest-control measures at a specific location. The information is collected and used to determine if mitigation is necessary to control an active pest problem or to prevent a future infestation.

In summary, the aims of this first component of IPM are to:

- develop a site-specific inspection process
- conduct the inspection to determine the presence of insects and rodents.

15.1.2.2. Identification

The purpose of identification is to accurately identify pests and conditions that can support pests present at a specific site.

The identification process is as follows. Insect and rodent specimens are collected and positively identified by field technicians, personnel with technical training or both. Information about where the specimen was collected and signs of an active infestation are documented and used to aid positive identification. Pest-control measures should not be implemented until a knowledgeable professional identifies the pest. Identifying the pest provides useful information about pest biology, preferred habitat and life-cycle. If positive identification is not possible in the field, a trained technical expert examines specimens. In addition, the field technician identifies conditions that support or encourage pest activity; the goal is to identify the root causes of pests and their activity that can be corrected by different control measures, to provide a permanent solution to the existing issue while acting to prevent infestations in the future.

In summary, the aims of this second component of IPM are to:

- obtain positive identification of pests of concern at the site; and
- understand the biology of the identified pest and identify conditions conducive to the survival and growth of the pest population.

15.1.2.3. Establishment of threshold levels

The purpose of establishing threshold levels is to provide a site-specific insect- and rodent-population level that can be tolerated on the basis of aesthetic, economic, legal and health concerns. Control measures are implemented when the population exceeds the established threshold.

The process of establishing threshold levels is as follows. Information collected in the inspection and identification steps is used to determine population levels and set action thresholds. When establishing a threshold, data on vector-borne infection and disease in the human population, along with information on the presence of pathogens in the rodent and insect populations, should also be considered. Depending on local circumstances, the action threshold for pests with the potential to cause concern, transmit disease or cause injury may be very low. For example, a single human case of HPS in an urban environment may cause public concern and require the implementation of community-wide control measures. It should be noted that in most settings where people are present, the tolerance of pests is zero, so the threshold level is one single specimen of a particular pest species or population being considered. This is also true of heavily regulated industries, such as restaurants and food plants, where regulations prohibit any evidence of pests, even if the pests have not infested human food.

In addition to health and safety thresholds, local legal restrictions on pest infestation, individual tolerance levels and economic concerns should be considered when developing a site-specific action threshold. Some economic concerns are residents being unable to afford widespread control measures or a municipality requesting preventative measures, rather than thorough inspection, due to cost concerns.

In summary, the aim of this third component of IPM is to establish pest threshold levels with the participation of community members, building residents, contact people and health authorities, where appropriate.

15.1.2.4. Employment of two or more control measures

The purpose of employing two or more control measures is to design and implement an IPM programme that uses as many control measures as necessary to suppress insect and rodent infestations. All practical, reasonable and effective control measures should be considered in the development of the plan.

The process of employing two or more control measures is as follows. Control measures may be grouped into five categories: sanitary, mechanical, cultural, biological and chemical. From these groups, two or more interventions should be selected that will deny pests access to food, water and harbourage and that interrupt the life-cycle of the targeted pest.

This integrated approach requires the participation of the people who live in infested premises. Residents may be asked to change housekeeping practices, clean interior and exterior areas, place trash in a refuse container equipped with a tight-fitting cover, and other measures designed to discourage pests from living at the site. Prevention is a critical aspect of an integrated programme, and such preventive measures as improved sanitation, proper waste storage and removal, and other cultural, mechanical or natural methods are used. Other effective measures include sealing cracks and crevices, stopping the intrusion and accumulation of moisture, using physical barriers to block pest entrance points and informing community members (particularly owners and occupants of infested premises) about actions they can take to control rodent and insect pests.

In summary, the aims of this fourth component of IPM are to:

- determine the best management practices for control of the identified pest;
- develop intervention options;
- select the most appropriate intervention option, with community and resident participation;
- develop an intervention plan; and
- obtain necessary approvals and implement the selected intervention option.

15.1.2.5. Evaluation of effectiveness

The purpose of evaluating effectiveness is to determine whether the IPM plan was implemented as designed (plan process objectives met) and whether the objectives of the plan were attained (pest reduction and exposure objectives met).

The process of evaluating effectiveness is as follows. The evaluation determines how effective the implementation of each of the first four steps has been. The process used to implement each step is described and evaluated to determine if any phase of the IPM plan could have been implemented with greater efficiency and effectiveness. The lessons learned from the evaluation are documented and used to improve the implementation process.

The methods used to evaluate the reduction of pest activity and exposure are determined and executed at the beginning of the IPM process. During implementation of the first four steps – inspection, identification, establishment of thresholds and employment of control measures – monitoring surveys are conducted to document pest numbers, pest activity and opportunities for human–pest interaction. These surveys determine whether the pest population has been reduced below the site-specific threshold level (pest reduction and exposure outcomes). Examples of pest and exposure reduction indicators include: frequency, number and time of visual sightings; insect activity following aerosol flushing; pests counts obtained from vacuuming; positive bait stations; flea egg counts; presence of rodent and cockroach droppings; positive sticky and snap traps; mosquito larval dipper counts, adult trapping counts, and adult landing and bite counts; tick counts; signs of rodent gnawing and food contamination, number of active burrows, runways and the presence of rodent tracks. The survey information is used to identify causative factors that have not been fully addressed during implementation. The evaluation of the effectiveness step documents the attainment of the process and pest reduction objectives established in the IPM plan.

In summary, the aims of this last component of IPM are to:

- evaluate the outcome of the interventions implemented and determine whether the process and pest reduction objectives have been achieved; and

- document actions and decisions and maintain clear and detailed records.

15.1.2.6. Develop an IPM plan

The key elements in developing an IPM plan are as follows.

- The management plan must have clear objectives and outcome measures.

- The plan should ensure the development of partnerships and collaborative efforts with all stakeholders, including building occupants, community members, political authorities, leaders, decision-makers and technical experts.

- Partners should actively participate in the development of the plan.

- The plan should include a community outreach and education component.
- The plan and actions should be revised as evaluation information becomes available.

15.1.3. Definitions of IPM

A number of definitions of IPM have been published. The following definition was published in 1959 (Stern et al., 1959).

> Applied pest control which combines and integrates biological and chemical control. Chemical control is used as necessary and in a manner which is least disruptive to biological control. Integrated control may make use of naturally occurring biological control as well as biological control affected by manipulated or induced biotic agents.

Since the release of this definition, more than 150 additional definitions of IPM have been proposed. A few of those definitions are provided here. A review of recent definitions shows that IPM is considered a multifactorial, long-term, systematic, and holistic or comprehensive approach to managing pests in ways that reduce their harmful effects below acceptable levels, while avoiding or minimizing environmental hazards.

FAO (1968) defined IPM as:

> ... a pest management system that, in the context of the associated environment and the population dynamics of the pest species, utilizes all suitable techniques and methods in as compatible a manner as possible, and maintains the pest population at levels below those causing economic injury.

Still another definition (Intersociety Consortium for Plant Protection, 1979) is as follows.

> IPM is the optimization of pest control in an economically and ecologically sound manner. This is accomplished by the use of multiple tactics in a compatible manner to maintain pest damage below the economic injury level while providing protection against hazards to humans, animals, plants, and the environment.

The NPMA (2006) describes IPM as:

> ... a process consisting of five basic steps. These include inspection, identification, the establishment of threshold levels, the employment of two or more control measures and the evaluation of effectiveness. To be acceptable, the control measures must be both environmentally compatible and economically feasible.

Additional IPM definitions can be found in a compendium of definitions put together by Bajwa & Kogan (1996).

15.1.4. Case studies and the benefits of IPM

Research by government agencies, academic institutions and the pest-control sector demonstrates that IPM is more effective than non-integrated pest-control approaches (Brenner et al., 2003; Miller & Meek, 2004). IPM has been effective in difficult pest-control environments, such as overcrowded urban centres with ageing and neglected buildings and infrastructure.

The application and outcome of IPM in public housing developments provide important evidence of the effectiveness and benefits of an integrated approach to pest control. In many public housing developments, pesticides are applied on a regularly scheduled basis to control cockroaches. The application of pesticides is the primary – and maybe the only – control measure implemented. It is often done without monitoring the pest population.

In the IPM approach, non-pesticide control measures are implemented and less pesticide is applied. Structural IPM programmes for the control of cockroaches limit their access to food and water. Measures taken to prevent such access include improvements in kitchen sanitation, proper food storage, improvements in general housekeeping, repairs of water leaks in supply and drain lines, and emptying condensation pans. Also, the movement of cockroaches is controlled so that food is inaccessible. Sealing cracks and openings, screening windows, and sealing doors prevent entry and ease of movement and have proven effective in reducing cockroach infestations (Hedges, 1999; CDC & HUD, 2006).

Several studies of IPM effectiveness are summarized below.

15.1.4.1. Case study 1

A prospective intervention trial conducted in New York City by the Mount Sinai Children's Environmental Health and Disease Prevention Research Center found urban IPM programmes effective in controlling cockroaches. This project reduced cockroach activity in apartment units. Apartment dwellers received information on the control of cockroach infestations and were able to control cockroach infestations in their own units without the use of pesticides. Similar work done in Chicago included the cleaning of vacant and occupied units, and using gel and paste pesticides in place of sprays. Caulking and the installation of screening also blocked access, and improvements in sanitation included proper storage and disposal of trash. The reduced cockroach activity from this IPM programme allowed an 83% reduction in the use of gel pesticides. In the New York City project and in the work done in Chicago, the community was involved in planning, implementing and evaluating the IPM programme (Brenner et al., 2003).

15.1.4.2. Case study 2

A year-long study conducted in a Virginia public housing development found IPM to be more effective than conventional pest control practices in controlling cockroaches. In the conventionally treated units, liquid and dust pesticides were applied. In contrast, IPM-treated homes were cleaned with HEPA-filtered vacuums and treated with baits and IGRs. Cockroaches were controlled in the IPM-treated units, while the cockroach populations in the apartments that received only monthly spray and dust applications did not change.

The study found that the IPM approach was more expensive to implement because of the on-site inspections and time needed to implement interventions. However, the IPM approach produced a significant reduction in the cockroach population in four months, and the population remained suppressed in the IPM apartments (Miller & Meek, 2004).

15.1.4.3. Case study 3

A cost assessment conducted by the University of Virginia found the cost of implementing an IPM programme in public housing to be US$ 4.06 per unit per month. The application of pesticides to conventionally control cockroaches in similar units cost US$ 1.50 per unit per month. The IPM programme was more effective in reducing the cockroach population, and residents reported they would be willing to pay for IPM services that reduced their exposure to pesticides while controlling cockroach infestations (Miller, 2006).

15.1.4.4. Case study 4

Cost and other IPM implementation considerations have been assessed and reported by the United States Department of Housing and Urban Development (HUD) Office of Public and Indian Housing in a guidance document. The work conducted by HUD identified the critical importance of involving housing development residents in IPM efforts. The HUD IPM guidance document also addresses IPM cost concerns by pointing out that well managed IPM programmes improve property maintenance and extend the useful life of a structure. The multiple benefits of IPM include improved building maintenance and the control of pests while reducing the application of pesticides. The HUD Office of Public and Indian Housing recommends considering these benefits when comparing the cost of implementing an IPM programme with the routine application of pesticides (HUD Office of Public and Indian Housing, 2006).

15.1.4.5. Case study 5

In the United States, the General Services Administration (GSA) is responsible for all federal office space. A study of 55 GSA-managed structures in the Washington, DC, area assessed both the quantities of insecticides applied indoors and requests for pest-control services by building occupants. Baseline measures of insecticide applications and pest service calls were established in 1988 for each of the 55 structures, and an IPM programme was initiated in 1989. When compared with similar data collected in 1994 and again in 1999, the pre-IPM baseline data collected in 1988 found that the amount of insecticides applied and the number of requests for pest-control services by building occupants decreased significantly (Table 15.2) (Greene & Breisch, 2002).

The objective of GSA IPM programmes is to "achieve long-term, environmentally sound pest suppression and prevention through a variety of technologic and management practices". Control strategies in GSA structural IPM programmes (GSA, 2005) included:

- reducing food, water, harbourage and access routes used by pests through structural and procedural modifications;
- using pesticide compounds, formulations and application methods that present the lowest potential hazard to people and the environment;

Table 15.2. GSA assessment of IPM practices in public buildings

Year	Service requests	Pesticide applications
1988 (baseline)	14 716	14 659
1994	3331	1674
1999	1581	954

Source: Greene & Breisch (2002).

- using technologies, such as trapping and monitoring, in place of pesticides; and
- coordinating efforts among all facility management programmes that have a bearing on pest-control outcomes.

15.1.5. Challenges in implementing IPM programmes

Cost, understanding the needs of the people being served, regulatory restrictions and emergencies have been identified as barriers to implementing IPM programmes.

The first challenge that frequently arises in implementing a programme is the cost of setting it up. Those responsible for paying for the treatment may want what are now considered outdated methods of pest management, because these methods are seemingly less expensive, especially in the short term. In this case, *outdated* may mean using pesticides without considering the IPM steps necessary to have an effective programme. It is generally accepted that IPM programmes may be more expensive initially when structural repairs may be needed to preclude the entry of vermin and eliminate their harbourage. An insufficient budget or an insufficiency in other resources may also restrict the full implementation of all IPM components (Norton & Mullen, 1994).

The second challenge is considering the needs of the people who require IPM intervention. If very young, elderly, ill or alleged chemically sensitive people are present in infested premises, the types of materials and measures that can be used will differ from those where such individuals are not present. Also, although the need for implementing pest control may seem clear to public officials, the people being served may not perceive the same level of concern and may object. A clear line of communication between the agency performing the service and those receiving it is very important.

The third challenge is limits established by law or regulations on types of pest-control measures that may be implemented or prohibited. For example, there are restrictions on the pest-control methods or materials that can be applied by a health care facility, a food plant or a susceptible population. Regulatory restraints will be a factor that must be considered when determining control measures that may be used.

The fourth challenge is emergencies, such as natural disasters and vector-related outbreaks. These situations may require that immediate treatment be implemented and that some phases of IPM – for example, removal of food sources before applying rodent control agents – may not be feasible. Stinging insects in a school environment are another example: such emergencies, however, do not preclude the use of IPM measures. Rather, appropriate and necessary action should be implemented quickly when pests present an imminent public health hazard.

Careful consideration of these four challenges will clarify what steps need to be taken when implementing an IPM programme. Also, open communication is vital, so that the strategies produce a programme agreeable to all parties.

Finally, some challenges may be unreasonable or impossible to address. For example, if *least toxic* is requested, care must be exercised: this is a relative and undefined term, and its use will vary with exposure and human use patterns and from site to site. It is improper to assume that the least toxic pest management method will always mean the least toxic product will be appropriate or will represent the least risk to people. Among the factors that should be considered in developing an IPM plan are the presence of an at-risk human population, site location, potential for exposure and impact of a vector-borne disease outbreak.

15.2. The future of IPM

The factors listed below influence the acceptance and use of IPM in urban environments.

15.2.1. Legal requirements

Authorities in the United Kingdom have recognized the need for the application of IPM practices. In 1988, the implementation of IPM was bolstered in the United Kingdom by the enactment of Control of Substances Hazardous to Health regulations (COSHH). These regulations imposed restrictions on the use of chemical pesticides. The regulations were further refined by amendments in 1994, 1996 and 1999 (Habgood, 1999; HMSO, 1999), and consolidated in 2002 (OPSI, 2002).

Before the enactment of COSHH, the control of pesticides regulations of 1986 and 1987 regulated pesticide use. These regulations were established to meet the requirements of the EU. An operational objective of the EU's sustainable development strategy is the replacement of pesticides of great concern with suitable alternative substances and technologies (Council of the European Union, 2006). The acceptance and use of IPM procedures can assist Member States in reaching this EU public health objective.

In the United States, the EPA is working with departments of education to develop and implement IPM programmes in and around schools (EPA, 2004, 2006). Also, local governments are adopting policies that reduce pesticide use and encourage the use of IPM. Moreover, pest-control operators are adopting IPM practice standards and offering IPM programmes to their clients. These trends will result in the wider application of IPM in urban environments.

15.2.2. Education and training of the workforce

Many pesticide-use training and permitting programmes offered in the United States now cover IPM principles. IPM training and certification programmes are also available from professional associations. For example, the New England Pest Management Association maintains a registry of IPM-certified individuals and pest-control companies. Individuals must attend IPM training and pass a written exam before they can be listed on the voluntary registry. Companies listed on the registry must provide evidence that they have developed and implemented IPM plans for clients. A better-trained IPM workforce will improve the delivery of pest-control services, and improved pest-control services in urban settings will reduce the opportunities for interactions between people and vectors. IPM training and professional certification programmes should be made available to pest-control and public health practitioners.

15.2.3. Demonstration of effectiveness and benefits

Evidence-based outcomes are needed to advance the acceptance and use of IPM in urban settings. Also, additional research is needed to measure the long-term health, environmental, social and economic benefits of IPM.

15.2.4. Advances in science and technology

New information on the behaviour, biology and ecology of pests should be shared with pest-control and public health authorities as it becomes available. Also, advances in biotechnology may provide additional tools and effective solutions for the IPM practitioner.

An example of an advance in biotechnology is the introduction of genetically altered mosquitoes that block the transmission of some mosquito-borne diseases (Gould, Magori & Huang, 2006). Available technology must also be applied to control mosquito pests in a broad setting. Global positioning systems (GPS) and geographic information systems (GIS) are now essential vector-control tools used to control mosquitoes. Data generated by GPS is combined increasingly with data from surveillance activities and then analyzed in GIS (Clarke, McLafferty & Tempalski, 1996). For example, to monitor high concentrations of a vector, GPS is used to pinpoint known breeding grounds, such as ponds, wetlands and favoured vegetation areas. GIS can then be used to overlay the different layers of information, so that mosquito-control measures can be targeted. The use of satellite imagery allows certain vector habitats to be identified and habitat changes to be monitored over time. Environmental factors can be used to predict the risk of contact between people and vectors.

GIS and GPS are used to map spatial and temporal variables. Mapping of spatial and temporal information allows IPM practitioners to target control measures and to monitor the progress in pest reduction. GPS and GIS have improved the monitoring of control measures for onchocerciasis in Guatemala (Richards, 1993) and trypanosomiasis in Africa (Rogers & Williams, 1993).

Understanding the environmental determinants of vector-borne disease transmission is critical to the development of an IPM plan (Glass et al., 2002). The most up-to-date technologies, knowledge and practices should be available so that public health practitioners and pest-control professionals are able to respond to and control urban pests in an ever-changing environment.

15.2.5. Community awareness and involvement

Successful IPM programmes actively engage residents and community members in plan design and implementation. Many local environmental health service programmes are familiar with community mobilization strategies that can facilitate community involvement. Residents are valuable resources: they are sources of information on vector activity, the environment and the local activities that expose people to pests. Community residents are also able to provide critical input on the development of locally appropriate action plans and the feasibility of implementing interventions. Community participation is a critical component of IPM (MacArthur, 2002; Environmental Health Project, 2003; Hubbard et al., 2005).

15.2.6. Urban planning and community design

Urban planners may not be aware of local and regional vector-borne disease issues. Disease vectors are present in old and new urban centres and could be newly established or reintroduced into these habitats by the movements of their human hosts.

The renewal of existing urban centres presents unique vector-control problems. Blighted urban areas oftentimes have high densities of rodents and insects. When insect and rodent habitats are disrupted during urban renewal projects, rodent and insect populations relocate to nearby areas. The construction of new urban centres and transportation systems provides pests with avenues for movement and with fresh habitats. For example, the extension of a rail system to the suburbs of an urban centre may provide rats with a means to enter new areas. Also, the development of parks and athletic fields may provide new favourable habitats for insects and rodents.

Environmental-health and vector-control professionals can also provide important guidance on the design and construction of healthy buildings and healthy homes. Construction methods and materials used in healthy buildings and homes prevent entry of insect and rodent pests. Healthy housing developments and landscapes are designed to control conditions conducive to the survival and reproduction of pests. Environmental-health and vector-control organizations can influence and play important roles in the promotion of healthy urban planning and community design.

15.2.7. Pest-control industry and public health authority collaboration

The control of disease vectors is an important function of public health. A strong partnership between public health agencies and the pest-control industry will advance the practice of IPM. This is especially true in urban environments where the interaction between people and pests is dynamic. Urban environments are complex systems that challenge those professionals responsible for the control of vector-borne diseases.

References[2]

AFPMB (1997). *Guidelines for preparing DoD pest control contracts using integrated pest management.* Washington, DC, Armed Forces Pest Management Board, Defense Pest Management Information Analysis Center (Armed Forces Pest Management Board Technical Information Memorandum No. 39; http://www.afpmb.org/pubs/tims/tim39.htm, accessed 21 December 2006).

AFPMB (2003). *Integrated pest management (IPM) in and around buildings.* Washington, DC, Armed Forces Pest Management Board, Defense Pest Management Information Analysis Center (Armed Forces Pest Management Board Technical Guide No. 29; http://www.afpmb.org/pubs/tims/tg29/tg29.htm, accessed 7 September 2006).

Bajwa WI, Kogan M (1996). Compendium of IPM definitions [web site]. Corvallis, OR, Integrated Plant Protection Center, Oregon State University (http://www.ippc.orst.edu/IPMdefinitions/defineI.html, accessed 3 January 2007).

Brenner BL et al. (2003). Integrated pest management in an urban community: a successful partnership for prevention. *Environmental Health Perspectives*, 111:1649–1653.

CDC (2006). *Integrated pest management: conducting urban rodent surveys.* Atlanta, GA, United States Department of Health and Human Services, Centers for Disease Control and Prevention:1–19 (http://www.cdc.gov/healthyplaces/publications/IPM_manual.pdf, accessed 21 December 2006).

CDC, HUD (2006). *Healthy housing reference manual.* Atlanta, United States Department of Health and Human Services, Centers for Disease Control and Prevention (http://www.cdc.gov/nceh/publications/books/housing/housing.htm, accessed 22 December 2006).

Clarke KC, McLafferty SL, Tempalski BJ (1996). On epidemiology and geographic information systems: a review and discussion of future directions. *Emerging Infectious Diseases*, 2: 85–92.

[2] The criteria for inclusion of peer-reviewed articles and studies were: inclusion of data relevant to the public health significance of pests, pest-management practices and programme implementation for IPM; specified dates for the data collection period, including data from 2000 onward; and a clear description of the methods used in the studies. The criteria for inclusion of government manuals and guidance documents developed by professional organizations were: inclusion of information, guidelines, best practices, principles, evaluation and measurement strategies for IPM practices and programmes; guidelines and best practices for pesticide reduction and mitigation strategies; specified dates for the data collection period (from 1987 onward); and data collection limited to officially cleared government documents from multiple governmental agencies. Additional criteria for manuals were the inclusion of data relevant to factors that affect vector distribution, exposure and personal protection.

Council of the European Union (2006). *Review of the European Union Sustainable Development Strategy (EU SDS) – Renewed Strategy*. Brussels, Council of the European Union (http://register.consilium.europa.eu/pdf/en/06/st10/st10117.en06.pdf, accessed 21 December 2006).

Dreyer H et al. (2005). IPM Europe: The European Network for integrated pest management in development cooperation. In: EFARD 2005 Conference, Zurich, 27–29 April 2005. Agricultural Research for Development: European Responses to Changing Global Needs. Eschborn, Germany, Deutsche Gesellschaft für Technische Zusammenarbeit GmbH:1–6 (http://www.ipmeurope.org/download/dreyer-et-al-2005.pdf, accessed 21 December 2006).

Earth Tech, Inc. (1996). *Model pesticide reduction plan, November 1996*. Brooks Air Force Base, TX, Air Force Center for Environmental Excellence, Environmental Quality Directorate (https://www.denix.osd.mil/denix/Public/Library/AF_P2/Pest/afpest.html, accessed 7 September 2006).

Environmental Health Project (2003). *Community-based environmental management for urban malaria control in Uganda – year 1*. Arlington, VA, Environmental Health Project (EHP Brief No. 20; http://pdf.dec.org/pdf_docs/PNACU780.pdf, accessed 21 December 2006).

EPA (2004). *Integrated pest management for schools: a how-to manual*. Washington, DC, United States Environmental Protection Agency (http://www.epa.gov/pesticides/ipm/schoolipm/index.html, accessed 13 September 2006).

EPA (2006). *Integrated pest management (IPM) in schools*. Washington, DC, United States Environmental Protection Agency (http://www.epa.gov/pesticides/ipm/, accessed 13 September 2006).

FAO (1968). *Report of the first session of the FAO Panel of Experts on Integrated Pest Control, Rome, 12–22 September 1967*. Rome, Food and Agriculture Organization of the United Nations.

Glass GE et al. (2002). Satellite imagery characterizes local animal reservoir populations of Sin Nombre virus in the southwestern United States. *Proceedings of the National Academy of Sciences of the United States of America*, 99:16817–16822.

Gould F, Magori K, Huang Y (2006). Genetic strategies for controlling mosquito-borne diseases. *American Scientist*, 94:238–246.

Greene A, Breisch NL (2002). Measuring integrated pest management programs for public buildings. *Journal of Economic Entomology*, 95:1–13.

GSA (2005). *The GSA integrated pest management program, 2005 revision*. Washington, DC, General Services Administration:4

(http://www.gsa.gov/gsa/cm_attachments/GSA_DOCUMENT/Structural_Pest_Control_Business_Practices_R2-uWAZ_0Z5RDZ-i34K-pR.pdf, accessed 24 July 2007).

Habgood V (1999). Pest control. In: Bassett WH, ed. *Clay's handbook of environmental health,* 18th ed. London, E & FN Spon:214–239.

Hedges SA (1999). The latest trends in cockroach control. *Pest Control Technology*, 27: 24–26, 32.

HMSO (1999). *Statutory Instrument 1999 No. 437: the control of substances hazardous to health regulations 1999*. London, Her Majesty's Stationery Office (http://www.opsi.gov.uk/SI/si1999/19990437.htm, accessed 22 December 2006).

Howard WE (1988). Rodent pest management: the principles. In: Prakash I, ed. *Rodent pest management*. Boca Raton, FL, CRC Press Inc.:285–294.

Hubbard B et al. (2005). Community environmental health assessment strengthens environmental public health services in the Peruvian Amazon. *International Journal of Hygiene and Environmental Health*, 208:101–107.

HUD Office of Public and Indian Housing (2006). *Guidance on integrated pest management.* Washington, DC, United States Department of Housing and Urban Development.

Intersociety Consortium for Plant Protection (1979). *Integrated pest management, a program of research for the state agricultural experimental stations and the colleges of 1980.* Washington, DC, USDA Science and Education Administration/Cooperative Forestry Research.

Jacobsen BJ (1997). Role of plant pathology in integrated pest management. *Annual Review of Phytopathology*, 35:373–391.

Kelly L (2005). *The Global Integrated Pest Management Facility: addressing challenges of globalization: an independent evaluation of the World Bank's approach to global programs.* New York, World Bank:1–3 (http://lnweb18.worldbank.org/oed/oeddoclib.nsf/DocUNIDViewForJavaSearch/210300C07054C81A852570A50076C0DC/$file/gppp_pest_management_facility.pdf, accessed 21 June 2007).

Kogan M (1998). Integrated pest management: historical perspectives and contemporary developments. *Annual Review of Entomology*, 43:243–270.

MacArthur I (2002). *Local environmental health planning: guidance for local and national authorities.* Copenhagen, WHO Regional Office for Europe (WHO Regional Publications European Series, No. 95; http://www.euro.who.int/document/e76436.pdf, accessed 22 December 2006).

Massachusetts Department of Food and Agriculture Pesticide Bureau (2006). *Integrated pest management for building managers. How to implement an integrated pest management program in your building(s)*. Boston, MA, Pesticide Bureau, Massachusetts Department of Food and Agriculture (http://www.mass.gov/agr/pesticides/publications/IPM_kit_for_bldg_mgrs.pdf, accessed 7 September 2006).

Miller DM (2006). Assessing the value of IPM in Virginia Public Housing [web site]. Blacksburg, VA, Department of Entomology, Virginia Tech (http://web.ento.vt.edu/ento/project.jsp?project=IPM+in+Public+Housing, accessed 21 December 2006).

Miller DM, Meek F (2004). Cost and efficacy comparison of integrated pest management strategies with monthly spray insecticide applications for German cockroach (Dictyoptera: Blattellidae) control in public housing. *Journal of Economic Entomology*, 97:559–569.

Mörner J, Bos R, Fredrix M (2002). *Reducing and eliminating the use of persistent organic pesticides: guidance on alternate strategies for sustainable pest and vector management*. Geneva, World Health Organization:25–42 (http://whqlibdoc.who.int/hq/2002/a76620.pdf, accessed 21 December 2006).

National Park Service (2004). Integrated pest management [web site]. Washington, DC, National Park Service (http://www.nature.nps.gov/biology/ipm/index.cfm, accessed 7 September 2006).

NPMA (2006). *Urban IPM handbook: an integrated approach to management of pests in and around structures*. Fairfax, VA, National Pest Management Association.

Norton GW, Mullen J (1994). *Economic evaluation of integrated pest management programs: a literature review*. Virginia Cooperative Extension Publication 448–120. Petersburg, VA, Virginia State University and Blacksburg, VA, Virginia Polytechnic Institute and State University:1–112.

OPSI (2002). *The dangerous substances and preparations (safety) (consolidation) (amendment no. 3) regulations 2002*. Norwich, United Kingdom, Office of Public Sector Information (Statutory Instruments No. 3010; http://www.opsi.gov.uk/si/si2002/uksi_20023010_en.pdf, accessed 13 August 2007).

Richards FO (1993). Use of geographic information systems in control programs for onchocerciasis in Guatemala. *Bulletin of the Pan American Health Organization*, 27:52–55.

Robinson WH (1996). Integrated pest management in the urban environment. *American Entomologist*, 42:76–77.

Robson MG, Hamilton GC (2005). Pest control and pesticides. In: Frumkin H, ed.

Environmental health: from global to local. San Francisco, CA, John Wiley and Sons, Inc.:544–580.

Rogers DJ, Williams BG (1993). Monitoring trypanosomiasis in space and time. *Parasitology*, 106(Suppl.):S77–S92.

Stern VM et al. (1959). The integrated control concept. *Hilgardia*, 29:81–101.

UNEP Chemicals (2000). *Proceedings of a UNFP/FAO/WHO workshop on sustainable approaches for pest and vector management and opportunities for collaboration in replacing POP's pesticides. Bangkok, March 6–10, 2000.* Geneva, United Nations Environment Programme Chemicals (http://portalserver.unepchemicals.ch/Publications/ProcBangkok19-22March01.pdf, accessed 22 December 2006).

United Nations Department of Economic and Social Affairs, Division for Sustainable Development (1992). *Agenda 21*. New York, United Nations Department of Economic and Social Affairs, Division for Sustainable Development (http://www.un.org/esa/sustdev/documents/agenda21/english/agenda21toc.htm, accessed 22 December 2006).

United Nations Educational, Scientific and Cultural Organization (1992). *Rio Declaration on Environment and Development (1992)*. New York, United Nations Educational, Scientific and Cultural Organization (http://www.unesco.org/education/information/nfsunesco/pdf/RIO_E.PDF, accessed 24 July 2007).

United States Fish and Wildlife Service (2004). *Integrated pest management: guidance for preparing and implementing integrated pest management plans*. Washington, DC, United States Fish and Wildlife Service (http://www.fws.gov/contaminants/Documents/GuidanceIPMPlan.pdf, accessed 22 December 2006).

WHO Strategy Development and Monitoring for Parasitic Diseases and Vector Control Team (2004). *Global strategic framework for integrated vector management*. Geneva, World Health Organization (document number: WHO/CDS/CPE/PVC/2004.10; http://whqlibdoc.who.int/hq/2004/WHO_CDS_CPE_PVC_2004_10.pdf, accessed 22 December 2006).

Annex 1. Abbreviations

Organizations, other entities and activities

AFPMB	Armed Forces Pest Management Board
APPMA	American Pet Products Manufacturers Association
CDC	United States Centers for Disease Control and Prevention
CEN	European Committee for Standardization
COSHH	Control of Substances Hazardous to Health Regulations
DFG	Deutsche Forschungsgemeinschaft
DG Environment	European Commission Environment Directorate-General
DG Sanco	European Commission Health and Consumer Protection Directorate-General
DPIF	State of Queensland, Australia, Department of Primary Industries and Fisheries
ECB	European Chemicals Bureau
ECDC	European Centre for Disease Prevention and Control
EHCS	English House Condition Survey
EMCA	European Mosquito Control Association
EPA	United States Environmental Protection Agency
EU	European Union
FAO	Food and Agriculture Organization of the United Nations
FIFRA	United States Federal Insecticide, Fungicide, and Rodenticide Act of 1972
FQPA	United States Food Quality Protection Act of 1996
GINA	Global Initiative for Asthma
GSA	United States General Services Administration
HSE	Health & Safety Executive
HUD	United States Department of Housing and Urban Development
IARC	International Agency for Research on Cancer
IFCS	WHO Intergovernmental Forum on Chemical Safety
INCHEM	Chemical Safety Information from Intergovernmental Organizations, also known as the International Programme on Chemical Safety (IPCS)
IPCS	WHO International Programme on Chemical Safety
IRRI	International Rice Research Institute
ISAAC	International Study of Asthma and Allergies in Childhood
JMPR	FAO/WHO Joint Meeting on Pesticide Residues
LARES	Large Analysis and Review of European housing and health Status (study/project)
NCICAS	National Cooperative Inner-City Asthma Study
NPMA	National Pest Management Association
OECD	Organisation for Economic Co-operation and Development
OPSI	United Kingdom Office of Public Sector Information
PCPA	Canadian Pest Control Products Act of 2006
PHED	Pesticide Handler's Exposure Database
PMRA	Canadian Pest Management Regulatory Agency
PPR Panel	Scientific Panel on Plant Health, Plant Protection Products and their Residues of the European Food Safety Authority
REACH	Registration, Evaluation, Authorisation and Restriction of Chemicals
UNEP	United Nations Environment Programme
USDA	United States Department of Agriculture
USDA-APHIS	United States Department of Agriculture, Animal and Plant Health Inspection Service
WHOPES	WHO Pesticide Evaluation Scheme

Technical terms

ACA	acrodermatitis chronica atrophicans
AChE	acetylcholine esterase
ADI	acceptable daily intake
AEL	acceptable exposure level
AI	active ingredient
AIDS	acquired immunodeficiency syndrome
ARfDs	acute reference doses
BR	breathing rate
BW	body weight
BZC	breathing zone concentration
CCHF	Crimean-Congo haemorrhagic fever
CCHFV	Crimean-Congo haemorrhagic fever virus
CEE	central European encephalitis
CEEV	central European encephalitis virus
CEH	critical equilibrium humidity

CTF	Colorado tick fever
DDT	dichlorodiphenyltrichloroethane
ED	exposure duration
EEEV	Eastern equine encephalitis viruses
EEG	electroencephalogram
ELISA	enzyme-linked immunosorbent assay
EM	erythema migrans
EsD	estimated duration
ETS	environmental tobacco smoke
FAD	flea allergy dermatitis
FMV	Fort Morgan virus
Freq	frequency
GABA	gamma-aminobutyric acid
GIS	geographic information system
GPS	global positioning system
HDMs	house dust mites
HEPA	high efficiency particulate air
HFRS	haemorrhagic fever with renal syndrome
HGA	human granulocytic anaplasmosis
HIV	human immunodeficiency virus
HME	human monocytic ehrlichiosis
HPAI	highly pathogenic avian influenza
HPS	hantaviral pulmonary syndrome
HTE	hand transfer efficiency
IgE	immunoglobulin E
IGRs	insect growth regulators
INR	international normalized ratio
IPM	integrated pest management
ISR	indoor surface residue
LB	Lyme borreliosis
LBRF	louse-borne relapsing fever
LC50	lethal concentration, 50%; for inhalation experiments, it is the concentration of a chemical in air that kills 50% of the test animals in a given time
LD50	lethal dose, 50%; this value is a statistical estimate of the number of milligrams of toxicant per kilogram of body weight required to kill 50% of test animals
LOAEL	the lowest observed adverse effect level
LPAI	low pathogenic avian influenza
mg/kg BW	milligram per kilogram body weight
MOE	margin of exposure
MOS	margin of safety
MRL	maximum residue limit
MSF	Mediterranean spotted fever
MUP	mouse urinary proteins

MVHR	mechanical ventilation with heat recovery
NOAEL	a dose level that causes no observed adverse effect
OPIDP	organophosphate-induced delayed polyneuropathy
OPs	organophosphates
OR	odds ratio
OspA	outer surface protein A
PM2.5	fine particulate matter with an aerodynamic diameter smaller than 2.5 µm
PMPs	pest management professionals
PPE	personal protective equipment
PPV	positive pressure ventilation
PSV	passive stack ventilation
PT/PTT	prothrombin/partial thromboplastin time
RMSF	Rocky Mountain spotted fever
RSSE	Russian spring-summer encephalitis
SA	surface area
SARS	severe acute reparatory syndrome
SEF	saliva extraction factor: the percentage of the dislodged chemical residue that is extracted from the fingers by the saliva of the child
SINV	Sindbis virus
SLEV	St. Louis encephalitis virus
SNV	Sin Nombre virus
SOPs	standard operating procedures
spp.	species
ssp.	subspecies
TBDs	tick-borne diseases
TBE	tick-borne encephalitis
TBEV	tick-borne encephalitis virus
TR	transferable residue
UF	uncertainty factor
ULV	ultra-low-volume
WEEV	Western equine encephalitis viruses
WNV	West Nile virus

Annex 2. Working Group

Steering Committee

ANTERO AITIO (retired), International Programme on Chemical Safety, WHO, Geneva, Switzerland

SHARUNDA D. BUCHANAN, Environmental Health Services Branch, Centers for Disease Control and Prevention, Atlanta, GA, United States

BOUÉ FRANCK, Agence Française de Sécurité Sanitaire des Aliments, Paris, France

ANDREW GRIFFITHS, Chartered Institute of Environmental Health, London, United Kingdom

JERRY M. HERSHOVITZ, Division of Emergency and Environmental Health Services, Centers for Disease Control and Prevention, Atlanta, GA, United States

FRANÇOIS HUBERT, Société François Hubert, Baugy, France

GRAHAM JUKES, Chartered Institute of Environmental Health, London, United Kingdom

NED KINGCOTT, Food Standards Agency, London, United Kingdom

ARNOLD E. LAYNE, Office of Administration, Resources and Management, U.S. Environmental Protection Agency, Washington, DC, United States, formerly with Office of Pesticide Programs, Registration Division, Insecticides Branch, U.S. Environmental Protection Agency, Washington, DC, United States

MARCO MARONI, International Centre for Pesticide Safety, Milan, Italy[1]

NELLY MOROSOVA, Kharkov Medical Academy of Postgraduate Education, Ukraine

JULIUS PTASHEKAS, Ministry of Health, Vilnius, Lithuania

MAGGIE TOMLINSON, Zoonotic and Emerging Infections and Biotechnology, Communicable Diseases Branch, Department of Health, London, United Kingdom

PEET TÜLL, European Centre for Disease Prevention and Control, Stockholm, Sweden

[1] Deceased.

Authors

Rob C. Aalberse, Department of Immunopathology, University of Amsterdam, Netherlands

Jens Amendt, Institute of Legal Medicine, University of Frankfurt, Germany

Brian R. Amman, Medical Ecology Unit, Centers for Disease Control and Prevention, Atlanta, GA, United States

Stephen Battersby, Robens Centre for Public and Environmental Health, University of Surrey, Guildford, and School of Law, University of Warwick, Coventry, United Kingdom

Gregory J. Baumann, National Pest Management Association, Fairfax, VA, United States

Ian Burgess, The Medical Entomology Centre, Shepreth, United Kingdom

Ginger L. Chew, Department of Environmental Health Sciences, Columbia University, New York, NY, United States

Anna Clara Fanetti, Department of Occupational Health, University of Milan, and Division of Occupational Medicine, International Centre for Pesticides and Health Risk Prevention, Milan, Italy

David Crowther, The Martin Centre, University of Cambridge, United Kingdom.

Frédéric de Blay, Service de Pneumologie, Hôpital Lyautey, Hôpitaux Universitaires de Strasbourg, France

Michael K. Faulde, Department of Medical Entomology/Zoology, Central Institute of the Federal Armed Forces Medical Service, Koblenz, Germany

Kenneth L. Gage, Bacterial Zoonoses Branch, Centers for Disease Control and Prevention, Fort Collins, CO, United States

Howard S. Ginsberg, United States Geological Survey Patuxent Wildlife Research Center, University of Rhode Island, Kingston, RI, United States

Harold J. Harlan, National Pest Management Association, Fairfax, VA, United States

Nancy C. Hinkle, Department of Entomology, University of Georgia, Athens, GA, United States

Randall B. Hirschhorn (retired), Department of Public Health, Philadelphia, PA, United States

Jerry R. Hogsette, Agricultural Research Service, United States Department of Agriculture, Gainesville, FL, United States

Zdenek Hubálek, Institute of Vertebrate Biology, Academy of Sciences, Brno, Czech Republic

Helge Kampen, Department of Epidemiology and Public Health, Yale University School of Medicine, New Haven, CT, USA, formerly with Institute for Medical Microbiology, Immunology and Parasitology, University of Bonn, Germany

Michael Y. Kosoy, Bacterial Zoonoses Branch, Centers for Disease Control and Prevention, Fort Collins, CO, United States

Marco Maroni, Department of Occupational Health, University of Milan, and Division of Occupational Medicine, International Centre for Pesticides and Health Risk Prevention, Milan, Italy[2]

Francesca Metruccio, Department of Occupational Health, University of Milan, and Division of Occupational Medicine, International Centre for Pesticides and Health Risk Prevention, Milan, Italy

Angelo Moretto, International Centre for Pesticides and Health Risk Prevention, Milan, Italy

David H. Oi, Center for Medical, Agricultural, and Veterinary Entomology, Agricultural Research Service, United States Department of Agriculture, Gainesville, FL, United States

Matthew S. Perzanowski, Department of Environmental Health Sciences, Columbia University, New York, NY, United States

Michael K. Rust, Department of Entomology, University of California, Riverside, CA, United States

John P. Sarisky, Environmental Health Services Branch, Centers for Disease Control and Prevention, Atlanta, GA, United States

Francis Schaffner, Institute for Parasitology, University of Zurich, Switzerland, formerly with EID Méditerranée, Montpellier, France

Kevin J. Sweeney, Office of Pesticide Programs, Registration Division, Insecticides Branch, U.S. Environmental Protection Agency, Washington, DC, United States

[2] Deceased.

Toby Wilkinson, The Martin Centre, University of Cambridge, and The Medical Entomology Centre, Shepreth, United Kingdom

Reviewers

Nida Besbelli, Technical Officer Chemical Safety, WHO European Centre for Environment and Health, Bonn Office, Germany

Jean Bousquet, Hôpital Arnaud de Villeneuve, Montpellier, France

Róbert Farkas, Department of Parasitology and Zoology, Szent István University, Budapest, Hungary

Barbara Hart, Royal Agricultural College, Cirencester, United Kingdom

Stephen A. Kells, Department of Entomology, University of Minnesota, St. Paul, MN, United States

John H. Klotz, Department of Entomology, University of California, Riverside, CA, United States

Boris Krasnov, Mitrani Department of Desert Ecology, Jacob Blaustein Institutes for Desert Research, Ben-Gurion University of the Negev, and Ramon Science Center, Israel

Klaus Kurtenbach, Department of Biology and Biochemistry, University of Bath, United Kingdom

Paolo M. Matricardi, Department of Pediatric Pneumology and Immunology, Charité Medical University, Berlin, Germany

Michael L. Perdue, WHO Global Influenza Programme, Geneva, Switzerland

Richard J. Pollack, Laboratory of Public Health Entomology, Harvard School of Public Health, Boston, MA, United States

Reiner Pospischil, Bayer CropScience, Monheim, Germany

Michael F. Potter, Department of Entomology, University of Kentucky, Lexington, KY, United States

Rudolf Schenker, Novartis Animal Health Inc., Basel, Switzerland

Mike W. Service, Department of Entomology, Liverpool School of Tropical Medicine, United Kingdom

ANGELIKA TRITSCHER, WHO Joint Secretary to the Joint FAO/WHO Expert Committee on Food Additives and the Joint FAO/WHO Joint Meeting on Pesticide Residues, Geneva, Switzerland

Observers

JONATHAN PECK, Killgerm Group Ltd., West Yorkshire, United Kingdom

ROBERT M. ROSENBERG, National Pest Management Association, Fairfax, VA, United States

Secretariat

NURIA AZNAR, Office Assistant Noise and Housing, WHO European Centre for Environment and Health, Bonn, Germany (2005–2007)

XAVIER BONNEFOY, Regional Advisor Noise and Housing, WHO European Centre for Environment and Health, Bonn, Germany[3]

MATTHIAS BRAUBACH, Technical Officer Housing and Health, WHO European Centre for Environment and Health, Bonn, Germany

BARBARA JUKES, Environmental Health Consultant, CIEH London, United Kingdom

HELGE KAMPEN, Project Officer Urban Pests, WHO European Centre for Environment and Health, Bonn, Germany (2006–2007)

KEVIN J. SWEENEY, Project Officer Urban Pests, WHO European Centre for Environment and Health, Bonn, Germany (2004–2005)

[3] Deceased.